J. Flury
R. Rummel
C. Reigber
M. Rothacher
G. Boedeker
U. Schreiber

Observation of the Earth System from Space

Jakob Flury
Reiner Rummel
Christoph Reigber
Markus Rothacher
Gerd Boedeker
Ulrich Schreiber
Editors

Observation of the Earth System from Space

with 249 Figures and 54 Tables

Springer

Dr. J. Flury
Professor Dr. R. Rummel
TU München
Institut für Astonomische
Physikalische Geodäsie
Arcisstr. 21
80290 München
Germany

Professor Dr. G. Boedecker
Bayerische Akademie
der Wissenschaften
Bayerische Kommission für die
Internationale Erdmessung
Marstallplatz 8
80539 München
Germany

Professor Dr. C. Reigber
Professor Dr. M. Rothacher
GeoForschungsZentrum Potsdam
Dept. 1, Geodäsie und Fernerkundung
Telegrafenberg
14473 Potsdam
Germany

Professor Dr. U. Schreiber
TU München
Forschungseinrichtung
Satellitengeodäsie
Fundamentalstation Wettzell
93444 Kötzting
Germany

Library of Congress Control Number: 2005933897

ISBN 10 3-540-29520-8 Springer Berlin Heidelberg New York
ISBN 13 978-3-540-29520-4 Springer Berlin Heidelberg New York

Springer is a part of Springer Science+Business Media
springeronline.com

Printed in The Netherlands

Cover design: E. Kirchner, Heidelberg
Production: Almas Schimmel
Typesetting: GOCE-Projektbüro Deutschland, TU München
Printing: Krips bv, Meppel
Binding: Stürtz AG-, Würzburg

Printed on acid-free paper 30/3141/as 5 4 3 2 1 0

Preface

In the recent years, space-based observation methods have led to a substantially improved understanding of Earth system. Geodesy and geophysics are contributing to this development by measuring the temporal and spatial variations of the Earth's shape, gravity field, and magnetic field, as well as atmosphere density. In the frame of the German R&D programme GEOTECHNOLOGIEN, research projects have been launched in 2002 related to the satellite missions CHAMP, GRACE and ESA's planned mission GOCE, to complementary terrestrial and airborne sensor systems and to consistent and stable high-precision global reference systems for satellite and other techniques.

In the initial 3-year phase of the research programme (2002-2004), new gravity field models have been computed from CHAMP and GRACE data which outperform previous models in accuracy by up to two orders of magnitude for the long and medium wavelengths. A special highlight is the determination of seasonal gravity variations caused by changes in continental water masses. For GOCE, to be launched in 2006, new gravity field analysis methods are under development and integrated into the ESA processing system. 200,000 GPS radio occultation profiles, observed by CHAMP, have been processed on an operational basis. They represent new and excellent information on atmospheric refractivity, temperature and water vapor. These new developments require geodetic space techniques (such as VLBI, SLR, LLR, GPS) to be combined and synchronized as if being one global instrument. In this respect, foundations have been laid for a substantial improvement of the reference systems and products of the International Earth Rotation and Reference Systems Service (IERS). Sensor systems for airborne gravimetry have been integrated and tested, and a particularly development is a laser gyro dedicated to the measurement of the rotational degrees of freedom of the motion caused by earthquakes. A total sum of about 10 million Euros has been spent by the German Federal Ministry of Education and Research (BMBF) and the German Research Foundation (DFG). The projects were carried out in close cooperation between universities, research institutes, and small and medium sized enterprises.

In this book the results of the first programme phase are collected in 30 scientific papers related to the six core programmes of the theme "Observation of the Earth system from space". The book provides an overview of the state-of-the-art of this research. At the same time it should provide inspiration for future work, since on many fields research is going on, and a number of projects will continue in the second programme phase. The editors are indebted to all authors and to the publisher for the excellent cooperation in the preparation of this book. The editing process and the compilation of the camera-ready manuscript were coordinated by J. Flury at the German GOCE project bureau at Technische Universität München. The support of the GEOTECHNOLOGIEN programme by BMBF and DFG is gratefully acknowledged as well as the continuous support by the GEOTECHNOLOGIEN coordination office.

Munich and Potsdam, August 2005
Jakob Flury
Reiner Rummel
Christoph Reigber
Markus Rothacher
Gerd Boedecker
Ulrich Schreiber

GEOTECHNOLOGIEN

Contents

Part I

CHAMP
CHAllenging Minisatellite Payload

CHAMP Mission 5 Years in Orbit

Christoph Reigber, Hermann Lühr, Ludwig Grunwaldt, Christoph Förste, Rolf König, Heiner Massmann and Carsten Falck

GeoForschungsZentrum Potsdam (GFZ), Dept. 1 'Geodesy and Remote Sensing', Telegrafenberg A 17, 14473 Potsdam, Germany, `reigber@gfz-potsdam.de`

Summary. In the summer of 2000 the geo-research satellite CHAMP was launched into orbit. Its innovative payload arrangement and its low injection altitude allow CHAMP to simultaneously collect almost uninterrupted measurement series relating to the Earth gravity and magnetic fields at low altitude. In addition, CHAMP sounds the neutral atmosphere and ionosphere using GPS observations onboard. After 60 months in orbit one arrives at a very positive conclusion for the CHAMP mission. The CHAMP satellite and its instruments have been operated almost uninterruptedly since launch. The great performance of the satellite subsystems and of the mission operation specialists has made it possible to keep CHAMP in the science operation mode for most of the time and in addition to lift its orbit two times. After a series of calibration and validation activities in the course of the mission, which included a number of onboard software updates and parameter adjustments, CHAMP has been providing excellent measurements from its state of the art instruments for now more than 4 years. The effective and steadily functioning of the CHAMP Science Data System and the supporting tracking networks has made it possible to provide large quantities of pre-processed data, precision data products and auxiliary information to hundreds of registered users in an almost uninterrupted manner. This was only possible due to the funding of the project DACH (CHAMP Data Acquisition and Data Use) within the 'GEOTECHNOLOGIEN' R+D programme of the BMBF. With the orbit altitude being presently about 60 km higher than originally planned for mid 2005, CHAMP will very likely orbit the Earth for another 3 years at quite low altitude. This mission extension at low altitude will make CHAMP a pioneering long-duration mission for geo-potential research and sounding of the atmosphere.

Key words: CHAMP, Mission overview, Science Data System achievements

1 Introduction

The geo-research mission CHAMP (**CHA**llenging **M**inisatellite **P**ayload), launched on July 15, 2000 from the Russian cosmodrome Plesetsk into a near polar, circular and 455 km altitude orbit, was established in 1997 as a Principal Investigator (PI) institution led project, with the PI (C. Reigber) and his institution (GFZ Potsdam) being fully responsible for the successful implementation and execution of the mission. During the various CHAMP mission phases, until the end of the commissioning phase, the project was funded by the German Federal Ministry of Education and Research (BMBF), the German Aerospace Centre (DLR) and the GFZ Potsdam. In mid 2001, 9 months after launch, the CHAMP overall system, consisting of the space and ground segment components, was commissioned and validated and ready to deliver high quality data and data products to the international science and application community. In order to stimulate additional calibration/validation activities and to trigger as many scientific studies and application investigations on the basis of CHAMP data and routinely generated products, an Announcement of Opportunity (AO) was issued in May 2001 for the international geo-science community. At this point of the mission timeline the operational phase of the CHAMP mission started, with the primary CHAMP Science Data System (SDS) funding being provided by the 'GEOTECHNOLOGIEN' R+D programme of the BMBF under grant 03F0333A for the first phase.

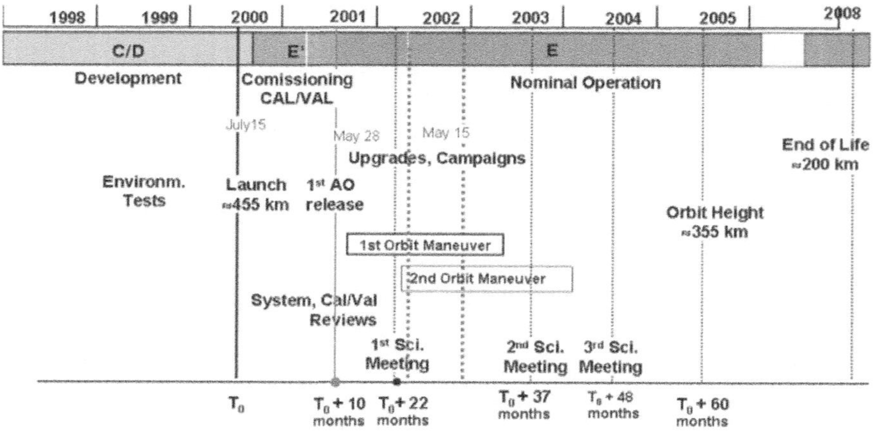

Fig. 1. CHAMP mission timeline.

The exceptionally good performance of all CHAMP system components over the past 5 years and the AO-triggered involvement of a large and still growing number of users around the globe has made it possible not only to provide unprecedented long, uninterrupted and well calibrated data series for various investigations, but also to apply new data reduction and analysis methods and to come up with new and value-added products besides those routinely generated in the CHAMP science data processing system. Many of the scientific achievements with CHAMP data are presented in the proceedings of the 1st CHAMP Science Meeting (Reigber et al., 2003) and the 2nd Science Meeting (Reigber et al., 2004).

The CHAMP mission, originally designed for a 5 years lifetime, will last a few years longer than initially planned, thanks to the smooth functioning of all mission elements, the successful execution of two orbit rises and the availability of still enough cold gas for the operation over a number of additional years. The purpose of this contribution is to shortly describe the status of the mission at this 5-year milestone and to elucidate the science instrument data and data products, which have been delivered in large quantities to the more than 500 scientists and application users worldwide.

2 Spacecraft, Instrumentation and Orbit Evolution

When the CHAMP spacecraft was designed it was optimized in the sense that it should best satisfy the requirements of the gravity and magnetic field objectives simultaneously, which are at times quite different. Design drivers in this respect were a well-determined and constant position of the centre of gravity, a three-axes stabilized attitude control causing only negligible lateral accelerations, a sizable boom for magnetic cleanliness and a long mission lifetime at low altitude.

In order to optimize the aerodynamic behaviour and magnetic field observation environment, the satellite was build as a relatively heavy trapezoid body of dimensions 430x75x162 cm^3 (l/h/w) with a 404 cm long deployable boom in flight direction (see Color Fig. I on p. 286). The spacecraft weighed 522 kg at the beginning of the mission, including 34 kg of cold gas for attitude control and orbit manoeuvres, of which nearly 21 kg have been consumed in the meantime. The average power consumption of 120 W (payload 46 W) is comfortably provided by 7 m^2 of solar cells and a 16 Ah NiH^2 battery. No degradation is detectable so far in the power system.

CHAMP is kept in an Earth-oriented attitude with the boom pointing in flight direction. For calibration experiments the spacecraft was steered in a number of occasions into quite different orientations, from perpendicular to the velocity vector to anti-flight direction. Three magnetic torquers are used to orient the spacecraft within a control band of ± 2 degrees. In case of dead-band exceedance, 12 cold gas thrusters restore the nominal attitude. Prime attitude sensors are star trackers and an onboard GPS receiver. Every 10 seconds the GPS receiver provides a new position and updates the onboard clock. A highly autonomous control and data handling system guarantees a save operation during longer periods (up to 12 h) of no contact with ground stations. Data are stored in a mass memory of 1.2 Gigabit capacity. The 4 m long boom, installed for magnetic cleanliness reasons, consists of three segments: the outer part with the scalar Overhauser magnetometer at the tip, the middle segment with the rigid optical bench on which two star sensor heads and two Fluxgate vector magnetometers are mounted, and the inner segment incorporating the deployment hinge.

In total CHAMP is equipped with seven different scientific instruments, the data of which are processed in an operational mode since May 2001 (see Color Fig. I on p. 286).

The NASA Jet Propulsion Laboratory (JPL) has provided the state-of-the-art "Blackjack" GPS space receiver. Accommodated for the first time onboard a LEO satellite as a mission control support and satellite-to-satellite (SST) gravity recovery instrument, it delivers NAV solutions accurate to about 6 m rms with an average availability of >99.5 %, the time tag for all science instruments within 1 ms and precision orbit ephemeris (POD) results for gravity recovery with phase residuals in the order of <3 cm (König et al., 2004). Since June 2001 radio occultation measurements have routinely been obtained with C/A measurements at high rates (50 Hz sampling frequency). The obtained profiles for atmospheric humidity and temperature (nearly 250 per day) reach close to the Earth surface and are in good agreement with operational meteorological analysis results (Wickert et al., 2004).

The STAR accelerometer, which was provided by the Centre National d'Études Spatial (CNES) and manufactured by the Office National d'Études et de Recherches Aérospatiales (ONERA), had its maiden flight on CHAMP. It meets the specified resolution of $<3 \times 10^{-9}$ m/s^2 for the two highly sensitive axes (Förste and Choi, 2004) and has been delivering since autumn 2000 valuable information on the surface forces accelerations, an information which is highly important for the accurate gravity field modelling and the development of air density models.

The GFZ-built CHAMP Laser Retro-Reflector (LRR) has demonstrated impressively the possibility to use a densely packed array with the minimum number of 4 prisms for a LEO satellite to obtain a sufficiently high return signal for easy target acquisition under both night and daytime conditions. Due to its compact design, the target signature of the CHAMP LRR is negligible and single-shot accuracies below 5 mm have been reported by the most advanced laser trackers (Grunwaldt and Meehan, 2003).

CHAMP was also the maiden flight for the Advanced Stellar Compass (ASC) used in dual-head configuration. Combined with the aberration correction capability – first time applied in orbit with CHAMP – this has led to a highly accurate attitude of approximately 15" of the raw data onboard. The instrument has been operating fully autonomously for 5 years already and directly outputs the final quaternions. On-ground post processing improves the accuracy to about 2" (Rother et al., 2003).

The Digital Ion Drift Meter and Langmuir Probe were provided by the Air Force Research Laboratory (AFRL) in Hanscom MA, USA. This newly developed instrument monitors the ion dynamics like the drift velocity, density and temperature along the orbit.

Since its first switch-on on the second day of the mission the Fluxgate magnetometer has been operating flawlessly. Thanks to the magnetic cleanliness of the spacecraft, the ambient magnetic field is measured at a high rate of 50 Hz and a resolution of 0.2 nT in all three axes. After having applied all necessary transformations and corrections to the vector field measurements on the basis of attitude and position observables, absolute vector accuracies of less than 2 nT have been reported (Rother et al., 2003).

The Overhauser magnetometer provides absolutely calibrated readings of the scalar field strength at a rate of 1 Hz and a resolution of 0.1 nT. It serves as measurement standard and calibration unit, and fully satisfies since the beginning of the mission the scientific requirements.

As stated, all CHAMP instruments are in a very good state and function even after 5 years in operation as foreseen. The only exception is the less sensitive radial component of the accelerometer, the observations of which cannot fully be used because of a malfunctioning of one of the six electrode pairs of the STAR accelerometer (Perosanz et al., 2004).

After a series of calibration and validation activities in the course of the mission, which included a number of software updates and parameter adjustments, and the scientific results obtained so far, it can be stated that CHAMP has been providing the best possible measurements from its state-of-the-art instruments for now almost 5 years, making CHAMP a pioneering mission in many respects.

In addition CHAMP is at the moment the lowest orbiting geo-research satellite, continuously tracked by GPS and continuously providing accelerometer and magnetic field data. CHAMP was injected into an almost circular ($e = 0.004$), near polar ($i = 87°$) orbit with an initial altitude of 454 km. This initial altitude was chosen as the best compromise to guarantee on one hand a five-year mission duration even under high solar activity conditions, predictable by models at the time prior to launch, and to account on the other hand for the requirements imposed by the scientific goals of the mission. Due to the extremely high solar flux and the corresponding high atmospheric drag acting on the satellite throughout the time period from mid 2001 to the end of 2002, the orbit decay was considerably faster than had been predictable, with the danger that the mission would have been finished already in 2004. To avoid this, a first orbit change manoeuvre was performed on June 10/11, 2002. Through a sequence of thruster firings at apogee the orbital altitude of CHAMP was increased by about 16 km. A second orbit change manoeuvre of the same type was carried out on December 9/10, 2002, resulting in a second rise of the orbit by about 20 km (see Color Fig. II on p. 286).

Now, in July 2005, CHAMP has lost almost exactly 100 km of its original orbital height and is orbiting at an altitude of about 355 km. After the two orbital manoeuvres the eccentricity e changed to the very small value of 0.0002, which means that CHAMP is now on an almost perfect circular path around the Earth.

The present orbital height is still 55 km above the originally for July 2005 planned height of about 300 km. With the solar flux predictions presently available, the 300 km altitude floor will be reached in autumn 2007 and this

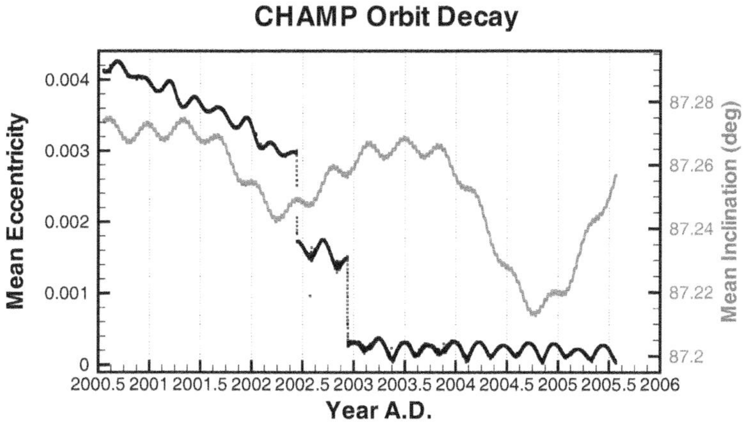

Fig. 2. Changes of mean eccentricity and inclination since launch.

will bring the CHAMP mission to a definite end in the spring to summer 2008 timeframe.

In the course of its free-drifting orbit periods CHAMP passed through many different commensurabilities and resonant regimes, with high sensitivity to 15th and 16th order terms of the geo-potential and overtones. Due to the orbit changes the satellite passed through a number of repeat cycles more than once (e.g., a 2-days repeat in May 2002, October 2002 and in May 2003) and will experience during the second mission part at low altitudes largely enhanced perturbations in the orbital motion.

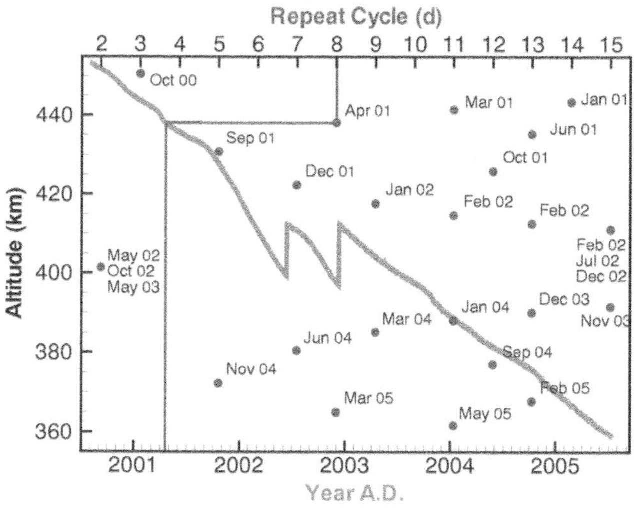

Fig. 3. Repeat cycles (in days) through which CHAMP passed since launch.

3 Ground System Performance

CHAMP's ground segment comprises all ground-based components which perform the operational control of the spacecraft and instruments, the data flow from the onboard memory and supporting ground tracking networks to the processors, the standard science product generation and the dissemination of data and products to the users. Color Figure III on p. 287 shows the general scheme of the ground segment for CHAMP.

DLR has been running for 5 years with great success the *Mission Operation System* (MOS) consisting of the Mission Control Centre (MCC) at the German Space Operation Centre (GSOC), Oberpfaffenhofen, and the Raw Data Centre (RDC) at DLR's German Remote Sensing Data Centre (DFD), Neustrelitz. The *Science Operation System* (SOS) at GFZ constitutes the interface between the science experimenters and satellite operation. It is responsible for mission scheduling, command preparation, and mission and orbit analysis.

CHAMP's on-board instruments continuously produce science and instruments' house-keeping data with an overall rate of 10.8 kbit/s, and the satellite adds 2.2 kbit/s of spacecraft house-keeping data, which makes a total of 141 MByte/d. These data are downloaded three to four times a day to the 7.3 m ground antenna of the DLR receiving station in Neustrelitz (53.5 N, 13 E), Germany, and for almost every pass to the GFZ/DLR 4 m receiving station in Ny Ålesund (78.9 N, 11.8 E), Spitsbergen. A third ground station, the DLR ground station in Weilheim (48 N, 11 E), is operated as the commanding and satellite control station. It also serves as a back-up station to Neustrelitz. It receives 'real-time' science and H/K data at a bit-rate of 32 kbit/s and sends commands at 4 kbit/s. A great number of command sequences were prepared and successfully transmitted to the spacecraft in the meantime. The number of commands executed by the CHAMP satellite since launch nears the 290,000 mark. After 5 years of science data gathering in orbit, approximately 6,700 times telemetry data sequences were downloaded to the three aforementioned ground stations.

CHAMP's *Raw Data Centre* is running, almost uninterruptible since launch, at the receiving station Neustrelitz with the following functions: telemetry data reception (transfer frames) and long-term storage in the *Raw Data Archive*, demultiplexing and extraction of science and H/K application packets (level-0 data), immediate transfer of H/K packets to GSOC, and temporary storage of all level-0 data in the *level-0 rolling archive* for access by the Decoding Centre of the Science Operation System (SOS-SD) at GFZ Potsdam. Here the *level-0 long-term archive* for CHAMP is located.

In addition to the spacecraft data, all CHAMP related ground station network data are accessed and archived at GFZ Potsdam: low rate (30 s, 10 s) and high rate (1 s), low latency GPS ground-based observations from individual GPS stations and the data centres of the International GPS Service (IGS), and CHAMP laser tracking data from the international laser data centres of the Inernational Laser Ranging Service (ILRS). The high-rate GPS ground-station data of the GFZ and JPL dedicated CHAMP GPS subnets, altogether about 25 stations, are mutually exchanged. All data transfer happens via the public Internet network.

The SOS-SD component is carrying out in a semi-automatic process all decoding of CHAMP level-0 data to level-1, that means the conversion from telemetry code into user-defined physical units.

The higher level scientific products are generated within the *Science Data System* (SDS) consisting of the

- Orbit and Gravity Field Processing System (SDS-OG),
- Magnetic and Electric Field Processing Systems and (SDS-ME)
- Neutral Atmosphere Profiling System (SDS-AP)

at GFZ Potsdam, and the

- Ionosphere Profiling System (SDS-IP)

at DLR's Institute for Communication and Navigation (IKN), Neustrelitz.

Data and data product archiving, administration and retrieval is managed by the CHAMP *Information System and Data Centre* (ISDC), located at GFZ Potsdam, which is also the users' www- and ftp-based interface for access to CHAMP data and scientific products. The number of users and user groups, registered at ISDC and retrieving data, data products and ancillary information from the archive, has continuously grown with time. Four years after having issued the Announcement of Opportunity, this number has reached the value of about 560, with more than 50% of these users originating from Germany, the USA and China (see Color Fig. IV on p. 287).

CHAMP's standard science products are labelled from level-1 to level-4 according to the number of processing steps applied to the original data. Decommutation and decoding of level-0 data results in level-1 products. These are daily files, associated with each individual instrument and source aboard CHAMP, with the data content being transformed from the telemetry format and units into an application software readable format and physical units. Level-1 products also include the ground station GPS and laser data. Level-2 products are pre-processed, edited and calibrated experiment data, supplemented and merged with necessary spacecraft housekeeping data and arranged in daily files. Level-3 products comprise the operational rapid products and fine processed, edited and definitely calibrated experiment data. Finally, level-4 leads to the geo-scientific models derived from the analysis of CHAMP experiment data, supported and value-added by external models and observations.

At the time of writing this contribution, the numbers of product files given in Table 1 have been reported by the ISDC to exist in the data base for each of the levels 1 to 4. Each additional year of CHAMP operation adds about 1.4 Terabyte of data to the total amount.

Table 1. Total amount of stored data/product files since launch

	number of files	total
Level-1	3570365	3109 GByte
Level-2	244017	599 GByte
Level-3	807744	1767 GByte
Level-4	7708	1723 GByte
total	7786533	7198 GByte

4 Mission Goals and Science Data System Achievements

The science goals of the CHAMP mission are to gain improved sources of information about the nature and composition of the Earth, about evolutionary processes continuing to shape it, as well as to gain information on dynamic processes taking place in the near Earth space, in the neutral atmosphere and the ionosphere. Precise global gravity and magnetic field models are of main importance for studying and understanding the structure and composition of the solid Earth, whereas evolutionary processes, influencing global change, express themselves either directly or indirectly through changes in gravity and magnetic field signals and changes of key parameters of the atmosphere and ionosphere.

The mission goals for CHAMP, as defined in the pre-launch period, were:

1. to acquire long-term, uninterrupted and well calibrated data series from CHAMP's gravity field, magnetic field and atmosphere sensors,
2. to produce on the basis of high-low SST and accelerometer observations a long-term mean estimate of the Earth's gravity field for the spectral components >1,000 km with an at least one order of magnitude improvement and to contribute to the determination of the time variability of the longest wavelength components of the field by comparing three-monthly models,
3. to measure and model the main and lithospheric magnetic fields of the Earth as well as secular variations and ionospheric currents with unprecedented spatial resolution and precision through high-precision scalar/vector magnetic field and electric field observations,
4. to probe the neutral atmosphere and ionosphere as global as possible, using GPS limb soundings with improved technology,
5. to give all interested science and application users free access to the CHAMP data and data products through a dedicated CHAMP data and information system.

After 5 years in orbit and after 51 months of routine operation it can be stated that the CHAMP mission succeeded in achieving the aforementioned mission goals. More than 98 % of all possible observations have been acquired and stored in the raw data archives. Within the three fields of research and application pursued with CHAMP, the following number of standard products have been made available to the general user community via the ISDC (see Color Figs. V and VI on p. 288) up to now:

(1) Orbit and Gravity Field Processing System (SOS-OG)

- level-1: 21 GByte of GPS to CHAMP satellite-to-satellite phase and code tracking observations (0.1 Hz),
- level-2: 8 GByte of preprocessed *accelerometer* observations (0.1 Hz) and linear and angular accelerations with attitude information plus the thruster-firing time events,
- level-3: 15 GByte of *predicted, ultra-rapid and rapid science orbits* of CHAMP and the GPS satellites in the Conventional Terrestrial System, and processed with a short time delay of a few hours to days after data download,
- level-4: global *Earth gravity field models*, represented by the adjusted coefficients of the spherical harmonic expansion: progressively accumulated solutions, named EIGEN-1S, EIGEN-2, EIGEN-3p and EIGEN-CHAMP03S (see http://www.gfz-potsdam.de/pb1/op/champ/results/index_RESULTS.html).

(2) Magnetic and Electric Field Processing System (SOS-ME)

- level-2: 38 GByte *magnetic field* observations, both scalar and vector field, in the sensor system as well as in local coordinates (North, East, Down), all at 1 Hz rate; 17 GByte precise attitude derived from Advanced Stellar Compass both for the spacecraft and for the boom instrumentation at a 1 Hz rate,
- level-4: *main field and lithospheric field models* by the spherical harmonic expansion coefficients, derived from spacecraft data and its secular variation coefficients from space and ground-based observations; recent models are named POMME 1.4 and MF3 (see: http://www.gfz-potsdam.de/pb1/op/champ/results/index_RESULTS.html).

(3) Atmosphere/Ionosphere Profiling Systems (SOS-AP/IP)

- level-1: 75 GByte *GPS-CHAMP radio occultation* measurements (50 Hz for AP and 1 Hz for IP),
- level-2: 272 GByte of *atmospheric excess path delays*; time-tagged atmospheric excess path of the occultation, link annotated with SNR and orbit (position and velocity), information of CHAMP and the occulting GPS satellite for each occultation event,
- level-3: 16 GByte of *vertical profiles* of atmospheric bending angle and geopotential, profiles of refractivity, *dry air*-density, -pressure and -temperature, and – adopting temperature from global analyses – specific and relative humidity, partial pressure and mixing ratios of *water vapour* in the troposphere. 9 GByte of occultation link related *Total Electron Content* data values and 0.1 GByte of *vertical TEC profiles*.

In addition, more than 2,000 GByte of High Rate GPS ground data are provided to the users via the ISDC.

The SDS team at GFZ has achieved a number of outstanding scientific results in the course of the 5 years operation of CHAMP and has made these results quickly available to the community:

- For the first time in space geodesy's history with the EIGEN solutions global gravity field models with full power up to degree/order 65 of the spherical harmonic expansion could be derived from observations of a single satellite and largest-scale temporal gravity variations could be extracted from 3 years worth of data (Reigber et al., 2004).
- With POMME, a series of field models for the accurate description of the main and external magnetic field has been introduced (Maus et al., 2004). Employing data of the CHAMP scalar and vector magnetometers, a detailed global model up to degree/order 90 of the crustal magnetic field was derived (Maus et al., 2005). This model MF3 is providing important information for studies of the crustal magnetisation. In addition, from two years of high-precision CHAMP satellite magnetic measurements it has been possible to map for the first time the magnetic signal of ocean tidal flow (Tyler et al., 2003).
- Unprecedented continuous long series of atmospheric and ionospheric profiles are derived by the SDS AIP team from CHAMP's GPS radio occultation data. More than 300,000 atmospheric occultation measurements are presently available as well as more than 200,000 ionospheric occultation data. Currently the delay time from data reception to the generation of key parameters of the neutral atmosphere and ionosphere is only a few hours and the quality of the data products as derived from inter-comparisons with independent observations and analyses is impressively high (Wickert et al., 2004; Jakowski et al., 2004).

Finally, with the CHAMP ISDC a modern tool for the management of system data of a space geodetic mission was introduced, which has found its extension into the GRACE era (see http://isdc.gfz-potsdam.de/champ/). More than 500 scientists and application users are registered at the moment, which are making intensive use of this service. With the continuous annual increase of CHAMP data users over the last four years, this number is likely to further grow in the next few years.

5 Conclusion and Outlook

After 5 years of mission operation the main conclusion is that the CHAMP mission fully meets the demands defined by the project team in the design and development phase for the space and for the ground segment. The CHAMP mission has already now provided an unprecedented set of data for geo-potential, atmospheric and ionospheric research and has marked a new era of LEO satellites with onboard GPS receivers, accelerometers and magnetometers. Many scientists from various fields of geosciences and the application area make intensive use of data and products provided by the CHAMP Science Data System for their own analyses and investigations. CHAMP has served in many respects as pathfinder for the GRACE mission and will do so for the next generation of magnetic field missions such as SWARM. CHAMP is likely to remain in orbit until mid 2008. With the decreasing orbital altitude and the extension of the observation period by additional three years, more sensitivity and precision will be gained in particular for the gravity field and magnetic field modelling. With its companion mission GRACE and a CHAMP observation period extended to seven or eight years, highly valuable information on the variability of the Earth gravity and magnetic fields and on long-term changes of key quantities of the atmosphere and ionosphere will be obtained. This information will support a better understanding of the mass balances in the Earth System and may help in future to early detect global changes and to understand their underlying mechanisms.

Acknowledgement. This is publication no. GEOTECH-162 of the program GEOTECHNOLOGIEN of BMBF and DFG. The successful operation and exploitation of the CHAMP mission over the last 5 years would not have been possible without the financial support or contributions in kind from various institutions and partner organisations and the engagement and enthusiasm of many individuals in the mission operation, the mission science data processing and the tracking and receiving station teams. We cannot thank everybody, but would like to acknowledge in particular the great efforts of the mission operation team at GSOC Oberpfaffenhofen and of the tracking stations of the ILRS and IGS as well as the technical support we have received from the experts of the science instrument providers NASA/JPL, CNES and AFRL. The exploitation of the mission data and their distribution to the users could not have been performed to the extent provided without BMBF's financial support of the DACH project (grant FKZ 0333A) within the GEOTECHNOLOGIEN geo-scientific R+D programme. Unforgotten are the outstanding contributions which our colleague Peter Schwintzer brought into the CHAMP mission before he suddenly passed away in December 2004. He was leading the development of the Science Data System in the early mission phases and was directing the SDS team throughout the exploitation phase until the last day of his too short life.

References

Förste C. and Choi, S. (2004) CHAMP accelerometer preprocessing at GeoForschungsZentrum Potsdam, in: Reigber C., Lühr H., Schwintzer P. and Wickert J. (Eds.) Earth Observation with CHAMP. Results from Three Years in Orbit, Springer-Verlag, 169–174.

Grunwaldt L. and Meehan T. (2003) CHAMP Orbit and Gravity Instrument Status, in: Reigber C., Lühr H. and Schwintzer P. (Eds.) First CHAMP Mission Results for Gravity, Magnetic and Atmospheric Studies, Springer-Verlag, 3–10.

Jakowski N., Tsybulya K., Mielich J., Belehaki A., Altadill D., Jodogne J.-C. and Zolesi B. (2004) Validation of GPS ionospheric radio occultation results onboard CHAMP by vertical sounding observations in Europe, in: Reigber C., Lühr H., Schwintzer P. and Wickert J. (Eds.) Earth Observation with CHAMP. Results from Three Years in Orbit, Springer-Verlag, 447–452.

König R., Michalak G., Neumayer K.H., Schmidt R., Zhu S., Meixner H., Reigber C. (2004) Recent developments in CHAMP orbit determination at GFZ, in: Reigber C., Lühr H., Schwintzer P. and Wickert J. (Eds.) Earth Observation with CHAMP. Results from Three Years in Orbit, Springer-Verlag, 65–70.

Maus S., Lühr H., Balasis G., Rother M., Mandea M. (2004) Introducing POMME, the POtsdam Magnetic Model of the Earth, in: Reigber C., Lühr H., Schwintzer P. and Wickert J. (Eds.) Earth Observation with CHAMP. Results from Three Years in Orbit, Springer-Verlag, 293–298.

Maus S., Rother M., Hemant K., Lühr H., Kuvshinov A. and Olsen N. (2005) Earth's lithospheric magnetic field determined to spherical harmonic degree 90 from CHAMP satellite measurements, *Geophys J Int*, revised submitted.

Perosanz F., Biancale R., Lemoine J.M., Vales N., Loyer S., Bruinsma S. (2004) Evaluation of the CHAMP accelerometer on two years of mission, in: Reigber, C., Lühr H., Schwintzer P. and Wickert J. (Eds.) Earth Observation with CHAMP. Results from Three Years in Orbit, Springer-Verlag, 77-82.

Reigber C., Lühr H. and Schwintzer P. (Eds.) (2003) First CHAMP Mission Results for Gravity, Magnetic and Atmospheric Studies, Springer-Verlag.

Reigber C., Lühr H., Schwintzer P. and Wickert J. (Eds.) (2004) Earth Observation with CHAMP. Results from Three Years in Orbit, Springer-Verlag.

Rother M., Choi S., Lühr H. and Mai W. (2003) CHAMP ME Data Processing and Open Issues, in: Reigber C., Lühr H. and Schwintzer P. (Eds.) First CHAMP Mission Results for Gravity, Magnetic and Atmospheric Studies, Springer-Verlag, 203–211.

Tyler R., Maus S., Lühr H. (2003) Satellite observations of magnetic fields due to ocean tidal flow, *Science* 299, 239–241.

Wickert J., Schmidt T., Beyerle G., Michalak G., König R., Kaschenz J. and Reigber C. (2004) Atmospheric profiling with CHAMP: Status of the operational data analysis, validation of the recent data products and future prospects, in: Reigber, C., Lühr H., Schwintzer P. and Wickert J. (Eds.) Earth Observation with CHAMP. Results from Three Years in Orbit, Springer-Verlag, 495–500.

Remarks on CHAMP Orbit Products

Rolf König, Grzegorz Michalak, Karl Hans Neumayer, and Shengyuan Zhu

GeoForschungsZentrum Potsdam, Telegrafenberg, D-14473 Potsdam, Germany, koenigr@gfz-potsdam.de

Summary. The GeoForschungsZentrum Potsdam (GFZ) runs an operational system for the CHAMP mission that provides precise orbits on a regular basis. Focus is put on recent analyses and achievements for the Rapid and Ultra-rapid Science Orbits.

Key words: CHAMP, GRACE, SAC-C, Precise Orbit Determination, Orbit Products

1 Introduction

Since the beginning of the CHAMP mission (Reigber, 2005) in 2000, the GeoForschungsZentrum Potsdam (GFZ) operationally provides precise orbits. These products comprise orbit predictions (the PreDicted Orbits or PDOs), rapidly available orbits (the Rapid Science Orbits or RSOs and the Ultra-Rapid Science Orbits or USOs), and offline generated orbits (the Post-processed Science Orbits or PSOs). All these routine orbits are dynamically integrated and differentially corrected for certain parameters to fit to the observations being available at the time of generation and being appropriate to meet the objectives the orbit is intended for. The orbits are provided at different frequencies, latencies, and accuracies depending again on their intention. And they are published at the CHAMP data center at GFZ (ISDC, 2001).

Developments in CHAMP Precise Orbit Determination (POD) have recently been discussed in König et al. (2005). The following concentrates therefore on newest improvements in accuracies and latencies, on new considerations regarding accuracy assessments of the RSOs of the GPS satellites, and on the accuracy of GRACE RSOs which have been invented newly to support radio occultation analysis with GRACE enhancing the CHAMP and SAC-C data set. Also given are some tests on the impact of ambiguity fixing and dense GPS clocks. These approaches are due next for the upgrade of the operational processing system.

The instruments of CHAMP provide data for use in POD, such as space-borne Global Positioning System (GPS) Satellite-to-Satellite Tracking (SST) observations, onboard accelerometer measurements, attitude, thruster firing and other POD relevant information from the housekeeping data. The ground based data are GPS data of the CHAMP low latency network, other ground GPS data from the International GNSS Service (IGS, see Beutler et al. (1999), IGS (2005)), and Satellite Laser Ranging (SLR) data from the International Laser Ranging Service (ILRS, see Pearlman et al. (2002), ILRS (2005)). The same holds true for the GRACE satellites, where however the SST observations only are exploited for the RSO. K-band intersatellite range observations as well as the attitude etc. data are omitted because they do not arrive in time. Also in case of SAC-C we must rely on space-borne GPS observations alone.

In all POD applications described in the following, the data are evaluated by GFZ's EPOS-OC (Earth Parameter and Orbit System - Orbit Computation) software system in version 5.4 at the time of writing this.

2 CHAMP Rapid and Ultra-Rapid Orbit Products

Modelling standards and earlier quality results for the CHAMP RSO and USO are given e.g. in Michalak et al. (2003). Recent efforts concentrated on improving and accelerating the pre-processing system. They resulted in more accurate GPS orbits with lower latency. Fig. 1 shows the comparison of the GPS RSO orbits to IGS Rapid Orbits (IGRs) after having applied a Helmert transformation in terms of Root Mean Square (RMS) values of position differences per axis, Fig. 2 the comparison of the GPS USOs to the IGRs. The IGRs are taken as a reference as IGS claims that their accuracy is better than 5 cm (IGS, 2005). Improvements concerned the optimization of the selection of approximately 50 stations of the GPS ground network. In effect since September 20, 2004, (marked by a dashed vertical line in Fig. 1 and 2), indeed less outliers can be noticed for both the RSO and the USO. Currently the GPS RSO shows 7.5 cm RMS versus IGR, the USO 8.5 cm. The USO is slightly less accurate because it is generated with a latency of approximately two hours after the last observation versus a latency of 17 hours for the RSO (the IGR also comes with a latency of 17 hours). Therefore the set of observations for the USO may lack data from some receivers, making the ground station network less optimal.

A validation of the RSOs of the GPS satellites PRN G05 and PRN G06 by SLR observations is performed for orbits since the beginning of year 2004. For that the GPS based orbits are fixed and compared to the SLR observations. Eventually the SLR residuals are compiled in Fig. 3. They exhibit a systematic bias of -5 cm, their standard deviation is 4.9 cm. The bias here is consistent with previously published results (e.g. Urschl et al. (2005)). Con-

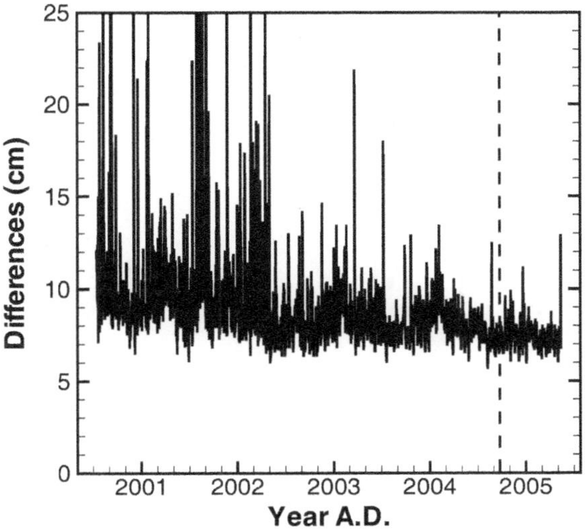

Fig. 1. Comparison of the GPS RSO to the IGR

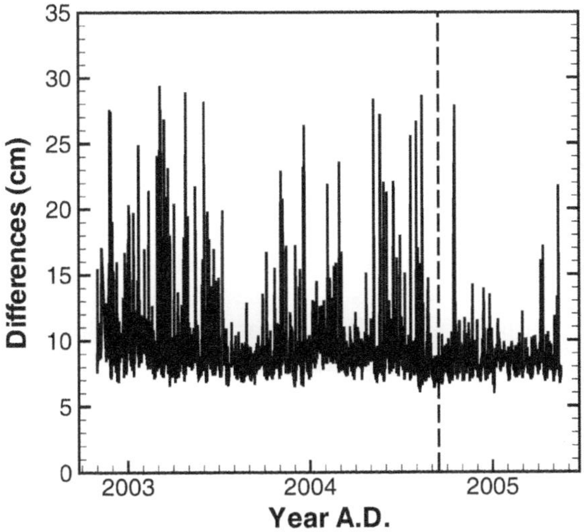

Fig. 2. Comparison of the GPS USO to the IGR

cluding from the SLR validation, a radial accuracy of 5 cm of the GPS RSOs can be assessed.

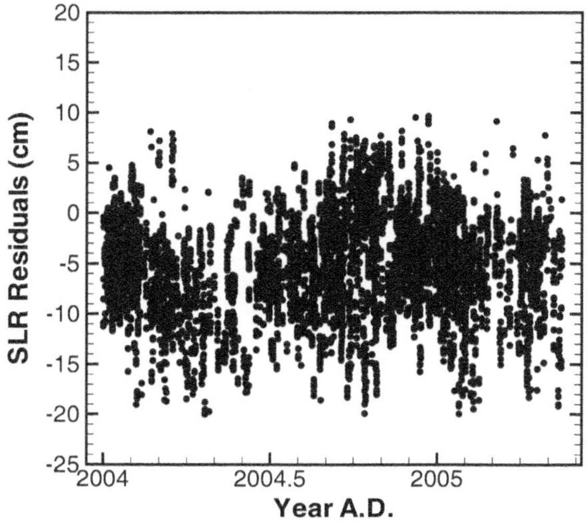

Fig. 3. SLR validation of the RSOs of PRN G05 and PRN G06

For the determination of CHAMP RSO and USO orbits, the respective GPS RSO and USO orbits and clocks are fixed. The resulting accuracies of the CHAMP RSO orbits can again be assessed by SLR validation. For the recent period the RMS is around 5.5 cm. It should be noted here in general, that the SLR data are taken as is, i.e. the RMS values can be contaminated by outliers. In addition, the SLR observations can be located at the beginning or at the end of an arc, which, due to the known dissipations of dynamical orbits at those periods, increases the RMS values as well.

A second assessment of CHAMP RSO accuracy is performed by sampling the position differences of subsequent orbits in the middle of the 2-hour period where the orbits overlap. The recently computed mean of the sampled position differences amounts to 5.0 cm. This is in good agreement with the SLR RMS and validates therefore the possibility to use the overlap analysis as accuracy assessment.

SLR validation and overlap analysis are also used to asses the accuracy of the CHAMP USO. The global SLR RMS is 7.4 cm. This is larger than in case of the CHAMP RSO due to its dependency on less accurate GPS USO orbits and because of more frequent occurences of gaps in the CHAMP SST observations. In Fig. 4 the position differences and their medians of overlapping arcs at epochs distant by 0 to 2 hours from the end of the preceding arc are given. The most critical part of the CHAMP USO orbit is its end, the last 15 minutes, where the median values are quite large, between 13 and 29 cm.

The main reason is found with poor accuracies of GPS USOs for the last 1 hour of the arc due to lacking data. Meanwhile an effort has been started to improve the acquisition of GPS ground data covering the last 1-2 hours of the GPS USO.

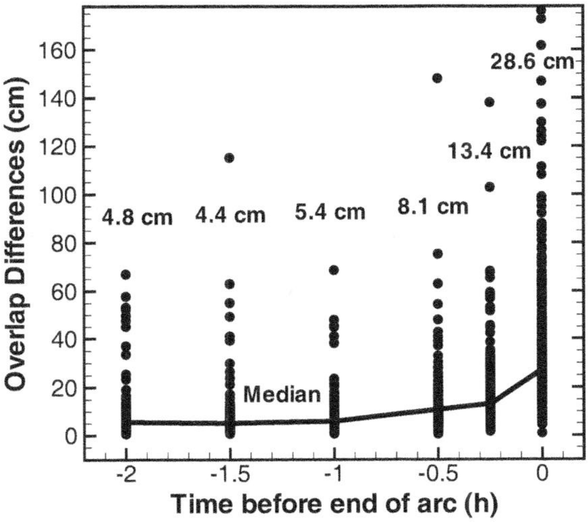

Fig. 4. CHAMP USO orbit accuracies

The GPS and CHAMP USOs are produced as pre-requisite for occultation data processing, which in turn generates atmospheric profiles or related products for use in Numerical Weather Prediction (NWP). The age of input data to NWP applications must not exceed three hours. The latencies of the CHAMP USO are given in Fig. 5. The recent improvement of pre-processing procedures by parallel acquisition and pre-processing of GPS ground data introduced on April 20, 2004, resulted in a reduction of the latency from 3.5h to 2.2h in mean. Further reductions are still possible by switching from a 3-hourly processing interval to dump-dependent processing. In case of CHAMP, the polar receiving station has view of the satellite during each revolution, i.e. approximately each 1.5 hours. Then the onboard data, the GPS SST observations etc., can be sent to the ground or dumped respectively.

3 SAC-C and GRACE Rapid Orbit Products

Recently the CHAMP RSO processing system was extended to generate orbits for three more occultation measuring satellites: SAC-C, GRACE A and GRACE B. The SAC-C satellit has no SLR reflector, so for accuracy assessment the overlap values only are available. The results are given in Table 1.

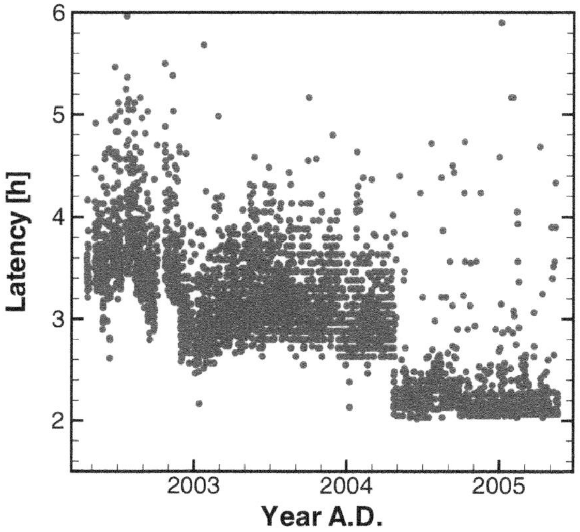

Fig. 5. CHAMP USO latencies

The mean overlap position difference 5.4 cm is close to the value for CHAMP, i.e. 5.0 cm. Since the modelling standards for both satellites are rather similar, it can be concluded from the overlap analysis that the accuracy of the SAC-C orbits is close to that of the CHAMP RSO.

In addition to CHAMP and SAC-C, the RSO for both GRACE satellites is produced since October 2004. Though, at the time being, the GRACE occultation measurements are switched off, permanent switch on is planned. Therefore the generation of the GRACE RSOs keeps going as long as resources allow. Recent accuracy assessments for both GRACE RSOs are compiled in Table 1, too. SLR RMS values are as large as those of CHAMP, but overlaps are about half as large as those of CHAMP and SAC-C. As the GPS receivers onboard the GRACE satellites deliver higher quality data, it can be concluded that the GRACE RSOs are of higher quality than the CHAMP and SAC-C RSOs.

Table 1. SAC-C RSO and GRACE RSO accuracies

	SLR RMS (cm)	Overlap Mean (cm)
SAC-C	-	5.4
GRACE A	5.2	2.8
GRACE B	4.8	2.9

4 Increasing the Accuracy of GPS and LEO Orbits

Ambiguity fixing (Mervat, 1995) for GPS observations is tested for a small sample of the GPS Post-processed Science Orbits (PSOs, 30 s ephemerides and clocks for sub-sequent gravity field processing). Table 2 summarizes the comparison of the standard and the ambiguity-fixed PSOs to the IGS final orbits for three 1.5-d arcs of May 2002. The IGS final orbits are considered as a reference because IGS claims, as in case of the IGR, that their accuracy is better than 5 cm (IGS, 2005). For further assessment, two out of all individual contributions to the combination of the IGS final orbits, the final orbits of the CODE and GFZ IGS analysis centers, are compared the same way as the PSOs to the IGS final orbits.

From Table 2 it can be conluded that ambiguity fixing improves the accuracy of the PSOs considerably. GFZ final and CODE final orbits should be as close as 2 cm to the IGS final orbits according to the IGS combination reports. However the values in Table 2 differ quite largely from this particularly for the GFZ finals. The reason being the weighting scheme applied in the combination whereas the results in Table 2 are derived from straightforward differences of all satellites being equally weighted.

Table 2. Impact of ambiguity fixing. Differences in position per axis for various orbits versus IGS final orbits

Arc	Standard PSO RMS (cm)	PSO with ambiguity fixing RMS (cm)	GFZ final RMS (cm)	CODE final RMS (cm)
2002.05.01	13.8	9.9	10.2	3.6
2002.05.03	11.4	6.9	8.5	3.2
2002.05.05	9.7	5.7	7.0	3.1
Mean	11.6	7.5	8.6	3.3

The GPS PSO (standard and with ambiguity fixing) was next used to generate CHAMP RSO type orbits for the period 2003.08.01 - 2003.08.14. Some arcs were excluded a priori because of gaps in the GPS clock solutions. Generally the CHAMP RSO is generated using the 5 minutely spaced ephemerides and clocks of the GPS RSOs. The 5-minute clocks are then being linearly interpolated to 30-second clocks. The impact of these different GPS orbits and clocks on CHAMP RSO accuracy can be seen in Table 3. The largest impact comes from proper 30-second clock solutions, case GPS PSO, for which the CHAMP SLR RMS drops drastically. The ambiguity fixed PSOs improve the CHAMP orbits additionally. Ambiguity fixing as well as improved interpolation of the 5-minute clocks of the GPS RSO will be implemented in the next future.

Table 3. Impact of GPS clocks and ambiguity fixing on CHAMP RSO type orbits measured by independent SLR residuals. CHAMP arcs where attitude and thruster data are missing, are marked by (*)

Arc	Standard GPS RSO RMS (cm)	Standard GPS PSO RMS (cm)	PSO with ambiguity fixing RMS (cm)	Number of SLR normal points
030801 10:00	3.68	2.08	2.00	167
030801 22:00	2.94	3.47	4.49	51
030802 10:00	3.81	2.46	2.54	139
030802 22:00	5.29	4.57	4.38	28
030803 10:00*	4.53	2.35	1.40	59
030803 22:00	0.13	1.10	0.04	1
030804 10:00	2.52	2.06	1.68	148
030804 22:00	2.90	2.72	3.59	45
030805 10:00	4.29	4.50	3.54	96
030805 22:00	5.11	3.17	3.47	122
030807 22:00	3.51	1.16	2.54	47
030808 10:00*	4.90	4.90	4.49	120
030808 22:00*	5.60	4.00	4.73	38
030810 10:00*	2.99	4.22	5.79	162
030810 22:00	4.94	5.67	4.33	66
030811 10:00	3.93	3.52	3.06	170
030811 22:00	3.51	3.89	2.93	81
030812 10:00	3.06	3.60	2.86	194
030812 22:00	4.19	3.05	2.25	49
030813 10:00	5.69	2.64	3.18	232
030813 22:00	6.02	3.29	3.48	52
030814 10:00	5.73	5.09	4.58	51
Global SLR RMS				
All arcs	4.24	3.48	3.45	2118
Arcs (*) excluded	4.24	3.30	3.08	1739

Another possibility for improving the LEO orbit accuracies is to use the integrated approach (Zhu et al., 2004) where all LEO and GPS orbits and the ground station coordinates are estimated in one step. Some results for a few GRACE 1.5-day arcs under different observation scenarios are given in the cited article. Here the integrated approach is applied for two months of GRACE A/B 1-day orbits and shown in Table 4. For comparison, also RMS values of SLR residuals are given for GRACE orbits produced during gravity field screening and for JPL reduced dynamic orbits. In the gravity screening runs, accelerometer data and empirical forces were used to achieve good initial orbits. For the integrated solution, solely accelerometer data were used. The independent SLR RMS for the integrated solutions is slightly larger than for the JPL solution. The difference can be deduced to gaps in the accelerometer data in the integrated solution. Therefore the integrated approach can produce LEO orbits accurate on the level of 2-3 cm.

Table 4. GRACE A and B orbit accuracies for three different solutions measured independently by 9872 SLR normal points for the period 2003.07.02-2003.08.31

Solution	RMS (cm)
Routine gravity screening (1.5d arcs, accelerometer + emp. coeff.)	5.15
Integrated (1-step) solution (1d arcs, accelerometer only)	2.92
JPL solution (reduced dynamic)	2.33

5 Summary and Conclusions

Rapid and ultra-rapid GPS, CHAMP, SAC-C and GRACE orbits generated operationally by GFZ e.g. for GPS radio occultation applications are accurate and reliable products. Recent improvements concern the optimized selection of a suitable GPS ground station network that resulted in more reliable GPS RSOs and USOs. Faster procedures for data acquisition and pre-processing led to considerable smaller latencies of the USOs. By applying ambiguity fixing and accurate GPS clock interpolation the LEO orbits can be generated on an operational basis with an anticipated accuracy of 2–3 cm versus the current 4–5 cm. Further accuracy improvements are possible by the integrated approach which, due to its large needs on computational resources, seems at this time to be of practical relevance only if it is applied offline. As demonstrated by the adoption of the SAC-C and GRACE A/B RSOs, the GFZ operational system is prepared to accomodate further future LEO missions carrying onboard GPS where fast and accurate orbits are required.

Acknowledgement. This is publication no. GEOTECH-186 of the program GEOTECHNOLOGIEN of BMBF and DFG, Grant 03F0333A. Proxies for Earth orientation parameters and ground station coordinates were provided by the International Earth Rotation and Reference Systems Service (IERS) and the IGS, part of the GPS ground data by the IGS, and part of the SLR data by the ILRS.

References

Beutler G, Rothacher M, Schaer S, Springer TA, Kouba J, Neilan RE (1999) The International GPS Service (IGS): An Interdisciplinary Service in Support of Earth Sciences. Adv Space Res 23:631-635

IGS (2005) International GPS Service Site. Internet: http://igscb.jpl.nasa.gov, modified April 02, 2005.

ILRS (2005) International Laser Ranging Service Site. Internet: http://ilrs.gsfc.nasa.gov, cited as of May, 2005.

ISDC (2001) CHAMP Information System and Data Center Site. Internet: http://isdc.gfz-potsdam.de/champ/, modified May 17, 2001.

König R, Michalak G, Neumayer KH, Schmidt R, Zhu SY, Meixner H, Reigber Ch (2005) Recent Developments in CHAMP Orbit Determination at GFZ. In: Reigber C, Luehr H, Schwintzer P, Wickert J (eds) Earth Observation with CHAMP, Results from Three Years in Orbit, pp. 65-70, Springer, Berlin Heidelberg

Mervat L (1995) Ambiguity Resolution Techniques in Geodetic and Geodynamic Applications of the Global Positioning System. Inauguraldissertation der Philosophisch-naturwissenschaftlichen Fakultät der Universität Bern

Michalak G, Baustert G, König R, Reigber C (2003) CHAMP Rapid Science Orbit Determination - Status and Future Prospects. In: Reigber C, Luehr H, Schwintzer P (eds) First CHAMP Mission Results for Gravity, Magnetic and Atmospheric Studies, pp. 98–103, Springer, Berlin Heidelberg

Pearlman MR, Degnan JJ, Bosworth JM (2002) The International Laser Ranging Service. Adv Space Res 30:135-143

Reigber Ch (2005) CHAMP Project Site. Internet: http://www.gfz-potsdam.de/pb1/op/champ/index_CHAMP.html, modified May 02, 2005.

Urschl C, Gurtner W, Hugentobler U, Schaer S, Beutler G (2005) Validation of GNSS orbits using SLR observations. Adv Space Res, in press

Zhu S., Reigber C., König R. (2004) Integrated adjustment of CHAMP, GRACE, and GPS data. J of Geodesy 78:103-108

Harmonic Analysis of the Earth's Gravitational Field from Kinematic CHAMP Orbits based on Numerically Derived Satellite Accelerations

Tilo Reubelt, Martin Götzelmann, and Erik W. Grafarend

Geodätisches Institut, Universität Stuttgart, Germany
reubelt@gis.uni-stuttgart.de

Summary. Based on very accurate kinematic CHAMP orbits, a new CHAMP gravitational field model was computed by means of a (point-wise) acceleration approach. In order to implement such an acceleration approach, the satellite's acceleration has to be derived from the kinematic CHAMP orbits by means of interpolation and subsequent numerical differentiation. The iterative method of preconditioned conjugate gradients is implemented to solve the large linear system of equations for the spherical harmonic coefficients. If appropriate preconditioning is applied, convergence can be reached within 7 – 15 iterations. An important topic concerning the accuracy of the gravity field solutions is the detection and filtering or down-weighting of spikes, jumps, outliers and inaccurate data in the kinematic orbits. These problems are adressed by data-preprocessing or robust estimation. Different gravity field solutions up to degree and order 90 were computed, where validation exhibits a signal-to-noise (S/N) ratio per degree of S/N $\geq$ 1 for coefficients up to degree 80 and S/N $\geq$ 2 for coefficients up to degree 70. Comparisons to different CHAMP-models, which were obtained by application of alternative algorithms, prove that the acceleration approach can compete with other methods of gravity field determination.

Key words: gravity field determination, CHAMP, low earth orbiting satellite (LEO), numerical differentiation, robust estimation, wavelet filter

1 Introduction

The classical, dynamic approach for the analysis of the high-low SST (satellite-to-satellite-tracking) measurements between the low Earth orbiting (LEO) CHAMP-satellite and GPS-satellites is based on the former methods developed for Laser-Ranging-observables and connects the GPS-observables directly with the gravity field parameters (Reigber, 1989). The relation between the satellite orbit and the gravity field parameters is achieved by the integration of the variational equations. With the proceeding CHAMP mission various models, e.g. EIGEN-1S (Reigber et al., 2002), EIGEN-2 (Reigber et al., 2003), EIGEN-3p and EIGEN-CHAMP03S (Reigber et al., 2005a), have been estimated with this method which indeed led to an improvement of state-of-the-art gravity field models. Besides this classical *one-step-method* alternative approaches have been developed in context with CHAMP. These are based on kinematic orbits and can be classified as *two-step-methods* (1. GPS-observations → kinematic orbits; 2. kinematic orbits → gravity field parameters). Applied alternative two-step-methods are the *energy-balance-approach* (Földváry et al., 2005), the *short-arc analysis* formulated as *boundary value problem* (Mayer-Gürr et al., 2005) and *acceleration approaches* (*point-wise*, see Austen et al. (2002), Reubelt et al. (2003a,b) and *average*, see Ditmar and van Eck van der Sluijs (2004)). The implementation of such algorithms is motivated by the fact, that kinematic orbits can nowadays be determined with an accuracy of less than 5 cm (Švehla and Rothacher, 2003, 2004), offering comparable results to the classical approach.

The two-step algorithms working with kinematic orbits can be classified as fast regarding the computation time, which is achieved by the underlying linear system of equations. Especially the acceleration approaches will be efficient, since any kind of integration of the force function is avoided and instead the comparable fast process of numerical differentiation is applied. Numerical differentiation, which normally increases the noise, is a less critical procedure within acceleration approaches, if the orbit is correlated. In this case, the noise may be reduced by numerical differentiation (Reubelt et al., 2003a,b). Indeed the correlation of kinematic orbits can be proven by either a comparison to dynamic orbits (Reubelt et al., 2003a,b) or by error-propagation of kinematic orbits (D. Švehla, personal communication).

Motivated by the explained advantages, detailed simulations and first results from real data analysis (Reubelt et al., 2003a,b), the CHAMP real data analysis within the subproject of the GIS in CHAMP-DACH was performed with a point-wise acceleration approach.

2 The point-wise acceleration approach

The acceleration approach is briefly outlined in this section, detailed explanations including mathematical formulas can be found in Austen et al. (2002) and Reubelt et al. (2003a). Normally, kinematic CHAMP orbits are given with respect to a Conventional Terrestrial System (CTS). In order to obtain accelerations free from frame accelerations, the CHAMP positions have to be transformed into the Conventional Inertial System (CIS). The satellite acceleration vector is derived from the positions by differentiating twice an interpolation function, which was fitted to the orbit. The Gregory-Newton-interpolation turned out to be an appropriated method for application in *numerical differentiation*. By reduction of non-conservative (non-gravitational) and gravitational (tides) disturbing accelerations the terrestrial gravitational acceleration vector can finally be obtained. While tidal effects as the direct attraction of sun, moon and the planets, the solid Earth and pole tides and the ocean tides can be modeled with sufficient accuracy, non-conservative effects caused by satellite surface forces as atmospheric lift and drag, solar radiation pressure and the Earth's albedo are measured with the in-situ CHAMP STAR accelerometer. Calibration parameters - determined by GFZ - for the correction of the bias, tilt and scale of the accelerometer instrument can be downloaded from CHAMP-ISDC at GFZ. Due to some major problems, as explained in Sect. 6, the satellite accelerations were not reduced by the accelerometer measurements. According to Newton's Law of motion, the "reduced" accelerations are balanced by the gradient of a spherical harmonic geopotential model in order to set up the linear system of equations for the determination of the spherical harmonic coefficients of the gravitational field. The gradient is naturally computed by the spherical partial derivatives in the local spherical system (normalized tangential system) and must therefore be transformed into the CIS (via the CTS). A method for the solution of the large system of equations and important aspects of data-preprocessing and weighting are addressed in Sects. 4 and 5.

3 Numerical differentiation

For derivation of the satellite's acceleration from kinematic orbit positions, numerical (double) differentiation has to be applied. In general, several possibilities exist for numerical differentiation, whereas the most important are the Fourier (FFT)-approach and differentiation based on interpolating splines and polynomials. The FFT-approach is not further considered since data gaps and outliers as well as sudden changes in the signal (orbit maneuvers,...) cause serious problems and the result is contaminated from edge-effects and aliasing-effects (Weigelt and Sneeuw, subm.). Due to a less oscillating nature, cubic splines instead of polynomials are generally suggested for interpolation. Problems in using spline-interpolation are the derivation of boundary values and the cause of edge-effects in the case of inaccurate boundary values. Compared with polynomials, a longer time-series is needed to apply spline-interpolation, which can be problematic in the presence of data-gaps and outliers. The aforementioned difficulties can be handled, if polynomial interpolation is adopted. Oscillations are avoided or marginal, when the order of the polynomial is not too high and the polynomial is shifted point-wise for interpolation. In our algorithm, numerical differentiation based on Gregory-Newton-interpolation is implemented. Details exceeding the following brief overview can be found in Reubelt et al. (2003a,b). According to Engeln-Müllges and Reutter (1987), Gregory-Newton-interpolation is a n-point-interpolation-scheme and can be expressed as a product-sum of *binomial coefficients*, containing the time t (sampling time Δt), and *forward-differences* of positions $\mathbf{X}_k$. By means of double-differentiation of the *binomial coefficients* with respect to time (2) and by expressing the *forward differences* $\mathbf{\Delta}^i_{1+i/2}$ in terms of *coordinate dif-*

ferences $\Delta\mathbf{X}_k^{k+1}$ (baselines) (3) the satellite accelerations $\ddot{\mathbf{X}}$ can be determined by numerical differentiation with (1).

second derivative of the Gregory-Newton-interpolation formula

$$\ddot{\mathbf{X}}(t) = \sum_{i=1}^{n-1} \binom{q}{i}^{//} \boldsymbol{\Delta}^i_{1+i/2} \tag{1}$$

with time difference quotient $q = \frac{t-t_1}{\Delta t}$

second derivative of binomial coefficients

$$\binom{q}{i}^{//} = \frac{1}{\Delta t^2\, i!} \sum_{j=0}^{i-1} \frac{\sum_{k=0}^{i-1} \left(\frac{q-j}{q-k}\right) - 1}{(q-j)^2} \prod_{l=0}^{i-1} (q-l) \tag{2}$$

forward differences in terms of baselines

$$\boldsymbol{\Delta}^i_{1+i/2} = \sum_{k=0}^{i-1} (-1)^{i-1+k} \binom{i-1}{k} \Delta\mathbf{X}_{k+1}^{k+2} \quad \text{with baselines} \quad \Delta\mathbf{X}_k^{k+1} = \mathbf{X}_{k+1} - \mathbf{X}_k \tag{3}$$

For reasons of accuracy the accelerations are only determined at the central point of the interpolation scheme. The resulting central difference filter is displayed in Fig. 1a.

(a)

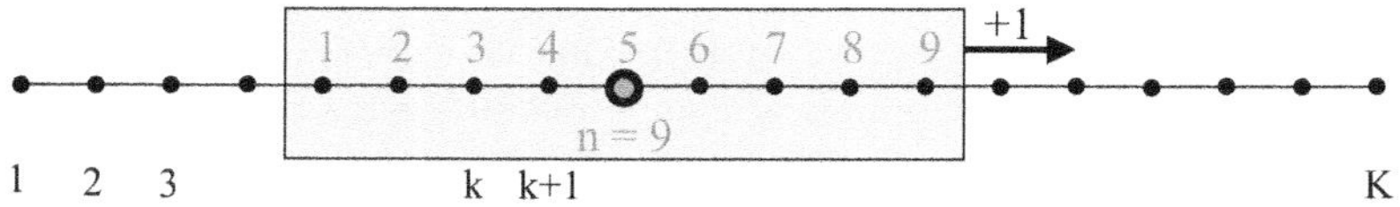

(b)

$-\frac{1}{560\Delta t^2}$	$\frac{8}{315\Delta t^2}$	$-\frac{1}{5\Delta t^2}$	$\frac{8}{5\Delta t^2}$	$-\frac{205}{72\Delta t^2}$	$\frac{8}{5\Delta t^2}$	$-\frac{1}{5\Delta t^2}$	$\frac{8}{315\Delta t^2}$	$-\frac{1}{560\Delta t^2}$

Fig. 1. (a) second derivative of Gregory-Newton-interpolation as filter and (b) its filter coefficients for a 9-point-scheme

To guarantee a good approximation, at least a 7-point-scheme must be used. On the other hand, the propagation of orbit-noise increases with a higher-order-interpolation scheme. The 9-point scheme (Fig. 1b) offers a good compromise between approximation and propagation of errors.

Generally, a major drawback involved with numerical differentiation is an increase of noise. This holds for white noise and it can be shown (Reubelt et al., 2003a,b), that the noise induced errors of the accelerations can be diminished if the coordinates are correlated. To demonstrate this, the *second order formula of Gregory-Newton-interpolation* (1) was expressed in baselines (relative coordinates) instead of absolute coordinates. Due to the correlation of positions, these baselines can be determined with higher accuracy than the absolute coordinates, similar to DGPS. By introduction of such baselines it can be explained why numerical differentiation with (1) enables the damping of noise. The effect of correlation is briefly outlined in Figs. 2a,b,c by means of a simulated erroneous orbit ($\sigma_X = 3\,\text{cm}$, $\rho = 0.97$), which was generated according to earlier real data investigations (Reubelt et al., 2003a,b) and an improved accuracy of real CHAMP kinematic orbits (Švehla and Rothacher, 2004). While the absolute orbit accuracy in Fig. 2a is 3 cm, the accuracy of the baselines is enhanced in Fig. 2b to about $\sigma_{\Delta X} = 7\,\text{mm}$. If the correlation of the orbit is assumed to $\rho_2 = 0.8$, the accuracy of the baselines only improves up to $\sigma_{\Delta X} = 1.9\,\text{cm}$ (not shown in the figure). The higher accuracy of the baselines for $\rho = 0.97$ leads to an accuracy of the accelerations in the level of $\sigma_{accl} = 1.6{\cdot}10^{-5}\ \text{m/s}^2$ (Fig. 2c) in contrast to an accuracy of $\sigma_{accl} = 4.5{\cdot}10^{-5}\ \text{m/s}^2$ for the lower correlated orbit ($\rho_2 = 0.8$). This means, that the level of accuracy of the accelerations and furthermore of the corresponding gravity solution is rather determined by the correlations and the accuracy of baselines than the absolute accuracy of the orbit.

Figs. 2d,e,f try to oppose the behavior and accuracy of a real CHAMP kinematic orbit (Švehla and Rothacher, 2003, 2004) to the simulations. The accuracy of a real kinematic orbit is difficult to judge since the truth is not known. Thus the quality and accuracy of the kinematic orbit, its baselines and accelerations are validated by a comparison to a reduced dynamic orbit (Švehla and Rothacher, 2003, 2004), which is based on the best present-day gravity field model EIGEN-GRACE02S (Reigber et al., 2005b) and which serves due to its smooth behavior as a good reference for evaluation. The orbit/baseline acceleration differences between the kinematic and reduced-dynamic orbit, illustrated in Figs. 2d,e,f, show RMS values of $2\,\text{cm}/6\,\text{mm}/1.5{\cdot}10^{-5}\ \text{m/s}^2$ (outliers and data gaps are neglected) and thus are in good agreement with the results of the simulated correlated orbit, displayed in Figs. 2d,e,f. Additionally the orbit-, baseline- and acceleration differences between two (smooth) reduced-dynamic-orbits (Švehla and Rothacher, 2003, 2004) based on EIGEN-GRACE02S and the less accurate EGM96 (Lemoine et al., 1998) are plotted, which show RMS-values of 2 cm, 3 mm and $3{\cdot}10^{-6}\ \text{m/s}^2$. The small differences between the baselines and accelerations of both reduced dynamic solutions prove that the reduced-dynamic orbits provide a good reference for validating the accuracy of kinematic baselines and accelerations. Only the estimation of the absolute accuracy of the kinematic orbit by a comparison to the reduced-dynamic orbit is difficult, since the absolute accuracy and long wavelength-behavior of the reduced-dynamic orbit may be in the same range,

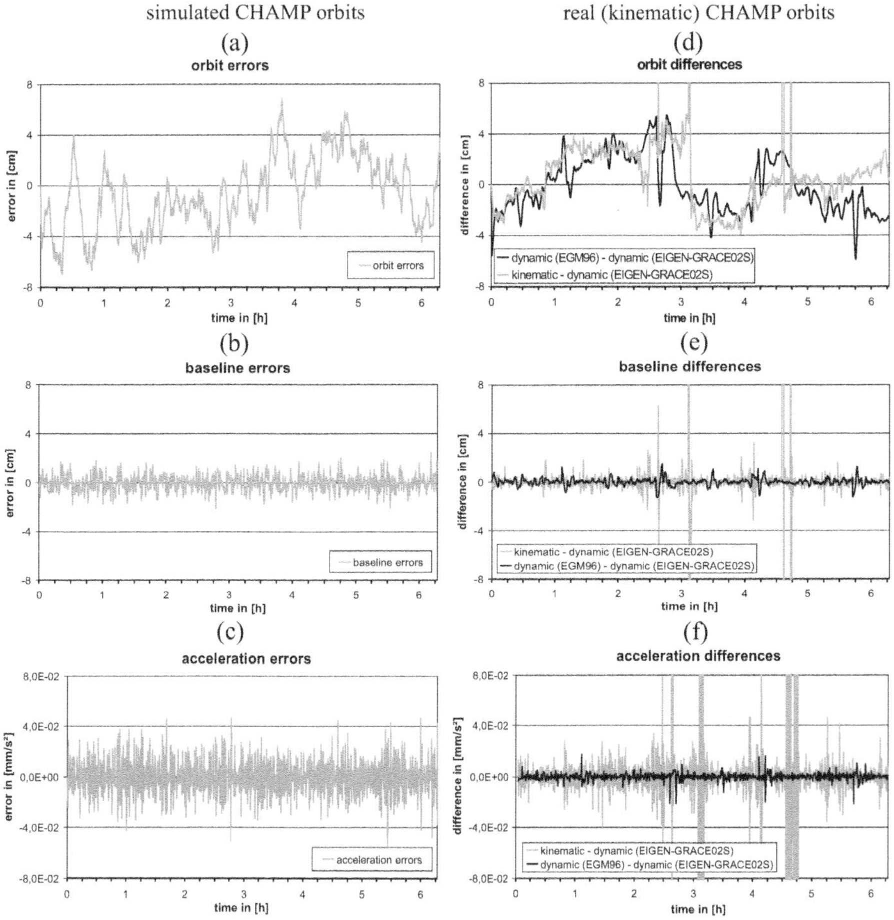

Fig. 2. (a),(b),(c) orbit/baseline/acceleration errors of a simulated CHAMP orbit (h = 470 km, $\Delta t = 30\,\mathrm{s}$, $\sigma_X = 3\,\mathrm{cm}$, $\rho = 0.97$) in the CIS (x-component); (d),(e),(f) orbit/baseline/acceleration differences between real CHAMP orbits (year 2002, day 200, GPS time, $\Delta t = 30\,\mathrm{s}$) in the CIS (x-component); bright: differences between kinematic and reduced-dynamic orbit (EIGEN-GRACE02S), dark: differences between 2 reduced-dynamic orbits (EIGEN-GRACE02S, EGM96)

as indicated by the orbit-difference between the two reduced-dynamic orbits. From external validation to SLR-data and orbit-adjustment-residuals (Švehla and Rothacher, 2003, 2004) we get the absolute accuracy of the kinematic orbit of about 2 - 3 cm. A comparison of this absolute accuracy to the baseline accuracy of $\leq 1\,\mathrm{cm}$ gives a hint, that kinematic orbit data is correlated, although the correlation may differ slightly from the simulations. At least the comparable accuracy of baselines and accelerations from simulations and real kinematic orbits indicate that real data processing will confirm the promising results from simulations (for instance in Reubelt et al. (2003a,b)).

The correlation of kinematic orbits originates from the natural correlation of GPS-measurements, from signal delays (ionosphere) and from the procedures (e.g. ambiguity and clock fixing) adopted in the orbit adjustment. Additionally to the comparison between kinematic and reduced-dynamic orbits, the variance-covariance-matrix of kinematic orbits clearly reveals such correlations (D. Švehla, personal contact).

For completeness it should be mentioned, that also smoothing methods like polynomial regression and smoothing splines have been tested in simulations in order to reduce the influence of noisy observations. These methods were not adopted to real data, since the accuracy of the estimated gravity decreased though the internal statistics of the accelerations improved. The reason for this might be, that not only the noise was reduced by smoothing, but also the signal.

4 Iterative solution of linear system of equations

For the determination of the gravity field parameters from the 2-years kinematic orbit, a system of equations consisting of 6 millions of observations and 8278 unknowns for the maximum degree $L = 90$ has to be solved. This may lead to two basic problems, namely the storage of the large design and normal matrices and the time-consuming computation of the normal matrix. The algorithm can be shifted to a super-computer or an iterative solution can be aimed concerning these problems. Iterative methods are able to deal with restricted memory, since the normal matrix must not be built up and the design matrix must not be stored. If iterative solvers in terms of preconditioned conjugate gradients (PCG) are implemented, the computations can be performed on a standard PC. Mathematical details on the method of preconditioned conjugate gradients, which led to a fast and stable convergence within 7 – 15 iterations in all computations, can be found in Ditmar and Klees (2002). In Reubelt et al. (subm.) the preconditioner, which is implemented as a block-diagonal approximation of the normal matrix consisting of one submatrix per order m, is examined in more details.

5 Data preprocessing and robust estimation

5.1 Data preprocessing

The gravity field determination in this contribution is based on a two years kinematic CHAMP orbit of the period March 2002 – March 2004, which was kindly provided by D. Švehla and M. Rothacher (Forschungseinrichtung Satellitengeodäsie at Technical University Munich). In contrast to (reduced) dynamic orbits, which depend on a model, kinematic orbits are generated purely from the geometric information of GPS phase-observations and pseudoranges. Kinematic orbits thus are not as smooth as (reduced) dynamic orbits and may contain data gaps, outliers and jumps, as visible in Fig. 2d. While the former two phenomena are mainly caused by an insufficient number of observed GPS satellites or a bad satellite-constellation, the latter can be assigned to a changing GPS-constellation.

Since the implemented method of the acceleration approach is very sensitive to outliers and jumps, as proved by simulations (Götzelmann et al., subm.), data-preprocessing is a very important aspect in gravity field determination. In the acceleration approach, the removal of outliers can mainly be applied at two levels. The first opportunity is the preprocessing of the given kinematic orbit (or its baselines) and, since errors propagate into numerical differentiation, the derived accelerations offer a second platform for outlier removal strategies.

The most natural way for data preprocessing would be to detect outliers by means of the variance-covariance matrix of the kinematic orbit, which was propagated from the orbit adjustment. By setting a threshold value for the orbit variances, inaccurate orbit observations can be removed. A weakness of this procedure is revealed by a comparison between orbit variances and orbit differences (kinematic - reduced dynamic): outliers and inaccurate positions do not always coincide with large orbit variances and vice versa. Moreover, jumps in the orbit may hardly be discovered from the orbit variances (only a reduced correlation between two positions may give a hint for the presence of jumps).

To overcome the mentioned problems, data preprocessing by comparisons to the smooth reduced dynamic orbits and accelerations computed from an existing model was tested. A very simple method would be to set a threshold value for the orbit and/or baseline differences between the kinematic and the reduced dynamic CHAMP orbit and remove all observations which exceed this limit. The comparison is more valuable on the level of baselines than on the level of positions, since reduced dynamic baselines provide a better reference for evaluation than absolute coordinates (as already explained in Sect. 3) and jumps can easier be detected. Thus, a kinematic baseline indicating a difference to a reduced dynamic baseline of 5 cm can already be classified as outlier whereas an orbit difference of 5 cm cannot necessarily be interpreted as an outlier. For orbit differences, a higher threshold-value, e.g. 10 cm or 20 cm must be set to ensure, that the difference is not caused by an inaccurate reduced dynamic orbit. In a second step, the determined accelerations can be compared with the accelerations of the reduced dynamic orbit or directly with accelerations computed from existing gravity field models, which were applied for the computation of the reduced dynamic orbits. Here, the direct comparison to accelerations estimated from a gravity field model was chosen. Due to gravity signal attenuation at the satellite orbit, degrees higher than $l \geq 90$ only contribute marginally to accelerations and thus the used gravity field model is only developed up to degree and order 90. From

Sect. 3 it is concluded, that a threshold value of 5 - $10 \cdot 10^{-5}$ m/s^2 for acceleration differences should be applied. This ensures that all outlying accelerations can be removed and the acceleration differences are not caused from errors in the applied gravity-field.

A more elegant and mathematically well-defined method to remove small, temporary occurring outliers from the input data set are wavelet filter techniques, which are based on fast discrete wavelet transformation. Due to their time localizing ability, these are very appropriate for detecting and removing local signal occurrences without effecting the remaining parts of the signal. By means of the fast discrete wavelet transformation the input signal is developed into a consecutive series expansion of approximation signals and detail signals of increasing scales. Fast wavelet transformation is applicable for orthogonal wavelets with compact support (finite number of corresponding filter coefficients). Daubechies wavelets of order 1 (Haar wavelet) and 2 were applied. All local spikes and outliers within the signal are solely mapped to the coefficients on the smallest scales. Considering multiples of the mean signal energy on these small scales, scale-dependent thresholds are computed. By localising the signal points, which correspond to the identified wavelet coefficients, outliers and bad data can be removed from the observation data set. Such wavelet techniques are applied first to the orbit differences between kinematic and reduced dynamic orbits and second to the difference of accelerations derived from the kinematic orbit and from a gravity field model. A comparison on the level of baselines is not necessary since the wavelet-filter enables already the detection of spurious data from orbit differences. For the wavelet-filtering of the accelerations, it proved to be sufficient to develop the reference gravity field model only up to degree 2, see Götzelmann et al. (subm.) Thus it can be ensured at this stage that the signal is not shifted to any structure (of resolution $l > 2$) of the reference field. More details about the applied wavelet-filters can be found in Götzelmann et al. (subm.).

5.2 Robust estimation

The alternative way for data preprocessing in terms of filtering is to use robust methods for parameter estimation, which are less sensitive to outlying observations than least squares estimation. Simultaneously, data weighting is addressed by means of robust estimation.

Gravity field modeling in the acceleration approach is posed as a least-squares parameter estimation problem within the Gauss-Markoff model, which presumes an underlying Gaussian normal distribution. Actually, the errors in the real data are rarely Gaussian normal distributed, and especially in the presence of outliers the probability function will distinctly depart from the Gaussian one. Common least squares is sensitive to gross errors and tends to smear outlying observations by averaging them into the solution. A natural step is to implement parameter estimation methods, whose robustness against spurious observations is superior. According to Huber (1981), robust estimation methods may, similar as least-squares adjustment, be interpreted as maximum-likelihood estimators with a different probability function. An opportunity to attain robust maximum likelihood estimations $\hat{\xi}^{(j)}$ is provided by iteratively reweighting the common least squares solution. Starting with the normal least squares estimation $\hat{\xi}^{(0)} = (\mathbf{A}^T\mathbf{A})^{-1}\mathbf{A}^T\mathbf{y}$ in the first step, the weights $w_{i,i}^{(j+1)}$ of the diagonal weight matrix $\mathbf{W}_d^{(j+1)}$ for each following iteration step are obtained from the residuals $\mathbf{v}^{(j)} = \mathbf{A}\hat{\xi}^{(j)} - \mathbf{y}$ of the preceding iteration step. The weights can for instance be determined with Huber's method: 1 if $|v_i| \leq a$; $a/|\mathrm{v_i}|$ if $|v_i| \geq a$. This means, that all observations, whose residuals lie within the boundary a, are assumed to be Gaussian normal distributed, observations with larger residuals underlie a different probability function. By means of the application of such robust methods, (i) no data preprocessing is necessary since spurious observations are iteratively downweighted and (ii) data weighting can be implemented easily by means of the estimated diagonal weight matrix in the PCG method. It must be mentioned, that the correlations among the accelerations are neglected in this procedure. However, as mentioned in (Koch, 1996; Xu, 1989), robust methods for uncorrelated data also work well for correlated data. This is confirmed by the following results of robust estimation in comparison to the results of standard least squares estimation including data-preprocessing. For completeness, it must be emphasised that robust methods work, in contrast to the outlier removal and filter strategies, without any other additionally data sources like reduced-dynamic orbits or reference models. Therefore it is guaranteed, that the solution is purely gained from the kinematic orbit data itself and does not display any dependency from the reference signal.

6 Results

Based on the two-years kinematic CHAMP orbit (version 6) of the period March 2002 – March 2004 different gravity field models with application of the strategies explained in the previous section were computed. The gravity field parameters were estimated without any application of regularisation to guarantee an unbiased solution, additionally the accelerations were not reduced from non-conservative disturbing effects. The reason for the latter are results from precedent investigations (Reubelt et al., subm.) of version 3 of the two years kinematic orbit, where the reduction of accelerometer-data worsened the results slightly. This might be due to the fact, that the provided accelerometer calibration parameters are not sufficient to remove the bias and tilt correctly. Here, in future, the inclusion of in-situ-estimation of calibration parameters within the acceleration approach should be investigated, which was not implemented so far.

For the evaluation of the estimated GIS-CHAMP models, comparisons (Fig. 3, Table 1) were drawn to the recently released GRACE gravity field model EIGEN-GRACE02S, which is of superior accuracy for degrees up to

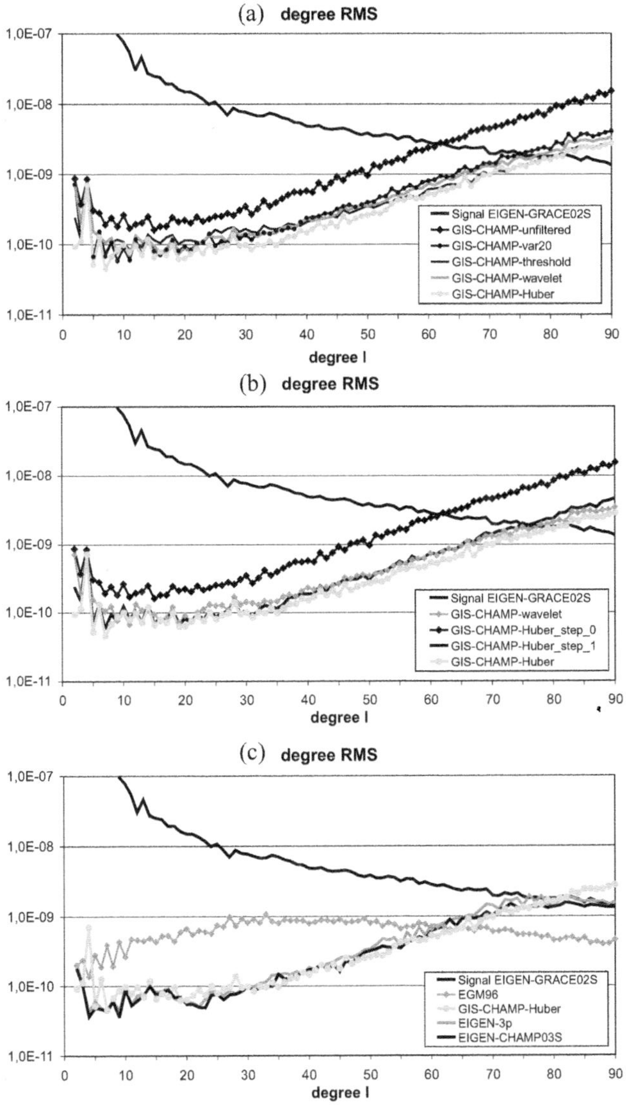

Fig. 3. Validation of different gravity field models GIS_CHAMP by means of a comparison to EIGEN-GRACE02S in terms of degree RMS; (a) comparison of models obtained from unfiltered data, various filtered data and robust estimation; (b) solutions of robust estimation (basic step, first step, final result) compared to wavelet-filtering; (c) result of robust estimation compared to models obtained with the classical approach (from a different observation period!)

100 due to the more sensitive measurement principle. To get an idea of the performance of the acceleration approach, validations of the CHAMP-models EIGEN-3P and EIGEN-CHAMP03S were added, which were estimated from GFZ-Potsdam by means of the classical approach from a time span from July 2000 – June 2003 and October 2000 – June 2003 respectively.

Fig. 3a shows the degree RMS (in comparison to EIGEN-GRACE02S) of different GIS-CHAMP models, which were obtained by application of the outlier removal and downweighting strategies described in the previous section. It can be concluded from these results, that methods for dealing with inaccurate data are very important, since the model computed from completely unfiltered data by means of least squares estimation, GIS-CHAMP-unfiltered, could be significantly improved by the procedures described in the previous section. Already by data-selection in terms of the orbit variances, where about 20% of the orbit data was filtered out (GIS-CHAMP-var20), the accuracy could be explicitly enhanced. A further advance of quality can be gained, especially for degrees > 40, if the wavelet filter is applied to the accelerations computed from the remaining orbit after preselection by means of the orbit variances. In GIS-CHAMP-wavelet, about 5% of the accelerations were additionally eliminated. The improvement by means of the wavelet-filter in contrast to detection by orbit variances can be explained by the fact, that orbit variances are not able to mark all outlying and inaccurate data. A similar result compared to wavelet-filtering can be received by the simple thresholding principle, where a higher accuracy was reached for degrees over 45 accompanied with a lower quality for some lower degrees. The model GIS-CHAMP-threshold (about 20% of the data were filtered out) was estimated by a threshold of coordinates and baselines compared to the EGM96-based reduced-dynamic orbit of 50 cm and 10 cm respectively and by a threshold-value of differences between computed accelerations and EGM96-accelerations of $5{\cdot}10^{-5}$ m/s^2. The best result, outreaching the methods of data preprocessing, especially for degrees 20 - 70, is obtained by means of robust estimation in terms of the Huber method (GIS-CHAMP-Huber, parameter $a = 1.5{\cdot}10^{-5}$ m/s^2). The superiority of robust estimation opposite to data-preprocessing can be explained by: (i) instead of a rigorous threshold all data is used and downweighted according to its accuracy and (ii) the weights are purely gained from the kinematic orbit itself without any additional information. Fig. 3b demonstrates the fast convergence of robust estimation. While the result of the basic step (GIS-CHAMP-Huber_step_0) coincides with the result from the unfiltered orbit (GIS-CHAMP-unfiltered), already in the first step (GIS-CHAMP-Huber_step_1) a comparable accuracy to data preprocessing (here: GIS-CHAMP-wavelet) is reached. With the second step, convergence is almost achieved and only marginal improvements, which won't be visible in the figure, can be gained by further iterations. GIS-CHAMP-Huber corresponds to the model estimated in step 5. Finally in Fig. 3c, a comparison of GIS-CHAMP-Huber and EIGEN-3P/EIGEN-CHAMP3S, estimated by means of the classical method, is displayed. All three models are of similar accuracy. GIS-CHAMP-Huber is slightly superior for degrees 45 – 75 and marginally worse for degrees < 30 compared to EIGEN-CHAMP03S, which is the final version of EIGEN-3P. For completeness, it must be kept in mind that EIGEN-CHAMP03S is estimated from a longer, but earlier observation period, where the satellite was in a higher orbit and thus less sensitive for higher degree terms. This might explain, why GIS-CHAMP-Huber is closer to EIGEN-GRACE02S for the higher spherical harmonic degrees. The remarkable worse accuracy for the very low degrees (2-6) of the GIS-CHAMP models in Figs. 3a,b,c was confirmed by other groups working with the same kinematic orbits, which leads to the assumption that this is an effect due to the data and not a problem related to the applied method. In addition to the CHAMP-models the degree RMS of the most accurate pre-CHAMP global geopotential model, EGM96, are illustrated in Fig. 3c. A clear increase of accuracy of the CHAMP-models in comparison to EGM96 is visible for the coefficients up to degree 60 or 65, which demonstrates the progress in gravity field determination achieved with CHAMP.

Table 1. RMS-value, area-weighted RMS (by $\cos(\phi)$) and maximum absolute value of the geoid-differences between various gravity field models and the reference model EIGEN-GRACE02S, developed up to degree and order 70

model / error (m)	EIGEN-		GIS-CHAMP-			
	CHAMP3p	CHAMP03S	var20	threshold	wavelet	Huber
geoid-RMS	0.356	0.233	0.281	0.211	0.253	0.189
weighted RMS	0.282	0.235	0.305	0.224	0.275	0.204
max. deviation	2.664	1.575	1.258	1.037	1.360	0.915

Table 1 displays the RMS values, the area-weighted RMS values and the maximum absolute values of the geoid differences between the GIS-CHAMP/EIGEN-CHAMP models and EIGEN-GRACE02S. The geoid differences were computed up to degree and order 70 (signal-to-noise ratio per degree is ≥ 1 for all models) on a 1° x 1° grid. Since the area of 1° x 1° grids decreases with shortening distance to the poles, additionally to the normal 1° x 1° - RMS the area-weighted RMS is regarded in order to diminish the influence of polar and near-polar data. If the normal and the area-weighted RMS are similar, the distribution of the geoid differences is quite similar, as it is the case for EIGEN-CHAMP03S. In contrast, the area-weighted RMS of EIGEN-3P is much higher than its normal RMS which is caused by larger differences around the poles. The normal RMS-value of the GIS-CHAMP models is slightly higher than their area-weighted RMS, which points to a lower accuracy at the equatorial areas. This can be explained as follows: (i) since the CHAMP groundtrack converges towards the poles more data per area is available. This means that an area in polar regions gets a higher weight in the solution if all observations are assumed to have equal quality. (ii) more kinematic orbit data was filtered out or downweighted at the equatorial areas.

The interpretation of Figs. 3a,b,c is confirmed by Table 1. The reached accuracy of the GIS-CHAMP-models with robust estimation (weighted RMS of 20.4 cm) is superior to data-preprocessing methods, the best outlier-removal strategy is the simple threshold method (weighted RMS of 22.4 cm). Concerning the weighted RMS, GIS-CHAMP-threshold is of similar accuracy and GIS-CHAMP-Huber is of slightly higher accuracy than EIGEN-CHAMP03S. A very interesting point is the maximum geoid differences. Obviously, the inclusion of outlier removal or downweighting strategies in the acceleration approach reduces larger geoid errors. Even the data selection by means of the orbit variances leads only to a maximum deviation of 1.258 m in comparison to 1.575 m at EIGEN-CHAMP03S. The high performance of robust estimation by means of downweighting inaccurate data is supported by a maximum deviation of only 0.915 m of the corresponding model.

7 Conclusions/Outlook

It has been demonstrated within this contribution and by other groups (Földváry et al., 2005; Mayer-Gürr et al., 2005), that models of the Earth's gravitational potential recovered by gravity field analysis based on kinematic CHAMP satellite orbit data can compete with those models generated by classical integration of the variational equations. A major reason for this is the outstanding quality of currently available kinematic orbits. The capability of kinematic orbit analysis in terms of a point-wise acceleration approach can further be enhanced, if methods are employed, which are either able to filter or to downweight single spurious observations. Exceedingly satisfying results are obtained if robust estimation is applied, which requires neither additional preparatory data preprocessing nor any reference information. Exploiting the full primary data set without rejecting any observation, the quality of models recovered by data-preprocessing was exceeded within a few iteration steps.

Reliable detection of outliers is also achieved by application of the wavelet filter or the simple threshold method. Although the accuracy of the resulting gravity field models remained inferior to robust estimation, filter methods are advantageous with respect to computational effort as no repeated iterative solution of the system of normal equations is required.

The second order numerical differentiation doesn't seem to be the weak point of acceleration approaches due to the correlation of the orbit data and the high accuracy of the baselines. The algorithm can be classified as fast, since the system of equations is linear and numerical differentiation is applied to the data instead of integration. The analysis can be carried out on a standard PC by means of the iterative PCG method, which guarantees fast convergence and is able to cope with restricted memory capacity.

It has been demonstrated in this contribution, that the implemented method of the acceleration approach is a well-suited procedure for gravity field determination. Future investigations and validations can address the following topics: (i) in-situ-estimation of calibration-parameters within the acceleration approach and (ii) external validations based on terrestrial data (GPS/leveling, gravity data, ...).

Acknowledgement. This is publication no. GEOTECH-145 of the program GEOTECHNOLOGIEN of BMBF and DFG, Grant No. 03F0333C. D. Švehla and M. Rothacher (FESG at TUM) are thankfully acknowledged for providing kinematic and reduced-dynamic CHAMP orbits. The CHAMP-data was kindly provided by GFZ Potsdam via the CHAMP ISDC data-center.

References

Austen G, Grafarend EW, Reubelt T (2002) Analysis of the Earth's Gravitational Field from Semi-Continuous Ephemeris of a Low Earth Orbiting GPS-Tracked Satellite of Type CHAMP, GRACE or GOCE. In: International Association of Geodesy Symposia, Vol. 125, Vistas for Geodesy in the New Millennium, J. Adam and K. P. Schwarz (Editors), Axel Springer Verlag.

Ditmar P, Klees R (2002) A method to compute the Earth's gravity field from SGG/SST data to be aquired by the GOCE satellite. Delft University Press, Delft, 2002.

Ditmar P, van Eck van der Sluijs AA (2004) A technique for modelling the Earth's gravity field on the basis of satellite accelerations. Journal of Geodesy, Vol. 78, 1-2: 12-33.

Engeln-Müllges G, Reutter F (1987) Numerische Mathematik für Ingenieure. 5th revised edition, BI Wissenschaftsverlag, Mannheim, Wien, Zürich.

Földváry L, Švehla D, Gerlach C, Wermuth M, Gruber T, Rummel R, Rothacher M, Frommknecht B, Peters T, Steigenberger P (2005) Gravity model TUM-2Sp based on the energy balance approach and kinematic CHAMP orbits. In: Reigber Ch, Lühr H, Schwintzer P, Wickert J (Eds.), Earth Observation with CHAMP - Results from Three Years in Orbit, Springer, Berlin, 13-16, 2005.

Götzelmann M, Keller W, Reubelt T (submitted) Gross error compensation for gravity field analysis based on kinematic orbit data. Journal of Geodesy.

Huber PJ (1981) Robust statistics. Wiley, New York, 1981.

Lemoine FG, Kenyon SC, Factor JK, Trimmer RG, Pavlis NK, Chinn DS, Cox CM, Klosko SM, Luthcke SB, Torrence MH, Wang YM, Williamson RG, Pavlis EC, Rapp RH, Olson TR (1998) The development of the joint NASA Goddard Space Flight Center and the National Imagery and Mapping Agency (NIMA) geopotential Model EGM96. NASA/TP-1998-206861, Greenbelt, Maryland, USA, 1998.

Koch KR (1996) Robuste Parameterschätzung. Allgemeine Vermessungsnachrichten AVN 1/1996, 103: 1-18, Wichmann, 1996.

Mayer-Gürr T, Ilk KH, Eicker A, Feuchtinger M (2005) ITG-CHAMP01: a CHAMP gravity field model from short kinematic arcs over a one-year observation period. Journal of Geodesy (2005) Vol. 78, Issue 7 – 8: 462-480.

Reigber C (1989) Gravity field recovery from satellite tracking data. In: Sanso F, Rummel R (Eds.), Theory of satellite geodesy and gravity field determination, Springer Verlag, Berlin, Heidelberg, New York 1989, pp. 197-234.

Reigber C, Balmino G, Schwintzer P, Biancale R, Bode A, Lemoine JM, König R, Loyer S, Neumayer H, Marty JC, Barthelmes KH, Perosanz F, Zhu SY (2002) A high quality global gravity field model from CHAMP GPS tracking data and accelerometry (EIGEN-1S). Geophysical Research Letters, 29(14), 10.1029/2002GL015064, 2002.

Reigber C, Schwintzer P, Neumayer H, Barthelmes KH, König R, Förste C, Balmino G, Biancale R, Lemoine JM, Loyer S, Bruinsma S, Perosanz F, Fayard T (2003) The CHAMP-only Earth Gravity Field Model EIGEN-2. Advances in Space Research 31(8): 1883-1888, 2003 (doi :10.1016/S0273—1177(03)00162-5).

Reigber C, Schmidt R, Flechtner F, König R, Meyer U, Neumayer KH, Schwintzer P, Zhu SY (2005a) An Earth gravity field model complete to degree and order 150 from GRACE: EIGEN-GRACE02S. Journal of Geodynamics 39(1):1-10.

Reigber C, Jochmann H, Wünsch J, Petrovic S, Schwintzer P, Barthelmes F, Neumayer KH, König R, Förste C, Balmino G, Biancale R, Lemoine JM, Loyer S, Perosanz F (2005b) Earth Gravity Field and Seasonal Variability from CHAMP. In: Reigber Ch, Lühr H, Schwintzer P, Wickert J (Eds.), Earth Observation with CHAMP - Results from Three Years in Orbit, Springer, Berlin, 25-30, 2005.

Reubelt T, Austen G, Grafarend EW (2003a) Harmonic analysis of the Earth's gravitational field by means of semi-continous ephemerides of a low Earth orbiting GPS-tracked satellite. Case study: CHAMP. J Geod 77: 257-278.

Reubelt T, Austen G, Grafarend EW (2003b) Space Gravity Spectroscopy-determination of the Earth's gravitational field by means of Newton interpolated LEO ephemeris; case studies on dynamic (CHAMP Rapid Science Orbit) and kinematic orbits. Advances in Geosciences, 1: 127-135, European Geosciences Union 2003.

Reubelt T, Götzelmann M, Grafarend EW (submitted) A new CHAMP gravitational field model based on the GIS acceleration approach and two years of kinematic CHAMP data. In Reigber C, Tapley B (Eds.): New Satellite Mission Results for the Geopotential Fields and Their Variations, special issue of Advances in Geosciences, EGU.

Švehla D, Rothacher M (2003) Kinematic and Reduced-Dynamic Precise Orbit Determination of Low Earth Orbiters. Advances in Geosciences, 1: 47-56, EGU, 2003.

Švehla D, Rothacher M (2004) Two years of CHAMP kinematic orbits for Geosciences. EGU, 1st General Assembly, Nice, France. Geophysical Research Abstracts, European Geophysical Society Vol. 6. ISSN: 1029-7006, 2004.

Weigelt M, Sneeuw N (submitted) Numerical Velocity determination and Calibration Methods for CHAMP Using the Energy Balance Approach. Proceedings of the 4th Meeting of the International Gravity and Geoid Commission "Gravity, Geoid and Space Missions – GGSM2004", Aug. 30 – Sept. 3, 2004, Porto, Portugal.

Xu P (1989) On robust estimation with correlated observations. Bull. Geod., 63: 237-252.

Earthquake Signatures in the Ionosphere Deduced from Ground and Space Based GPS Measurements

Norbert Jakowski[1], Volker Wilken[1], Konstantin Tsybulya[1], and Stefan Heise[2]

[1] Deutsches Zentrum für Luft- und Raumfahrt e.V. in der Helmholtz-Gemeinschaft, Institut für Kommunikation und Navigation, Kalkhorstweg 53, D-17235 Neustrelitz, Germany `Norbert.Jakowski@dlr.de`

[2] GeoForschungsZentrum Potsdam (GFZ), Telegrafenberg, D-14473 Potsdam, Germany

Summary. Atmospheric perturbations induced by weather fronts, nuclear explosions, volcano eruptions, and earthquakes can generate signatures in the ionospheric plasma density by atmospheric–ionospheric coupling processes. Because of their sensitivity to the ionospheric ionization, ground and space based GPS measurements offer a unique opportunity for detecting earthquake signatures in the ionosphere. Although numerous case studies and statistical analyzes were made, the GPS radio occultation measurements on CHAMP did not show a clear ionospheric response to earthquakes. On the other hand the retrieved total electron content (TEC) data along numerous ray paths between ground based receivers and GPS satellites has shown clear earthquake related signals for selected earthquakes of magnitudes larger than 6. By using the dense GPS network in North America, earthquake related structures have been found after the Denali earthquake on November 3, 2002 and during the California earthquake on December 22, 2003. Single station observations revealed also typical earthquake signatures after the Sumatra earthquake on December 26, 2004. It is assumed that these significant structures are generated by upward propagating atmospheric acoustic waves which are excited by seismic surface waves. Detection techniques and wave propagation features are discussed.

Key words: ionosphere, radio occultation, TEC, GPS, earthquake, Rayleigh wave

1 Introduction

The ionospheric plasma is subjected to a number of quite different forces from both, below and above the ionosphere. The major impact is due to solar forcing. Both electromagnetic and particle radiation are extremely variable and modify the ionospheric plasma directly, via ionization processes, and indirectly, via interactions with the magnetosphere and thermosphere, thus generating electric fields, composition changes and neutral winds. Forces from below, e.g. from the troposphere or lithosphere, are principally very small compared with the solar control. Nevertheless, ionospheric response to earthquakes has been reported in the literature for many years, e.g. (Row, 1967; Wolcott et al., 1984; Calais and Minster, 1996). Hereafter earthquakes may cause pressure waves that propagate upward and may generate so-called "Travelling Ionospheric Disturbances" (TID's) in the ionosphere. Since the atmospheric density decreases almost exponentially with altitude, energy conservation implies that the wave amplitude also increases exponentially. Such amplification of the waves can reach a factor of 10^5 to 10^6. Due to the absorption and acoustic cut-off, the atmosphere filters the acoustic waves. Only waves with periods between about 2 and 6 minutes can reach the ionospheric layers. The sensitivity of GPS based ionospheric measurements to earthquakes was first reported by Calais and Minster (1996). While analyzing the Denali earthquake on November 3, 2002, Ducic et al. (2003) and Wilken et al. (2004) have provided evidence that seismic wave induced upward propagating atmospheric acoustic waves significantly modify the ionospheric plasma measurable by differential GPS phases. Acoustic waves generated at the ground need approximately 10 minutes to reach the F-layer of the ionosphere, where the close coupling between the neutral atmosphere and ionized plasma may result in a wavelike modulation of the electron density. According to the PREM (Preliminary Reference Earth Model, (Dziewonski and Anderson, 1981)) the group velocity of Rayleigh waves ranges between 3.8 km/s at a period of 100 s and 3.5 km/s at a period of 240 s. The aim of the paper is to provide further evidence that severe earthquakes may cause wavelike electron density variations in the ionosphere and that these are detectable by integral GPS measurements. Following this line we have analyzed both space based radio occultation measurements obtained since April 2001 on board CHAMP as well as numerous ground based GPS measurements obtained during and after earthquakes.

2 Space Based Measurements on CHAMP

Principally, the GPS measurements take advantage of the dispersive nature of the ionosphere, i.e. of the frequency dependent refractivity. In the first order approximation the differential phase of $L1$ and $L2$ carrier frequencies of GPS satellites is directly proportional to the integrated electron density along the ray path. Hence, the computation of differential phases along all available satellite tracks provides a sensitive tool to get information about the ionospheric ionization. For analyzing ground based measurements we used data of globally distributed GPS stations available via SOPAC (Scripps Orbit and Permanent Array Center).

Space based GPS measurements on board CHAMP are very effective when deriving vertical electron density profiles by Ionospheric Radio Occultation (IRO) measurements (Jakowski et al., 2002) and reconstructing the 3-D electron density distribution of the topside ionosphere/plasmasphere (Heise et al., 2002). Space based radio occultation measurements on CHAMP cover the entire globe. On the other hand, the coincidence of IRO measurements with earthquake events in space and time is rather rare when using only one satellite.

The search algorithm for earthquake signatures is based on the analysis of dual frequency GPS measurements. The ionosphere related term - the ionospheric range error in positioning d_I - is related to

$$d_I \approx \frac{40.3}{f^2} \int n_e \, ds \,. \tag{1}$$

The integral over the electron density n_e along the ray path s represents the total electron content TEC. Comparing the phase measurements at two coherent frequencies, all frequency independent terms vanish. So the differential GPS phase $\Delta\phi$ at frequencies $f_1 = 1.575\,\text{GHz}$ and $f_2 = 1.228\,\text{GHz}$ is approximately given by

$$\Delta\phi = 40.3 \frac{f_1^2 - f_2^2}{f_1^2 f_2^2} TEC + const \,. \tag{2}$$

After extracting the relative TEC from differential GPS phases according to (2), ionospheric irregularities may be detected easily. TID intensity, amplitude and wavelength of wavelike phenomena are estimated for each occultation as described by Tsybulya and Jakowski (2005). Estimates of the TID intensity from IRO measurements in the temporal and spatial vicinity of the earthquakes are considered both in case studies as well as statistical analyzes.

2.1 Case Studies

The case studies are based on individual earthquakes for which all available IRO measurements were collected over 24 hours within a cross section radius of 30° around the epicenter. To demonstrate this procedure, we briefly discuss the extremely strong Sumatra earthquake on December 26, 2004. Obviously there is no perfect coincidence of the occultation sites with the epicenter (Fig. 1). The plotted measurements do not indicate any wavelike structure worth to be mentioned (Fig. 2). Acoustic waves should in principle be detectable. Due to the long ray path through the ionosphere in the radio occultation mode, wavelike variations may be averaged out. Precursor effects during the 24 hour interval before the earthquake were also not observed. In agreement with this, further case studies from other earthquakes did not indicate any significant anomaly worth to be reported here.

2.2 Statistical Analysis

To enhance the significance of the observations, a superposition of ionospheric wavelike effects up to 7 days before and after an earthquake onset is computed

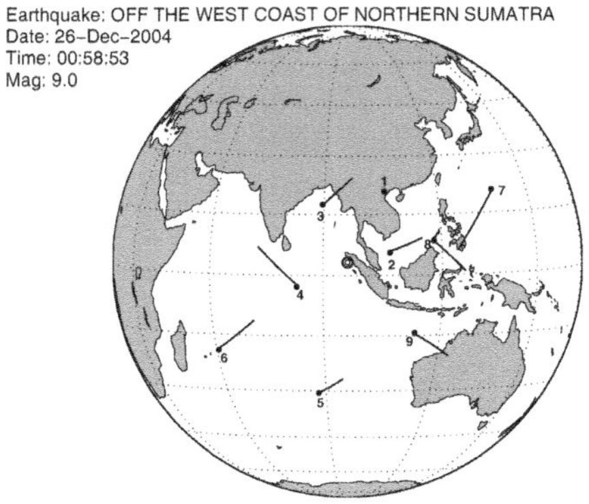

Fig. 1. Earthquake in North-Sumatra on December 26, 2004. Indicated are the locations of the IRO measurements on board CHAMP up to 24 hours after the earthquake

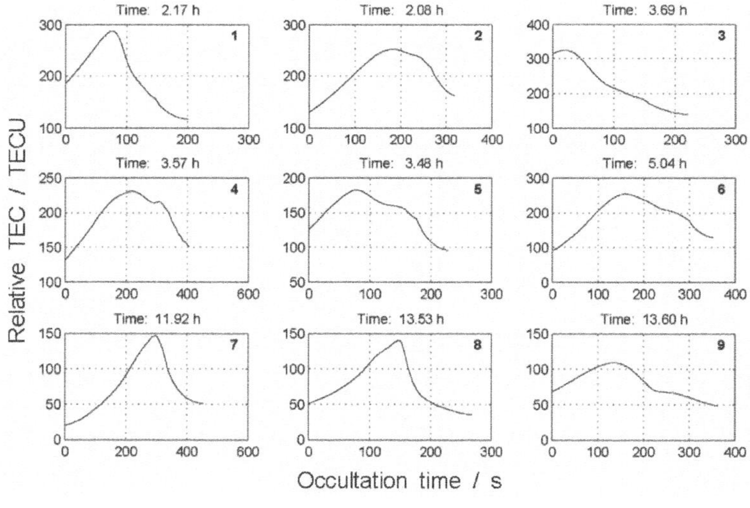

Fig. 2. IRO measurements at geographical locations marked in Fig. 1

for all earthquakes in different groups characterized by specific selection criteria, e.g. the magnitude (Fig. 3). Due to the more extended data base, the coincidence radius was reduced to 10° around the epicenter. The histogram does not show serious anomalies in the data. Although the TID–activity level shows a maximum at the day after the earthquake for magnitudes less than 7.5, the statistical significance is obviously not sufficient (Fig. 3). To draw a clear conclusion, more IRO data are needed for the analysis.

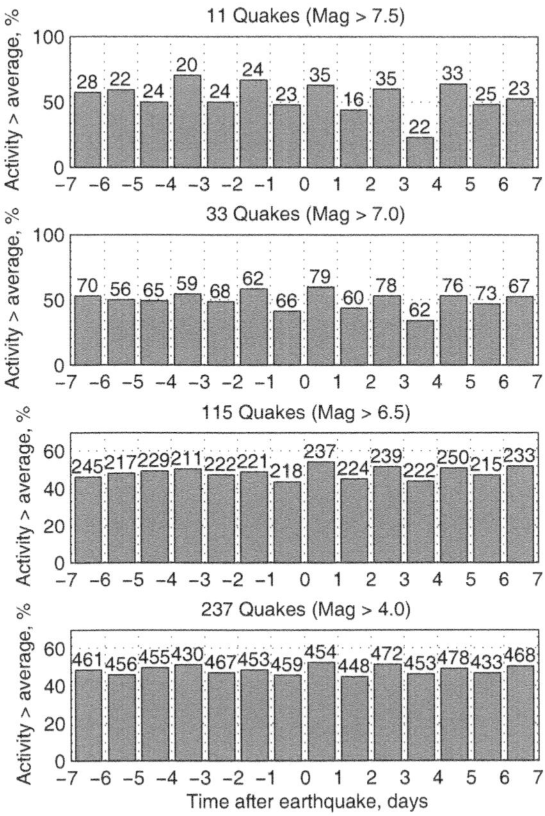

Fig. 3. Superposition of the TID activity obtained from IRO measurements on CHAMP within the time window of ±7 days around the onset of the earthquakes. The different groups distinguish by different threshold values for the magnitude. The measure is given by the deviation from the mean over a full day. The numbers on the top of the columns indicate the number of radio occultations involved

3 Ground Based GPS Measurements

A big advantage of the ground based GPS measurements compared with the discontinuous IRO measurements is the 24 hours continuity. Since the mostly used geodetic data have a time resolution of 30 s, wave periods of the expected acoustic waves in the 2-6 min interval may just be analyzed from the data. We took benefit from the fact that in some earthquake regions dense receiver networks exist which guarantee a high statistical significance of the results. Therefore, our analysis is focused on specific earthquakes in the California and Japan region. To find earthquake signatures in the ionosphere it is useful to follow the trace of seismic surface waves (Ducic et al., 2003). As already mentioned, these waves propagate with a group velocity of approximately 3.5 km/s at a period of about 240 s.

3.1 Data Processing

Ground based GPS measurements are powerful in monitoring the total electron content of the ionosphere, e.g. (Jakowski, 1996), (Jakowski et al., 1998), (Jakowski, 1998). Since the absolute accuracy is limited to a few TEC units (1 TECU $= 10^{16}$ el. / m^2), special care has to be taken to detect extremely small structures in the TEC data. The GPS navigation and observation data of selected stations are provided in RINEX-format (Receiver Independent Exchange) with a sampling rate of 30 s. For North-American stations we used the "Scripps Orbit and Permanent Array Center" (SOPAC). The estimation of TEC from GPS measurements is based both on code as well as carrier phase measurements on the L1/L2 frequencies. Since the differential carrier phases do not provide absolute TEC values, the calibration is achieved by including the differential code phases which are much noisier due to multipath effects. To overcome this problem, the accurate but relative phase measurements are leveled to the absolute code measurements by a least squares method for elevation angles greater than 20 degrees (Jakowski, 1996). The problem is then reduced to the estimation of the calibration constant or instrumental biases of the satellite and receiver hardware which are estimated by using a Kalman filter over a full 24 hours run for all considered satellite–receiver links (Sardon et al., 1994). The estimated TEC along the satellite–receiver links is then mapped to the corresponding equivalent vertical TEC (vTEC) at the so-called subionospheric point. This location is defined by the ground geographic coordinates of the point where the ray path crosses the ionospheric shell height; assumed is a single layer approximation of the ionosphere at 400 km height. Systematic trends in vTEC, such as the diurnal variation (Fig. 4), are eliminated by recursively computing the backward differences of the time–series of satellite–receiver links. Subsequently, a band pass filter with periods of 150-350 s was applied, covering the Airy phase of Rayleigh waves at 240 s.

3.2 Detection of Earthquake Signatures

Next we present the results of the analysis of three earthquakes which are briefly described here:

• Central Alaska, USA (63.517° N/ -147.444° E) at 22:12:41 UTC on November 03, 2002 (DOY: 307) with a Magnitude of M = 7.9 ($M_S = 8.5$) in a depth of 5 km. The Ap-index on that day was 35 (one day before/after: 28/23) and the F10.7–index was 166.5 (one day before/after: 162.1/174.4).

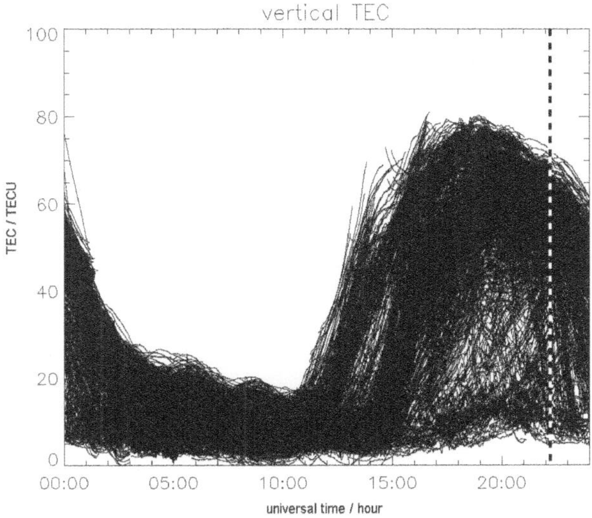

Fig. 4. Vertical TEC on November 3, 2002. The dashed line marks the onset time of the earthquake at the epicenter indicating a high ionization level during the earthquake. Although large differences appear between the electron content values from different links, a clear diurnal variation of TEC is observed

• Central California, USA (35.706° N/ -121.102° E) at 19:15:56 UTC on December 22, 2003 (DOY: 356) with a Magnitude of M = 6.5 (M_S = 6.4) in a depth of 8 km. The Ap-index on that day was 18 (one day before/after: 24/7) and the F10.7-index was 133.2 (one day before/after: 129.1/137.4).

• Sumatra, Indonesia (3.295° N/ 95.982° E) at 00:58:53 UTC on December 26, 2004 with a magnitude of M = 9.0 (M_S = 8.8) in a depth of 30 km. The Ap-index on that day was 11 (one day before/after: 14/8) and the F10.7-index was 88.7 (one day before/after: 90.0/93.7).

While Fig. 4 illustrates the ionospheric total electron content, Color Fig. VII on p. 289 shows the propagation and observation geometry of the Central Alaska Earthquake on November 3, 2003. The red point marks the epicenter where the earthquake began in a depth of about 5 km at 22:12:41 UT. The outer dashed circle represents the position of the Rayleigh wave front which travels with a speed of 3.5 km/s. The inner full circle illustrates the position of ionospheric wave front generated by the Rayleigh–wave 11 minutes before. The positions of the GPS receivers are indicated by small yellow colored square symbols. The wave amplitudes derived from band pass filtered vTEC values are shown by color coded pixels at the subionospheric points. When considering subsequent snapshots, the formation and propagation of the ionospheric wave can be observed. The close correlation of the ionospheric wave with the seismic wave can be seen in the path–time diagram of the observed amplitudes of both the seismogram as well as the ionospheric wave in Fig. 5. The parallel structures of both phenomena separated by a time interval of about 10 minutes needed for upward propagation of the excited acoustic waves to ionospheric altitudes provide strong evidence for the continuous generation of earthquake signatures in the ionosphere by the Rayleigh wave. Similar re-

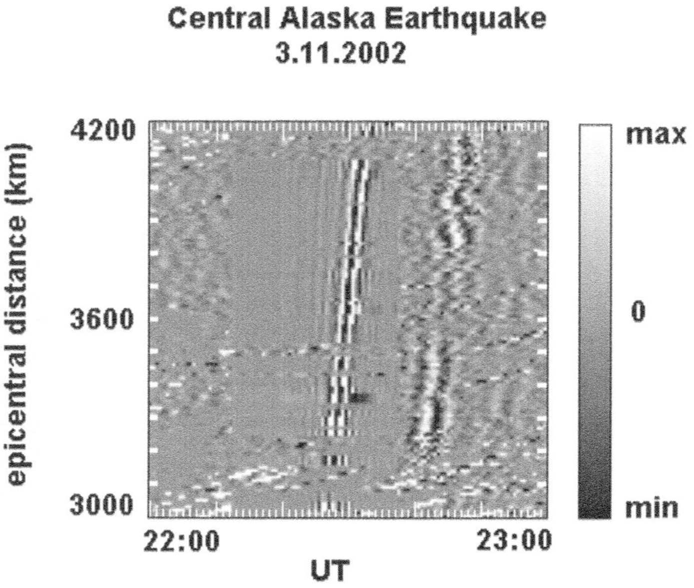

Fig. 5. Path-time diagram of seismogram amplitudes and band pass filtered vertical TEC values referring to the distance from the epicenter of the Alaska earthquake on November 3, 2002

sults were obtained when analyzing the GPS data of the Central California earthquake on December 22, 2003.

As already mentioned, the time development and propagation of the excited ionospheric wave can be traced by mapping the amplitudes of band pass filtered vTEC for each observation. To overcome problems related to the motion of the subionospheric points and the corresponding non–uniform data distribution we applied for visualization the assimilation technique that is used for the creation of TEC maps in DLR Neustrelitz since many years (Jakowski, 1996), (Jakowski et al., 1998), (Jakowski, 1998). For reasons of simplicity, the model used for assimilation is defined by zero amplitude values for all grid points (green color). As Color Fig. VIII on p. 289 convincingly documents, a clear wave structure is formed about 420 s after the beginning of the earthquake. Surprisingly the wave does not simply travel in radial direction; the propagation follows mainly the coast line. Probably, this phenomenon is due to the vicinity of the epicenter, where the radial propagation of the ionospheric wave is not yet dominant.

The path–time diagram referring to the distance from the epicenter again indicates a radial propagation velocity in the order of 3.5 km/s, as shown in Color Fig. IX on p. 290.

The huge Sumatra earthquake of magnitude 9 that has caused large tsunami waves in the Indian Ocean has been clearly detected in the GPS signals. Unfortunately, the number of GPS stations was not large enough to reconstruct clear wave like structures in the ionosphere. Tsunami waves with periods of 15-50 minutes cannot produce acoustic waves. They will surely excite gravity waves whose ionospheric response is not yet clear. As the band pass filtered vTEC data of the Sampali station (3.62° N, 98.71° E) shows, there appears a strong wave like signal with amplitudes of up to 2×10^{15} electrons per sqm in the ionosphere about 10 minutes after the earthquake onset lasting about 100 minutes (Fig. 6).

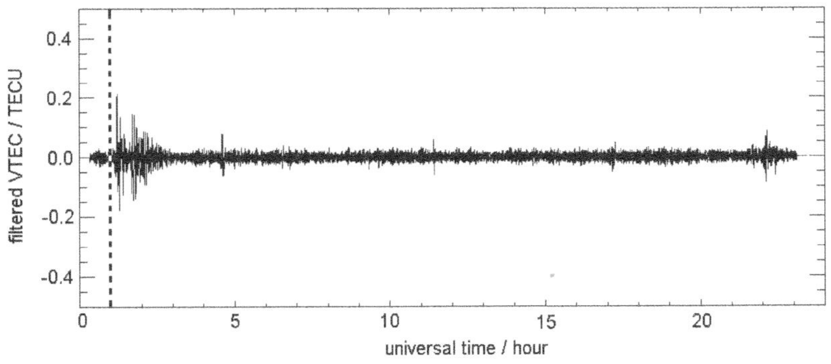

Fig. 6. Time series of band pass filtered vertical TEC data over the GPS station Sampali: (3.62° N, 98.71° E) on December 26, 2004. The onset of the earthquake is marked by the dashed line at 00:58:53 UT

4 Summary and Conclusions

The analysis of ground based GPS measurements obtained during three earthquakes, provides clear evidence for the excitation of earthquake signatures in the ionosphere. It is assumed that seismic waves at the Earth's surface excite acoustic waves propagating upward and modifying the ionospheric plasma density after about 10 minutes.

We have shown that the ionospheric wave phenomena, obtained after a careful band pass filtering process, are closely correlated with corresponding seismograms measured at different distances from the epicenter of the Alaska earthquake on November 3, 2002.

The analysis of space based GPS data measured on CHAMP does not provide a clear conclusion with respect to the detection of earthquake signatures in the ionosphere up to seven days before and after the earthquake's onset. Principally, Traveling Ionospheric Disturbances should be measurable on CHAMP although density structures are considerably smoothed by the GPS radio occultation measurements (Tsybulya and Jakowski, 2005).

As we learned from the ground based measurements, the waves are critically in time and space and furthermore, have very small amplitudes. Thus, if coincidence of space and time is required within small spatial and temporal windows, single satellite RO missions like CHAMP cannot fulfill such strict requirements in most cases. Multi satellite missions will be able to produce more dense spatial and temporal data coverage. The statistics indicate a slight maximum of TID activity at the day after the earthquake onset. But this result is statistically not significant.

Further intensive research is required to improve our understanding of coupling processes between the lithosphere and atmo-/ionosphere. If tsunami excited gravity waves are clearly detectable, ionospheric measurements have the potential to contribute to an early warning system of tsunamis.

Acknowledgement. This is publication no. GEOTECH-144 of the programme GEOTECHNOLOGIEN of BMBF and DFG, Grant no. 03F0333B. We are grateful to the data centers Scripps Orbit and Permanent Array Center (SOPAC), the Southern California Earthquake Data Center (SCEDC) and the Northern California Earthquake Data Center (NCEDC), the United States Geological Survey's (USGS) / National Earthquake Information Center (NEIC), and the National Geophysical Data Center (NGDC)/ National Oceanic and Atmospheric Administration (NOAA) for providing numerous GPS and seismic data.

References

Blanc, E. (1985) Observations in the Upper Atmosphere of Infrasonic Waves from Natural or Artificial Sources - A summary, Annales Geophysicae, Vol. 3, pp. 673-687

Blewitt, G. (1990) An Automatic Editing Algorithm for GPS Data, Geophysical Research Letters, Vol. 17, No. 3, pp. 199-202, doi:10.1029/90GL00294

Calais, E., Minster, J. B. (1996) GPS, Earthquakes, the Ionosphere, and the Space Shuttle, Physics of the Earth and Planetary Interiors, Vol. 105, No. 3-4, pp. 167-181

Ducic, V., Artru, J., Lognonne, P. (2003) Ionospheric Remote Sensing of the Denali Earthquake Rayleigh Surface Waves, Geophysical Research Letters, Vol. 30, No. 18, pp. SDE 8-1, doi:10.1029/2003GL017812

Dziewonski, A. M., Anderson, D. L. (1981) Preliminary Reference Earth Model. Physics of the Earth and Planetary Interiors, Vol. 25, No. 4, pp. 297-356

Heise, S., Jakowski, N., Wehrenpfennig, A., Reigber, Ch., Lühr, H. (2002) Sounding of the topside ionosphere/plasmasphere based on GPS measurements from CHAMP: Initial results, Geophysical Research Letters, Vol. 29, No. 14, p. 44, doi:10.1029/2002GL014738

Jakowski, N. (1996) TEC Monitoring by using Satellite Positioning Systems. In: Kohl, H., Rüster, R., Schlegel, K.: Modern Ionospheric Science. Katlenburg-Lindau: ProduServ GmbH Verlagsservice, Berlin, pp. 371-390

Jakowski, N., Sardon, E., Schlüter, S. (1998) GPS-Based TEC Observations in Comparison With IRI95 and the European TEC Model NTCM2, Adv. in Space Res., Vol. 22, pp. 803-806

Jakowski, N. (1998) Generation of TEC Maps over the COST251 Area Based on GPS Measurements, Proceedings of the 2nd COST 251 Workshop, (Ed. A. Vernon), March 30-31, pp. 51-57

Jakowski, N., Wehrenpfennig, A., Heise, S., Reigber, Ch., Lühr, H., Grunwaldt, L., Meehan, T. K. (2002) GPS radio occulation measurements of the ionosphere from CHAMP: Early results, Geophysical Research Letters, Vol. 29, No. 10, p. 95, doi:10.1029/2001GL014364

Row, R. V. (1967) Acoustic-Gravity Waves in the Upper Atmosphere Due to a Nuclear Detonation and an Earthquake, J. Geophys. Res., Vol. 72, pp. 1599-1610

Sardon, E., Rius, A., Zarraoa, N. (1994) Estimation of the Transmitter and Receiver Differential Biases and the Ionospheric Total Electron Content from Global Positioning System Observations, Radio Science, Vol. 29, No. 3, pp. 577-586

Tsybulya, K. and Jakowski, N. (2005) Medium- and Small-Scale Ionospheric Irregularities Detected by GPS Radio Occultation Method, Geophysical Research Letters, Vol. 32, No. 9, doi:10.1029/2005GL022420

Wilken, V., Jakowski, N., Saito, A., Heise, S. (2004) Earthquake Signatures in the Ionosphere Detected by GPS Measurements, Poster presentation at the International Beacon Satellite Symposium October 18-22, Trieste, Italy

Wilken, V., Jakowski, N., Kind, R., Saito, A., Heise, S. (2005) Ionospheric Wave Like Perturbations Related to Rayleigh Surface Waves Measured by GPS, Poster presentation at the European Geosciences Union, General Assembly April 24-29, Vienna, Austria

Wolcott, J. H., Simons, D. J., Lee, D. D., Nelson, R. A. (1984) Observations of an Ionospheric Perturbation Arising From the Coalinga Earthquake of May 2, 1983, J. Geophys. Res., Vol. 89, pp. 6835-6839

Global Atmospheric Sounding with GPS Radio Occultation aboard CHAMP

Jens Wickert, Torsten Schmidt, Georg Beyerle, Stefan Heise, and Christoph Reigber

GeoForschungsZentrum Potsdam (GFZ), Department 1, Geodesy & Remote Sensing, Potsdam, Germany, *wickert@gfz-potsdam.de*

Summary. The German CHAMP (CHAllenging Minisatellite Payload) satellite provides continuously GPS radio occultation data since February 2001. The measurements are analyzed by an operational satellite orbit and occultation processing system at GFZ. In total, more than 200,000 precise globally distributed vertical profiles of refractivity, temperature and water vapor are provided as of June 2005. The operational ground infrastructure from GFZ allows for the demonstration of a rapid data analysis since February 2003. A mean delay between measurement and provision of atmospheric excess phase data of ~4 hours is continuously reached. Results of various validation studies with data from meteorological analyzes of the European Centre for Medium-Range Weather Forecasts (ECMWF) and the global radiosonde network indicate an excellent quality of the CHAMP data. But in the lower troposphere systematic deviations are observed, the known negative refractivity bias of the occultation data. It is most pronounced in the Tropics and is also observed by other GPS occultation instruments. The CHAMP data stimulated a number of studies to investigate the observed bias and to improve the GPS occultation data quality in the lower troposphere. First radio occultation measurements from the GRACE-B (Gravity Recovery And Climate Experiment) satellite are available for a 25 h period on July 28/29, 2004.

Key words: GPS, radio occultation, CHAMP, GRACE, atmospheric profiling

1 Introduction

Atmospheric profiling aboard the German CHAMP (Reigber et al., 2005) satellite was activated on February 11, 2001 (Wickert et al., 2001). First ionospheric occultations were recorded on April 11, 2001 (Jakowski et al., 2002). CHAMP's atmospheric measurements brought significant progress (Hajj et al., 2004; Kuo et al., 2004; Wickert et al., 2004b) for the innovative GPS (Global Positioning System) radio occultation (RO) technique (e.g. Kursinski et al., 1997) in relation to the pioneering GPS/MET (GPS/METeorology) mission

(Ware et al., 1996; Rocken et al., 1997). Main advantages of the calibration-free RO method are global coverage, high vertical resolution and all-weather capability combined with high accuracy. These properties allow for various applications in atmospheric/ionospheric research (e.g. Hajj et al., 2000; Wickert et al., 2004a; v. Engeln et al., 2005; Kuo et al., 2005; Wang et al., 2004), weather forecast (e.g. Kuo et al., 2000; Healy et al., 2005; Healy and Thepaut, 2005) and climate change detection (e.g. Randel et al., 2003; Schmidt et al., 2004, 2005a; Foelsche et al., 2005). Several upcoming RO missions will provide thousands of occultations per day in the near future and will increase the potential of the CHAMP occultation measurements (e.g., EQUARS (EQUatorial Atmosphere Research Satellite, Takahashi et al. (2004), launch scheduled for 2006); COSMIC (Constellation Observing System for Meteorology, Ionosphere and Climate, Rocken et al. (2000), launch scheduled for 2006) or Metop (Meteorology Operational, Loiselet et al. (2000); Larsen et al. (2005), launch scheduled for 2006)). We briefly review results from the CHAMP RO experiment and include results from the first short activation of GPS RO aboard the U.S. American/German GRACE mission (Tapley and Reigber, 2004; Dunn et al., 2003).

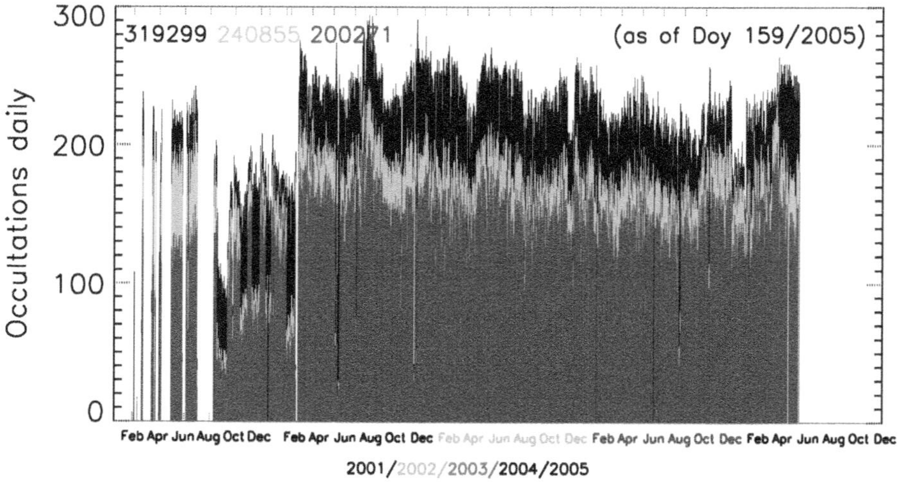

Fig. 1. Number of daily CHAMP occultations (duration > 20 s) as of June 8, 2005. The total height of the columns corresponds to the number of daily measurements. The black and light grey color indicate occultations with less quality atmospheric atmospheric excess phase and vertical atmospheric profiles, respectively. The height of the dark grey columns corresponds to the number of vertical atmospheric profiles provided to the CHAMP data center (ISDC, Information System and Data Center) at GFZ

2 Status of the CHAMP occultations

Occultation measurements were performed during 1,456 days since February 2001 as of June 8, 2005; giving a total of 319,299 recorded events (~219 per day). For ~75.4 % of the occultations (240,855) atmospheric excess phases are available (see Fig. 1). Vertical profiles of atmospheric parameters were derived for 200,271 occultations (~62.7 %). The CHAMP data are also processed by JPL (Jet Propulsion Laboratory) and UCAR (University Corporation for Atmospheric Research) and analysis results are provided via their data centers. ROSE (Radio Occultation Sensor Evaluation), a joint initiative of GFZ, JPL and UCAR, is aimed to compare the retrievals of these three analysis centers and to evaluate and optimize the quality of CHAMP's occultation data analysis (Ao et al., 2003b; Wickert et al., 2005a).

3 Operational data analysis

The atmospheric sounding instrument aboard CHAMP is the "BlackJack" GPS flight receiver provided by JPL. The operational ground infrastructure of GFZ is used to demonstrate an automated occultation data analysis. The infrastructure and the orbit and occultation processing systems are described in detail by, e.g., König et al. (2005b); Schmidt et al. (2005b); Wickert et al. (2004b, 2005b), a schematic overview is given in Color Fig. X on p. 291. A near real time (NRT) provision of atmospheric excess phases is continuously demonstrated by GFZ since February 2003. An average delay of ~5 hours between each measurement and provision of corresponding analysis results was reached. Optimized GPS ground station data handling for the precise orbit determination reduced this delay to ~4 hours since mid April 2004. The average minimum delay has been reduced from 3.5 to 2.5 hours accordingly (see Fig. 2). A further reduction of the average delay is currently planned within a joint research project of GFZ, ECMWF, and the German Weather Service with the MetOffice (U.K.) as external partner. The challenging goal of this project is, to reach ~2 hrs delay in average. This is feasible due to the use of a polar satellite receiving antenna at Ny-Ålesund, Spitsbergen (access to CHAMP's data every ~1.5 h), and a global low latency GPS ground network (access to the ground data every ~15 min), but requires further optimization concerning precise satellite orbit generation and the occultation processing. The demonstration of NRT data analysis will be the precondition for a continuous assimilation of the occultation data to global numerical weather forecast models. A positive impact of CHAMP data on global weather forecasts was already demonstrated by Healy et al. (2005) and by Healy and Thepaut (2005).

Details on the GFZ retrieval algorithms for the derivation of atmospheric parameters and related references are given by, e.g., Wickert et al. (2004b, 2005b). Basics of the GPS RO technique are described, e.g., by Kursinski et al. (1997). The automatically generated data products (refractivity and dry

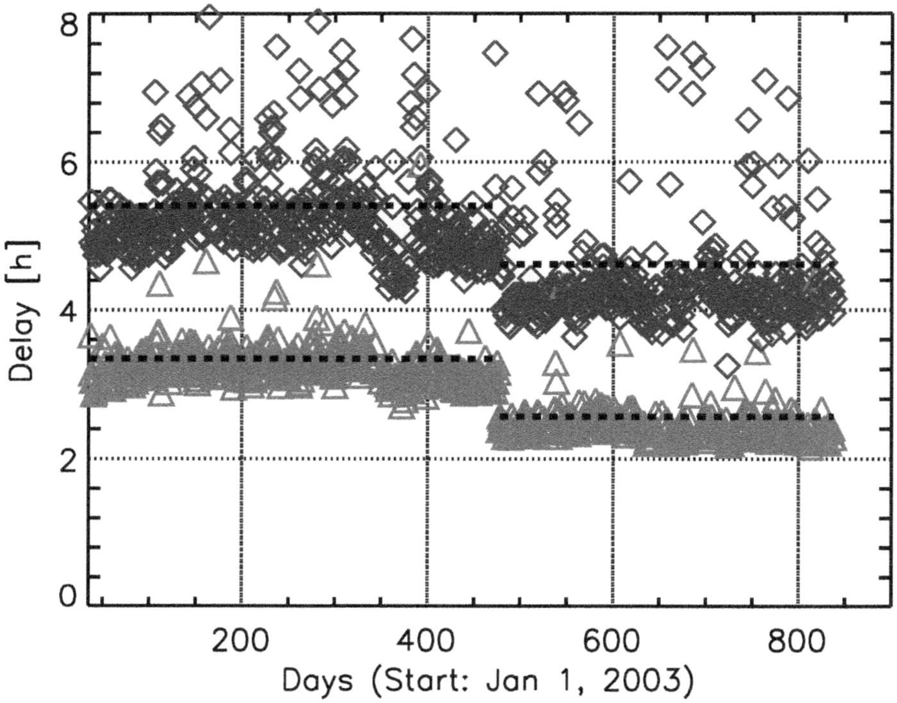

Fig. 2. Time delay between CHAMP occultation measurements and availability of analysis results at GFZ from February 2003 until mid May 2005. Black diamonds indicate the daily mean of the time delay between each measurement and the availability of the corresponding calibrated atmospheric excess phases. An average of ~5 hours for nearly the entire period is reached. The minimum time delays are marked by grey triangles. Due to improvements in the satellite orbit provision the mean delay was reduced to ~4 hours since end April 2004

temperature profiles; product: CH-AI-3-ATM) are provided via the CHAMP data center ISDC at GFZ (*http://isdc.gfz-potsdam.de/champ/*).

Vertical profiles of specific humidity are provided on demand. Background information from ECMWF is used to derive the humidity profiles from the CHAMP refractivities. Two methods for the water vapor derivation were implemented to the operational data analysis. In addition to a standard 1Dvar retrieval (Healy and Eyre, 2000) a direct method (DWVP), introduced by Heise et al. (2005), is implemented. Both methods come to statistically comparable results and reveal a bias of less than 0.1 g/kg and a standard deviation of less than 1 g/kg specific humidity in relation to radiosonde measurements in the mid troposphere. Fig. 3 shows the seasonal mean of the global water vapor distribution for Northern summer (2004) and winter (2003/2004) at 500 hPa derived from operational CHAMP water vapor retrieval, according to a grid of 5° resolution in latitude and 10° in longitude respectively.

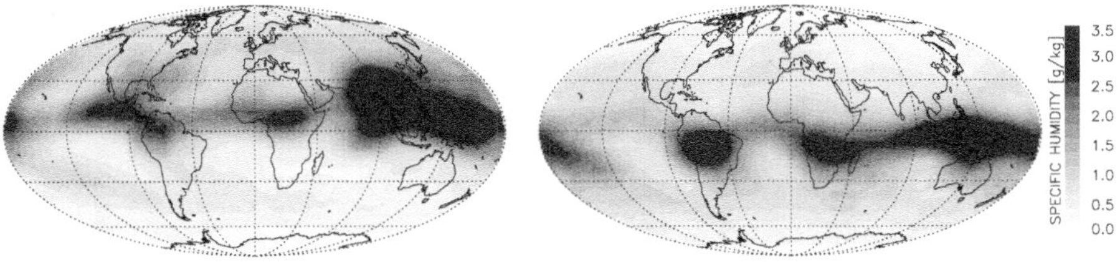

Fig. 3. Mean global distribution of specific humidity at 500 hPa derived from CHAMP occultation data. Left: Northern summer 2004, Right: Northern winter 2003/2004. The humidity was derived using the DWVP method (for details see text and (Heise et al., 2005))

4 Validation of atmospheric profiles

The quality of the CHAMP data was evaluate within several validation studies using independent meteorological data from meteorological analyzes, radiosondes and other satellite data (e.g. Marquardt et al., 2003; Wickert, 2004; Kuo et al., 2005; Wickert et al., 2005c; Gobiet et al., 2005). As an example for these studies Fig.4 shows a comparison of CHAMP refractivity profiles (GFZ product version 005) with corresponding analysis data from ECMWF (Gaussian grid with 0.5° x 0.5° resolution at the Equator, 60 altitude levels) between 10 and 30 km. The comparison shows nearly bias-free refractivity (see Fig. 4). The standard deviation is ~1 %.

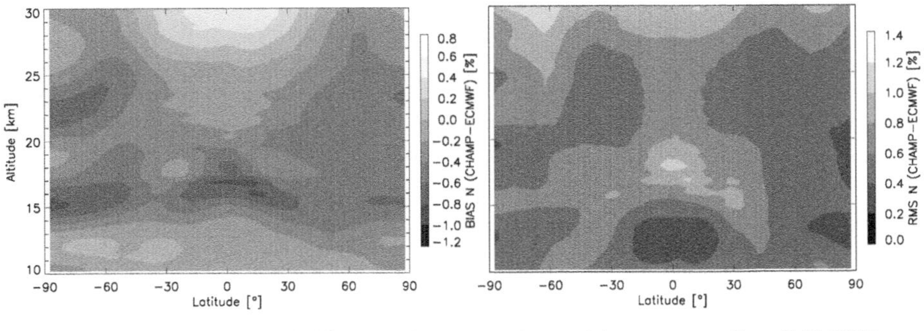

Fig. 4. Comparison of CHAMP refractivity data with corresponding ECMWF analyzes (CHAMP-ECMWF) in the upper troposphere/stratosphere (left: bias; right: rms) between May 14, 2001 and June 8, 2005 (~ 200, 000 profiles)

5 Lower troposphere refractivity bias

In contrast to the excellent agreement of CHAMP with ECMWF between 10 and 30 km, systematic deviations are observed in the lower troposphere below

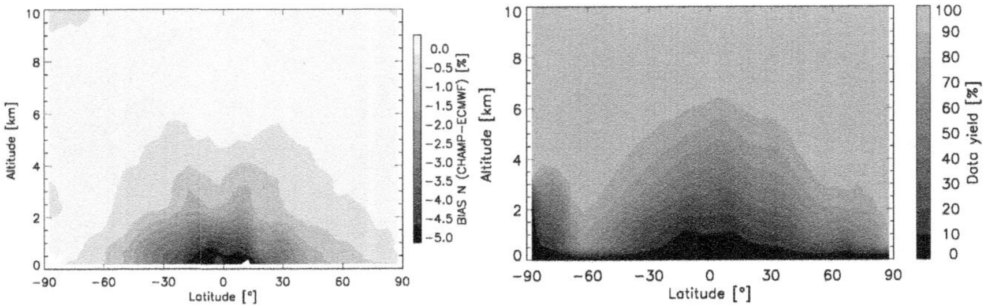

Fig. 5. Comparison of CHAMP refractivity data with corresponding ECMWF analyzes in the troposphere (left: bias; right: data yield, related to data at 10 km) between May 14, 2001 and June 8, 2005 ($\sim 200,000$ profiles)

5 km (Fig. 5), which is combined with a decreasing data yield. A negative refractivity bias of the CHAMP data can be observed, which depends on latitude and is most pronounced in the tropics, where it reaches a value of 5 % at 1 km. However, in mid latitude and polar regions the CHAMP data are nearly bias free throughout the entire troposphere. This bias is a known phenomenon of the CHAMP data and is discussed in more detail by Ao et al. (2003a); Beyerle et al. (2003a,b, 2005b). Causes of the bias are, beside multi-path propagation, also signal tracking errors of the GPS receiver and superrefraction, a physical limitation of the RO technique. Further progress in reducing the bias and increasing the data yield is expected by the application of advanced signal tracking methods (Open Loop (OL) technique, see, e.g. (Sokolovskiy, 2001; Beyerle et al., 2005b)) and improved signal strength due to the use of more advanced occultation antenna configuration (foreseen, e.g., for COSMIC or Metop). As an example for these investigations and to illustrate the potential of the advanced tracking techniques for the improvement of the occultation data quality, Fig. 6 shows end-to-end occultation simulation results according to Beyerle et al. (2005b). An improved GPS receiver model was used (OL, four quadrant carrier phase extractor) for the simulations based on realistic atmospheric conditions, derived from 1,309 radiosonde profiles from the tropics (with occurrence of superrefraction in $\sim$ 40% of the profiles). The results were compared with those of a "CHAMPlike" receiver (Closed Loop, two quadrant phase extractor, flywheeling). Hereby the GPS receiver parameters should not be an accurate representation of the CHAMP receiver. Fig. 6 shows the corresponding comparisons of output and input refractivity profiles. A significant improvement of the analysis results can be observed when using the OL GPS receiver model. The OL retrievals show significantly reduced bias and standard deviations in relation to the "CHAMPlike" receiver model results. This corresponds with significant increased data yield in the lower troposphere. The OL results are bias free down to $\sim$2 km, with standard deviations from $\sim$0.1 to $\sim$ 0.5%. Below 2 km the physical limitation of the GPS RO technique, the superrefraction, is the reason for a deterioration of the data quality, which is represented by significant deviations between

output and input profiles up to ~1.4 % negative bias and ~2.5 % standard deviation at 0.5 km for an "ideal" receiver, representing the truth. But even in this altitude range the results of the "real" OL receiver model match nearly perfect the "truth" and is less sensitive for low SNR (signal to noise ratio). More details on these investigations are given by Beyerle et al. (2005b).

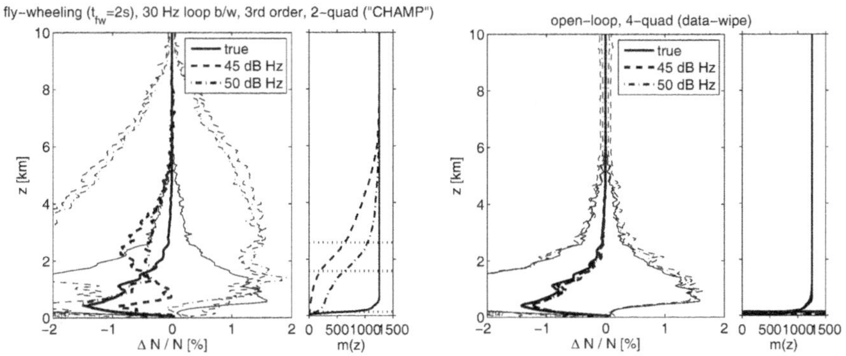

Fig. 6. Results of GPS receiver simulations. Thick lines indicate relative mean and thin lines relative one-sigma standard deviation of $N_{output} - N_{input}$ related to N_{input} and the corresponding number of compared data per altitude for two different GPS receiver models (for details see text and (Beyerle et al., 2005b). Solid lines indicated the "truth", which shows a negative bias caused by superrefraction, dotted lines indicate the results including the receiver for two different carrier signal-to-noise density ratios of 45 and 50 db (equivalent to SNR of ~ 141 and 251 V/V, respectively)

6 Application of CHAMP's data for atmospheric research

CHAMP refractivity and temperature data in the upper troposphere/lower stratosphere region are not affected by background temperature fields and are most accurate in that altitude region (Kursinski et al., 1997). Therefore they can be used to investigate climate change processes. Global temperature changes can be detected by analyzing surface temperatures, but also using measurements in the free atmosphere. Hereby the amplitude of the observed temperature variations can be higher in distinct zonal regions as observed variations at the surface (e.g. Tett et al., 2002) and trends in the tropopause region can be visible earlier and more clear as at the surface (Randel et al., 2003; Sausen and Santer, 2003). CHAMP based temperatures climatologies (e.g., seasonal zonal means) are derived and analyzed within the CHAMP-CLIM project (Foelsche et al., 2005). First results on climatic variations of the Earth's atmosphere can be expected from CHAMPCLIM after 7 years of CHAMP measurements in 2008. This data set will be extended by the measurements from COSMIC (launch probably early 2006) and Metop (planned launch 2005).

Another point of interest with respect to the detection of climate change are tropopause parameters. Especially tropopause temperature, altitude and pressure have received attention in the recent years to describe climate variability (e.g. Sausen and Santer, 2003). The accuracy and high vertical resolution of GPS RO measurements are nearly perfect preconditions for precise monitoring of tropopause characteristics on a global scale. The tropopause was the main focus of several studies using CHAMP or GPS/MET data (e.g. Randel et al., 2003; Schmidt et al., 2004, 2005a).

To illustrate the potential of these investigations, Color Fig. XI on p. 292 shows the global tropopause structure derived using RO data from CHAMP and the Argentine SAC-C satellite (Satelite de Aplicaciones Cientificas-C, (Hajj et al., 2004)). The observed regional patterns are consistent with climatologies based on radiosonde measurements and meteorological analyzes, as discussed in more detail by Schmidt et al. (2005a).

Another application example to investigate atmospheric phenomena is illustrated by Fig. 7. It shows vertical temperature anomalies in the equator region, detected with CHAMP RO data. These anomalies can be related to the Quasi-Biennal-Oscillation (Randel et al., 2003; Schmidt et al., 2004), a well known atmospheric wave phenomenon of the tropical stratosphere.

7 First results from GRACE

The U.S.-German GRACE satellites were launched on March 17, 2002. The GPS receiver ("BlackJack", JPL) aboard the GRACE-B satellite (aft-looking antenna to observe setting occultations) was activated for the first time in atmospheric sounding mode from July 28 06:00 UTC until July 29 07:00 UTC, 2004 (Beyerle et al., 2005a; Wickert et al., 2005b). 120 occultations (parallel tracking of occultation and reference satellite $\geq$ 20 s) were recorded during this 25 h interval. The data were analyzed using the orbit and occultation processing system for CHAMP (see Sect.. 3). Fig. 8 shows retrieved profiles of dry temperature, specific humidity of the first occultation measurement from GRACE (55.31°N, 22.32°E) on July 28, 2005 06:10 UTC, and corresponding ECMWF profiles.

The significantly improved stability of the satellite clock from GRACE-B in relation to CHAMP allows for the application of a zero-differencing technique to avoid the disturbing influence of the additional link to a referencing GPS satellite and to minimize the RO data amount, recorded aboard GRACE-B. The method was first applied and is described in detail by Beyerle et al. (2005a). For the operational GRACE data analysis we apply an implementation of the zero-difference technique using the 30 s clock solutions provided by

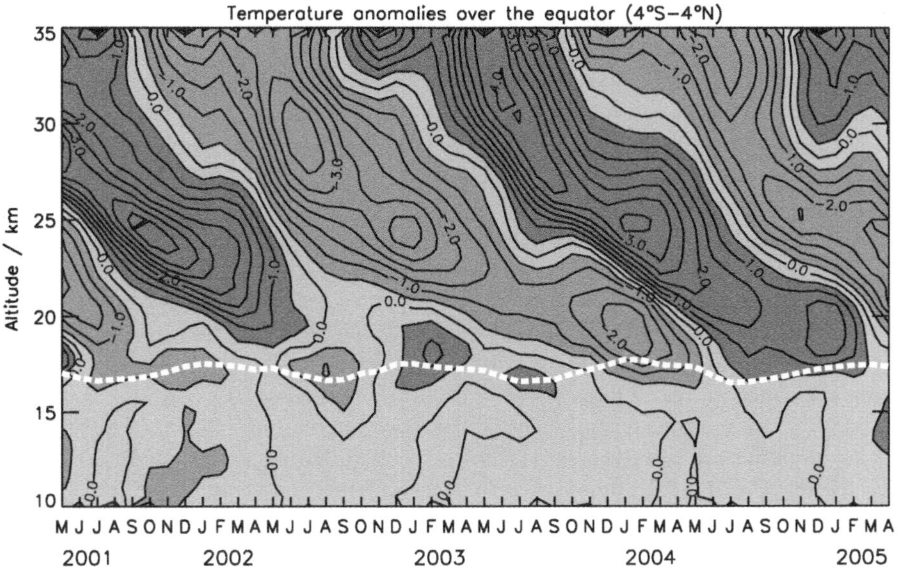

Fig. 7. Temperature anomaly above the equator region (zonal mean; 4°N-4°S), derived from CHAMP RO data (May 2001-April 2005). Contour interval ±0.5 K. The dotted white line indicates the monthly means of the cold point tropopause altitude (update from (Schmidt et al., 2004))

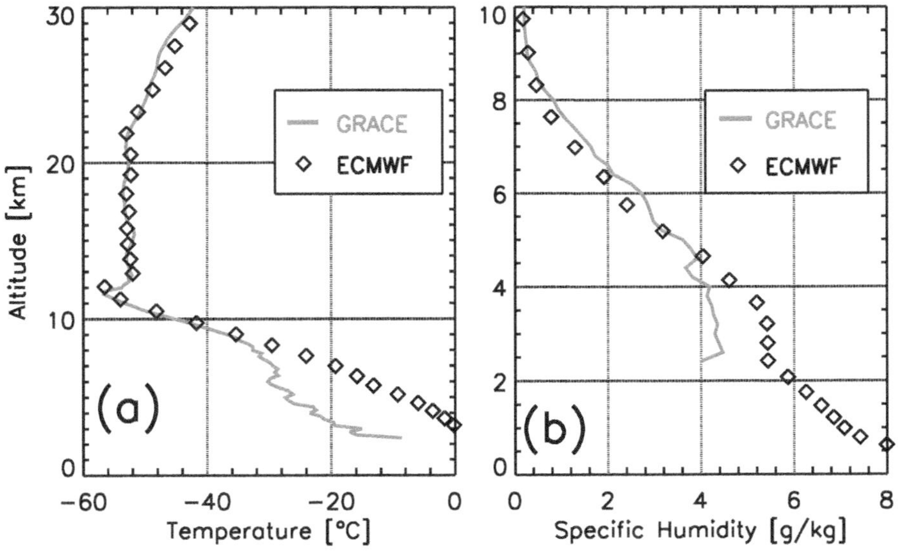

Fig. 8. Vertical profiles of (a) dry temperature and (b) specific humidity, derived from the first occultation measurement aboard GRACE, compared with corresponding ECMWF analysis (22.32°E, 55.31°N), July 28, 2004, 06:10 UTC; from (Wickert et al., 2005b)

the precise orbit determination facility from GFZ (König et al., 2005a,b). The zero differencing results show statistically equivalent behavior as the standard double difference profiles with respect to ECMWF (Wickert et al., 2005c). This finding is in contrast to the early results from Wickert et al. (2005b), which indicated a better agreement of the zero difference results with ECMWF.

8 Summary and outlook

More than 300,000 occultation measurements are available after four and a half years of GPS radio occultation with CHAMP. The mission is currently expected to last at least until 2008, the first long-term data set of GPS RO data is anticipated. Occultation data and analysis results are available at the GFZ data center (*http://isdc.gfz-potsdam.de/champ*). A rapid occultation data transfer and analysis is continuously demonstrated since February 2003. Currently an average delay of ~4 hours between each measurement aboard CHAMP and data product provision is continuously reached. The data from CHAMP are the base for the preparation of future occultation missions and related processing systems, for impact studies to improve the global weather forecasts, for studies with relevance to detect climate change and for other applications in atmospheric/ionospheric research. Validation results indicate an excellent data quality in the upper troposphere and lower stratosphere. A negative refractivity bias is observed in the lower troposphere. End-to-end simulations indicate, that improved GPS tracking techniques have the potential to significantly increase the lower troposphere data quality. First results from the GRACE-B satellite are available, which indicate the potential of GRACE to enhance the daily number of globally distributed occultation measurements after continuous activation of the GRACE occultation experiment. Recent information on the status of the RO experiments aboard CHAMP and GRACE, the operational data analysis at GFZ and current validation results can be obtained via WWW (*http://www.gfz-potsdam.de/gasp*).

Acknowledgement. This is publication GEOTECH-146 of the research programme GEOTECHNOLOGIEN of BMBF and DFG, Grant 03F0333A, which supported the provision of CHAMP's occultation raw data. The analysis of the RO measurements from CHAMP were partly funded by the German Research Foundation (DFG), grant WI2634/2-1/503976. We thank all members of the CHAMP team for their excellent work, which is the basis for our investigations. The German Weather Service provided ECMWF analysis data. Radiosonde data are from Alfred-Wegener-Institut for Polar and Marine Research Bremerhaven.

References

Ao, C. O., T. K. Meehan, G. A. Hajj, A. J. Mannucci, and G. Beyerle, Lower-troposphere refractivity bias in GPS occultation retrievals, J. Geophys. Res.,

108(D18), doi:10.1029/2002JD003,216, 2003a.

Ao, C. O., W. B. Schreiner, and J. Wickert, First report on the CHAMP radio occultation intercomparison study, JPL Publication, 03-016, 2003b.

Beyerle, G., M. E. Gorbunov, and C. O. Ao, Simulation studies of GPS radio occultation measurements, Radio Sci., 38(5), doi:10.1029/2002RS002,800, 2003a.

Beyerle, G., J. Wickert, T. Schmidt, and C. Reigber, Atmospheric sounding by GNSS radio occultation: An analysis of the negative refractivity bias using CHAMP observations, J. Geophys. Res., p. doi:10.1029/2003JD003922, 2003b.

Beyerle, G., T. Schmidt, G. Michalak, J. Wickert, and C. Reigber, GPS radio occultation with GRACE: Atmospheric profiling utilizing the zero difference technique, Geophys. Res. Lett., in print, 2005a.

Beyerle, G., T. Schmidt, J. Wickert, S. Heise, C. Reigber, and G. König-Langlo, An analysis of refractivity biases detected in GPS radio occultation data: Results from simulation studies, aerological soundings and CHAMP satellite observations, paper available via www.arxiv.org/physics/0502052, 2005b.

Dunn, C., et al., Instrument of GRACE - GPS Augments Gravity Measurements, GPS World, Feb. 1, 2003.

Foelsche, U., G. Kirchengast, A. Gobiet, A. Steiner, A. Löscher, J. Wickert, and T. Schmidt, The CHAMPCLIM project: An overview, in Earth Observation with CHAMP: Results from Three Years in Orbit, edited by C. Reigber, P. Schwintzer, H. Lühr, and J. Wickert, pp. 615–620, Springer Verlag, 2005.

Gobiet, A., U. Foelsche, A. Steiner, M. Borsche, G. Kirchengast, and J. Wickert, Validation of stratospheric temperatures in ECMWF operational analyses with CHAMP radio occultation data, Geophys. Res. Lett., 32(12), doi:10.1029/2005GL022,617, 2005.

Hajj, G. A., L. C. Lee, X. Pi, L. J. Romans, W. S. Schreiner, P. R. Straus, and C. Wang, COSMIC GPS ionospheric sensing and space weather, Terrestrial, Atmospheric and Oceanic Science, 11, 2000.

Hajj, G. A., et al., CHAMP and SAC-C atmospheric occultation results and intercomparisons, J. Geophys. Res., 109(D06109), doi:10.1029/2003JD003,909, 2004.

Healy, S., and J. Eyre, Retrieving temperature, water vapor and surface pressure information from refractive-index profiles derived by radio occultation: A simulation study, Quart. J. Roy. Meteorol. Soc., 126, 1661–1683, 2000.

Healy, S., A. Jupp, and C. Marquardt, Forecast impact experiment with gps radio occultation measurements, Geophys. Res. Lett., 32(3), doi: 10.1029/2004GL020,806, 2005.

Healy, S. B., and J.-N. Thepaut, Assimilation experiments with CHAMP GPS radio occultation measurements, Q. J. R. Meteorol. Soc., subm., 2005.

Heise, S., J. Wickert, G. Beyerle, T. Schmidt, and C. Reigber, Global monitoring of tropospheric water vapor with GPS radio occultation aboard CHAMP, Adv. Space Res., in print, 2005.

Jakowski, N., A. Wehrenpfennig, S. Heise, C. Reigber, H. Lühr, L. Grunwaldt, and T. K. Meehan, GPS radio occultation measurements of the ionosphere from CHAMP: Early results, Geophys. Res. Lett., 29(10), doi:10.1029/2002RS002763, 2002.

König, R., G. Michalak, and K. Neumayer, Remarks on champ orbit products, in Observation of the Earth System from Space, edited by R. Rummel, C. Reigber, M. Rothacher, G. Boedecker, U. Schreiber, and J. Flury, Springer Verlag, 2005a.

König, R., G. Michalak, K. Neumayer, R. Schmidt, S. Zhu, H. Meixner, and C. Reigber, Recent developments in CHAMP orbit determination at GFZ, in Earth Observation with CHAMP: Results from Three Years in Orbit, edited by C. Reigber, P. Schwintzer, H. Lühr, and J. Wickert, Springer Verlag, 2005b.

Kuo, Y.-H., S. V. Sokolovskiy, R. A. Anthes, and F. Vandenberghe, Assimilation of GPS radio occultation data for numerical weather prediction, Terrestrial, Atmospheric and Oceanic Sciences, 11(1), 157–186, 2000.

Kuo, Y.-H., T.-K. Wee, S. Sokolovskiy, C. Rocken, W. Schreiner, D. Hunt, and R. A. Anthes, Inversion and error estimation of GPS radio occultation data, J. Meteorol. Soc. Jpn., 1B(82), 507–531, 2004.

Kuo, Y.-H., W. S. Schreiner, J. Wang, D. L. Rossiter, and Y. Zhang, Comparison of GPS Radio occultation soundings with radiosondes, Geophys. Res. Lett., 32, l05817, doi:10.1029/2004GL021443, 2005.

Kursinski, E. R., G. A. Hajj, J. T. Schofield, R. P. Linfield, and K. R. Hardy, Observing Earth's atmosphere with radio occultation measurements using Global Positioning System, J. Geophys. Res., 19(D19), 23,429–23,465, 1997.

Larsen, G., K. Lauritsen, F. Rubek, and M. Sørensen, Processing of CHAMP radio occultation data using GRAS SAF software, in Earth Observation with CHAMP: Results from Three Years in Orbit, edited by C. Reigber, P. Schwintzer, H. Lühr, and J. Wickert, pp. 543–548, Springer Verlag, 2005.

Loiselet, M., N. Stricker, Y. Menard, and J. Luntama, GRAS – MetOps GPS based atmospheric sounder, ESA Bulletin, May, 102, 38–44, 2000.

Marquardt, C., K. Schöllhammer, G. Beyerle, T. Schmidt, J. Wickert, and C. Reigber, Validation and data quality of CHAMP radio occultation data, in First CHAMP Mission Results for Gravity, Magnetic and Atmospheric Studies, pp. 384–396, Springer Verlag, 2003.

Randel, W. J., F. Wu, and W. R. Rios, Thermal variability of the tropical tropopause region derived from GPS/MET observations, J. Geophys. Res., 108(4024), doi: doi:10.1029/2002JD002595, 2003.

Reigber, C., P. Schwintzer, H. Lühr, and J. Wickert (Eds.), Earth Observation with CHAMP: Results from Three Years in Orbit, Springer Verlag, 2005.

Rocken, C., Y.-H. Kuo, W. Schreiner, D. Hunt, S. Sokolovskiy, and C. McCormick, COSMIC system description, Terrestrial, Atmospheric and Oceanic Sciences, 11, 21–52, 2000.

Rocken, C., et al., Analysis and validation of GPS/MET data in the neutral atmosphere, J. Geophys. Res., 102(D25), 29,849–29,866, 1997.

Sausen, R., and B. Santer, Use of changes in tropopause height to detect human influences on climate, Met. Zeitschrift, 12(3), 131–136, 2003.

Schmidt, T., J. Wickert, G. Beyerle, and C. Reigber, Tropical tropopause parameters derived from GPS radio occultation measurements with CHAMP, J. Geophys. Res., 109(D13105), doi:10.1029/2004JD004566, 2004.

Schmidt, T., S. Heise, J. Wickert, G. Beyerle, and C. Reigber, GPS radio occultation with CHAMP and SAC-C: global monitoring of thermal tropopause parameters, Atmosph. Chemistry and Physics, 5, 1473–1488, 2005a.

Schmidt, T., J. Wickert, G. Beyerle, R. König, R. Galas, and C. Reigber, The CHAMP atmospheric processing system for Radio occultation measurements, in Earth Observation with CHAMP: Results from Three Years in Orbit, edited by C. Reigber, P. Schwintzer, H. Lühr, and J. Wickert, Springer Verlag, 2005b.

Sokolovskiy, S. V., Tracking tropospheric radio occultation signals from low Earth orbit, Radio Sci., 36(3), 483–498, 2001.

Takahashi, H., et al., Scientific satellite EQUARS for remote sensing of the equatorial atmosphere, 7th Latin-American Conference on Space Geophysics, Atibaia, April, 2004.

Tapley, B., and C. Reigber, GRACE (Gravity Recovery and Climate Experiment), in McGraw-Hill Yearbook of Science & Technology, 2004.

Tett, S., et al., Estimation of natural and anthropogenic contributions to twentieth century temperature change, JGR, 107, 4306, 10.1029/2000JD000,028, 2002.

v. Engeln, A., J. Teixeira, J. Wickert, and S. Buehler, Detecting the planetary boundary layer using CHAMP data, Geophys. Res. Lett., 32, doi:10.1029/2004GL022,168, 2005.

Wang, D. Y., et al., Cross-validation of MIPAS/ENVISAT and GPS-RO/CHAMP temperature profiles, J. Geophys. Res., 109, doi:10.1029/2004JD004963, 2004.

Ware, R., et al., GPS sounding of the atmosphere from low Earth orbit: Preliminary results, Bull. Am. Meteorol. Soc., 77(1), 19–40, 1996.

Wickert, J., The CHAMP radio occultation experiment: Algorithms, Processing system, and First results (in German), Scientific Technical Report 02/07, GFZ Potsdam, 2002.

Wickert, J., Comparison of vertical refractivity and temperature profiles from CHAMP with radiosonde measurements, Scientific Technical Report 04/19, GFZ Potsdam, ISSN 1610-0956, 2004.

Wickert, J., A. Pavelyev, Y. A. Liou, T. Schmidt, C. Reigber, A. Pavelyev, K. Igarashi, and S. Matyugov, Amplitude variations in the GPS signals as a possible indicator of the ionospheric structures, Geophys. Res. Lett., 31, l24801, doi:10.1029/2004GL020607, 2004a.

Wickert, J., T. Schmidt, G. Beyerle, R. König, C. Reigber, and N. Jakowski, The radio occultation experiment aboard CHAMP: Operational data processing and validation of atmospheric parameters, J. Meteorol. Soc. Jpn., 82(1B), 381–395, 2004b.

Wickert, J., C. Ao, and W. Schreiner, GPS based atmospheric sounding with CHAMP: Comparison of data analysis and results from GFZ, JPL and UCAR, GFZ Scientific Technical Report, in preparation, Potsdam, 2005a.

Wickert, J., G. Beyerle, R. König, S. Heise, L. Grunwaldt, G. Michalak, C. Reigber, and T. Schmidt, GPS radio occultation with CHAMP and GRACE: A first look at a new and promising satellite configuration for global atmospheric sounding, Ann. Geophysicae, 23(653-658), 2005b.

Wickert, J., T. Schmidt, G. Beyerle, G. Michalak, R. König, S. Heise, and C. Reigber, GPS radio occultation with CHAMP and GRACE: Recent results, in Proc. OPAC-2 workshop, Graz September 2004, edited by G. Kirchengast, U. Foelsche, and A. K. Steiner, Springer Verlag, 2005c.

Wickert, J., et al., Atmosphere sounding by GPS radio occultation: First results from CHAMP, Geophys. Res. Lett., 28(17), 3263–3266, 2001.

Part II

GRACE
The Gravity Recovery And Climate Experiment

Design and Operation of the GRACE ISDC

Bernd Ritschel, Andrea Bendig, Hartmut Palm, Ronny Kopischke, Sebastian Freiberg, Frank Flechtner, and Ulrich Meyer

GeoForschungsZentrum Potsdam, Telegrafenberg A3, 14473 Potsdam, Germany
rit@gfz-potsdam.de

Summary. The GRACE Information System and Data Center designed, implemented and operated for the management of all scientific GRACE products is one fixed part of the GRACE Science Data infrastructure. The main objectives for the long term archiving and the worldwide dissemination of GRACE data and metadata driven by user requirements as well as the underlying software development workflow from the business modelling, analysis and design up to the implementation and deployment steps are described in the following article.

Key words: Information System and Data Center (ISDC), geoscience products, data archive

1 Introduction

Daily up to 200 Mbyte of data are generated by the different sensors mounted onboard the two GRACE satellites. These sensor data, collected and encoded by computers onboard of the satellites, are transmitted as raw data as often as possible to DLR's (Deutsches Zentrum für Luft- und Raumfahrt) receiving stations located in Neustrelitz and Weilheim.

These level-0 raw data, together with ancillary GPS ground station and meteorological data, are processed by the GRACE Science Data System (JPL Pasadena, University of Texas Center for Space Research (UTCSR) and GFZ Potsdam) to various level-1 and level-2 products according to the objectives of the GRACE mission. All products are stored at two different archives which are a fixed part of the overall GRACE Science Data System (GRACE SDS) infrastructure (Flechtner et al., 2003): the GRACE Information System and Data Center (ISDC) at GFZ and the Physical Oceanography Distributed Active Archive Center (PO.DAAC) at JPL. While the GRACE-part of the PO.DAAC system is designed as a simple GUI front end to data directories managed by a FTP service, the GRACE ISDC allows users to search for unique products using product-specific retrieval forms.

At the GRACE ISDC the products are divided into the main categories orbit/gravity (OG) and atmosphere/ionosphere (AI). The GRACE-A (GA) and GRACE-B (GB) OG and AI products are categorized in different product types by the level of processing (0, 1a, 1b and 2) and by a unique product name consisting of a fixed and a variable part separated by a "+" sign (e.g. GB-OG-1B-GPSDAT+JPL-GPS1B_2002-07-31_B_00).

Up to now 169 different OG and 2 AI products are defined and can be accessed as public or (GFZ-only) internal data sets (Table 1).

Table 1. Number of GRACE public and internal* product types at ISDC

Product type	Level 0	Level 1a	Level 1b	Level 2	Total
Orbit/Gravity	10*	40*	45	7+4*	52+54*
Atmosphere/Ionosphere			1*	1*	2*
GPS			6*		6*
Meteorological				57*	57*
Total					58+113*

At present more than 123,000 unique products with a summarized value of 1.3 TB are stored in the GRACE data archive. According to data policy reasons, only 30 % of all product types are accessible for the public science community. The remaining 70 % are for internal processing purposes only.

More than 170 public users and user groups from the worldwide scientific community and 14 internal users from the GFZ are registered at the GRACE ISDC. Many of them use the GRACE ISDC regularly for product retrieval, product requests and download of data, products and documents.

In the following, the design and operation of the GRACE ISDC, which is used for the management of the scientific GRACE products, is described.

2 Objectives and business modeling

Making use of the experience gained during the development and operation of the CHAMP ISDC being responsible for the management of the scientific products of the CHAMP satellite mission (Ritschel et al., 2002, 2003b,c; Reigber et al., 2003), a similar system for the GRACE satellite mission has been designed and implemented by the GFZ Data Center.

The overall process for the development and the implementation of the GRACE ISDC is based on IBM's Rational Unified Process®(RUP) (http://www-306.ibm.com/software/awdtools/rup/). RUP is an iterative and configurable software development process platform that delivers proven best practices and a configurable architecture for the development of distributed and sophisticated applications (Fig. 1).

The main scientific and technical objectives for the development of the GRACE ISDC are:

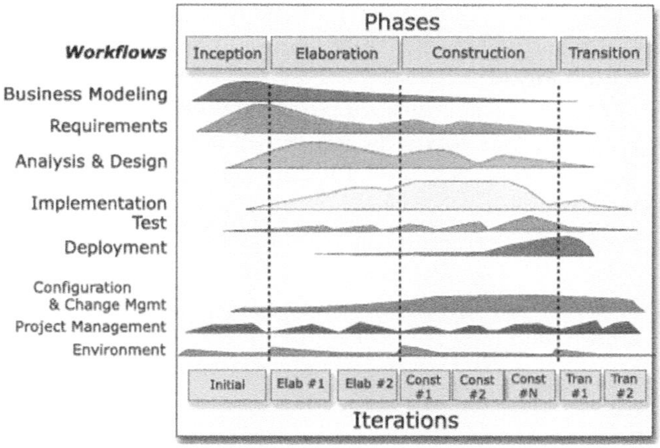

Fig. 1. Rational Unified Process, Copyright ©1987 - 2000 Rational Software Corp

- overall management of the input and output data and product stream,
- long-term archiving and backup of all GRACE products,
- user-controlled and user-specifically granted access to the GRACE products and related documents and
- periodical check of the ISDC GRACE data stock and harmonization with the content of the PO.DAAC data archive.

Based on these main objectives, an appropriate business process was identified and the appropriate model has been developed. This business model comprises different general and some specific business cases. The main business case deals with the general purpose of the ISDC, including the overall management of the products as well as the use of the ISDC by the different user types. This takes into account the user searching for products, the SDS placing products into the ISDC and the administrator responsibilities for the control and operation of the overall system. Besides the identification of the business process, an overall business process realization represented by a business-intended functional use-case model has been designed.

3 Requirements and constraints

Starting from this business model, the detailed user requirements for the GRACE ISDC can be defined under consideration of technical system constraints. Whereas in the business model the scope of the overall ISDC system is described, the requirements phase of the RUP is used for the detailed description of any type of request a stakeholder (scientific user, project manager and decision maker, administrator, etc.) might have on the GRACE ISDC system. These stakeholder requests are transformed in business rules and described on a technical and formalized basis using appropriate use-case scenarios (Flechtner et al., 2003) according to the Unified Modelling Language (UML) standard (http://www.uml.org/).

In general, the use-case model is a model of the system's intended functions and its environment, and serves as a contract between the customer and the developers. Thus, the use-case model is an essential input for analysis, design, and test activities.

Based on the main scientific and technical objectives, the different user requirements can be classified according to the product management, the product retrieval/access, the user management and the provision of general information and documents.

Concerning the product management, the GRACE ISDC requirements apply to the

- continuous product input provided by the different scientific producers at UTCSR, JPL and GFZ,
- product archiving and backup on tape for the creation of a long-term archive and on disc for online product access purposes and
- user-controlled product output using different access methods.

The product retrieval/access requirements consist of different product search and access features using online retrieval, batch mode and express delivery according to data policies (depending on user grants).

The user management requirements based on the GRACE project data policies require the consideration of different user categories (public, extended and internal) with different product access grants. Finally, there are requirements concerning the provision of general information, documents and tools.

The different stakeholder requirements to the GRACE ISDC are documented in verbal descriptions of the associated use-cases. The main use-case parties involved in the GRACE ISDC are scientific users, scientific producers, GRACE project managers and ISDC system administrators.

The overall GRACE ISDC software development workflow containing various iteration loops is shown in Fig. 2. The process starts with the use-case generation based on the requirements analysis. The formalized presentation of use-cases using the UML consists of parties involved (scientific user, scientific producer and system administrator for the GRACE ISDC) in different roles visualized by a stickman symbol interacting with a feature or interface of the ISDC system symbolized by an oval sign.

The main use-cases for the scientific user (roles) are:

1. searching for general information → browsing for documents, software tools and FAQs
2. searching for metadata → browsing and retrieval of product defining data
3. online product retrieval → online browsing and retrieval of requested products

4. offline product download → batch processing using product request list files

The main use-cases for the scientific producer (roles) are:

1. product announcer → delivering of product description
2. product supplier → transfer of products into interface directory
3. product input manager → processing of products
4. announcer of new features → delivering of new requests

The main use-cases for the system administrator (roles) are:

1. system administrator → system monitoring and controlling
2. user administrator → management of users and user grants
3. maintenance administrator → change management and system maintenance
4. advanced administrator → providing of new features

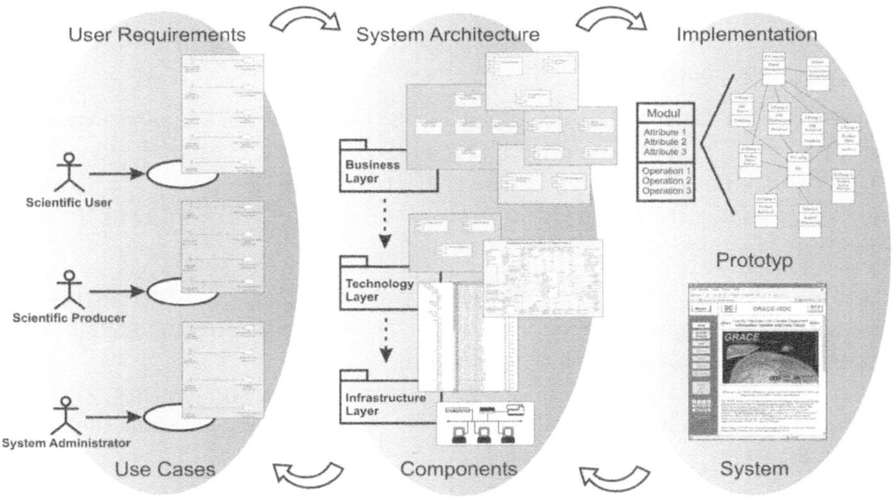

Fig. 2. GRACE ISDC software development workflow based on the RUP

Figure 2 shows the workflow starting with the implementation process using object-oriented software development techniques, the generation of different state prototypes and finally the system deployment and operation of the target platform.

In addition to the functional requirements on the GRACE ISDC entailed by the stakeholders, the technical constraints concerning the IT infrastructure have to be taken into account. For the GRACE ISDC such technical constraints mainly apply to the technical environment (hardware and software) of the production system. These constraints are reflected in the structure of the system architecture of the GRACE ISDC (Fig. 2) and in the deployment of the main components related to basis software components as well as application software components.

The detailed transformation of the business rules of the GRACE ISDC into the main use-cases supplemented by technical realization descriptions are the basis for the creation of the overall system architecture of the GRACE ISDC.

4 Analysis and software design

The objective of the crucial analysis and design phase of the software development workflow is the creation of the overall software architecture of the GRACE ISDC.

The use-cases and the appropriate technical realization documents as well as technical, financial and project organizational constraints are the fundament for the design of a functionally structured system architecture. For the development of the GRACE ISDC, a three-layer system architecture (Fig. 2) has been used. The business layer contains the mapping and the realization of the functional features of the ISDC. In the technology layer the main basic software components

- Sybase RDBMS (RDBMS = Relational Database Management System),
- Apache HTTP web server (HTTP = Hyper Text Transfer Protocol),
- UMN map server (UMN = University of Minnesota),
- WU-FTPD (Washington University-File Transfer Protocol Domain) and
- OPA (Online Product Archive)

are integrated. The infrastructure layer symbolizes the basic services provided by the Computing Center of the GFZ such as mass storage management, e-mail and firewall, and network in general.

Derived from the scope of the three-layer system architecture, all layers themselves are split into smaller parts. The business layer contains the following subsystems and components:

- product retrieval, near real-time product access, scheduled product access,
- system boundary (interface), product import, product access, system administration, product saving (interface),
- DIF (Directory Interchange Format) meta data document import, DIF document normalization, metadata storage, product storage and
- maintenance and controlling, user management.

The technology layer is split into the subsystems product archive and data base and other application services, whereas the infrastructure layer contains the Computing Center basic services.

Figure 3 shows the main components of the GRACE ISDC which are responsible for the realization of the functions and features corresponding to the business model. The left and right handed arrows present the GRACE products input and output stream, whereas the top sided mark the main user interaction processes with the GRACE ISDC.

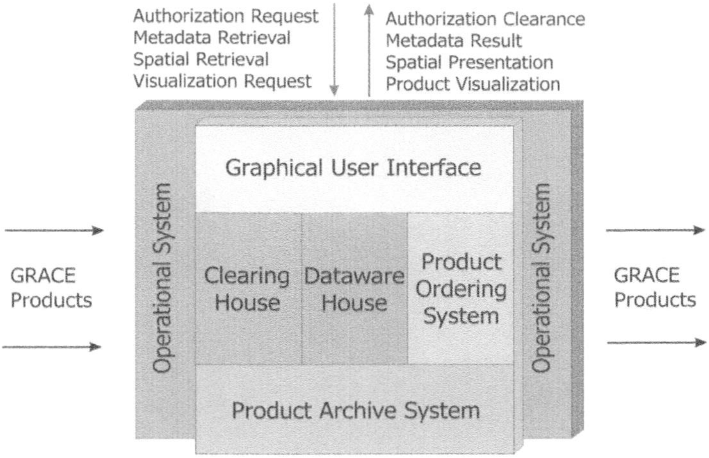

Fig. 3. Basic components of the GRACE ISDC (functional view)

The operational system component in the background containing the basic interfaces to the other components is mainly responsible for the technical realization of the product input and output data stream. The clearinghouse, the heart of the GRACE ISDC, has been designed as a powerful catalogue system based on the mandatory and product-related metadata (Fig. 4). A smaller part of the clearinghouse is also responsible for the user management. All GRACE products are stored and archived in the product archive system consisting of an online RAID (Redundant Array of Inexpensive Disks/Redundant Array of Independent Disks) hard disc area as part of the ISDC technology layer and a tape area for long-term archiving and backup as part of the infrastructure layer. The product ordering system controls the user-granted access to the products, depending on the different product request methods. Spatial retrieval and presentation as well as product visualization are features of the data warehouse component (Braune et al., 2002). The interactive communication between the ISDC users and the system, which is mainly used for online product retrieval, is realized by the graphical user interface component (Ritschel et al., 2003a).

In order to handle all the different scientific product types within one unique information system, a unique standard for the definition of ISDC products had to be introduced. According to the ISDC product philosophy, first applied for the CHAMP ISDC, each product consists of one data file and one corresponding metadata file (Fig. 4). As the data file format is unrestricted, all metadata files correspond to the extended Directory Interchange Format (DIF) standard (http://gcmd.gsfc.nasa.gov/User/difguide/whatisadif.html).

Using the standardized DIF files as data source for the metadata database of the clearinghouse, only one application is necessary in order to manage all different product types. Whereas for the input of metadata only the DIF format is used, the output of metadata is designed for the ISO 19115 standard for spatial metadata too (Braune et al., 2002; Czegka et al., 2003).

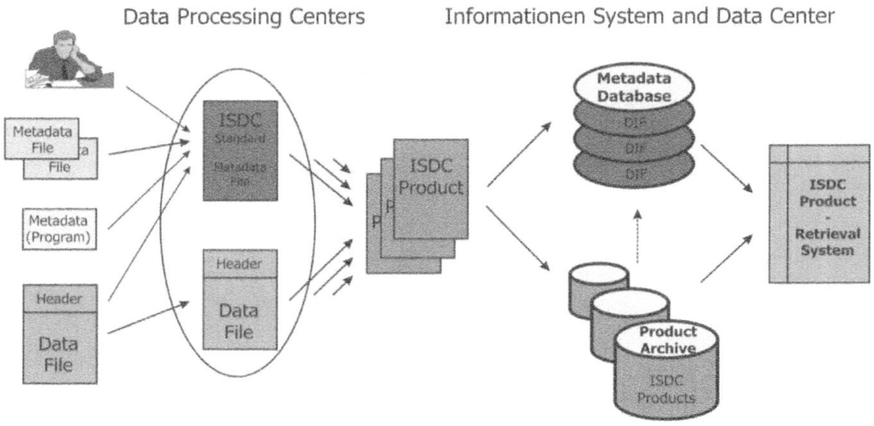

Fig. 4. ISDC product philosophy and metadata processing

5 Implementation and deployment

In the implementation process, the organization of the program code has to be defined in terms of coding the subsystems which are organized in layers. Subsystem-inherent classes and objects (source files, binaries, executables and others) are generated and tested both isolated and in interaction with other objects. For the implementation of the GRACE ISDC, existing but modified and advanced CHAMP ISDC components for the clearinghouse and the product ordering system could be used (Ritschel et al., 2002, 2003c). The main components of the technology and infrastructure layer (cf. Fig. 2), like Sybase RDBMS, Apache HTTP web server, WU-FTPD server and others, are used with small adjustments for the GRACE ISDC.

The deployment of the different software subsystems and components of the GRACE ISDC is shown in Fig. 5. Here it is important to distinguish between the technical and organizational environment for the actual operation of the overall system and a separate area for the development and maintenance of unique system components. Due to a firewall with a so called demilitarized zone (DMZ), major parts of the GRACE ISDC are implemented in the DMZ as well as inside the intranet. The main services based on commercial and open-source software products of the technology layer are dedicatedly assigned to unique workstations. This is realized for the database, the WWW, the FTP and the map service. A major part of the business logic of the GRACE ISDC is running on the WWW/Visualization workstation. The management of the input and output product stream as well as the storage of the data in the long-term and tape-based HSM archive (Hierarchical Storage Management) and the hard-disc-based OPA (Online Product Archive) is performed by special data pumps running on a dedicated workstation. Due to a limited bandwidth (100 MB/s) of the general network and in order to avoid bottlenecks during the transfer of huge data amounts, a high-speed (2 GB/s) shared file system is used for the communication between the FTP and OPA servers.

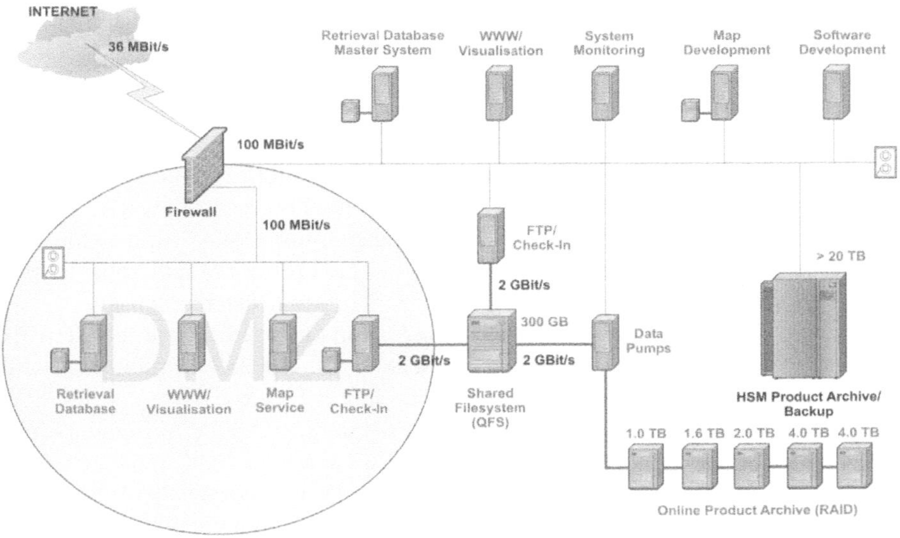

Fig. 5. Deployment of ISDC components

6 Operation and outlook

The internal operational phase of GRACE ISDC had already started before the launch of the GRACE satellites in March 2002. Because of a longer than expected validation phase of the GRACE satellites, sensors and products, the GRACE ISDC became operational not before December 2003, when a first set of GRACE products was provided to the GRACE Science Team (ST) for validation purposes. Important milestones of the operation of the GRACE ISDC were January 26, 2004, the release of a second set of data to the GRACE ST, and August 8, 2004, the release of 2 years of data to the public science community.

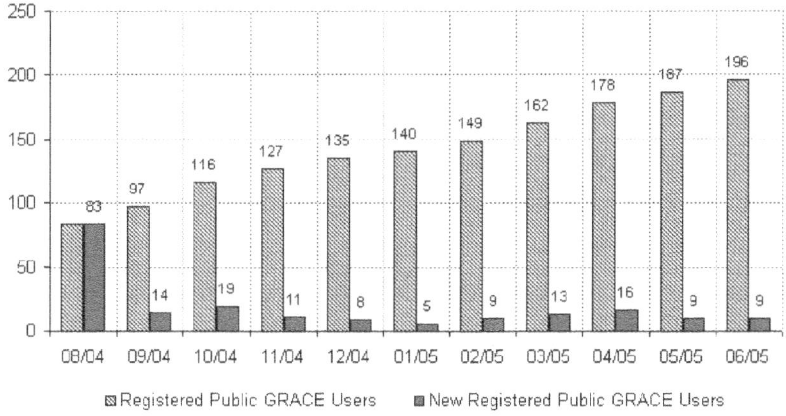

Fig. 6. Public GRACE ISDC users (08/2004 – 06/2005)

A permanently increasing number of international users and user groups visit the ISDC daily for the retrieval and download of scientific GRACE products (cf. Fig. 6). Additionally the GFZ scientific processing groups share the ISDC for the storage of and the access to internal products. For routine work the batch mode interface based on the ISDC product list file is used.

Many years of experience with the operation of the ISDC for the CHAMP and the GRACE satellite missions (Ritschel et al., 2003a,b) validate the usefulness of the adopted RUP workflow process model for the design of the ISDC systems. The feedback from the ISDC user community, new requirements from the project management and technological progress are the main factors for improvements and further development of the ISDC systems.

The main goal for a future system is the development of an innovative internet portal for the management of a continuously growing input of satellite products as well as different geodetic and geophysical ground station data and products. The integration of the already existing CHAMP and GRACE ISDC, the ISDC for data of the worldwide Global Geo Dynamic (GGP) project network of super-conducting gravimeters as well as the future TSX ISDC designed for the TerraSAR-X satellite mission will be an important step towards a unique and universal access to geo-scientific data and information (http://gesis.gfz-potsdam.de).

A modular and well-scaled system shall be developed and implemented within the next two years (Ritschel et al., 2005). This new portal system will allow research not limited by product groups or satellites. The content management system will be adjusted to the present data stock but will be expandable to data amounts to be expected in the future. This includes a re-structuring of the product archiving system. The two most complex applications for user and product administration – which are currently separated – shall be replaced by a new integrated solution. A focal point will be the development and realization of a comprehensible and particularly user-friendly user interface.

Acknowledgement. This is publication GEOTECH-152 of the research programme GEOTECHNOLOGIEN of BMBF and DFG. The German Ministry of Education and Research (BMBF) supports the GRACE project within the GEOTECHNOLOGIEN geoscientific R+D programme under grant 03F0326A.

References

Braune, S., Ritschel, B., Palm, H. (2002), MapServer und ArcInfo als Werkzeuge zum räumlichen Retrieval und zur Online-Präsentation von Metadaten zu geowissenschaftlichen CHAMP-Satellitenprodukten, in: Strobl, J., Blaschke, T., Griesebner, G. (eds.): Angewandte Geographische Informationsverarbeitung XIV: Beiträge zum AGIT-Symposium Salzburg 2002, Wichmann, 58–63

Czegka, W., Braune, S., Palm, H., Ritschel, B., Klump, J., Lochter, F. (2003), Beispiele ISO 19115 DIS konformer Metadaten in Katalogservices. Zwei Anwendungen aus dem Bereich umwelt- und geowissenschaftlicher Geofachdaten im Rahmen der Metadatencommunity der "GIB", 1. UNIGIS-Update-Konferenz, Salzburg, 14-15 Apr 2003, http://www.unigis.ac.at/club/u2/2003/UP_Beitrag_Czegka_Braune.pdf

Flechtner, F., Ackermann, C., Meixner, H., Meyer, U., Neumayer, K.-H., Ritschel, B., Schmidt, A., Schmidt, R., Zhu, S., Reigber, Ch. (2003), Development of the GRACE Science Data System, in: Observation of the System Earth from Space, Status Seminar, Bavarian State Mapping Agency (BVLA), Munich, 12 June 2003, GEOTECHNOLOGIEN Science Report No. 3, 48–50

Reigber, Ch., Schwintzer, P., Lühr, H., Massmann, F.-H., Galas, R., Ritschel, B. (2003), CHAMP Mission Science Data System Operation and Generation for Scientific Products, in: Observation of the System Earth from Space, Status Seminar, Bavarian State Mapping Agency (BVLA), Munich, 12 June 2003, GEOTECHNOLOGIEN Science Report No. 3, 129–131

Ritschel, B. and CHAMP-ISDC Team (2002), CHAMP - From Data Acquisition to Innovative Services, 27th General Assembly European Geophysical Society, Nice, 2002, Abstract EGS02-A-06459

Ritschel, B., Behrends, K., Braune, St., Freiberg, S., Kopischke, R., Palm, H., Schmidt, A. (2003a), CHAMP/GRACE-Information System and Data Centre (ISDC) - The User Interfaces for Scientific Products of the CHAMP and GRACE Mission, in: Observation of the System Earth from Space, Status Seminar, Bavarian State Mapping Agency (BVLA), Munich, 12 June 2003, GEOTECHNOLOGIEN Science Report No. 3, 132–133

Ritschel, B., Behrends, K., Braune, St., Freiberg, S., Kopischke, R., Palm, H., Schmidt, A. (2003b), Two Years of CHAMP Product Management - Experiences and Conclusions, EGS-AGU-EUG Joint Assembly, Nice, 2003, *Geophysical Research Abstracts*, 5, 11612

Ritschel, B., Braune, S., Behrends, K., Freiberg, S., Kopischke, R., Palm, H., Schmidt, A., Schneider, M. (2003c), CHAMP-ISDC - Informationssystem und Datenzentrum für geowissenschaftliche Produkte des CHAMP-Satellitenprojekts, *Zeitschrift für Geologische Wissenschaften*, 31(1), 21–30

Ritschel, B. and CHAMP-ISDC Team (Behrends, K., Bendig, A, Bruhns, Ch., Freiberg, S., Kopischke, R., Palm, H.) and Colleague Students (Buchta, St., Domann, P., Glosa, M., Lowisch, St., Volgmann, O.) (2005), Managing the workflow from data acquisition to knowledge generation concerning the satellite missions CHAMP and GRACE, European Geosciences Union, General Assembly 2005, Vienna, Austria Geophysical Research Abstracts, 7, 07296, SRef-ID: 16077962/gra/EGU05-A-07296, 24–29 April 2005, Abstract, Poster

De-aliasing of Short-term Atmospheric and Oceanic Mass Variations for GRACE

Frank Flechtner, Roland Schmidt, and Ulrich Meyer

GeoForschungsZentrum Potsdam (GFZ), Dept. 1 'Geodesy and Remote Sensing', Telegrafenberg A 17, 14473 Potsdam, Germany, *flechtne@gfz-potsdam.de*

Summary. GFZ is responsible for routine calculation of atmospheric and oceanic mass variations which have to be considered during GRACE precise orbit determination and calculation of gravity field partial derivatives. This Level-1B Atmosphere and Ocean De-aliasing product (AOD1B) is made available to the GRACE Science Data System and user community in terms of spherical harmonic coefficients with a maximum time delay of about 3-4 days dependent on the availability of required ECMWF meteorological fields. The spatial and time-variable vertical structure of the atmosphere is taken into account by vertical integration of the atmospheric masses. Oceanic mass variations are derived from a barotropic ocean model (PPHA) which was provided by JPL. The individual atmospheric and oceanic contributions as well as the processing strategy to derive the combined AOD1B product are described in the first part of this paper.

The PPHA model has some deficiencies such as the exclusion of the Arctic Ocean or reduced level of energy compared to in-situ ocean bottom pressure data. Thus, the influence of different non-tidal ocean models on GRACE gravity field solutions has been investigated for a seasonal cycle. It turned out that the barotropic MOG2D and the baroclinic OMCT models, both providing global output and based on more complex algorithms and parameterization, produce slightly better agreement when compared to NIMA gravity anomalies or to an altimeter-derived geoid. Similar results are obtained when comparing daily times series of 10x10 degrees gravity field models, which have been derived without correcting short-term mass variations, with the candidate non-tidal ocean models.

These tests indicate that the PPHA model shall be substituted by OMCT or MOG2D. Nevertheless, a dramatic improvement of the monthly gravity field solutions towards the pre-launch simulated baseline accuracy will not be reached. Instead, future work should primarily concentrate on the improvement of the temporal resolution and the inclusion of short-term (daily) hydrological mass variations.

Key words: De-aliasing, Mass Variation, Ocean Model

1 Introduction

The primary objective of the GRACE (Gravity Recovery And Climate Experiment) mission is to provide precise monthly estimates of the global Earth's gravity field for a period of up to five years. The temporal sequence of these gravity field estimates will yield the time history of its variability (Tapley and Reigber, 2001). GRACE gravity field processing is performed in a joint Science Data System between the GeoForschungsZentrum Potsdam (GFZ) and the University of Texas Center for Space Research (UTCSR). Both centers have already generated dramatically improved mean Earth gravity field models based on GRACE mission data only (Reigber et al., 2005; Tapley et al., 2003). The GFZ EIGEN-GRACE02S model is about one order of magnitude more accurate than the latest CHAMP-derived global gravity models and two orders of magnitude more accurate than the latest pre-CHAMP satellite-only gravity models (at 1000 km half-wavelength). Additionally numerous GRACE data users have already demonstrated that GRACE monthly gravity field solutions give a clear indication that temporal variations in the Earth's mass distribution are detectable at 1500 km resolution or better (Han et al., 2005; Schmidt et al., 2005; Tapley et al., 2004; Wahr et al., 2004).

During gravity field determination temporal variations of the Earth and ocean tidal potential are removed by use of appropriate models (IERS2000 standards or FES2004 (LeProvost, 2002)). Special attention has to be given to the short-term (weekly to hourly) mass variations causing time-variant gravity forces that act on the GRACE satellites and which have to be corrected during GRACE instrument data processing. This process is called "de-aliasing" and is realized by taking into account 6-hourly spherical harmonic coefficients which are interpolated and added to the background static gravity field. These spherical harmonic coefficients are routinely generated at GFZ Potsdam as the so-called Level-1B Atmosphere and Ocean De-aliasing product (AOD1B) and have to be provided to the GRACE Science Data System and to the user community on a day by day basis with a maximum delay of 12 days. Consequently, to calculate the atmosphere and ocean mass variations meteorological input data have to be acquired and an ocean model has to be operated on a routine basis.

The sensitivity of CHAMP, GRACE and GOCE to atmospheric, oceanic and hydrological mass variations in terms of degree standard deviations is shown in Fig. 1 (taken from Gruber (2001)). It is clearly visible that the signal of the high-frequency sub-daily atmospheric mass variations is above the expected sensitivity of the monthly GRACE gravity field solutions up to degree 35 to 40 (spatial resolution of 500 to 570 km). High-frequency mass variations in the oceans are much smaller than in the atmosphere, but still have an impact on the GRACE observations. Additionally, the monthly continental hydrological mass variation (precipitation minus evaporation minus run-off), which is the major outcome from monthly GRACE gravity field comparisons, is shown for Europe. Here, the variation should be detectable up to degree and order 25 to 30 (spatial resolution of 666 to 800 km). But it should be noted that the hydrological signal at world's largest river basins such as Amazon, Ganges or Congo is much larger and therefore GRACE should detect signals of even higher spatial resolution.

GOCE will measure directly the second derivative of the gravity potential. Thus, compared to GRACE, the sensitivity of the gradiometer to long-wavelength mass variations will be much smaller but still similar compared to CHAMP because of the pure high-low GPS link for orbit positioning. Nevertheless, due to GOCE's mission profile of twice 6 months, the monthly gravity variability derived by GRACE has to be used to remove the so-called seasonal (hydrological) bias of the GOCE observations.

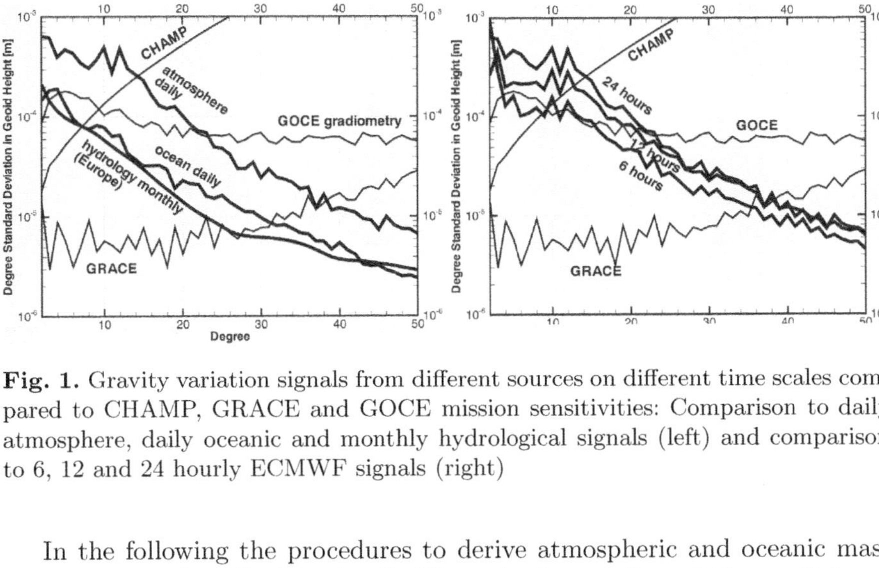

Fig. 1. Gravity variation signals from different sources on different time scales compared to CHAMP, GRACE and GOCE mission sensitivities: Comparison to daily atmosphere, daily oceanic and monthly hydrological signals (left) and comparison to 6, 12 and 24 hourly ECMWF signals (right)

In the following the procedures to derive atmospheric and oceanic mass variations are described in detail. First, the integration of the spatial and time-variable vertical structure of the atmospheric mass is described (Sect. 2). Then, the non-tidal ocean model used to derive the oceanic mass variation is introduced. Section 4 summarizes the meteorological input data which are necessary to force the integration and the ocean model. The processing strategy and the combination of the atmospheric and oceanic contributions are described in Sect. 5. Finally, the influence of the non-tidal ocean model on the gravity field solution is analysed.

2 Vertical Integration of the Atmosphere

For precise applications the spatial and time-variable vertical structure of the atmosphere has to be taken into consideration. Thus, a vertical integration of the atmospheric masses has to be performed. Fundamental basis is the gravitational potential V at a point outside the Earth which is expressed by a spherical harmonic expansion using normalized coefficients C_{nm} and S_{nm} of degree n and order m (Heiskanen and Moritz, 1967):

$$V = \frac{kM}{r} \sum_{n=0}^{\infty} \sum_{m=0}^{n} \left(\frac{a}{r}\right)^n P_{nm}(\cos\theta)(C_{nm}\cos m\lambda + S_{nm}\sin m\lambda) \tag{1}$$

where

$$\begin{Bmatrix} C_{nm} \\ S_{nm} \end{Bmatrix} = \frac{1}{(2n+1)Ma^n} \iiint_{Earth} r^n P_{nm}(\cos\theta) \begin{Bmatrix} \cos m\lambda \\ \sin m\lambda \end{Bmatrix} \lambda dM \tag{2}$$

and

k = gravity constant,
a = radius of the sphere,
M = mass of the Earth,
P_{nm} = normalized associated Legendre polynomials and
r, θ, λ = spherical coordinates of a mass element dM.

The relation between mass elements dM and volume elements dV is defined by the density ρ

$$dM = \rho\, dV = \rho\, r^2 dr \sin\theta\, d\theta\, d\lambda \tag{3}$$

Substitution of (3) into (2) and application of the hydrostatic equation

$$\rho\, dr = -\frac{dP}{g_r} \tag{4}$$

results in

$$\begin{Bmatrix} C_{nm} \\ S_{nm} \end{Bmatrix} = \frac{1}{(2n+1)Ma^n} \iint_{Earth} \int_{P_S}^{0} \frac{r^{n+2}}{g_r} dP P_{nm}(\cos\theta) \begin{Bmatrix} \cos m\lambda \\ \sin m\lambda \end{Bmatrix} \sin\theta\, d\theta\, d\lambda. \tag{5}$$

Here P, P_s, g and g_r are air pressure, surface pressure and gravity acceleration (mean and at height r) which can be approximated by

$$g_r = g\left(\frac{a}{r}\right)^2. \tag{6}$$

The radial coordinate r is composed of (see Fig. 2 and Swenson and Wahr (2002))

$$r = r_s + \delta r = a + \xi + h + \delta r = a + \xi + z \tag{7}$$

where ξ is the height of the mean geoid above the mean sphere and h is the elevation of the Earth's surface above the mean geoid (Earth's surface topography).

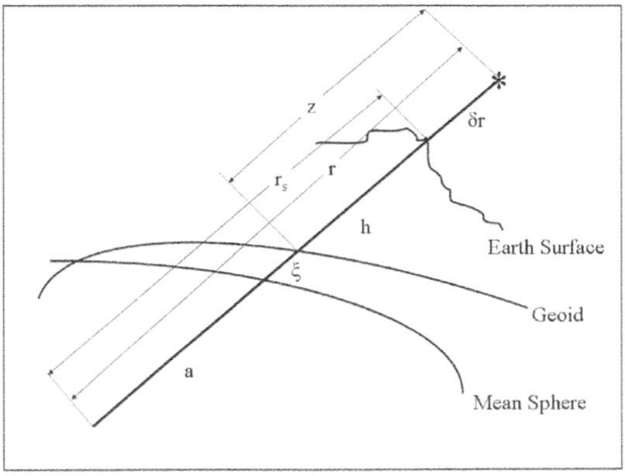

Fig. 2. Radial component r as used in vertical integration

The geopotential height ϕ at a point above the Earth's surface r is defined by:

$$\phi = \frac{1}{g}\int_0^z g_r dz = a\left(\frac{z}{a+z}\right) \Longleftrightarrow z = \frac{\phi}{1-\frac{\phi}{a}} \tag{8}$$

Substitution of (8) into (7) and the result into (5) yields

$$\left\{ \begin{array}{c} C_{nm} \\ S_{nm} \end{array} \right\} = -\frac{1}{(2n+1)Ma^{n+2}} \iint_{Earth} \left[\int_{P_S}^{0} \left(\frac{a}{1-\frac{\phi}{a}} + \xi \right)^{n+4} dP \right]$$

$$P_{nm}(\cos\theta) \left\{ \begin{array}{c} \cos m\lambda \\ \sin m\lambda \end{array} \right\} \sin\theta \, d\theta \, d\lambda. \tag{9}$$

After including the degree dependent term into the integral (for numerical reasons), introducing the elastic deformation of the solid Earth under variable load via the degree dependent load Love number k_n, and subtracting a corresponding mean field $\overline{P}_{VI}$ covering at least one year of data (in order to eliminate seasonal effects in the mean field) from the inner integral we get the final equation for the determination of atmospheric mass variations using the vertical integration approach:

$$\begin{Bmatrix} C_{nm} \\ S_{nm} \end{Bmatrix} = -\frac{a^2(1+k_n)}{(2n+1)Mg} \iint_{Earth} \left(\int_{P_S}^{0} \left(\frac{a}{a-\phi} + \frac{\xi}{a} \right)^{n+4} dP - \overline{P}_{VI} \right)$$

$$P_{nm}(\cos\theta) \begin{Bmatrix} \cos m\lambda \\ \sin m\lambda \end{Bmatrix} \sin\theta \, d\theta \, d\lambda \tag{10}$$

Meteorological analysis centers usually do not provide geopotential heights ϕ at certain model levels, but temperature and specific humidity. Thus, before the integration with (10) can be performed numerically, the geopotential heights for all levels have to be computed. This computation can be done according to White (2001) and Schrodin (2000), where N_{level} represents the lowest level.

$$\phi_{k+1/2} = \phi_S + \frac{1}{g} \sum_{j=k+1}^{N_{level}} R_{dry} T_v ln \frac{P_{j+1/2}}{P_{j-1/2}} \tag{11}$$

where

$\phi_{k+1/2}$ = geopotential height at half level (layer interfaces),
ϕ_S = geopotential height at surface,
R_{dry} = gas constant for dry air = 287 m^2/s^2K,
$P_{k+1/2}$ = pressure at half level (layer interface)
= $a_{k+1/2} + b_{k+1/2}P_S$,
$a_{k+1/2}, b_{k+1/2}$ = model dependent coefficients provided with ECMWF meteorological data,
T_v = virtual temperature = (1 + 0.608 S)T,
S = specific humidity and,
T = temperature.

Finally, the geopotential heights at pressure levels can be used to compute the inner integral in (10). In the second term (ξ/a), the mean geoid above the mean sphere can be approximated by the geopotential height at the Earth's surface (orography) which is available at ECMWF too.

3 Non-tidal Ocean Model

The ocean's response to atmospheric forcing can be divided into two classes: barotropic and baroclinic. A barotropic ocean model is one in which the whole water column has the same density and it is forced by wind and pressure only. A baroclinic ocean model includes vertical density changes and their effects, and requires additional forcing such as evaporation minus precipitation or radiation fluxes to handle thermodynamic effects. Barotropic motions are fast (fraction of a day to a few weeks), while baroclinic motions are slow (weeks to centuries). The tides are the best example of barotropic motion, even though they include some baroclinic energy in special places. El Niño is a predominantly baroclinic phenomenon. A barotropic model is simpler, has fewer parameterizations, and runs faster on a computer than a baroclinic model. Tierney et al. (2000) showed that the difference between barotropic and baroclinic models in terms of sea surface height change is negligible in a global average (< 0.1 mm of sea water) for periods shorter than 100 days, and at those short periods only noticeable in some steep topography regions (Zlotnicki, 2003).

Thus, the oceanic mass variation for the AOD1B product is derived from a barotropic ocean model, which was provided by JPL. This model is denoted PPHA, because it was developed by Pacanowski, Ponte, Hirose and Ali (Hirose et al., 2001). PPHA computes the component of oceanic mass redistribution ("barotropic sea level") due to wind stress and atmospheric pressure for an area between 65° N and 75° S including the Mediterranean Sea, Hudson Bay, North Sea and shallow waters on a 1.125° grid and lacks baroclinic dynamics and surface buoyancy. Further details are described in Flechtner (2003).

4 Meteorological Input Data

Vertical integration of the atmospheric mass distribution and forcing of the PPHA barotropic ocean model require different meteorological input data which have to be available on a routine basis, with short time delay and with sufficient temporal and spatial resolution in order to derive the operational AOD1B GRACE product (<12 day requirement). Data provided by the National Center for Environmental Predictions (NCEP) and by the German Weather Service (DWD) do not fulfil these requirements due to insufficient spatial resolution or non-guaranteed permanent access. Therefore GFZ has signed a contract with DWD to regularly acquire the operational analysis data from the European Center for Medium-range Weather Forecast (ECMWF) Integrated Forecast System (IFS) at synoptic times 0:00, 6:00, 12:00 and 18:00. These products are usually available within 2-3 days. The spatial resolution is defined on a Gaussian n160 grid which corresponds to a resolution of 0.5°. The temperature and specific humidity data are provided for 60 layers from surface up to 0.1 hpa. Further details on the IFS products, services and research can be found in ECMWF (2004).

Nine different products have to be downloaded each 6 hours for vertical integration of the atmosphere (surface pressure, geopotential height at surface, multi-level temperature and specific humidity data) and to run PPHA (wind speed in u and v direction, surface pressure, sea surface temperature, temperature and dew point temperature at 2m level). The data require about 125 MB/d disk space and are stored in the GFZ Information System and Data Center (ISDC).

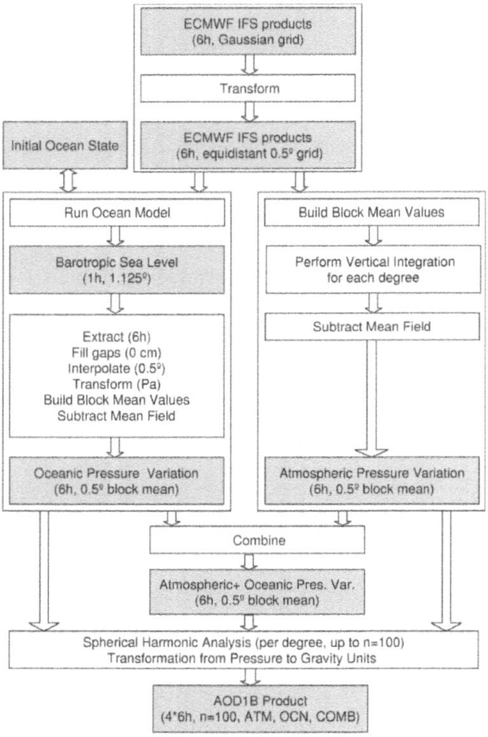

Fig. 3. Processing strategy for atmospheric and oceanic mass variations

5 Processing Strategy and Combination of Atmospheric and Oceanic Mass Variations

In a first step, all meteorological ECMWF data are transformed from Gaussian n160 to equidistant 0.5° grid. Then, the PPHA relevant products and an initial ocean model state are used to run the ocean model which produces 24 hourly, 1.125° grided files of barotropic sea level [cm] as well as an updated ocean model state which is used to process the following day. For the later combination with the atmosphere the results for epochs at 0, 6, 12 and 18 hours are extracted, transformed to pressure units by multiplication with the density of salt water and gravity acceleration of the mean sphere (4), interpolated to 0.5°, filled with zero values (corresponding to a pure inverse barometric response (IB) of the sea surface) for undefined ocean areas in polar regions and finally averaged to block mean values. In a second step the vertical integration is performed for 0, 6, 12 and 18 hours for each 0.5° grid point. The resulting integrated pressure data are then averaged to block mean values, combined with the oceanic contribution, both reduced by a 2001 mean field, and expanded to spherical harmonics. Finally, the atmospheric, oceanic

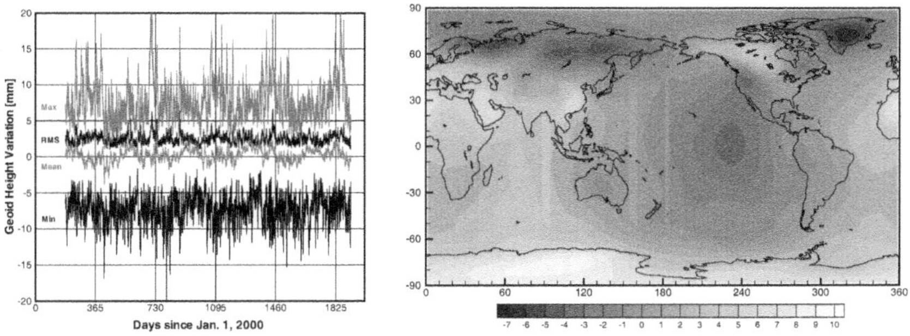

Fig. 4. Geoid height variability [mm] caused by atmospheric and oceanic mass variation for April 2000 until March 2005 (mean, rms, minimum and maximum; left) and for January 1, 2005 at 0:00 UTC(right)

and combined Stokes coefficients up to degree 100, reconverted from pressure unit to dimensionless gravity, are stored as the AOD1B (Level-1B Atmosphere and Ocean De-aliasing) RL01 (release 01) product (Fig. 3).

The geoid variability (root mean square), derived from the AOD1B product is at the level of 3-4 mm, minimum and maximum values can reach up to 15 mm (Fig. 4).

The AOD1B spherical harmonic coefficients have to be added to the background mean gravity field during GRACE precise orbit determination to calculate the gravity field partial derivatives. Thus, the resulting monthly mean gravity field products represent primarily the leftover monthly continental hydrological mass variation (Schmidt et al., 2005) and long-period barotropic fluctuations of the ocean. In order to compare the GRACE monthly solutions over the oceans with ocean model derived bottom pressure, the corresponding mean atmospheric and oceanic mass variation has to be added back again. Therefore, the GRACE level-2 processing centers average all AOD1B products which have been used to derive the individual monthly and mean gravity field solutions and provide these mean spherical harmonic coefficients as the GAC (GRACE Average of non-tidal atmosphere and ocean Combination) product along with the corresponding GSM (GRACE Satellite only Model) products. In case of comparison with in-situ ocean bottom pressure data the 6-hourly AOD1B products can be used directly.

6 Influence of the non-tidal Ocean Model on Gravity Field Solutions

The PPHA model has deficiencies which might influence the quality of the GRACE gravity field solutions: Due to the exclusion of the Arctic Ocean a pure inverse barometric response of the sea surface (no mass variation) has to be applied north of 65° latitude, it has less variance in bottom pressure than other comparable models (V. Zlotnicki, personal communication 2004) and shows a reduced level of energy compared with in-situ ocean bottom pressure data (Kanzow et al., 2005). To investigate the influence of the non-tidal ocean model on GRACE gravity field solutions two experiments have been performed for three different de-aliasing products (Flechtner et al., 2005). First, sets of monthly gravity field solutions have been derived which only differ by the applied short-term mass variations. Four months (May 2003, August 2003, November 2003 and February 2004) were selected for processing, to investigate also their seasonal variability. Second, daily gravity field solutions have been calculated for July 2003 until September 2003 without correcting for short-term mass variations and have been correlated with the candidate de-aliasing products.

All three de-aliasing products are based on the same input data and modeling assumptions to guarantee a fair comparison: the same 6-hourly ECMWF meteorological fields were used, the atmospheric and oceanic mass variations were derived by subtracting a 2001+2002 mean field, the atmospheric contribution was calculated by vertical integration (quasi-identical for all products) and the S2 atmospheric tide was filtered from surface pressure data before forcing the non-tidal ocean model. The later is necessary to avoid double book-keeping with the ocean tide model during orbit and gravity field parameter estimation. Thus, the dominating difference between these products is the non-tidal ocean model, used to calculate the oceanic mass variation. Consequently, the de-aliasing products of this study are denoted w.r.t. their applied non-tidal ocean models:

PPHA: This product is identical to the standard AOD1B RL01 product, except that the mean field period has been changed from 2001 to 2001+2002 and the S2 atmospheric tide has been filtered before forcing the non-tidal ocean model PPHA using a strategy described in Ponte and Ray (2002).

MOG2D: This model was provided by GRGS (Groupe de Recherche de Géodésie Spatiale) and is based on the same assumptions as AOD1B RL02 (see "PPHA" above), but the non-tidal ocean model is MOG2D, a barotropic, non-linear and time stepping model with global output (including the Arctic Ocean, Hudson Bay, Bering Strait and the Weddell and Ross Seas). The ice cover impact is taken into account through the diminution of the water column height, due to the submerged ice thickness. MOG2D is governed by shallow water continuity and momentum equations. The model can include tides and its main originality is a finite element space discretisation which allows to increase the resolution e.g. from 400 km in the open sea down to 20 km in areas with strong topographic gradients or in shallow waters (Carrère et al., 2003).

OMCT: The atmospheric contribution is identical to AOD1B RL02 (see "PPHA" above), but the oceanic mass variations were provided by the Technical University of Dresden (TUD). They are derived from the baroclinic Ocean Model for Circulation and Tides (OMCT) which is a further development of the Hamburg Ocean Primitive Equation Model (HOPE) by adjustment to the weather time-scale and coupling with an ephemeral tidal model (Thomas et al., 2001). In contrast to the HOPE model, OMCT allows to take into account effects arising from loading and self-attraction, atmospheric pressure forcing or the consideration of continental freshwater fluxes. Effects due to sea-ice are accounted for by means of an implemented thermodynamic sea-ice model which allows a prognostic calculation of ice-thickness, -drift and -compactness. As MOG2D, the OMCT output is provided on a global scale.

Monthly gravity field solutions have been calculated for May 2003, August 2003, November 2003 and February 2004 up to degree and order 150 using the standard GFZ level-2 GRACE processing strategy based on GPS code, GPS phase and K-band range-rate observations and background models (initial gravity field, ocean, atmospheric and Earth tides, etc.) (Reigber et al., 2005). The only difference was the de-aliasing model (PPHA, MOG2D or OMCT).

Additionally, daily time series of gravity field models complete up to degree and order 10 have been produced for the time span July 2 to September 30, 2003 using an integrated adjustment method (Zhu et al., 2004). The standard GFZ level-2 GRACE background models (see above) have been accounted for during data processing, while variations in the gravitational potential produced by non-tidal atmosphere and ocean mass transport were deliberately omitted. The recovered GRACE gravity changes represented by the daily time series of spherical harmonics were then compared to the corresponding coefficients of the de-aliasing products PPHA, MOG2D and OMCT, respectively. Hydrological mass variations are assumed to change slower than days and can therefore be accounted for by the subtraction of a bias and a trend. Detailed results can be found in (Hu et al., 2005). The major findings of these comparisons are as follows (see Figures and Table below):

- The mean geoid height differences of the three investigated de-aliasing products are in the order of ± 1.5 mm. Maximum differences are found in the Arctic (no PPHA output), in the Southern ocean and in the North Pacific (see Color Fig. XII on p. 293, left).
- The GPS phase and K-band range-rate observation residuals are nearly identical for OMCT and PPHA and slightly smaller (1-2 %) when using MOG2D.
- The degree amplitudes of gravity field solution differences are clustered for all degrees which is a sign that all solutions provide nearly the same signal.
- The mean geoid height differences between corresponding gravity field solutions reflect nearly one by one the mean background AOD differences spatial structure, but scaled by a factor of about 1.5 (see Color Fig. XII on p. 293, right). The opposite sign is due to the subtraction of the AOD product from the background gravity field during orbit determination.
- Additionally, previously observed striping which is due to other aliasing effects (e.g. insufficient instrument parameterization, ocean tide mismodeling or GRACE orbit configuration), is still visible.

- Comparisons with NIMA gravity anomalies and CLS01 minus ECCO oceanic geoid heights show for short-wavelengths (2.5° and 5° resolution) a better mean agreement with MOG2D (~3-8 %) and OMCT (~2-8 %) than when using the "standard" PPHA model (see Table 1).
- The recovered daily gravity field coefficients are highly correlated with the model predictions for low order coefficients, but the agreement degrades when the order is increasing. The highest correlation and smallest residual geoid height differences were derived with MOG2D, followed by OMCT and PPHA, which both give quite similar results. A pure IB response of the sea surface assumption produced the worst results (Fig. 5).

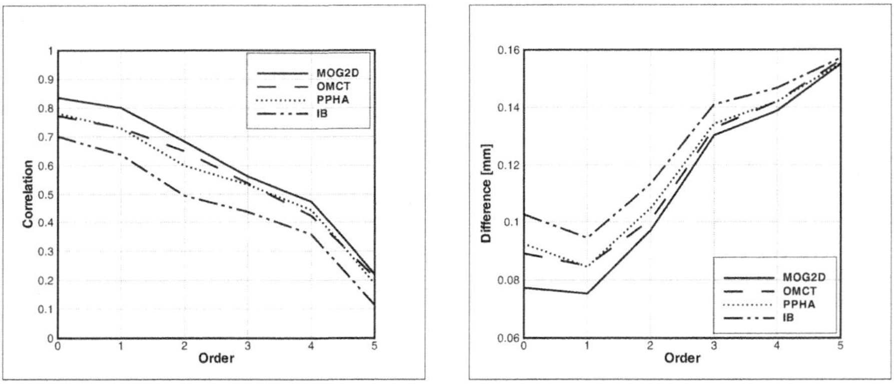

Fig. 5. Mean correlation coefficients as a function of order (left) and residual geoid height differences [mm] (right) between the daily GRACE gravity field solutions and different de-aliasing products (degree 1 and 2 ignored)

Table 1. Comparison (weighted root mean square) of monthly gravity field products derived from MOG2D, PPHA and OMCT de-aliasing products for May 2003, August 2003, November 2003 and February 2004 with altimeter-derived geoid heights (N, CLS01 minus ECCO oceanic geoid, top) and gravity anomalies (Δg, NIMA marine gravity anomalies, bottom) for different grid spacing (5°~n=36; 2.5°~n=72)

	May 2003		August 2003		November 2003		February 2004	
	2.5°x2.5°	5°x5°	2.5°x2.5°	5°x5°	2.5°x2.5°	5°x5°	2.5°x2.5°	5°x5°
	Weighted RMS of Geoid Height Differences about Mean [m]							
MOG2D	1.797	0.406	1.100	0.272	1.173	0.283	0.796	0.219
OMCT	1.814	0.411	1.111	0.275	1.194	0.287	0.797	0.218
PPHA	1.947	0.438	1.147	0.282	1.219	0.296	0.865	0.232
	Weighted RMS of Gravity Anomaly Differences about Mean [mgal]							
MOG2D	38.771	8.235	23.700	5.210	25.219	5.442	17.130	3.911
OMCT	39.034	8.330	23.928	5.274	25.696	5.533	17.253	3.920
PPHA	41.980	8.933	24.664	5.432	26.135	5.718	18.648	4.251

The study shows, in agreement with Tierney et al. (2000), that the barotropic MOG2D and the baroclinic OMCT models provide similar results when used for monthly gravity field determination and compared to external NIMA gravity anomalies and CLS01 minus ECCO derived geoid heights. Both models behave consistently better than PPHA which is believed to be caused primarily by the polar gaps of the PPHA model. Thus, PPHA should be substituted in future GRACE (re)processing by MOG2D or OMCT. Nevertheless, a significant step to derive the GRACE baseline accuracy will not be made because the resulting formal gravity field errors are still comparable and about one order of magnitude too big.

7 Conclusions

GFZ has developed a software package which routinely calculates atmospheric and oceanic mass variations with a maximum time delay of about 3-4 days dependent on the availability of required ECMWF meteorological fields. A vertical integration of the atmospheric masses is performed to take into account the spatial and time-variable vertical structure of the atmosphere. The oceanic mass variations are derived from the barotropic ocean model PPHA which was provided by JPL. Both constituents are combined and made available as the GRACE Level-1B Atmosphere and Ocean De-aliasing product (AOD1B) to the GRACE Science Data System and user community in terms of spherical harmonic coefficients which have to be added to the background static gravity field during GRACE precise orbit and gravity field partial derivatives determination.

The PPHA ocean model has deficiencies which might influence the quality of the GRACE gravity field solutions: exclusion of the Arctic Ocean requiring a pure inverse barometric response assumption of the sea surface north of 65°, less variance in bottom pressure than other comparable models and reduced level of energy compared with in-situ ocean bottom pressure data. Thus, the influence of different non-tidal ocean models on GRACE gravity field solutions has been investigated. It turned out that the barotropic MOG2D model and the baroclinic OMCT model, both providing global output and based on more complex algorithms and parameterization, produce slightly better results when compared to NIMA gravity anomalies or to an altimeter-derived geoid. Similar results are obtained when comparing daily times series of 10x10 degrees gravity field models, which have been derived without correcting short-term mass variations, with the candidate de-aliasing products. Again, best correlation and smallest residuals are obtained for MOG2D, followed by OMCT and PPHA. Nevertheless, a dramatic improvement of the monthly gravity field solutions towards the pre-launch simulated baseline accuracy will not be reached by the substitution of the non-tidal ocean model.

Future work has to concentrate on the improvement of the temporal resolution and the inclusion of short-term (daily) hydrological mass variations.

First is necessary to improve the half-daily S2 atmospheric tide representation inside the vertical integration of the atmosphere which cannot be precisely taken into account by pure linear interpolation of 6-hourly spherical harmonic coefficients. At least 3-hourly time series are necessary, which are available at ECMWF for a short time. Neglecting short-term hydrological mass variations during GRACE precise orbit determination may cause aliasing effects, which are not yet investigated or taken into account. This should be possible soon because hydrological models with improved temporal and spatial resolution will become available within the near future. As an example it is planned within the Geotechnologien II Program of the German Ministry of Education and Research (BMBF) starting mid of 2005, to assimilate GRACE mission data into the WaterGAP Hydrological Model (WGHM, Döll et al. (2003)) and to extract daily output.

Remark. The AOD1B products can be downloaded at the GRACE Information System and Data Center (ISDC) at GFZ Potsdam: http://isdc.gfz-potsdam.de/grace.

Acknowledgement. This is publication no. GEOTECH-147 of the GEOTECHNOLOGIEN programme of BMBF and DFG, grant 03F0326A. The authors would like to thank Victor Zlotnicki and Ahmed Ali from JPL who provided the PPHA ocean model plus valuable background information.

References

Carrère, L, and F Lyard (2003) Modeling the barotropic response of the global ocean to atmospheric wind and pressure forcing-comparisons with observations, *Geophys. Res. Let.*, 30(6), 1275, doi: 10.1029/2002GL016473

Döll, P, F Kaspar, and B Lehner (2003) A global hydrological model for deriving water availability indicators: model tuning and validation, *J. Hydrol.*, 270, 105-134

ECMWF (2004): http://www.ecmwf.int/research/ifsdocs/index.html

Flechtner, F (2003) AOD1B product description document, GRACE project document JPL 327-750, rev. 1.0, JPL Pasadena, Ca

Flechtner, F, R. Schmidt, S Y Zhu, and Ul Meyer (2005) GRACE gravity field solutions using different de-aliasing models, poster EGU05-A-04815 presented at the European Geosciences Union General Assembly 2005, Vienna, Austria, 24-29 April 2005

Gruber, Th (2001) Identifcation of Processing and Product Synergies for Gravity Missions in View of the CHAMP and GRACE Science Data System Requirements, Proceedings of the 1st International GOCE User Workshop, ESA Publication Division, WPP-188, 45-50

Han, S C, C K Shum, C Jekeli, and D Alsdorf (2005) Improved estimation of terrestrial water storage changes from GRACE, *Geophys. Res. Let.*, 32, doi: 10.1029/2005GL022382

Heiskanen W A, and H Moritz (1967) Physical Geodesy; W.H. Freeman Publications Co., San Francisco

Hirose, N, I Fukumori, V Zlotnicki, and R Ponte (2001) High-frequency barotropic response to atmospheric disturbances: Sensitivity to forcing, topography, and friction, *J. Geophys. Res.*, 106, 30987

Hu, X G, C Shi, F Flechtner, R König, P Schwintzer, R Schmidt, Ul Meyer, F H Massmann, Ch Reigber, and S Y Zhu (2005) High frequency temporal Earth gravity variations detected by GRACE satellites, article in this book

Kanzow, T, F Flechtner, A Chave, R Schmidt, P Schwintzer, and U Send (2005) Seasonal variation of ocean bottom pressure derived from GRACE: Local validation and global patterns, accepted by *Geophys. Res. Let.*

LeProvost, C (2002) FES2002 - A New Version of the FES Tidal solution Series, Abstract Volume, Jason-1 Science Working Team Meeting, Biarritz, France

Ponte, R, and R Ray (2002) Atmospheric pressure corrections in geodesy and oceanography: A strategy for handling tides, *Geophys. Res. Lett.*, 29(24), L2153, doi: 10.1029/2002GL016340

Reigber, Ch, R. Schmidt, F Flechtner, R König, Ul Meyer, K H Neumayer, P Schwintzer, and S Y Zhu (2005) An Earth gravity field model complete to degree and order 150 from GRACE: EIGEN-GRACE02S, J. Geodynamics, 39, 1-10, doi: 10.1016/j.jog.2004.07.001

Schmidt, R., P. Schwintzer, F. Flechtner, Ch. Reigber, A. Güntner, P. Döll, G. Ramillien, A. Cazenave, S. Petrovic, H. Jochmann, and J. Wünsch (2005), GRACE observations of changes in continental water storage, accepted by *Global Planet. Change*

Schrodin R. (Ed.) (2000), Quarterly Report of the Operational NWP-Models of the Deutscher Wetterdienst; No. 22, Dec. 1999 – Feb. 2000

Swenson S., and J. Wahr (2002), Estimated Effects of the Vertical Structure of Atmospheric Mass on the Time-Variable Geoid; Paper, *J. Geophys. Res. - Solid Earth*, 107(B9), 2194, doi: 10.1029/2000JB000024

Tapley, B., and Ch. Reigber (2001), The GRACE mission: Status and future plans, *EOS Trans AGU*, 82(47), Fall Meet. Suppl., G41 C-02

Tapley, B., D. Chambers, S. Bettadpur, and J. Ries (2003), Large Scale Ocean Circulation from the GRACE GGM01 Geoid, *Geophys. Res. Lett.*, 30(22), 2163, doi: 10.1029/2003GL018622

Tapley, B., S. Bettadpur, J. Ries, P. Thompson, and M. Watkins (2004), GRACE measurements of mass variability in the Earth system, *Science*, 305, 503-505, doi: 10.1126/science.1099192

Tierney, C., J. Wahr, F. Bryan and V. Zlotnicki (2000), Short-period oceanic circulation: implications for satellite altimetry, *Geophys. Res. Lett.*, 27(9), pp 1255-1258

Thomas, M., Sündermann, J., and E. Maier-Greiner (2001), Consideration of ocean tides in an OGCM and impacts on subseasonal to decadal polar motion excitation, *Geophys. Res. Let.*, (12), 2457

Wahr, J., S. Swenson, V. Zlotnicki, and I. Velicogna (2004), Time-variable gravity from GRACE: First results, *Geophys. Res. Lett.*, 31, L11501, doi: 10.1029/2004GL019779

White P.W. (Ed.) (2001), IFS Documentation Part III: Dynamics and Numerical Procedures (CY21R4); ECMWF Research Department

Zhu, S.Y., Ch. Reigber, and R. König (2004), Integrated adjustment of CHAMP, GRACE, and GPS data, *J. Geodesy*, 78, 103-108, doi: 10.1007/s0019000403790

Zlotnicki, V. (2003), Barotropic Ocean Model for GRACE Dealising, GRACE Science Data System internal document

Integrated Sensor Analysis GRACE

Björn Frommknecht, Ulrich Fackler, and Jakob Flury

Institute for Astronomical and Physical Geodesy, Technische Universität München, Arcisstrasse 21, 80333 Munich, frommknecht@bv.tum.de, fackler@bv.tum.de, flury@bv.tum.de

Summary. In this article performance estimates for the accelerometers, the star sensors and the K-Band ranging system onboard the GRACE satellites are given. It is shown that the accelerometers perform slightly worse than specified and that the star sensor and K-Band ranging system performances agree with the specifications. It is also demonstrated that mainly for the accelerometers performance assessment further investigations are needed as effects of unknown origin affect the measurements. For each instrument the results from the highrate L1a to the filtered and downsampled L1b data processing are shown and discussed. Concerning the accelerometer processing, good agreement with the data provided by JPL has been reached, concerning the star sensors and the K-Band system differences remain.

Key words: GRACE, sensor analysis, performance estimation, data processing

1 Introduction

The GRACE mission provides gravity field estimates with unprecedented accuracy. The complex and highly sophisticated sensor system onboard the GRACE satellites is responsible for this increase in quality. At the IAPG (Institute for Astronomical and Physical Geodesy) an integrated sensor analysis of the GRACE sensor system has been conducted during the last three years. A data set consisting of three days of data was used for the studies. Main fields of interest were:

- The derivation of mathematical models of the individual gravity relevant sensor systems of the GRACE satellites and the integration into a complete model that also allows to understand the interactions and couplings of the sensor systems
- The estimation of the sensor performance from real data, the comparison to the anticipated performance and the explanation of occurring differences
- The derivation of processing methods to high rate instrument data (called L1a data) to the level of low rate filtered 5s-data (called L1b data).

In this article performance estimates of the three key instruments of the GRACE mission, the accelerometer, the star sensor and the K-Band ranging system are derived and discussed. The processing of each instrument's data from high rate instrument data level to low rate filtered and downsampled level is briefly described. The results are compared to the data set provided by the JPL.

2 SuperSTAR Accelerometer

2.1 Performance Estimation

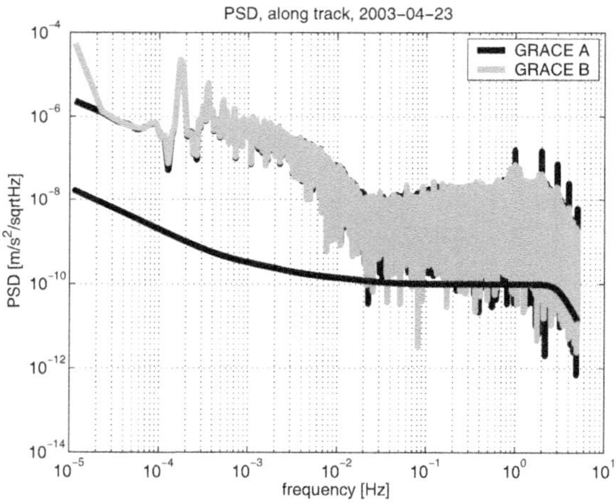

Fig. 1. Power Spectral Density (PSD) of the along-track component of the L1a accelerometer measurements of GRACE A & B

To investigate the performance of the linear acceleration measurements of the accelerometers onboard the GRACE satellites, only the raw data (L1a) is suited as the processed (L1b) data is treated with a low-pass filter that makes it impossible to estimate the high frequency noise. The accelerometers have two sensitive axes (along track and radial) and one less sensitive axis (cross-track). The low-frequency noise is not accessible through the analysis of a single accelerometer's measurements as the low-frequency noise is superposed by the measurement signal. The measurement signal are the non-gravitational accelerations acting on the satellites. At a certain frequency the signal to noise ratio become one and noise dominates the measurement. In Fig. 1 the comparison between the theoretical noise level and the apparent noise level of the real measurements is shown. Apparently the measurements of GRACE A and B show the same spectral characteristics, dominant frequencies are once and twice per revolution at about $2 \cdot e^{-4}$ Hz and $4 \cdot e^{-4}$ Hz. The signal decreases until at about $3 \cdot e^{-2}$ Hz it levels out into white noise. The level of this white noise is about a factor of 10 higher than the specified noise level for the accelerometers. A speciality of the GRACE A accelerometer are the prominent peaks on 1 Hz and multiples, their origin is unknown. Three possible causes of the higher noise level have been identified:

1. thrusterevents
2. twangs
3. spikes and peaks

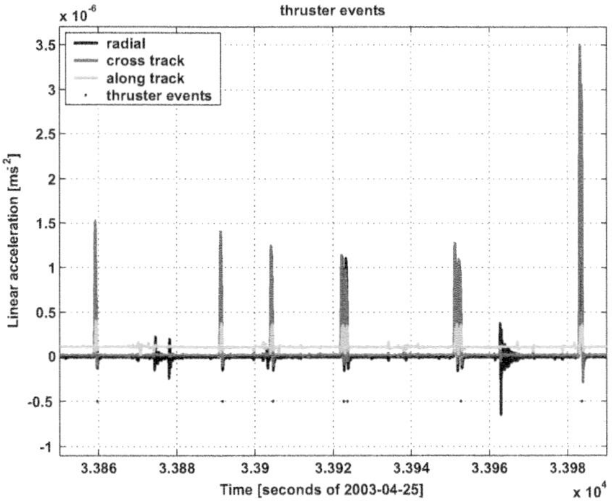

Fig. 2. Series of thruster events on GRACE A

Thrusterevents

A series of thruster events is shown in Fig. 2. Although the purpose of the thruster firings is to exert angular accelerations on the satellites, a thruster event also shows up in the measurement of the linear accelerations because of thruster misalignment and the displacement of the center of mass from the center of the proof mass of the accelerometer. The amplitudes range from 1 up to $3.5 \cdot e^{-6}\mathrm{m/s}^2$. As the linear accelerations due to thruster misalignment are real linear accelerations acting on the satellites it is correct that they are measured and show up in the linear accelerations measurements. The measured linear accelerations due to the center of mass offset however have to be considered as errors in the measurement, as they represent no real linear acceleration acting on the satellites. For both satellites the center of mass offset is kept so small through mass trim maneuvers that this effect can be neglected. As during the periods of thruster events only the effects caused by the misalignment of the thrusters are measured, it is sensible to exclude periods of thruster events for the performance estimation.

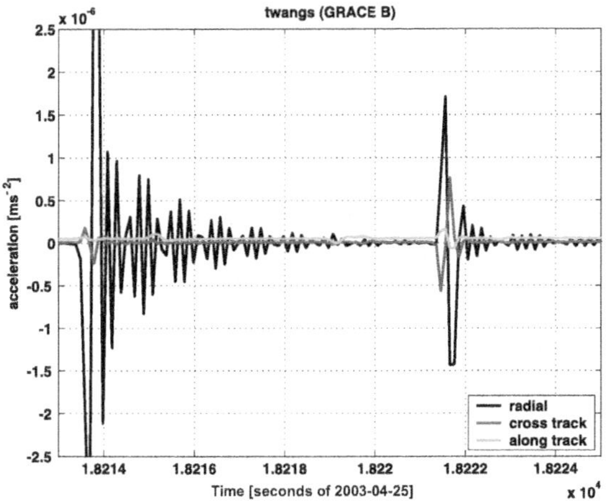

Fig. 3. Typical example of 'twangs' in the linear acceleration measurements

Twangs

Typical examples of a so called 'twang' are shown in Fig. 3. A twang is very similar to the reaction of a damped harmonic oscillator to a step impulse. It consists of an oscillation with decreasing amplitude, the typical duration is about 4 to 5 seconds (cf. (Hudson, 2003)). The amplitudes reach over $3 \cdot e^{-6}\mathrm{m/s^2}$. It is assumed that the reason for the twangs is the flattering of a thin foil on the bottom of the GRACE satellites, that could explain the massive occurrence in the radial component of the linear accelerations. Another possible reason is that they are a feature of the SuperSTAR accelerometer. If the latter was true, the twangs would have to be added to the error budget. If the twangs are caused by the foil, they are real non-gravitational accelerations acting on the satellites and these periods should be excluded from the performance estimation.

Peaks and Spikes

The third effect that adds to the apparent noise level are regular peaks and spikes occurring in all three components of the linear accelerations. A typical example is shown in Fig. 4. The peaks consist of a deflection from the mean in an arbitrary direction, followed by a deflection with a magnitude of one third of the original one in the opposite direction (cf. (Flury, 2004)). The amplitude reach a level of about $1 \cdot e^{-7}$ $\mathrm{m/s^2}$ for the along-track component and $1 \cdot e^{-8}$

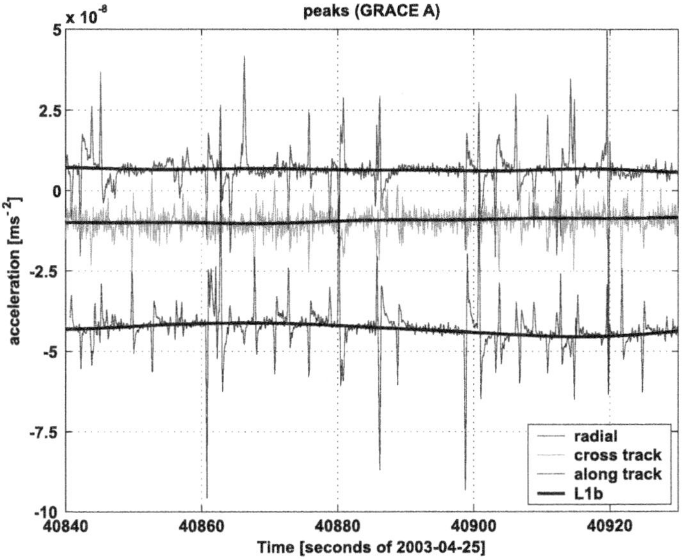

Fig. 4. Typical example for the spikes and peaks affecting the linear acceleration measurements

m/s^2 for the cross-track and radial components. Regarding the L1b data it seems that the effect is completely removed by the applied low-pass filter, however it is unclear if low-frequency effects remain in the filtered data. As the peaks and spikes are regular, it is assumed that they are a feature of the accelerometer.

Performance Assessment

The performance of the accelerometer was assessed in two ways:

1. Analysis of undisturbed periods of single accelerometer data
2. Spectral analysis of the difference of the accelerometer measurements of both satellites during undisturbed periods using L1b and L1a data.

Analysis of the signal of a single accelerometer

For the analysis of the single accelerometer data, only periods without thruster events, twangs or peaks have been used. According to (Flury, 2004) the following standard deviations are derived for undisturbed periods:

$$\begin{aligned} \text{along-track } & 3 \cdot e^{-10} \text{m/s}^2 \\ \text{cross-track } & 7 \cdot e^{-10} \text{m/s}^2 \\ \text{radial } & 2.1 \cdot e^{-10} \text{m/s}^2 \end{aligned}$$

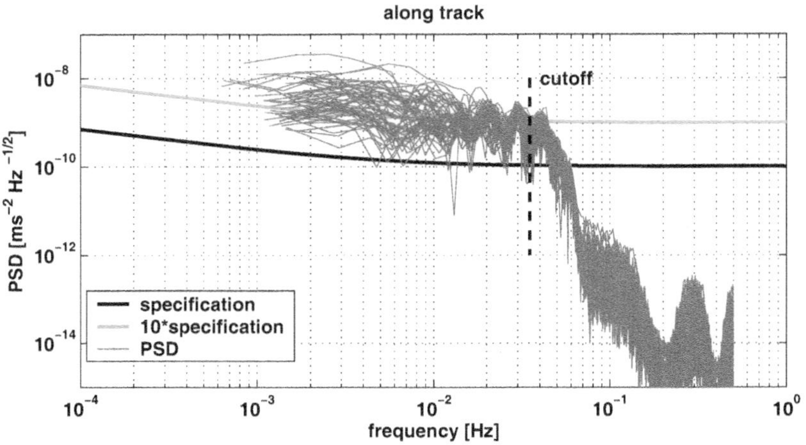

Fig. 5. PSD of differences of L1B accelerometer data between GRACE A and B during thruster free periods

Comparing this with the specifications given in (Stanton et al., 1998), the following conclusions can be drawn:

1. the noise level of the along-track and the radial component is about 2-3 times higher than the specified level of $1 \cdot e^{-10}$ m/s^2.
2. the noise level of the cross-track component is about 70% of the specified noise level of $1 \cdot e^{-9}$m/s^2.

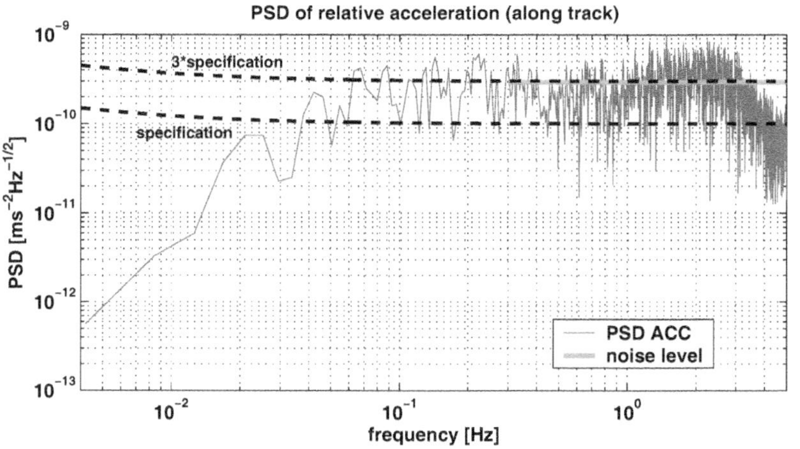

Fig. 6. PSD of differences of L1a accelerometer data between GRACE A and B during periods free of thruster events, twangs and peaks

Analysis of the differences between accelerometer measurements

The GRACE mission with its identical twin satellites offers the unique possibility to assess the performance of its accelerometers by using their redundancy: due to the orbit configuration, the trailing satellite arrives after about 27 s at the same location as the leading one. During this short time the atmosphere does not change significantly, thus the accelerometer measurements should be comparable. The difference between to accelerometer measurements includes the following effects:

- errors in the attitude information of both satellites
- scale factors and higher order error terms
- the sum of the measurement noise of both satellites

Figure 5 shows the PSD of the difference of L1b accelerometer measurements. Only periods without thruster events were used, but periods with twangs and peaks were not excluded to get longer timer series to eventually assess an upper limit of the low frequency part of the accelerometers noise. The differences fit well to an upper limit of 10 times the specification. The analysis of longer periods would yield more exact results, but because of the frequent thruster firings this is not possible.
In Fig. 6 the difference PSD of L1a accelerometer measurements is shown. Here periods without thruster events, twangs or peaks were used. The difference is following an error model that is three times above the specification. This analysis confirms the findings from the analysis of the single accelerometer measurement data analysis.
Comparing these results with the specifications given in (Stanton et al., 1998), the following conclusions can be drawn:

1. the noise characteristics of the accelerometer on GRACE A and B are very similar.
2. the noise level of the along-track and the radial component is about 2-3 times higher than the specified level of $1 \cdot e^{-10}\mathrm{m/s}^2$.

2.2 L1a to L1b Data Processing

The processing of the accelerometer data from L1a to L1b data has been conducted in accordance with (Wu and Kruizinga, 2004). Initial differences were reported to JPL where they helped to discover a glitch in the official processing software that led to the replacement of 30% of the data with interpolated values. The differences between IAPG and JPL processing results are now on the level of numerical accuracy. The effects of the strategy of the JPL to apply a low pass filter of 35 mHz to the accelerometer data instead of the same filter as used for the downsampling of the K-Band data should be investigated. It is unclear how the various effects like twangs, spikes and thruster events should be treated correctly. Although the applied low-pass filter removes the major effect of them, it should be investigated if long-term effects remain.

3 Star Sensor

3.1 Performance Estimation

The star sensor measurements are polluted by mainly white noise, not colored noise as the accelerometer. The orientation is derived from pictures taken of the sky in the field of view of the star tracker. The orientation w.r.t. an inertial system is derived by comparison with a star catalogue. The elements of the orientation matrix, which is basically a rotation matrix, can not be measured with an homogeneous accuracy: rotations about the line of sight of the star tracker are measured with an accuracy about 8 times worse the accuracy of rotations perpendicular to the line of sight of the star trackers (cf. (Wu and Kruizinga, 2004) and (Stanton et al., 1998)). Figure 7 shows the PSD of the

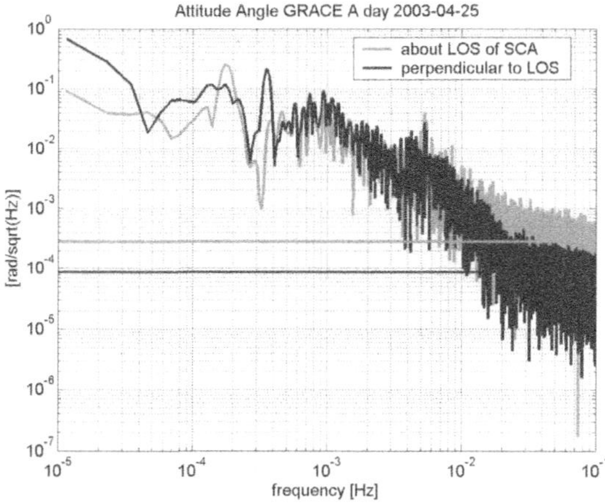

Fig. 7. PSD of attitude angles of GRACE A in star sensor reference frame. The horizontal lines show the estimated level of the white measurement noise

attitude angles of GRACE A. The estimated noise level for the rotations perpendicular to the line of sight is about $1 \cdot e^{-4}$ rad/$\sqrt{\text{Hz}}$ which is about three times the specification of $3 \cdot e^{-5}$ rad/$\sqrt{\text{Hz}}$. The noise level for rotations about the line of sight is about $2 \cdot e^{-4}$rad/$\sqrt{\text{Hz}}$ which agrees well with the specification of $2.4 \cdot e^{-4}$ rad/$\sqrt{\text{Hz}}$. Figure 7 shows the PSD of the attitude angles of GRACE B. The estimated noise level for the rotations perpendicular to the line of sight is about $3 \cdot e^{-5}$rad/$\sqrt{\text{Hz}}$ which agrees well with the specification. The noise level for rotations about the line of sight is about $2 \cdot e^{-4}$rad/$\sqrt{\text{Hz}}$ which also agrees well with the specification of $2.4 \cdot e^{-4}$ rad/$\sqrt{\text{Hz}}$.
The following conclusion concerning the star sensor performance can be drawn:

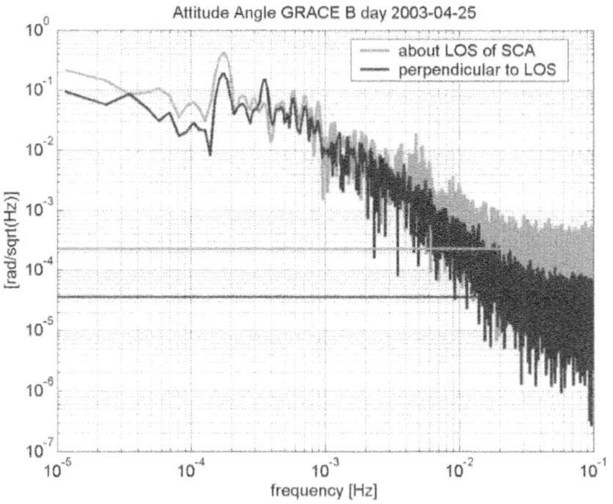

Fig. 8. PSD of attitude angles of GRACE B in star sensor reference frame. The horizontal lines show the estimated level of the white measurement noise

- The star sensor on GRACE A performs slightly worse than the star sensor on GRACE B concerning the rotation perpendicular to the line of sight.
- The performance of the star sensor on GRACE B agrees well with the specification.
- The performance of the star sensor on GRACE A concerning the orientation perpendicular to its line of sight is about three times worse than the specification. The performance concerning the rotation about its line of sight agrees well with the specifications.

3.2 L1a to L1b Data Processing

Onboard each GRACE satellite two star cameras are available for attitude determination. Apart from sun or moon intrusions the star sensors operate in dual mode, i.e. it is possible to combine the data of the two sensor heads. The combination takes into account that the quality of the derived rotation about the line of sight of the sensor heads is worse than the quality of the rotation derived perpendicular to the line of sight (from (Wu and Kruizinga, 2004)):

$$\mathrm{Q_{comb}} = Q_1 \cdot (1, \mathrm{M}\Delta_{12}) \tag{1}$$

where:

$$\mathrm{M} = \frac{1}{2} \cdot \begin{pmatrix} 1 & 0 & 0 \\ 0 & 1 & -\lambda \\ 0 & -\lambda & 1 \end{pmatrix} \tag{2}$$

$\lambda = \frac{\kappa^2-1}{\kappa^2+1}$ and $\kappa = 8$ taking into account that the attitude information for rotations about the line of sight of the star trackers is about eight times worse

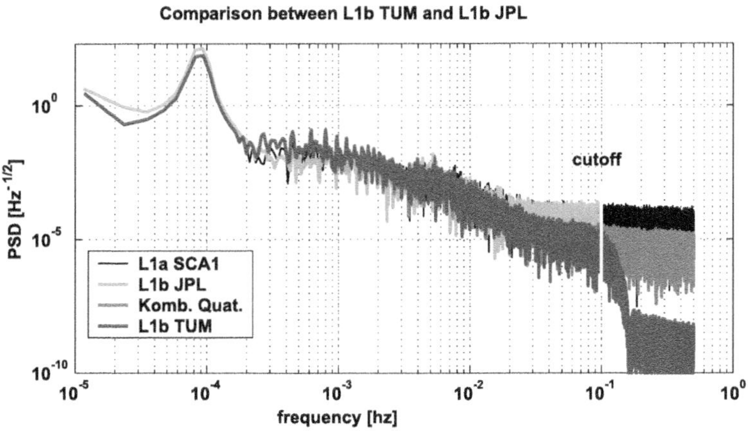

Fig. 9. Comparison of the L1b Data provided by JPL and the L1b data generated at IAPG. Differences are noticeable. It is interesting that the combined solution from JPL shows no effect of smoothing but the solution from IAPG does

than for rotations perpendicular to the line of sight. Δ_{12} is the observed difference between the two sensor heads. Basically the combination is a weighted mean. Figure 9 shows a comparison of the L1b star sensor data derived at IAPG and the star sensor data provided by JPL. Significant differences on all frequencies are noticeable. It is interesting that the spectrum of the combined quaternions from JPL on L1b compared to L1a shows no significant difference. In contrast the L1b quaternions derived by IAPG show smoothing effects visible as a lower noise level towards the higher frequencies. A possible explanation for this effect is that probably for the derivation of the JPL data a different value for κ was used. Taking a closer look at the differences in Fig. 10 reveals that the differences are small except from certain spikes appearing in a regular pattern. Closer investigation shows that these epochs are epochs where the star sensor switches from dual to single head operation. More detailed investigations are needed to identify the reasons for these deviations.

4 K-Band

The measurements of the K-Band ranging system are affected by two kinds of noise:

1. the instability of the local oscillator
2. the phase noise of the measurements

As Fig. 11 shows, the error source 1 can be avoided if GPS timing is used, which is the case for the K-Band processing. Towards the lower frequency the

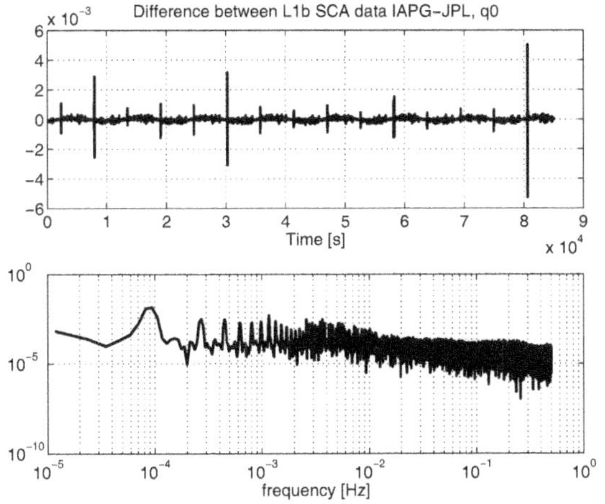

Fig. 10. Difference between IAPG derived L1b star sensor data and the data provided by the JPL

error is dominated by the inaccuracy of the synchronization of the measurements from both satellites, which shows up as a drift. Towards the higher frequencies the error is dominated by white noise with an magnitude of 1 micron. The range measurement is the original measurement, the range rate and the range acceleration are derived from the range via differentiation. The noise model is extracted from (Thomas, 1999). Figure 12 shows the PSD of

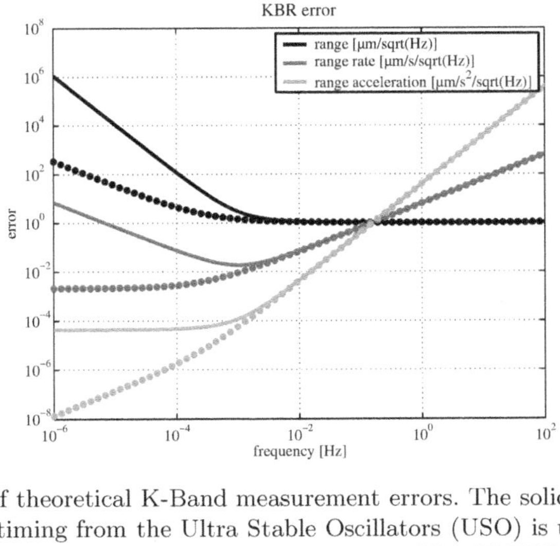

Fig. 11. PSD of theoretical K-Band measurement errors. The solid lines represent the error if the timing from the Ultra Stable Oscillators (USO) is used, the dotted lines represent the error if GPS timing is used. For the GRACE KBR data processing GPS timing is used

the L1b range measurement. Prominent peaks are on the once and twice per revolution frequencies. The signal decreases until at about $2.5 \cdot e^{-2}$Hz noise dominates the signal. Note that the noise is not white, it is colored as the range rate signal is shown. The dashed line represents the specified error psd. The level of the noise agrees well with the specified error level. Is is about $2\ \mu\mathrm{m/s}/\sqrt{\mathrm{Hz}}$. Note the peaks on $4.5 \cdot e^{-2}$ Hz and multiples. Their origin is unknown, it could be correlated with the peaks observed in the accelerometer measurements on GRACE A.

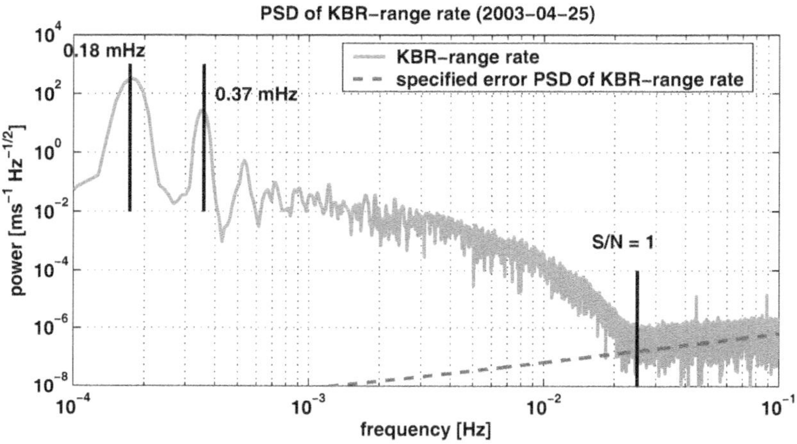

Fig. 12. PSD of KBR range rate signal and specified error level

4.1 L1a to L1b Data Processing

Again the processing was conducted according to (Wu and Kruizinga, 2004). Figure 13 shows the difference between the range rate derived at IAPG and the range rate provided by JPL. The difference is significant. It has a sinusoidal form, the dominant frequencies are once per revolution and multiples. The amplitudes reach the level of about 2 μm/s, about twice the specified noise level. Further investigations are needed to determine the reason for the difference, a first guess would be differences in the time tagging, resp. the derivation of the clock correction for the onboard time, as even a small phase shift could lead to the difference that has been derived. It seems that numerical problems can be excluded, as the difference shows a regular structure, and doesn't behave irregularly due to round-off errors.

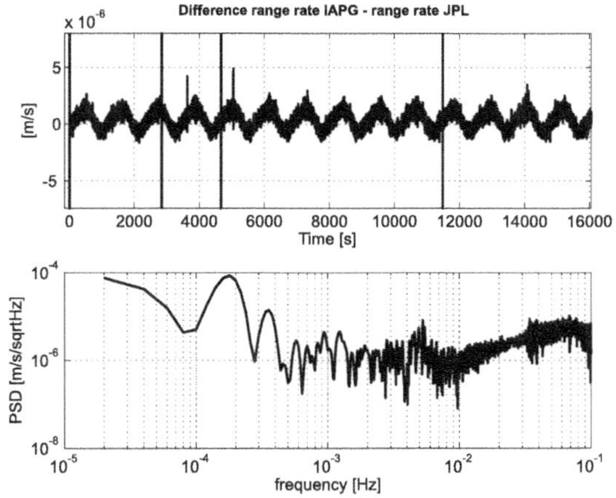

Fig. 13. Difference between the Range Rate derived at IAPG and the Range Rate provided by JPL

5 Conclusions and Outlook

Concerning the performance assessment of the sensor system the following conclusions have been derived:

1. SuperSTAR accelerometers:
   - the noise level of the along-track and the radial component is about 2-3 times higher than the specified level of $1 \cdot e^{-10}$ m/s$^2/\sqrt{\text{Hz}}$.
   - the noise level of the cross-track component is about 70% of the specified noise level of $1 \cdot e^{-9}$m/s$^2/\sqrt{\text{Hz}}$.

   The given noise levels are derived from periods without twangs, peaks or thruster events.
2. Star sensors:
   - The star sensor on GRACE A performs slightly worse than the star sensor on GRACE B concerning the rotation perpendicular to the line of sight.
   - The performance of the star sensor on GRACE B agrees well with the specification (30 μrad$/\sqrt{\text{Hz}}$ relative to line of sight (LOS) and 240 μ rad$/\sqrt{\text{Hz}}$ around LOS).
   - The performance of the star sensor on GRACE A concerning the orientation perpendicular to its line of sight is about three times worse than the specification. The performance concerning the rotation about its line of sight agrees well with the specifications.
3. K-Band Ranging System
   - The performance of the range measurements agrees well with the specifications of 1 μm $/\sqrt{\text{Hz}}$.

Concerning the L1a to L1b data processing the following status has been reached:

- The results from the processing of the accelerometer data agree well with the results provided by JPL
- Concerning the Star Sensor processing, significant differences remain to be investigated.
- The results from the K-Band ranging system show large discrepancies compared to the noise level of the K-Band measurements (about a factor of 1e3 higher on range level and a factor of 2 on range rate level) The reasons for this discrepancies have to be investigated.

Outlook

The following future studies are planned:

- Accelerometers:
  1. Investigation if the twangs and peaks that show up in the accelerometer measurements are caused by real physical accelerations on the satellites or if they are a feature of the accelerometers
  2. Investigation of the effect of different low-pass filters to be applied for the downsampling from L1a to L1b.
- Star Sensors:
  1. Investigation of different strategies for the combination of the observations from the two sensor heads of each star sensor.
  2. Investigation of the possibility to combine star sensor measurements with accelerometer measurements.
  3. Investigation of the reasons for the differences in the L1b data derived by IAPG and the data provided by JPL.
- K-Band-Ranging System:
  1. Investigation of the reasons for the differences in the L1b data derived by IAPG and the data provided by JPL.

Acknowledgement. This work was funded by the German Ministry of Education and Research, grant number 03F0326B. We are very thankful to JPL and GFZ/Potsdam for the provision of the L1a and L1b data from April 25th to 27th 2003 to Prof. Rummel as a member of the GRACE Science Team. This is publication GEOTECH-148 of the research programme GEOTECHNOLOGIEN of BMBF and DFG.

References

Flury J (2004) Beiträge zur Signalanalyse der GRACE L1a Akzelerometerdaten, Internal report, Institute for Astronomical and Physical Geodesy, Technische Universität München

Hudson D (2003) In-flight characterization and calibration of the SuperSTAR accelerometer, MSc thesis, Center of Space Research, University of Texas, Austin

Stanton R, S Bettadpur, C Dunn (1998) Science & Mission Requirements Document, Technical report, Jet Propulsion Laboratory, Pasadena

Thomas JB (1999) An Analysis of the Gravity Field Estimation based on Dual-1-Way Intersatellite Biased Ranging, Technical report, Jet Propulsion Laboratory, Pasadena

Wu SC, G Kruizinga (2004) Algorithm theoretical basis document for the grace l1b data processing, Technical report, Jet Propulsion Laboratory, Pasadena

Static and Time-Variable Gravity from GRACE Mission Data

Roland Schmidt, Frank Flechtner, Ulrich Meyer, Christoph Reigber, Franz Barthelmes, Christoph Förste, Richard Stubenvoll, Rolf König, Karl-Hans Neumayer, and Shengyuan Zhu

GeoForschungsZentrum Potsdam (GFZ), Dept. 1 Geodesy and Remote Sensing, Telegrafenberg A17, 14473 Potsdam, Germany, rschmidt@gfz-potsdam.de

Summary. Based on the GRACE mission data, a new era of static and time-variable gravity models with unprecedented resolution and accuracy have been generated by the GRACE Science Data System teams. In general, the spatial resolution of the field from pre-CHAMP satellite only models of about 1000 km can be increased by a factor of 5 - 6 thanks to the micrometer-precise K-band intersatellite link. The currently obtained gain in accuracy reaches one to two orders of magnitude, compared to the most advanced combination gravity pre-CHAMP models, but is still one order of magnitude away from the projected GRACE baseline accuracy.

In this article we highlight the advances in gravity recovery with GRACE, based on recent results from GFZ Potsdam for a new GRACE-only medium-wavelength gravity model, called EIGEN-GRACE03S, a new combined high-resolution model complete up to degree and order 360, called EIGEN-CG03C, and the derivation of time-variable gravity signals from monthly GRACE-only gravity models.

Evaluation of EIGEN-GRACE03S and EIGEN-CG03C shows that both models benefit in its long-to-medium wavelength part from an extended data base for GRACE, an augmented processing of the GRACE data as well as a meanwhile more complete and homogeneous compilation of surface data. The progress in resolution and accuracy with respect to earlier GRACE-based gravity models is moderate but visible at the level of 1 - 2 percent for standard comparisons.

The derivation of time-variable gravity signals from a time series of 16 monthly GRACE-only gravity solutions reveals the mission's sensitivity to hydrology-induced surface mass variations. The annual-varying signal on global and regional scales can be resolved down to spatial scales of a few hundred kilometers and the estimates are well above the assumed error level of the GRACE gravity solutions. Observable discrepancies with respect to the signal amplitudes, phases and spatial distribution indicate the potential contributions from GRACE to hydrological modelling, but also reveal systematic errors in the GRACE monthly fields.

Key words: GRACE, static gravity, time-variable gravity, dynamic gravity recovery

1 Introduction

Since its launch in March 2002 the US-German twin-satellite mission GRACE (Gravity Recovery And Climate Experiment, Tapley and Reigber (2001)) provides nearly continuous, highly precise instrument data of the spacecrafts' positions (BlackJack GPS receiver), attitude (star cameras), non-gravitational forces (SuperSTAR accelerometer) and the inter-satellite range and its rate of change (K-band link) for the determination of the Earth's gravity field. Based on these novel data the groups of the joint US-German Science Data System (SDS) at the Center for Space Research at the University of Texas in Austin (UTCSR) and at the GeoForschungsZentrum Potsdam (GFZ) have generated unique global gravity models with unprecedented accuracy and resolution. For the static field various global gravity models solely from GRACE data (so-called GRACE-only models, GGM01S (Tapley et al., 2004a), GGM02S (Tapley et al., 2005), EIGEN-GRACE01S (Reigber et al., 2003), EIGEN-GRACE02S (Reigber et al. , 2005a)) as well as combination models using additional high-resolution altimeter-derived and terrestrial gravity data (GGM01C (Tapley et al., 2004a) or EIGEN-CG01C (Reigber et al., 2005b)) have been computed and released to the scientific community.

However, the primary objective of the mission is the determination of time-variable changes in the Earth's gravity field caused by geophysically and climatoligcally driven processes. These are derived from time series of global gravity models in terms of spherical harmonics estimated from monthly batches of GRACE data thus representing the evolution of the changing gravity field at a monthly resolution. Although the anticipated accuracy of GRACE-based gravity models (the so-called *GRACE baseline* accuracy) has not yet been fully reached, the mission's sensitivity and capability of resolving time-variable gravity has been widely demonstrated. In particular seasonal mass redistributions in the continental water cycle are traceable in GRACE data (see e.g. Tapley et al. (2004b), Wahr et al. (2004), Han et al. (2005), Schmidt et al. (2005)).

In this contribution we compile an overview on recent results for the determination of the static and time-variable gravity field from GRACE obtained at GFZ Potsdam. Section 2 gives a description of the procedure applied for determination of GRACE-only gravity models at GFZ. In Sect. 3 current versions of a new long-term GRACE-only gravity model and a new high-resolution combination model derived from CHAMP, GRACE and surface data are presented. In Sect. 4 we discuss methods to assess the accuracy of monthly GRACE-only models as a preparatory step for the derivation of time-variable signals from GRACE in Sect. 5. The final Sect. 6 gives a summary and an outlook.

2 Gravity Field Model Determination

At GFZ GRACE-based gravity field models, consisting of the coefficients of the spherical harmonic expansion of the Earth's gravity field, are derived from the mission's data using the dynamic orbit determination and gravity recovery method implemented in the GFZ-owned Earth Parameter and Orbit System (EPOS) software. The method is based on the satellite's perturbated equation of motion around the Earth's geocenter using a complete set of models for gravitational and non-conservative forces. The solution of the dynamic motion equation is obtained by means of a numerical integration procedure starting from an initial state for the satellite's position and velocity as well as parameters for the force models. In case of GRACE the conservative force models comprise the static gravity field, third body perturbations from the Sun, Moon and planets, accelerations from luni-solar tidal effects on the solid Earth and oceans and short-term atmospheric and oceanic mass variations. Non-conservative forces are measured by the SuperSTAR accelerometers onboard each GRACE spacecraft. After correction of the instrument specific biases and scale factors, the accelerometer data given at 5 s intervals is used as true non-conservative forcing in the integration.

In order to estimate geometric and dyamic parameters (such as gravity coefficients) the integration method is combined with a least-squares adjustment procedure. In case of GRACE, such parameters are estimated from the GPS and K-band Satellite-to-Satellite Tracking (SST) data, where the micrometer-precise range-rate K-band SST data is the primary observable for gravity recovery. For gravity recovery the processing is performed in two stages. In the first step a reference orbit is computed from the SST data to be used for linearisation of observational equations for the estimation of the gravity coefficients in the second step.

Because of the huge amount of satellite data (about 1 million GPS- and 400,000 K-band- SST data for one month) and the large number of gravity unknowns (about 23,000 parameters for a gravity model complete to degree and order 150), the processing is split into batches (*arcs*) of nominally 1.5 days length to reduce the computational effort. This value is used as a compromise between the need for a short arc in order to prevent an increase of modeling errors and to keep the problem tractable on computers, and a longer arc to cover at least one half of GRACE's primary gravitational orbit resonance period. Thus arc-wise normal equation systems relating the observational residuals to the parameters are set up. After some manipulations (e.g. reduction of arc-dependent parameters like GPS phase ambiguities or intial elements) the arc-wise normal equation systems are accumulated to one global normal equation system which is eventually solved by matrix inversion. Monthly estimates of the gravity field are determined from the accumulation of arc-wise normal equation systems covering one calendar month, long-term static gravity field models are based on multiple monthly batches covering one year or longer. Further details on the method and the applied models can be found in Reigber et al. (2005a).

3 Static GRACE Gravity Models

Following the procedure of the previous section, in a recent processing GRACE data in the period February 2003 to July 2004 has been exploited. Not included are June 2003 and July 2004 due to larger instrument data gaps. The arc-wise normal equations were accumulated to one global system covering a 376 days period. The system was solved for spherical harmonic coefficients complete to degree and order 150. The resulting model, called EIGEN-GRACE03S hereafter, is a successor of the GFZ-generated models EIGEN-GRACE01S and EIGEN-GRACE02S.

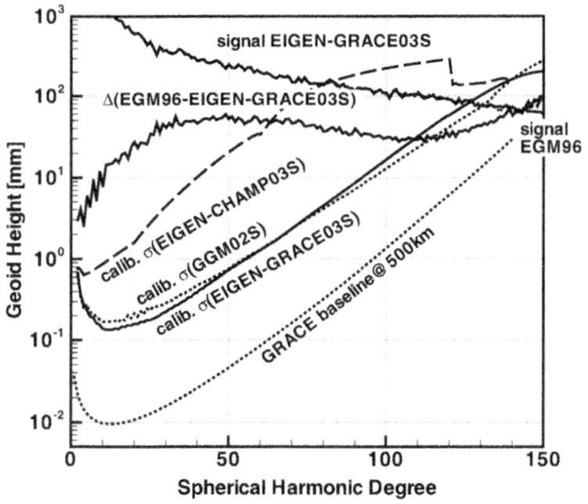

Fig. 1. Signal and error amplitudes per degree in terms of geoid heights

The gain in resolution and accuracy from GRACE-based models in general and EIGEN-GRACE03S in particular is shown in the spectral domain in Fig. 1. It depicts degree signal and degree error amplitudes in terms of geoid heights. EIGEN-GRACE03S seems to have full power up to degree 125 as is inferred by comparing the errors of EIGEN-GRACE03S to the high-resolution model EGM96 (Lemoine et al., 1996). The errors of EIGEN-GRACE03S[1] are about one to two orders of magnitude smaller than corresponding errors from

[1] The GRACE errors were aposteriori calibrated based on differences of subset solutions for GRACE and comparisons to the independently determined models GGM01S and GGM02S from UTCSR. The CHAMP errors are calibrated by comparisons to GRACE-only models

EGM96 and CHAMP. Accumulating the errors of EIGEN-GRACE03S over the spherical harmonic degrees gives a 1 cm error in geoid heights around degree 77 corresponding to $\lambda/2 \approx 260$ km. For the CHAMP-only model EIGEN-CHAMP03S this error level is obtained already at degree 30 respectively $\lambda/2 \approx 680$ km. For EGM96 the 1 cm error is surpassed significantly even around degree 10. On the other hand it can be seen that the anticipated GRACE baseline accuracy has not yet been reached by about one order of magnitude.

Table 1. Comparison of satellite only gravity models with altimeter-derived geoid heights (N, CLS01-ECCO oceanic geoid) and gravity anomalies (Δg, NGA (former NIMA) marine gravity anomalies) for a grid spacing of $5° \times 5°$ and $2.5° \times 2.5°$ (degree and order 36 and 72, respectively) in terms of root mean square (RMS) of cos(latitude) weighted differences about mean

Model (days of data)	RMS (N) [cm] $5° \times 5°$ / $2.5° \times 2.5°$	RMS (Δg) [mgal] $5° \times 5°$ / $2.5° \times 2.5°$
EGM96S (pre-CHAMP)	36/70	1.85/5.39
GRIM5-S1 (pre-CHAMP)	44/76	2.00/5.40
GGM01S (111)	14/18	0.29/1.88
EIGEN-GRACE01S (39)	14/17	0.28/1.55
EIGEN-GRACE02S (110)	14/16	0.28/1.25
EIGEN-GRACE03S (376)	14/15	0.28/1.19

Application of EIGEN-GRACE03S in orbit determination of the geodetic satellites (not shown) and comparisons to surface geoid and gravity data further reveal the strength and homogeneity of the new GRACE-only gravity model. Table 1 lists the statistics for EIGEN-GRACE03S and for pre-CHAMP era models such as EGM96S (Lemoine et al., 1996), GRIM5-S1 (Biancale et al., 2000) and earlier GRACE-only solutions. Comparisons to the pre-CHAMP satellite only models indicate the striking gain in resolution due to the K-band link. Intercomparisons to the GRACE-only models show the moderate but visible improvements in the EIGEN-GRACE03S solution with respect to the earlier models. EIGEN-GRACE03S clearly benefits from the large data base, augmentations in the prepocessing of the GRACE instrument data as well as advances in the background modeling (e.g. ocean tides).

In addition to the GRACE-only model EIGEN-GRACE03S a new combination model, called EIGEN-CG03C, based on CHAMP, GRACE and surface gravity data has been computed. The major differences to its predecessor EIGEN-CG01C are the reprocessed GRACE data used for EIGEN-GRACE03S and a more complete and updated surface data compilation. The CHAMP data is identical in EIGEN-CG01C and EIGEN-CG03C. The combination technique closely follows the procedure described in Reigber et al. (2005b):

Table 2. Surface Data used for the generation of EIGEN-CG03C

No.	Description Surface Data Set
1	Arctic Gravity Project (ArcGP) gravity anomalies (Forsberg and Kenyon, 2004).
2	NRCan gravity anomalies (Véronneau, 2003).
3	AWI and LDO gravity anomalies (Bell et al., 1999), over two small areas of Antarctica and adjacent sea ice.
4	NGA (former NIMA) altimetric gravity anomalies over the ocean, including standard deviations.
5	Geoid undulations over the oceans by using CLS01 altimetric Sea Surface Heights (Hernandez et al., 2001) and Sea Surface Topography from the ECCO simulation (Stammer et al., 2002).
6	NGA (former NIMA) terrestrial gravity anomalies (if not covered by data sets 2 or 3) including standard deviations with almost worldwide continental coverage, except for Antarctica and some smaller data gaps.
7	NGA (former NIMA) ship-borne gravity anomalies over water depths less than 200 m.

- The normal equation systems for the spherical harmonic expansion of the geopotential for CHAMP (available up to degree and order 120 and within CHAMP-resonant orders up to degree 140) and GRACE (complete up to degree and order 150) were combined to give an intermediate normal equation system EIGEN-CG03S complete up to degree and order 150. Because of the obviously decreasing sensitivity of GRACE and CHAMP for the spectral components beyond degree 120, the contributions from degree 121 up to degree 150 were stabilized in this normal equation system and kept separately in the subsequent combination with the surface data.
- The surface data were averaged to $1° \times 1°$ block mean values and for each data set an individual normal equation system complete up to degree and order 120 was generated (the upper limit for degree and order of 120 was chosen due to restricted computer resources). Then, these normal equations were combined taking into account individual weighting depending on the individual data accuracies.
- The two satellites and the ground based normal equation systems were then combined and solved to get a gravity model complete up to degree and order 120 under the following conditions:
  - the long-wavelength part up to degree and order 70 was based on the CHAMP/GRACE satellite data only and
  - the satellite and terrestrial contributions between degree 71 and 120 were overlapped, whereas the terrestrial normal equation system was strongly down-weighted relative to the satellite-only system by an empirically found factor.
- For degree 121 up to 359 a block diagonal normal equation system was created and solved based on surface data, which are given as $30' \times 30'$ block mean gravity anomaly values.

Table 3. GPS-leveling minus model-derived geoid heights weighted root mean square (wrms) about mean (cm, number of points in brackets)

Combination Model	USA (6169)	Canada (1930)	Europe (186)	Germany (675)
EIGEN-CG03C	43	35	38	20
EIGEN-CG01C	44	32	40	22
EGM96	47	38	45	28

- The spherical harmonic coefficients of degree 360 were derived from numerical integration of the gridded gravity anomalies.
- Finally, the obtained three gravity model components (for degree 1-120, 121-359 and 360 respectively) were summed up to get the full combination model.

Color Fig. XIII on p. 293 depicts the global distribution of free air gravity anomalies derived from EIGEN-CG03C. Improvements with EIGEN-CG03C become visible by comparison with external data such as geoid heights determined point-wise by GPS-leveling. Table 3 shows the results for EIGEN-CG03C, EIGEN-CG01C and EGM96. Compared to EGM96, the EIGEN-models benefit in its long-to-medium wavelength part from the unprecedented performance of the CHAMP and GRACE satellite-only gravity models and at short wavelengths from a meanwhile more complete and updated surface data compilation.

For the derivation of time-variable gravity signals 16 monthly GRACE-only gravity models in the period of EIGEN-GRACE03S, i.e. February 2003 to July 2004 have been computed from the accumulation of corresponding monthly batches of the arc-wise normal equation systems. As with the long-term field spherical harmonic coefficients complete to degree and order 150 were solved for each system. The resulting time series of monthly sets of spherical harmonic coefficients is thus the basis for the evolution of the gravity field investigated in Sect. 5. Prior to that, results of the accuracy assessment of monthly GRACE-only gravity models are presented in the next section.

4 Accuracy Assessment of Monthly GRACE-only Gravity Models

The formal errors of the spherical harmonic coefficients of the monthly and of the long-term GRACE-only gravity solutions as obtained by the least-squares adjustment process are known to be too optimistic. This originates from the fact that spurious gravity features given in the GRACE-only gravity models are significantly larger in amplitude than the formal coefficients uncertainties predicted. Such errors, showing up as meridional-oriented stripping features of gravity functionals in the space domain, have been explained to some extend by deficiencies of apriori models of time variable gravity signals such as ocean tides and short-term mass variations causing spatio-temporal aliasing (see e.g. Han et al. (2004) or Wünsch et al. (2005) and references cited therein). For error propagation of satellite-based functionals (e.g. surface mass fluctuations), realistic estimates of the GRACE gravity model errors are needed, however.

One possibility is to apply a degree-dependent scaling on the original variance-covariance matrices of the spherical harmonic coefficients obtained from the least squares adjustment process. However, since no independent data set of comparable global distribution, strength and homogeneity exist, an approximate calibration has to be determined from the GRACE data internally. For long-term GRACE-only solutions such individual degree-dependent calibration factors were obtained by comparing differences of signal amplitudes of GRACE-only subset solutions that cover different periods.

For the assessment of monthly solutions this approach has been applied in a similar manner. In contrast to static models where unreduced time-variable gravity is thought to average out if the processing covers a sufficient long time period, substracting the solutions with a year apart removes such signals in the monthly solutions. The basic idea is to reduce the dominant time-variable gravity signal from hydrology that has a strongly seasonal variation. To this end, differences of signal degree amplitudes are computed for all inter-annual combinations possible in the given period. The distribution of the residual degree amplitudes is thought to represent the uncertainty of the monthly models. In order to obtain a somewhat smoother error curve all available sets of differences are averaged. Next, degree-dependent scaling factors are determined by a degree-wise comparison of the formal error degree amplitudes to the averaged difference degree amplitudes. Eventually, the resulting degree-wise scaling factors are applied per degree to the formal variance-covariance matrices giving the calibrated matrices that can be used for error propagation. Since the error level of the monthly solutions is represented by the single set of the averaged difference degree amplitudes, it is sufficient to derive one calibrated variance-covariance matrix to be valid for all monthly solutions.

The degree amplitudes of this calibrated error of the monthly solutions is shown in Fig. 2, together with the EIGEN-GRACE03S errors and the GRACE baseline accuracy. It can be seen that the error of time-variable signal derived from the monthly solutions relative to EIGEN-GRACE03S will be dominated by the error of the monthly solutions. Just like for EIGEN-GRACE03S the errors of the monthly solutions are still about one order of magnitude higher than the GRACE baseline. Hence, the resolution of time-variable gravity signals such as caused by hydrology will be limited, as the comparison to the signal amplitudes of monthly hydrology from the WGHM (WaterGAP Hydrology Model, Döll et al. (2003)) illustrates. From Fig. 2 one would infer a maximum resolution of about degree 13 (i.e. $\lambda/2 \approx 1500$ km) instead of degree 35 as expected from the GRACE baseline. However, such comparisons on the basis of degree amplitudes are not too instructive because the full spatial dependency is obscured in degree amplitudes. In Sect. 5 it will be shown that the actual resolution can be much higher in areas of large surface mass variability.

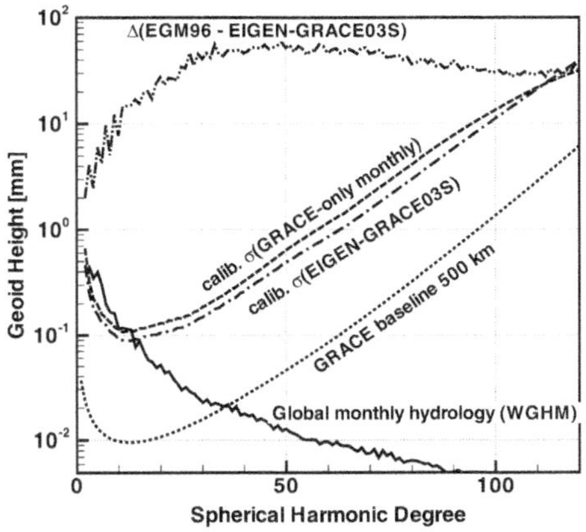

Fig. 2. Degree amplitudes of the calibrated errors for EIGEN-GRACE03S monthly and mean solutions using approach 1. Additionally the GRACE baseline accuracy and the global water storage change derived from the WGHM model are shown for comparison

As an alternative, one can estimate the accuracy level of the monthly solutions when looking at the RMS of surface mass variability in the space domain after the reduction of time-variable signal. For this purpose global grids ($1^\circ \times 1^\circ$) of surface mass variability are computed from the monthly gravity models with respected to the long-term mean field EIGEN-GRACE03S. Then, for each point a model with a bias, a drift and an annual periodic term is fitted and removed from the time series. Finally, the global, weighted RMS of the residual signal of the time series of all grid points is taken as a measure for the accuracy of the monthly solutions.

Table 4 lists results for Gaussian averages of the accuracy of surface mass variability (in terms of the thickness of an equivalent mass of water) propagated from the scaled variances and covariances of the first approach (column 2) and from the alternative approach (column 3). The displayed values are the weighted RMS (cosine latitude weigthing) of the $1^\circ \times 1^\circ$ grid points. For approach 1 the complete scaled covariance matrices (labelled cov in column 2) and the scaled variances only (labelled var in column 2) have been used for error propagation. Comparison between these two indicates the good decorrelation in the determination of the gravity coefficients by GRACE towards the long wavelengths of the gravity field, giving more or less identical RMS values for both cases. Consequently, it would be sufficient to consider only the coefficient variances for error propagation when looking at long-wavelength features. Towards the shorter wavelengths the RMS values are larger for the *cov*-case indicating an increasing correlation between the solved spherical harmonic coefficients. This is expected since the resolution will be limited due to ground track coverage, observation's sensitivity and so on. Therefore the study of the short wavelength features has to consider in principle full variance-covariance matrices rather than variances only.

Table 4. Accuracy of the surface mass estimates derived from EIGEN-GRACE03S monthly solutions using approach 1 and 2 in terms of thickness of an equivalent mass of water for different spatial averages

Gaussian Filter Radius km	Approach 1 RMS (cov/var) cm	Approach 2 RMS cm
1500	1.4/1.3	0.9
1000	1.6/1.6	1.4
750	2.0/1.8	2.1
500	3.1/2.5	4.0
400	4.8/3.7	6.7

Comparing the results of approach 1 and approach 2 in Table 4 reveals a good agreement between the two approaches at a wavelength of about 750 km. Towards longer wavelengths approach 2 has smaller statistics indicating that approach 1 could be too pessimistic there. For short wavelengths a reverse picture is obtained, suggesting that approach 1 gives too optimistic values in that region.

Since both approaches reduce only purely annual signals at best, but non-annual periodic and secular signals may remain, both approaches are likely to give too pessimistic estimates in general. In this way the obtained values could be seen as an upper bound for the models' accuracy. On the other hand one should note that above values are global averages and significantly smaller or larger errors can occur when looking at an actual spatial distribution.

5 Time-Variable Gravity from GRACE and Hydrology

As outlined in Sect. 2 the monthly GRACE-only gravity models are corrected for time-variable gravity signals such as ocean tides, tides of the solid Earth, atmospheric and oceanic short-term mass variations and secular changes in zonal coefficients of degree 2, 3 and 4 due to global isostatic adjustment processes via apriori models. Consequently the observed residual time-variable signal derived from differences of the monthly solutions versus a mean should basically represent the non-modelled gravity changes. Among these, mass redistributions due to the global water cycle cause the largest variations and are clearly detected by GRACE.

In the sequel results for an annual periodic time-variable gravity based on the 16 monthly gravity field solutions and the corresponding long-term static model EIGEN-GRACE03S are presented. Instead of changes in gravity, the surface mass variability in terms of the thickness of an equivalent mass of water has been derived. Following the method in the previous section, for each month a global $1^\circ \times 1^\circ$ grid of surface mass variability with respect to the long-term mean has been computed. Then the annual signal is determined by fitting a bias, a trend and an annual periodic sine (amplitude and phase) to the times series of each grid point. The left column of the Color Fig. XIV on p. 294 shows the amplitudes of the annual surface mass variability derived from GRACE for three different Gaussian filter radii (1000, 750 and 500 km, respectively). For comparison the corresponding results for the amplitudes derived from monthly maps of changes in the continental water storage from WGHM are displayed in the right column of Color Fig. XIV.

Visual inspection indicates a high spatial correlation between GRACE and WGHM at all selected resolutions. In particular the surface mass variability inside large drainage basins in South America (Amazon), South-East Asia (Ganges), Africa (Congo, Niger) and Siberia (Lena, Ob) are detected by GRACE. In other areas, e.g. Central America or the Labrador Peninsula GRACE and WGHM differ significantly. However, such discrepancies are not unexpected because hydrological models are known to have deficiencies in representing related mass variations at large scales and, to the contrary, are expected to be improved by GRACE. In this way observed deviations rather highlight the potential contributions of GRACE for hydrological model.

On the other one can see that the estimates of the amplitudes are superimposed by the aforementioned meridional-oriented stripping features that are due to the spatio-temporal aliasing and represent errors. The effect is pronounced when increasing the spatial resolution and in particular visible over the oceans.

The phases of the annual signal estimated by GRACE (in days relative to January 1st) are depicted in the left column of Color Fig. XV on p. 295. On the right the absolute differences of the phases from GRACE minus the phases from WGHM are shown. As before the same three averaging filter lengths (1000, 750 and 500 km, respectively) are used. In accordance to the results for the amplitudes, for large drainage basins such as the Amazon or the two independent data sets are well in phase. But as above, areas of disagreement are evident and the deviations may be explained by deficiencies in the WGHM model, but also the GRACE gravity model errors, again. Over the oceans, where no annual periodic signal can be detected by GRACE, the model errors dominate the estimated phase values and give a quite noisy distribution. Therefore the oceans have been omitted in the plots on the right of Color Fig. XV on p. 295.

Along with global estimates of surface mass variability regional investigations for river basins have been carried out. In this document results for four river basins of different size are summarized: the Amazon, the Ganges, the Congo and the Danube basin. The corresponding basin masks are taken from the WGHM data base. To extract the time-variable gravity signal from sets of monthly of spherical harmonics from GRACE and WGHM we use the method proposed in Swenson and Wahr (2003). This method allows the construction of regional averaging kernel functions in terms of spherical harmonics where the effect of signal leakage from outside the region of interest and the contributions of gravity model errors are minimized using some constraint on the acceptable gravity model error. For the gravity model error the calibrated variances of the coefficients as obtained in Sect. 4 from approach 1 are used in the sequel. Correlations of the solved spherical harmonics were neglected. For the constraint on the accepted error from the gravity model a value of 2 cm in terms of the thickness of an equivalent mass of water is selected here.

The left column of Color Fig. XVI on p. 296 shows the resulting averaged surface mass variability using the approximate basin functions displayed in the right column. The averaged values from GRACE are shown as red dots (including the 2 cm uncertainty as vertical bars), the corresponding values from WGHM are plotted as blue dots. The estimated annual signal is given as solid lines, red for GRACE and blue for WGHM. For the Amazon and Ganges the GRACE and WGHM signal are well in phase. The GRACE-based variability has a larger amplitude as it has been revealed already in the global estimates, indicating the potential contributions on an improved estimate of the total mass variability from GRACE. For the Congo basin the comparison shows a less stringent agreement. This may be due to the fact that the given Congo basin function is not well adopted to the actual signal maxima North and South of the Congo as seen by GRACE (cf. global plots of annual amplitudes in Color Fig. XIV on p. 294 left column). For the Danube basin, although a rather small basin, GRACE detects a plausible annual surface mass variability, again. However, the observed phase shift of about 2 months with respect to WGHM is still unexplained.

6 Summary and Outlook

The GRACE mission has demonstrated its capability of resolving the static and the time-variable gravity field with unprecedented accuracy. As evident from the recent GRACE-only solution EIGEN-GRACE03S generated at GFZ Potsdam, the gain in the resolution of the static field is a factor of 6 (from 1000 km to 160 km) compared to pre-CHAMP gravity models and a factor of 2.5 (from 400 km to 160 km) compared to CHAMP gravity models. The accuracy can be increased by one to two orders of magnitude in comparison to the pre-CHAMP gravity models. EIGEN-GRACE03S provides a 1 cm accuracy geoid with a spectral resolution up to degree and order 77, being slightly above the resolution of its predecessor EIGEN-GRACE02S (1 cm accuracy geoid with a spectral resolution up to degree and order 75, cf. Reigber et al. (2005a)). The moderate improvements of EIGEN-GRACE03S with respect to earlier GRACE-only solutions are also manifest by comparisons to external surface geoid and gravity data and orbit computation tests.

In addition to EIGEN-GRACE03S, a new combination, high-resolution model, EIGEN-CG03C, combining CHAMP, GRACE and surface gravity data has been computed. The model benefits in the long-to-medium wavelength part also from the improvements in the EIGEN-GRACE03S solution and at the short wavelengths from a further augmented surface data base and compilation.

Derivation of time-variable gravity signals from time series of monthly GRACE-only gravity models on global and regional scales gives access to surface mass variations caused by mass redistributions in the global water cycle. In particular in the world's largest river basins seasonal hydrological mass redistributions are detected by GRACE down to wavelengths of a few hundred kilometers. On the other hand the estimates are degraded by systematic effects from spatio-temporal aliasing and other causes, preventing the anticipated resolution.

Although the mission's baseline accuracy has not yet been fully reached, current static and time-variable gravity models from GRACE nevertheless provide substantial information for various geoscientific applications. An important example is the accurate recovery of the sea surface topography on the basis of a 1 cm accuracy geoid from GRACE-only models like EIGEN-GRACE03S for the determination of large scale circulations (see e.g. Tapley et al. (2003)). New generation combination models like EIGEN-CG03C will be of benefit to geophysical applications concerning the interpretation of the static gravity field in relation to the structure of the Earth's interior and geodynamic processes in the Earth's mantle/lithosphere. Finally, the successfull, though still limited resolution of hydrology-induced surface mass variations from monthly GRACE-only gravity models is the first step into the recovery of third dimension of gravity needed for the understanding of climatologically and geopysically driven processes in the context of a comprehensive view on the Earth system.

In future work GRACE gravity models shall be improved further to the ultimate precision possible. Different aspects need to be treated in this context. One concern is a more detailed investigation of the spatio-temporal aliasing in combination with a possibly improved parametrization and/or the usage of updated and more complete apriori models. Another should be concentrated on a further potential refinement of the processing respectively a definitive assessment of the GRACE instrument data. A third should be dedicated to an integrated analysis of CHAMP and GRACE data (see e.g. Zhu et al. (2004)). Parallel to that possible benefits from methods alternative to the dynamic approach should be investigated. Last but not least, the results on the static and time-variable gravity models should be applied, evaluated and discussed in close cooperation between the gravity modelers and the scientific users.

Remark. The models EIGEN-GRACE03S, the EIGEN-CG03C and the monthly GRACE-only gravity models can be downloaded at the GRACE Information System and Data Center (ISDC) at GFZ Potsdam: *http://isdc.gfz-potsdam.de/grace.*

Acknowledgement. This is publication no. GEOTECH-153 of the GEOTECHNOLOGIEN programme of BMBF and DFG, grant 03F0326A. We are grateful to one anonymous reviewer for the helpful comments and suggestions.

References

Bell RE, Childers VA, Arko RA (1999) Airborne and precise positioning for geologic applications. *J. Geophys. Res.*, 104 (B7), 15281 - 18292

Biancale R, Balmino G, Lemoine JM, Marty JC, Moynot B, Barlier F, Exertier P, Laurain O, Gegout P, Schwintzer P, Reigber Ch, Bode A, König R , Massmann FH, Raimondo JC, Schmidt R, Zhu SY (2000) A new global Earth's gravity field model from satellite orbit perturbations: GRIM5-S1. *Geophys. Res. Lett.*, 27, 3611-3614

Döll P, Kaspar F, Lehner B (2003) A global hydrological model for deriving water availability indicators: model tuning and validation. *J. Hydrol.*, 270, 105-134

Flechtner F, Schmidt R, Zhu SY, Meyer UL (2005) GRACE gravity field solutions using different de-aliasing models. Poster EGU05-A-04815 presented at the European Geosciences Union General Assembly 2005, Vienna, Austria, 24-29 April 2005

Forsberg R, Kenyon S (2004) Gravity and geoid in the Arctic region - The nortern gap now filled. *Proceedings of 2nd GOCE User Workshop, ESA SP-569*, ESA Publication Division, Noordwijk, The Netherlands

Han SC, Jekeli C, Shum CK (2004) Time-variable aliasing effects of ocean tides, atmosphere, and continental water mass on monthly mean GRACE fields. *J. Geophys. Res.*, 109, B04403, doi:10.1029/2003JB002501

Han SC, Shum CK, Jekeli C, Alsdorf D (2005) Improved estimation of terrestrial water storage changes from GRACE. *Geophys. Res. Let.*, 32, doi: 10.1029/2005GL022382

Hernandez FP, Schaeffer MH, Calvez J, Dorandeu Y, Faugére Y, Mertz F (2001) Surface Moyenne Oceanique: Support Scientifique à la mission altimetrique Jason-1, et à une mission micro-satellite altimetrique. Contract SSALTO 2945-Ot2-A1. Rapport final no. CLS/DOS/NT/00.341, CLS, Remonville St Agne

Lemoine FG, Kenyon S, Factor JK, Trimmer RG, Pavlis NK, Chinn DS, Cox CM, Klosko SM, Luthcke SB, Torrence MH, Wang YM, Williamson RG, Pavlis EC, Rapp RH, Olsen TR (1998) The development of the joint NASA GSFC and the National Imagery and Mapping Agency (NIMA) geopotential model EGM96. NASA Technical Paper NASA/TP-1998-206861, Goddard Space Flight Center, Greenbelt

Reigber CH, Schmidt R, Flechtner F, König R, Meyer UL, Neumayer KH, Schwintzer P, Zhu SY (2003) First GFZ GRACE gravity field model EIGEN-GRACE01S from 39 days of GRACE data. *http://www.gfz-potsdam.de/pb1/op/grace/results/index_RESULTS.html*

Reigber CH, Schmidt R, Flechtner F, König R, Meyer UL, Neumayer KH, Schwintzer P, Zhu SY (2005a) An Earth gravity field model complete to degree and order 150 from GRACE: EIGEN-GRACE02S. *J. Geodynamics*, 39, 1-10, doi: 10.1016/j.jog.2004.07.001

Reigber CH, Schwintzer P, Stubenvoll R, Schmidt R, Flechtner F, Meyer UL, König R, Neumayer KH, Förste CH, Barthelmes F, Zhu SY, Balmino G, Biancale R, Lemoine JM, Meixner H, Raimondo JC (2005b) A high resolution global gravity field model combining CHAMP and GRACE satellite mission and surface data: EIGEN-CG01C. accepted by *J. of Geodesy*

Schmidt R, Schwintzer P, Flechtner F, Reigber CH, Güntner A, Döll P, Ramillien G, Cazenave A, Petrovic S, Jochmann H. Wünsch J (2005) GRACE observations of changes in continental water storage, accepted by *Global and Planetary Change*

Tapley BD, Reigber CH (2001) The GRACE mission: Status and future plans. *EOS Trans AGU*, 82 (47), Fall Meet. Suppl., G41 C-02

Tapley BD, Chambers D, Bettadpur S, Ries J (2003) Large Scale Ocean Circulation from the GRACE GGM01 Geoid. *Geophys. Res. Lett.*, 30 (22), 2163, doi: 10.1029/2003GL018622

Tapley BD, Bettadpur S, Watkins MM, Reigber CH (2004a) The Gravity Recovery and Climate Experiment: Mission Overview and Early Results. *Geophys. Res. Lett.*, 31, L09607, doi:10.1029/2004GL019920

Tapley BD, Bettadpur S, Ries J, Thompson P, Watkins MM (2004b) GRACE measurements of mass variability in the Earth system. *Science*, 305, 503-505, doi: 10.1126/science.1099192

Tapley BD, Ries J, Bettadpur S, Chambers D, Cheng M, Condi F, Gunter B, Kang Z, Nagel P, Pastor R, Pekker T, Wang F (2005) GGM02 - An improved Earth gravity field model from GRACE. In review, *J. of Geodesy*

Stammer D, Wunsch C, Giering R, Eckert C, Heinbach P, Marotzke J, Adcraft A, Hill CN, Marshall J (2002) Global ocean circulation during 1992-1997 estimation from ocean observations and a general circulation model. *J. Geophys. Res.*, 107 (C9): 3118, doi:10.1029/2001JC000888

Swenson S, Wahr J (2003) Methods for inferring regional surface-mass anomalies from Gravity Recovery and Climate Experiment (GRACE) measurements of time-variable gravity. *J. Geophys. Res.*, 107 (B9), 2193, doi: 10.1029/2001JB000576

Véronneau (2003), pers. commun.

Wahr J, Swenson S, Zlotnicki V, Velicogna I (2004) Time-variable gravity from GRACE: First results. *Geophys. Res. Lett.*, 31, L11501, doi: 10.1029/2004GL019779

Wünsch J, Schwintzer P, Petrović S (2005) Comparison of two different ocean tide models especially with respect to the GRACE satellite mission. *Scientific Technical Report*, STR05/08, GeoForschungsZentrum Potsdam

Zhu SY, Reigber CH, König R (2004) Integrated adjustment of CHAMP, GRACE, and GPS data. *J. Geodesy*, 78, 103-108, doi: 10.1007/s0019000403790

Gravity Field Recovery from GRACE-SST Data of Short Arcs

Torsten Mayer-Gürr, Annette Eicker, and Karl Heinz Ilk

Institute of Theoretical Geodesy, University of Bonn, D-53115 Bonn, Germany
tmg@geod.uni-bonn.de

Summary. The signal content in the low-low SST observables of the gravity field twin-satellite mission GRACE (Gravity Recovery And Climate Experiment) varies in the space domain depending on the roughness of the gravity field features. On the one hand, the maximum degree of the spherical harmonic expansion has to be selected as high as possible to bring out the maximum of gravity field information out of the data. On the other hand, an increasing maximal degree deteriorates the stability of the normal equations to solve for the gravity field parameters. Therefore, a trade-off is necessary between the selection of a maximal degree adequate for representing the signal content in the observables, on the one hand, and a maximal degree which can still be recovered without causing instabilities, on the other hand. We propose to integrate the global gravity field recovery with regional gravity field refinements tailored to the specific gravity field features in these regions: In a first step, the gravity field only up to a moderate safely determinable degree is recovered; the specific analysis features tailored to the individual gravity field characteristics in areas of rough gravity field signal will be modelled subsequently by space localizing base functions in a second step. In a final third step, a spherical harmonic expansion up to an (in principle) arbitrary degree can be derived based on a numerical Gauss - Legendre - quadrature procedure without any stability problems. The procedure will be applied in a first example to observations of a GRACE simulation scenario to test the potential capabilities of the approach. A second application demonstrates the determination of a global gravity field model and regional refinements based on low-low SST data of the GRACE twin satellite mission for the August 2003 observations.

Key words: GRACE, SST, low-low, high-low, global gravity field recovery, gravity field refinement, gravity field zoom-in, space localizing base functions, Gauss-Legendre-quadrature

1 Introduction

As a result of the dedicated space-borne gravity field mission GRACE (Gravity Recovery And Climate Experiment – Tapley et al. 2004), in orbit since

2002, a breakthrough in accuracy and resolution of gravity field models has been achieved. Subsequent solutions by using the observations collected over a period of time of, e.g., one month, enables the derivation of time dependencies of the gravity field parameters. The innovative character of this mission lies in the continuous observation of the twin satellites by the Global Positioning System (GPS) and the highly precise line-of-sight range and range-rate K-band measurements between the twin satellites. In addition, the surface forces acting on these satellites are measured and can be considered properly during the recovery procedure. As a result of this mission, the presently best combination static model, EIGEN-CG03C, has been derived from 376 days of GRACE observations and three years of CHAMP (Förste et al., 2005) as well as monthly snap-shots of the gravity field, showing clearly temporal variations of the gravity field closely correlated to the hydrological water cycle. Another GRACE gravity field model is GGM02C, based on the analysis of 363 days of GRACE in-flight data (Centre of Space Research, Austin, - UTEX CSR, 2004, http://www.csr.utexas.edu/grace/gravity). It is represented by a spherical harmonic expansion up to degree 200 and constrained with terrestrial gravity information.

These gravity field models are represented, as usual, in terms of spherical harmonics. A disadvantage of this kind of gravity field modelling is the missing flexibility. Because of the inhomogeneous structure of the gravity field in the space domain the signal to noise ratio varies in the satellite-to-satellite tracking observables depending on the geographical region the satellite is actually passing. The heterogeneity of the gravity field cannot be properly taken into account in case of global solutions based on spherical harmonic expansions. The reason is that the recovery of the gravity field by satellite techniques is an improperly posed problem which requires a proper regularization that influences especially the high frequent spectral part of the gravity field. In most cases, a Tichonov-type regularization is applied which acts globally in case of a gravity field representation by base functions of global support. The regularization parameter can be derived by the L-curve procedure, by cross validation or by a recently proposed method of variance component estimation (Koch and Kusche, 2003; Mayer-Gürr et al., 2005). The establishment of the regularization matrix usually is based on Kaula's rule of thumb or derived from the degree variances of an available gravity field model. The disadvantage of this sort of uniform global regularization is that the regularization factor is selected such that an overall filtering of the observations leads to a mean damping of the global gravity field features. Depending on the procedure for selecting the regularization factor, the consequences are either an over-damping of the rough gravity field features, while the smoother parts would need a slightly stronger regularization to avoid a contamination of the recovery results by observation noise, or vice-versa.

This disadvantageous property suggests a hybrid modelling of the gravity field: the long wavelength features of the gravity field should be represented by a series of spherical harmonics up to a properly selected degree and the gravity field details should be modelled by series of space localizing base functions such as spherical wavelets or harmonic spline functions. The maximum degree of the spherical harmonic representation has to be selected depending on the measurement scenario; the space localizing gravity field parameters have to be recovered by a regionally adapted recovery procedure and an individual regularization which is tailored to the roughness of the gravity field in specific regions of the Earth. There are various possibilities to adapt the space localizing base functions to the individual gravity field features such as the customization of the resolution properties of the base function to enable an optimal fit to the gravity field spectrum to be recovered and the definition of the nodal point distribution necessary to model the individual gravity field structures properly. It is also very easy to include additional conditions for the regional gravity field solutions, such as inclinations of the geoid in special parts, or more general, arbitrary functionals of the gravity field parameters. Furthermore, several regional solutions with global coverage can be merged to obtain a global solution. A spherical harmonic representation, in principle up to an arbitrary maximum degree and only limited by the maximal signal content of the observations, can be obtained by means of numerical quadrature methods now in a direct stable way avoiding the instable inverse procedure. It should be pointed out that in the approach presented here, the above mentioned disadvantages of a global solution by spherical harmonics, derived by an inversion procedure, are avoided, as demonstrated by Eicker et al. (2005).

The mathematical model of our recovery approach is sketched in Sect. 2. Details to the method as far as the global recovery by spherical harmonics are concerned are presented, e.g., in Mayer-Gürr et al. (2005). In Sect. 3 the combination of normal matrices based on individual short arcs is explained as well as the weighting of the normal matrices within a variance component estimation procedure including the computation of the regularization parameter. In Sect. 4, a simulation scenario is presented and the potential capabilities of the method are demonstrated based on a global gravity field recovery from SST range-rate observations. The solutions of a global gravity field recovery from the GRACE low-low SST data of August 2003 with a global coverage of regional refinement patches are presented in Sect. 5. The results are compared to the gravity field models EIGEN-CG03C and GRACE-GGM02C and demonstrate the high quality in the spectral band above degree n=60 of a spherical harmonic expansion. Sect. 6 contains a summary and an outlook for future investigations.

2 Setup of the mathematical model

The gravity field recovery approach tailored to a twin satellite gravity mission of the GRACE type as presented here, is based on Newton's equation of motion, formulated as a boundary value problem in the form of a Fredholm type integral equation. This idea has been proposed as a general method for orbit determination by Schneider in 1967 (Schneider, 1968), modified for gravity field determination by Schneider and Reigber (1969), investigated in detail by Reigber (1969) and successfully applied subsequently. In the following, the idea has been applied to the satellite-to-satellite tracking (SST) problem by Ilk (1984) and later to the satellite gravity gradiometry (SGG) analysis. After that, the method has been developed and tested based on various simulation scenarios, e.g. see Ilk et al. (1995). As first real data applications, the gravity field models ITG-CHAMP01 and ITG-CHAMP02 have been derived based on this method, applied to the analysis of kinematical short arcs (Mayer-Gürr et al., 2005). The basic characteristic of this method is the use of short arcs for regional and global gravity field recovery applications. The complete recovery procedure consists of three steps which can be applied independently as well:

- Global gravity field recovery based on a spherical harmonic expansion up to a moderate degree to provide a first global reference model as a basis for further refinements,
- Regional refinements of the gravity field by spherical splines as space localizing base functions, adapted to the specific gravity field features, if possible covering the globe,
- Determination of a global gravity field model by merging the regional refinement solutions and deriving potential coefficients by a numerical quadrature technique.

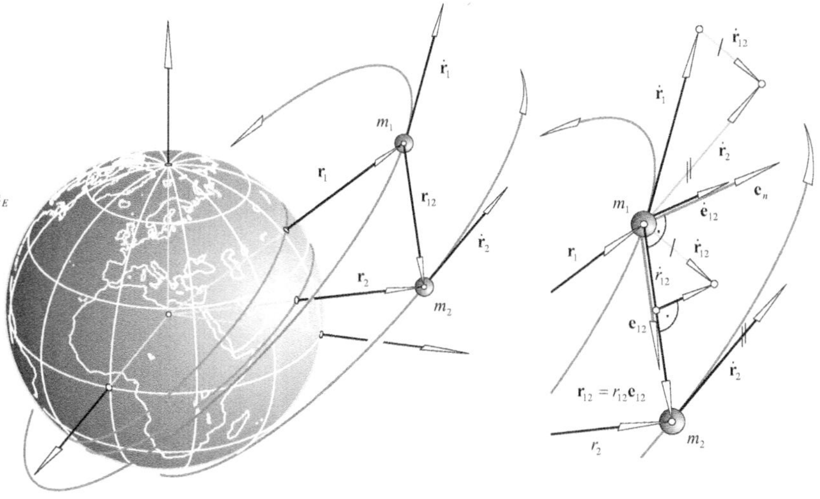

Fig. 1. Low-low satellite-to-satellite tracking experiment

2.1 The equation of relative motion for twin satellites

If precise intersatellite functionals as line-of-sight ranges or range-rate measurements are available, as in case of the GRACE mission, the mathematical model can be based on Newton's equation of motion for the line-of-sight distance (Fig. 1),

$$\ddot{r}_{12}(t) = \frac{\dot{\mathbf{r}}_{12}^2 - \dot{r}_{12}^2}{r_{12}} + \mathbf{e}_{12} \cdot \mathbf{g}(t; \mathbf{r}_{12}, \mathbf{r}_1, \dot{\mathbf{r}}_1, \dot{\mathbf{r}}_2; \mathbf{x}), \tag{1}$$

formulated as a boundary value problem,

$$\begin{aligned} r_{12}(t) &= (1-\tau)\, r_{12,A} + \tau r_{12,B} - \\ &- T^2 \int\limits_{\tau'=0}^{1} K(\tau,\tau') \left(\frac{\dot{\mathbf{r}}_{12}^2 - \dot{r}_{12}^2}{r_{12}} + \mathbf{e}_{12} \cdot \mathbf{g}(t; \mathbf{r}_{12}, \mathbf{r}_1, \dot{\mathbf{r}}_1, \dot{\mathbf{r}}_2; \mathbf{x}) \right) d\tau', \end{aligned} \tag{2}$$

satisfying the boundary values,

$$r_{12,A} := r_{12}(t_A), \quad r_{12,B} := r_{12}(t_B), \quad t_A < t_B. \tag{3}$$

The quantity $K(\tau, \tau')$ is the integral kernel,

$$K(\tau,\tau') = \begin{cases} \tau\,(1-\tau'), & \tau \le \tau', \\ \tau'\,(1-\tau), & \tau' \le \tau, \end{cases} \tag{4}$$

with the normalized time variable,

$$\tau = \frac{t - t_A}{T} \quad \text{with} \quad T = t_B - t_A, \quad t \in [t_A, t_B]. \tag{5}$$

The solution can be given in spectral form as follows,

$$r_{12}(t) = (1-\tau)\, r_{12,A} + \tau r_{12,B} + \sum_{\nu=1}^{\infty} r_{12,\nu} \sin(\nu\pi\tau), \tag{6}$$

with the coefficients $r_{12,\nu}$, $\nu = 1, 2, ..., \infty$,

$$\begin{aligned} r_{12,\nu} = -\frac{2T^2}{\pi^2\nu^2} \int\limits_{\tau'=0}^{1} \sin(\nu\pi\tau') \cdot \\ \cdot \left(\frac{\dot{\mathbf{r}}_{12}^2 - \dot{r}_{12}^2}{r_{12}} + \mathbf{e}_{12} \cdot \mathbf{g}(\tau'; \mathbf{r}_{12}, \mathbf{r}_1, \dot{\mathbf{r}}_1, \dot{\mathbf{r}}_2; \mathbf{x}) \right) d\tau'. \end{aligned} \tag{7}$$

The coefficients $\mathbf{r}_\nu$, $\nu = 1, 2, ..., \infty$, of (6) can be derived from the right hand side of (1) (for details see Ilk et al., 1995). The specific force function,

$$\begin{aligned} \mathbf{g}(\tau'; \mathbf{r}_{12}, \mathbf{r}_1, \dot{\mathbf{r}}_1, \dot{\mathbf{r}}_2; \mathbf{x}) = \mathbf{g}_d(\tau'; \mathbf{r}_1, \mathbf{r}_2, \dot{\mathbf{r}}_1, \dot{\mathbf{r}}_2) + \nabla V_{(12)E}(\tau'; \mathbf{r}_{12}, \mathbf{r}_1; \mathbf{x}_0) + \\ + \nabla T_{(12)E}(\tau'; \mathbf{r}_{12}, \mathbf{r}_1; \Delta\mathbf{x}), \end{aligned} \tag{8}$$

with the gravity field parameters $\mathbf{x}$ can be separated in a disturbance part $\mathbf{g}_d$, which represents the non-conservative disturbing forces, in a reference part $\nabla V_{(12)E}$, modelled by the tidal potential of the Earth (E) acting on the satellites 1 and 2,

$$\nabla V_{(12)E}(\tau';\mathbf{r}_{12},\mathbf{r}_1;\mathbf{x_0}) = \nabla\left(V\left(\mathbf{r}_1+\mathbf{r}_{12}\right) - V\left(\mathbf{r}_1\right)\right), \tag{9}$$

representing the long-wavelength gravity field features and in an anomalous part $\nabla T_{(12)E}$,

$$\nabla T_{(12)E}(\tau';\mathbf{r}_{12},\mathbf{r}_1;\Delta\mathbf{x}) = \nabla\left(T\left(\mathbf{r}_1+\mathbf{r}_{12}\right) - T\left(\mathbf{r}_1\right)\right), \tag{10}$$

modeling the high frequent refinements and parameterized either by corrections $\Delta\mathbf{x}$ to the global gravity field parameters $\mathbf{x_0}$ or by parameters $\Delta\mathbf{x}$ of a linear approximation with space localizing base functions.

The coefficients $r_{12,\nu}$, $\nu = 1, 2, ..., \infty$, of (6) can be derived by inter-satellite measurements of different types, e.g., in case of relative accelerations,

$$r_{12,\nu} = -\frac{2T^2}{\pi^2\nu^2}\int\limits_{\tau'=0}^{1} \sin\left(\nu\pi\tau'\right)\ddot{r}_{12}(\tau')\,d\tau', \tag{11}$$

and/or in case of range-rate measurements,

$$r_{12,\nu} = \frac{2T}{\pi\nu}\int\limits_{\tau'=0}^{1} \cos\left(\nu\pi\tau'\right)\dot{r}_{12}(\tau')\,d\tau', \tag{12}$$

and/or in case of inter-satellite range observations,

$$r_{12,\nu} = 2\int\limits_{\tau'=0}^{1} \sin\left(\nu\pi\tau'\right)\left(r_{12}(\tau') - (1-\tau')\,r_{12,A} - \tau' r_{12,B}\right)\,d\tau', \tag{13}$$

respectively.

There is a space domain model based on (2) with the force function according to (8) or a spectral domain model based on (7) and the spectral observations according to (11) to (13). An alternative to this approach starts with Newton's equation of relative motion as follows,

$$\ddot{\mathbf{r}}_{12}(t) = \mathbf{g}(t;\mathbf{r}_{12},\mathbf{r}_1,\dot{\mathbf{r}}_1,\dot{\mathbf{r}}_2;\mathbf{x}). \tag{14}$$

The formulation as a boundary value problem reads, analogously to (2)

$$\mathbf{r}_{12}(t) = (1-\tau)\,\mathbf{r}_{12,A} + \tau\mathbf{r}_{12,B} - T^2\int\limits_{\tau'=0}^{1} K\left(\tau,\tau'\right)\mathbf{g}(t;\mathbf{r}_{12},\mathbf{r}_1,\dot{\mathbf{r}}_1,\dot{\mathbf{r}}_2;\mathbf{x})\,d\tau'. \tag{15}$$

The relative velocity can be derived by differentiation with respect to the time,

$$\dot{\mathbf{r}}_{12}(t) = \frac{1}{T}\left(\mathbf{r}_{12,B} - \mathbf{r}_{12,A}\right) - T \int\limits_{\tau'=0}^{1} \frac{dK(\tau,\tau')}{d\tau}\, \mathbf{g}(\tau'; \mathbf{r}_{12}, \mathbf{r}_1, \dot{\mathbf{r}}_1, \dot{\mathbf{r}}_2; \mathbf{x})\, d\tau'. \quad (16)$$

The mathematical model for range observations can be derived by projecting the relative vector to the line-of-sight connection in combination with (15),

$$r_{12}(\tau) = \mathbf{e}_{12}(\tau) \cdot \mathbf{r}_{12}(\tau). \quad (17)$$

Analogously, the mathematical model for range-rate measurements in combination with (16) reads as follows,

$$\dot{r}_{12}(\tau) = \mathbf{e}_{12}(\tau) \cdot \dot{\mathbf{r}}_{12}(\tau). \quad (18)$$

In both equations, $\mathbf{e}_{12}$ is the unit vector in the line-of-sight direction (Fig. 1). This vector is known with high accuracy, assuming that the satellite positions are measured with an accuracy of a few cm and taking into account the distance of approximately 200km between the two satellites.

For all model alternatives the normal equations can be established and solved by a regularized solver of Tichonov type, where the regularization parameter is preferably computed according to the variance component estimation procedure of Koch and Kusche (2003).

2.2 Gravity field representation

The reference potential according to (9) can be formulated in the usual way as follows,

$$V = \frac{GM_E}{r} \sum_{n=0}^{n_{\max}} \sum_{m=0}^{n} \left(\frac{R_E}{r}\right)^n \left(c_{nm} C_{nm}(\vartheta, \lambda) + s_{nm} S_{nm}(\vartheta, \lambda)\right), \quad (19)$$

with the surface spherical harmonics,

$$C_{nm}(\vartheta, \lambda) = P_{nm}(\cos\vartheta)\cos m\lambda, \quad S_{nm}(\vartheta, \lambda) = P_{nm}(\cos\vartheta)\sin m\lambda. \quad (20)$$

The anomalous potential $T(\tau'; \mathbf{r}, \Delta\mathbf{x})$ according to (10) reads for a global gravity field recovery,

$$T = \frac{GM_E}{r} \sum_{n=2}^{N_{\max}} \sum_{m=0}^{n} \left(\frac{R_E}{r}\right)^n \left(\Delta c_{nm} C_{nm}(\vartheta, \lambda) + \Delta s_{nm} S_{nm}(\vartheta, \lambda)\right), \quad (21)$$

with the corrections $\Delta c_{nm}, \Delta s_{nm} \in \Delta\mathbf{x}$ to the reference potential coefficients $c_{nm},\ s_{nm} \in \mathbf{x_0}$. In case of a regional recovery the anomalous potential $T(\mathbf{r})$ is modelled by parameters of space localizing base functions,

$$T(\mathbf{r}) = \sum_{i=1}^{I} a_i \, \varphi(\mathbf{r}, \mathbf{r}_{Q_i}), \tag{22}$$

with the unknown field parameters a_i arranged in a column matrix $\Delta\mathbf{x} := (a_i,\ i = 1, ..., I)^T$ and the base functions,

$$\varphi(\mathbf{r}, \mathbf{r}_{Q_i}) = \sum_{n=0}^{N_{max}} k_n \left(\frac{R_E}{r}\right)^{n+1} P_n(\mathbf{r}, \mathbf{r}_{Q_i}). \tag{23}$$

The coefficients k_n are the difference degree variances of the gravity field spectrum to be determined minus the reference gravity field ($\Delta\bar{c}_{nm}, \Delta\bar{s}_{nm}$ are the fully normalized potential coefficients),

$$k_n = \sum_{m=0}^{n} \left(\Delta\bar{c}_{nm}^2 + \Delta\bar{s}_{nm}^2\right). \tag{24}$$

R_E is the mean equator radius of the Earth, r the distance of a field point from the geo-centre and $P_n(\mathbf{r}, \mathbf{r}_{Q_i})$ are the Legendre polynomials depending on the spherical distance between a field point P and the nodal points Q_i of the set of base functions. The maximum degree N_{max} in (23) should correspond to the envisaged maximum resolution expected for the regional recovery; in the following examples this maximum degree is selected as $N_{max} = 120$. With the definition in (23) the base functions $\varphi(\mathbf{r}, \mathbf{r}_{Q_i})$ can be interpreted as isotropic and homogeneous harmonic spline functions (Freeden et al., 1998). The nodal points are defined on a grid generated by a uniform subdivision of an icosahedron of twenty equal-area spherical triangles. In this way the global pattern of spline nodal points Q_i shows approximately uniform nodal point distances. Details can be found e.g. in Eicker et al. (2005).

3 Solution of the combined normal equations

For the analysis of GRACE observations not only the gravity field parameters have to be estimated, but also arc-related parameters as for example the two boundary position vectors of each arc. These parameters sum up to about 27000 additional unknowns for an analysis period of one month in case of short arcs with a mean arc length of approximately 30 minutes. To reduce the size of the normal equation matrices, the arc-related parameters are eliminated before the arcs are merged to the complete system of normal equations. Every short arc i builds a (reduced) partial system of normal equations $\mathbf{N}_i$. To combine the normal equation matrices for the short arcs, separate variance factors σ_i for each arc will be determined, to consider the variable precision of the range and range-rate observations. Furthermore, because of the intrinsic stability problems of the gravity field recovery process, an additional regularization factor σ_x and a regularization matrix $\mathbf{N}_x$ will be introduced into the

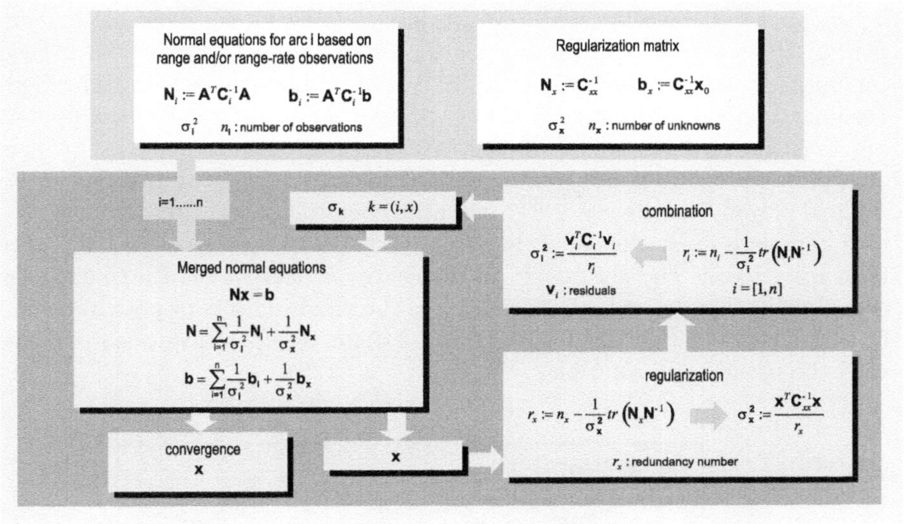

Fig. 2. Merging of normal equations for the short arcs within the iterative variance-covariance computation procedure including the determination of the regularization parameter

gravity field recovery procedure. The variance factors are computed by means of a variance component estimation procedure described by Koch and Kusche (2003). The iterative combination scheme combined with a variance component estimation and the computation of the regularization factor is shown in Fig. 2; for details of the iterative procedure refer to Mayer-Gürr et al. (2005).

4 Gravity field recovery within a simulation scenario

4.1 Simulation scenario

The recovery procedure has been tested based on a simulation scenario. It shall demonstrate the capability of the recovery approach using controlled error measures of the orbits and the inter-satellite observations. Nearly circular orbits of the GRACE twin satellites with a mean altitude of 490km and a mean distance between the two satellites of approximately 230km are generated. A pseudo real gravity field has been used for the orbit computations represented by a spherical harmonic expansion up to degree n=180. The potential coefficients have been taken over from the EGM96 gravity field model (Lemoine et al., 1998). The satellite positions are generated every 5 seconds covering a 30 days mission period. Each position coordinate is corrupted by white noise with an RMS of 3cm. Two different error scenarios for the observables have been investigated: For ***case 1***, the range-rate measurements between the twin satellites are corrupted by white noise with an RMS of

$0.2\mu m/s$ and the accelerations are considered to be measured with an accuracy of $10^{-9}m/s^2$. For ***case 2***, only the range-rate observations have been corrupted by white noise with an RMS of $1\mu m/s$, while the accelerations have been considered as error-free. For this test only range-rates have been used as observations, no ranges. The 30-days-orbit has been split into 1500 short arcs of approximately 30 minutes arc length. The total number of unknown gravity field parameters are the 19877 potential coefficients beginning from degree n=2 complete up to degree n=140. The first four potential coefficients have been fixed to one and zero, to force the centre of mass of the Earth onto the origin of the Earth-fixed reference frame. The resulting sets of potential coefficients have been truncated at degree n=110 for the subsequent comparisons with the pseudo-real solution.

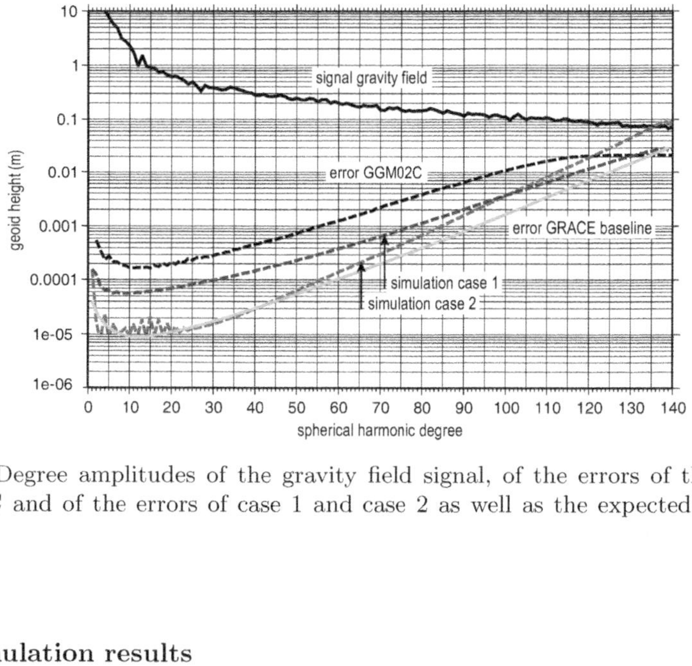

Fig. 3. Degree amplitudes of the gravity field signal, of the errors of the model GGM02C and of the errors of case 1 and case 2 as well as the expected GRACE baseline

4.2 Simulation results

The mathematical model (18) with (16) based on a spherical harmonic gravity field representation according to (8) to (10) with (19) and (21) has been applied for the gravity recovery. Fig. 3 shows the degree amplitudes of the gravity field signal, of the errors of the model GGM02C and of the errors of case 1 and case 2 as well as the expected GRACE baseline. The error degree variances of case 2 approximate the baseline quite well in the lower spectral range while the error behaviour of case 1 comes closer to the baseline in the higher frequencies. This means that the long wavelength features of the recovery result are strongly affected by the accuracy of the acceleration measurements while the effect in the higher frequencies are increasingly less influenced. A high range-rate accuracy can be exploited only if the accelerations are known adequately. The range-rate accuracy of case 1 seems to be realistic for the GRACE mission. Therefore, one can state that the baseline has been achieved under simplified error model assumptions. It is interesting to have a closer look at the geoid differences for both cases in the space domain. Obviously, the stripe pattern of the differences to the pseudo-real solution in Color Fig. XVII on p.297 is caused mainly by the noise of the acceleration measurements, while Color Fig. XVIII on p. 297 shows the typical instability effects of the downward continuation process caused by the noise in the range-rate measurements. This holds even in case of a moderate maximum spherical harmonic degree of n=110. Nevertheless, it becomes obvious that the resolution of the recovered solution can be extended to a higher degree than the selected one (degree n=110) as Fig. 3 shows. The maximum resolution would be reached at the point where the error degree amplitude graph would intersect the signal degree amplitude graph.

5 Gravity field recovery from GRACE range-rate measurements of August 2003

5.1 Data set

The following recovery results refer to the K-band range-rate measurements of the GRACE twin satellite mission for the month August 2003. The observations are corrected for the tides caused by Sun, the Moon and the planets. The ephemerides are taken from the JPL405 data set. Effects originating from the deformation of the Earth caused by these tides are modelled following the IERS 2003 conventions. Ocean tides are computed from the FES2004 model. Effects of high frequency atmosphere and ocean mass redistributions are removed prior to the processing by the GFZ AOD dealiasing products. The 30-days-orbit has been split into 1500 short arcs of approximately 30 minutes arc length. For each arc the coordinates of the boundary vectors have been determined as well as an accelerometer bias.

5.2 Global solution

In the first step, a global spherical harmonic solution up to degree n=90 beginning from degree n=3 has been determined for the month August 2003 from the GRACE range-rate measurements, in the following designated as gravity field model "ITG-GRACE-2003-08". The mathematical model (18) with (16) based on a pure spherical harmonic gravity field representation according to (8) and (9) with the spherical harmonic model (19) has been applied. The arc-related parameters are eliminated before merging the normal equations for each short arc to the total system of normal equations as outlined in Sect. 3. The results are compared to the gravity field models EIGEN-CG03C (GeoForschungsZentrum Potsdam - Förste et al. 2005) and GGM02C (Centre of Space Research, Austin, - UTEX CSR, 2004, http://www.csr.utexas.edu/grace/gravity).

The geoid height differences of our model ITG-GRACE-2003-08 and the CSR model GGM02C are shown in Color Fig. XIX on p. 298: RMS: 2.6cm, avg: 2.0cm, min/max: -12.7/14.7cm. The geoid height differences with the GFZ combination model EIGEN-CG03C show similar results: RMS: 2.7cm, avg: 2.1cm, min/max: -12.4/12.5cm while the GFZ and the CSR models coincide slightly better: RMS: 2.0cm, avg: 1.6cm, min/max: -12.8/11.1cm. But one has to keep in mind that the models EIGEN-CG03C and GGM02C contain much more data covering a considerably longer mission period while our solution has been derived from only one month range-rate observations. Furthermore, the model EIGEN-CG03C has been derived as a spherical harmonic expansion up to degree n=360, including GRACE and CHAMP data as well as terrestrial gravity and altimetry data. Similarly, the CSR model has been determined as a spherical harmonic expansion up to degree n=200 from GRACE data and constrained by terrestrial data. Despite the fact that the comparisons between all these models have been performed only up to degree n=90, the quality of our model is remarkably well. In some areas such as in the Central Asian region, our solution coincides better with the model GGM02C than the model EIGEN-CG03C.

5.3 Regional solutions

For the regional refinement solutions the same mathematical model as used for the global solution and formulated in (18) with (16) has been applied except for the gravity field representation. Based upon the global solution the additional gravity field refinements are represented according to (8) with (10) represented by spherical spline functions according to (22). To avoid geographical truncation effects at the region boundaries, gravity field parameters defined in an additional strip around the specific regions have to be taken into account; the width of the strip depends on the approximation of the real field by the global reference field; in most cases a strip of 10 ° is sufficient. The nodal points are located at a regular grid with a mean distance between the nodal points of approximately 160km. This grid is generated by a uniform subdivision of an icosahedron of twenty spherical equal-area triangles. The regional refinement resolution corresponds approximately to the degree n=120 of a spherical harmonic representation of the residual gravity field. Fig. 4 shows the recovered gravity anomalies of the Himalayan region with its extreme rough gravity field features. A comparison with the gravity anomalies derived from the GFZ combination model EIGEN-CG03C, evaluated at a uniform grid of 1 ° resolution, results in an RMS of 1.32mGal, an average value of 1.06mGal and minimum/maximum differences of -5.45/5.73mGal. The corresponding values with respect to the CSR model GGM02C are slightly better as follows: RMS 1.18mGal, avg 0.97mGal and min/max -3.62/4.00mGal.

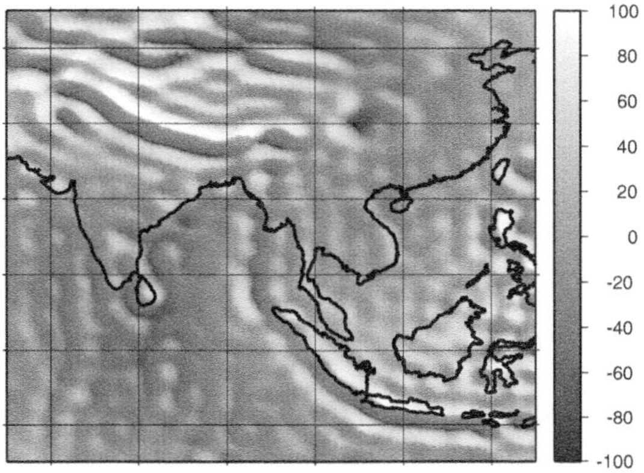

Fig. 4. Regional solution in the Himalayan area, gravity anomalies in mGal

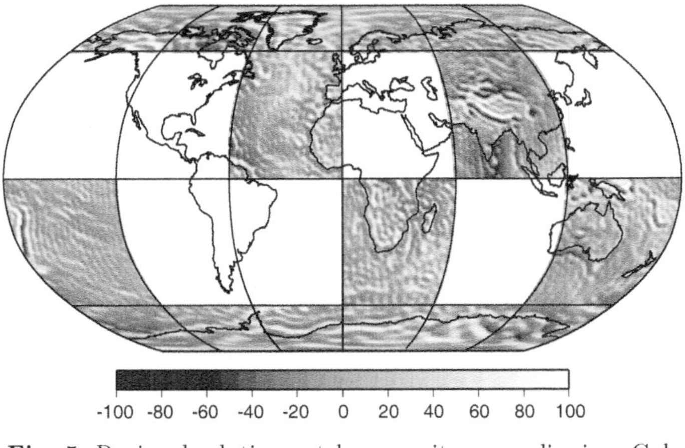

Fig. 5. Regional solution patches, gravity anomalies in mGal

5.4 Merged solutions

To combine the advantages of the satellite mission GRACE to cover almost perfectly the complete Earth with the advantages of regional focussing techniques - as pointed out earlier - the surface of the Earth has been divided into patches defined by the latitudes 60 ° and -60 °, as well as by the meridians 0 °, 60 °, 120 °, 180 °, 240 ° and 300 ° (Fig. 5). For the regional gravity field recovery only the satellite data over the respective regions have been used. To avoid geographical truncation effects at the boundaries of the regions, an additional strip of 10° has been taken into account. For each patch the residual gravity field has been recovered individually applying a tailored regularization where the regularization factor as well as the variance factors for the short arcs crossing the regions are determined by a variance component estimation procedure as outlined in Sect. 3.

To derive a global gravity field model represented by spherical harmonics without losing the details of a regional zoom-in, the refined disturbing potential values are calculated at points of a specific grid, the so-called Gauss–Legendre–Grid. It has equi-angular spacing along circles of latitude; along the meridian the nodes are located at the zeros of the Legendre polynomials of degree $N+1$. Then the potential coefficients of the spherical harmonic expansion are calculated by means of the Gauss-Legendre-Quadrature (see for example Stroud and Secrest, 1966). This quadrature method has the advantage of maintaining the orthogonality of the Legendre functions despite the discretization procedure, which allows an exact calculation of the potential coefficients,

$$\begin{Bmatrix} c_{nm} \\ s_{nm} \end{Bmatrix} = \frac{R_E}{GM4\pi} \sum_{k=1}^{K} T_k P_{nm}(\cos\vartheta_k) \begin{Bmatrix} \cos(m\lambda_k) \\ \sin(m\lambda_k) \end{Bmatrix} w_k, \tag{25}$$

with the area weights

$$w_k = \frac{2}{(1-t_k^2)\left(P'_{N+1}(\cos\vartheta_k)\right)^2} \quad . \tag{26}$$

The functionals T_k are the disturbing gravitational potential values at the K nodes of the quadrature, P_{nm} are the associated Legendre functions and P'_{N+1} the first derivatives of the Legendre polynomials of degree $N+1$ with respect to ϑ. N is the maximum spherical harmonic degree to be determined.

The geoid height differences of our merged model ITG-GRACE-2003-08 and the CSR model GGM02C are shown in Color Fig. XX on p. 298. The differences show an RMS of 7.27cm, an average value of 5.61cm and minimum/-maximum values of -28.7/30.6cm. The corresponding values with respect to the GFZ combination model EIGEN-CG03C (Color Fig. XXI on p. 298) are slightly worse with an RMS of 8.92cm, an average value of 6.88cm and min/-max values of -44.2/42.7cm. The GFZ and the CSR model coincide slightly better as Color Fig. XXII on p. 299 demonstrates. The RMS is 6.20cm, the average value 4.81cm and the minimum/maximum values are -35.9/36.1cm. But again, the models EIGEN-CG03C and GGM02C are not directly comparable to our one-month solution. Despite the fact that the comparisons between all these models have been performed only up to degree n=110, the quality of our model is remarkably well. In some areas is the coincidence of our solution with GGM02C better than the coincidence of GGM02C with EIGEN-CG03C, e.g., in the Central Asian region and in the polar areas.

Difference degree variances of different 2003-08-solutions with respect to the CSR gravity field model GGM02C are shown in Fig. 6 and with respect to the GFZ model EIGEN-CG03C in Fig. 7. These two gravity field models can be considered superior in quality with respect to the monthly 2003-08-solutions ITG-GRACE-2003-08, CSR-GRACE-2003-08 and GFZ-GRACE-2003-08. The difference degree variance graphs of our solution ITG-GRACE-2003-08 with respect to GGM02C and EIGEN-CG03C show slightly better results in the spectrum from degree n=70 upwards than the CSR and GFZ monthly solutions, more or less identical results in the spectral band from n=50 to 70 and still slightly worse results in the long and medium wavelength features.

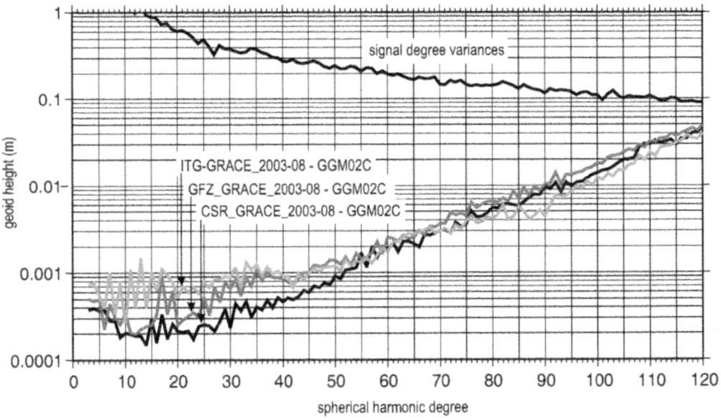

Fig. 6. Difference degree variances of different 2003-08-solutions with respect to GGM02C

This result and also the stripe pattern of the differences in Color Fig. XX and Color Fig. XXI on p. 298 could indicate some deficiencies in the bias parameter determination of the accelerometer measurements. This can be concluded from the simulation examples. Another reasons are inaccurate dealiasing products used for this analysis. But there are still numerous further reasons which have to be investigated in more details. The additional use of range measurements and the analysis of the kinematic arcs of the GRACE twin satellites may have the potential to improve the long and medium wavelength features of our recovery results.

6 Conclusions and outlook

The use of short arcs for gravity field recovery, based on the solution of Newton's equation of relative motion, formulated as a boundary value problem of Fredholm type is an adequate recovery technique for the processing of SST observations of the low-low types range and range-rates. The results achieved in

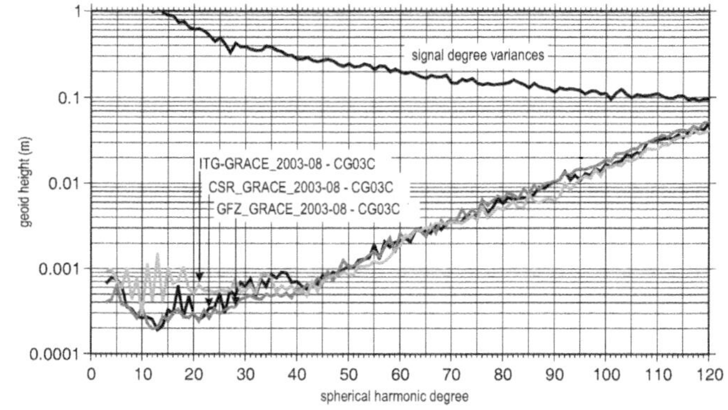

Fig. 7. Difference degree variances of different 2003-08-solutions with respect to EIGEN-CG03C

this investigation but also the successful application for the processing of kinematical arcs as an alternative to the SST high-low gravity field determination modus (cf. Mayer-Gürr et al. 2004) underline the usefulness of our gravity field recovery approach. This approach is not only very flexible in terms of various observation types of the new gravity satellite missions, moreover, it combines the advantages of a global gravity field determination with a regional gravity field zoom-in. The three computation steps a) global gravity field recovery based on a spherical harmonic expansion up to a moderate degree to provide a first global reference model as a basis for further refinements, b) regional refinements of the gravity field by spherical splines as space localizing base functions, adapted to the specific gravity field features, and c) determination of a global gravity field model by merging the regional refinement solutions and deriving potential coefficients by a numerical quadrature technique fulfil all expectations for a flexible gravity field recovery technique. Furthermore, the method is modest in terms of computation costs, as the complete global recovery problem is split up into much smaller partial problems. The computation of global gravity field models up to an arbitrary degree (only limited by the signal content of the observables) is possible on a single PC, because the stability properties of the numerical quadrature procedure do not limit the resolution to an upper degree. Together with the determination of a global solution regional zoom-ins of the gravity field are computed during the recovery procedure as well. This assures that all signal information of the observations is extracted in a best possible way.

Further improvements are expected with respect to the following aspects: a) joint use of ranges and range-rates combined with the analysis of the kinematical arcs of the GRACE twin satellites, b) refining the regularization strategy to enable smoother transition zones between the zoom-in-regions, c) tailoring the zoom-in areas more accurately to the demand of the gravity field features in the specific regions, d) more precise selection of the base functions and the nodal point distribution adapted to the roughness of the gravity field, possibly combined with a multi-resolution strategy, e) careful investigation of the aliasing effects originating from the patching of several regional solutions, f) homogeneization of the regional solutions to avoid long wavelength errors. Besides these topics, the transformation of the mathematical model into spectral domain may enable further improvements.

Acknowledgement. We are grateful to GFZ Potsdam for providing the data for this investigation and to the Center of Space Research (Austin, USA) for providing the gravity field models. The support of BMBF (Bundesministerium für Bildung und Forschung) and DFG (Deutsche ForschungsGemeinschaft) of the GEOTECHNOLOGIEN programme is gratefully acknowledged. This is publication GEOTECH-150 of the programme GEOTECHNOLOGIEN of BMBF and DFG, Grant 03F0326C.

References

Eicker A, Mayer-Gürr T, Ilk, KH (2005) An integrated global/regional gravity field determination approach based on GOCE observations, this volume

Förste C, Flechtner F, Schmidt R, Meyer U, Stubenvoll R, Barthelmes F, Neumayer KH, Rothacher M, Reigber C, Biancale R, Bruinsma S, Lemoine JM, Raimondo JC (2005) A New High Resolution Global Gravity Field Model Derived From Combination of GRACE and CHAMP Mission and Altimetry/Gravimetry Surface Gravity Data, Poster presented at EGU General Assembly 2005, Vienna, Austria, 24-29, April 2005

Freeden W, Gervens T, Schreiner M (1998) Constructive Approximation on the Sphere, Oxford University Press, Oxford

Ilk KH (1984) On the analysis of satellite-to-satellite tracking data, Proceedings of the International Symposium on Space Techniques for Geodesy, pp. 59, Sopron

Ilk KH, Rummel R, Thalhammer M (1995) Refined Method for the Regional Recovery from GPS/SST and SGG, CIGAR III/2, ESA contract No. 10713/93/F/FL, European Space Agency

Koch KR, Kusche J (2003) Regularization of geopotential determination from satellite data by variance components, Journal of Geodesy 76(5):259-268

Lemoine FG, Kenyon SC, Factor JK, Trimmer RG, Pavlis NK, Chinn DS, Cox CM, Klosko SM, Luthcke SB, Torrence MH, Wang YM, Williamson RG, Pavlis EC, Rapp RH, Olson TR (1998) The development of the joint NASA GSFC and the National Imagery and Mapping Agency (NIMA) geopotential model EGM96, NASA/TP-1998-206861, Goddard Space Flight Center, Grenbelt, MD

Mayer-Gürr, T., Ilk, K.H., Eicker, A., (2004): ITG-CHAMP02: An Improved Gravity Field Model from a Two-Year Observation Period, New Satellite Mission Results for the Geopotential Fields und Their Variations, Proceedings Joint CHAMP/-GRACE Science Meeting, GFZ Potsdam, July 6-8

Mayer-Gürr T, Ilk, KH, Eicker A, Feuchtinger M (2005) ITG-CHAMP01: A CHAMP Gravity Field Model from Short Kinematical Arcs of a One-Year Observation Period, Journal of Geodesy (2005) 78:462-480

Reigber C (1969) Zur Bestimmung des Gravitationsfeldes der Erde aus Satellitenbeobachtungen, DGK, Reihe C, Heft Nr. 137, München

Schneider M, (1968) A General Method of Orbit Determination, Royal Aircraft Translation No. 1279, Ministry of Technology, Farnborough Hants, England

Schneider M, Reigber C (1969) On the Determination of field parameters using a generalized Fourier-Analysis, In: B. Morando (ed.) Dynamics of Satellites, Symposium Prague COSPAR - IAU - IAG/IUGG - IUTAM, Springer-Verlag Berlin 1970

Stroud AH, Secrest D (1966) Gaussian Quadrature Formulas, Prentice –Hall, Englewood Cliffs, N.J.

Tapley BD, Bettadpur S, Watkins M, Reigber Ch (2004) The gravity recovery and climate experiment: mission overview and early results. Geophys Res Lett 31, L09607: doi10.1029/2004GL019920

Mapping Earth's Gravitation Using GRACE Data

Pavel Novák, Gerrit Austen, Mohammad A. Sharifi, and Erik W. Grafarend

Stuttgart University, Department of Geodesy and Geoinformatics,
Geschwister-Scholl-Strasse 24/D, 70174 Stuttgart, Germany
pnovak@gis.uni-stuttgart.de

Summary. This article describes an approach for global mapping of the Earth's gravitational field developed, tested and successfully implemented at the Geodetic Institute of the Stuttgart University. The method is based on the Newtonian equation of motion that relates satellite-to-satellite tracking (SST) data observed by the two satellites of the Gravity Recovery And Climate Experiment (GRACE) directly to unknown spherical harmonic coefficients of the Earth's gravitational potential (geopotential). Observed values include SST data observed both in the low-low (inter-satellite range, velocity and acceleration) and the high-low (satellites' positions) mode. The low-low SST data specific for the time being to the GRACE mission are available through a very sensitive K-band ranging system. The high-low SST data are then provided by on-board Global Positioning System (GPS) receivers. The article describes how the mathematical model can be modified. The geopotential is approximated by a truncated series of spherical harmonic functions. An alternative approach based on integral inversion of the GRACE data into the geopotential is also formulated and discussed. The article also presents sample numerical results obtained by testing the model using both simulated and observed data. Simulation studies suggest that the model has a potential for recovery of the Stokes coefficients up to degree and order 120. Intermediate results from the analysis of actual data have a lower resolution.

Key words: GRACE mission, geopotential, satellite-to-satellite tracking, Stokes's coefficients, Green's integrals

1 Introduction

Current satellite missions dedicated to the global mapping of the Earth's gravitational field provide data that can be inverted into global models of the gravitational potential (geopotential). The US-German satellite mission Gravity Recovery And Climate Experiment (GRACE) launched on March 17, 2002 is based on two spacecraft following each other along almost circular (approximate eccentricity of 0.001), low (decaying altitude during the lifetime of the mission from 500 to 300 km) and polar orbits (approximate inclination of 89°). The satellites fly approximately 220 ± 50 km apart. During the expected five-year lifetime of the mission, the relative motion of the two satellites is measured using a very sensitive K-band ranging system that senses changes in separation between the two satellites. Moreover, the GRACE satellites are equipped with geodetic-quality Global Positioning System (GPS) receivers allowing for accurate orbit determination as well as accelerometers measuring non-gravitational forces affecting the satellites. The mission is thus based on satellite-to-satellite tracking (SST) in both the low-low (LL) and high-low (HL) mode. More information on the mission can be found for example in (Tapley and Reigber, 1998).

This article discusses the inversion of the GRACE SST data into the geopotential and its parameters. The method is based on their direct relationship through the Newtonian equation of motion (Reubelt et al., 2003). Three modifications of this equation are proposed: (i) balancing the vector of the inter-satellite acceleration differences with corresponding differences of the geopotential gradients, both projected into the inter-satellite direction, and representing the geopotential in a form of a finite spherical harmonics series expansion (e.g. Rummel et al. , 1978; Austen and Grafarend, 2005); (ii) Approach (i) modified by expanding the gradient of the geopotential at the barycentre of the two GRACE satellites, i.e., dealing with the GRACE satellites as a large one-directional gradiometer (Keller and Hess, 1998); (iii) replacing the gradient difference of the geopotential by a boundary integral of the Abel-Poisson type and solving the geopotential at a reference surface of known geometry. While the first two approaches aim at recovery of spectral description of the geopotential (Stokes's coefficients), the last approach allows for description of the geopotential on a known boundary surface outside Earth's masses.

2 SST data of type GRACE

In this section, the GRACE observables are shortly reviewed. Starting with the HL-SST, positions of the GRACE satellites in the terrestrial (Earth-fixed) geocentric coordinate system can be derived from GPS code and carrier beat phase observations. Corresponding position vectors are defined in terms of the geocentric terrestrial Cartesian coordinates (vectors and matrices are bold)

$$\mathbf{x} = [\, x \quad y \quad z \,]^T \,. \tag{1}$$

Note that the argument of time related to all values in (1) was omitted. The space segment of the GRACE mission consists of the two satellites following each other along a similar orbit. Let us assign the leading satellite with the index 1 and the trailing satellite with the index 2. Positions of the satellites and their derivatives will be then labeled with corresponding indices. Besides the position of the satellites, also their velocity vector as well as the acceleration vector

$$\dot{\mathbf{x}} = [\, \dot{x} \quad \dot{y} \quad \dot{z} \,]^T \;, \qquad \ddot{\mathbf{x}} = [\, \ddot{x} \quad \ddot{y} \quad \ddot{z} \,]^T \;, \tag{2}$$

are being used in the following derivations. An operator δ for differences between parameters corresponding to the positions of the two satellites is used throughout the article. Vector differences in position, velocity and acceleration are then given as follows:

$$\delta\mathbf{x} = \mathbf{x}_2 - \mathbf{x}_1 \;, \qquad \delta\dot{\mathbf{x}} = \dot{\mathbf{x}}_2 - \dot{\mathbf{x}}_1 \;, \qquad \delta\ddot{\mathbf{x}} = \ddot{\mathbf{x}}_2 - \ddot{\mathbf{x}}_1 \;. \tag{3}$$

The LL-SST provides the inter-satellite range, e.g. (Blaha, 1992),

$$\varrho = \sqrt{< \delta\mathbf{x} \mid \delta\mathbf{x} >} = \mid \delta\mathbf{x} \mid \;. \tag{4}$$

Considering the unit vector of the inter-satellite direction

$$\mathbf{e} = [\, e_x \quad e_y \quad e_z \,]^T = \frac{\delta\mathbf{x}}{\varrho} \;, \tag{5}$$

the LL-SST data can be used for derivation of the first-order time derivative of the range called herein inter-satellite velocity e.g. (Blaha, 1992)

$$\dot{\varrho} = < \delta\dot{\mathbf{x}} \mid \mathbf{e} > + < \delta\mathbf{x} \mid \dot{\mathbf{e}} > = < \delta\dot{\mathbf{x}} \mid \mathbf{e} > \;, \tag{6}$$

since

$$< \delta\mathbf{x} \mid \dot{\mathbf{e}} > = 0 \;. \tag{7}$$

Finally, the LL-SST also provides the second-order time derivative of the range called inter-satellite acceleration, e.g. (Rummel, 1980),

$$\ddot{\varrho} = < \delta\ddot{\mathbf{x}} \mid \mathbf{e} > + < \delta\dot{\mathbf{x}} \mid \dot{\mathbf{e}} > = < \delta\ddot{\mathbf{x}} \mid \mathbf{e} > + \frac{\mid \delta\dot{\mathbf{x}} \mid^2 - \dot{\varrho}^2}{\varrho} \;. \tag{8}$$

The second term on the right-hand side of (8) is sometimes referred to as a *velocity correction*. The velocity correction accounts for the fact that the GRACE assembly with its inter-satellite direction $\mathbf{e}$ forms a moving observational system in inertial space.

3 Mathematical model

In the inertial reference frame, acceleration of a unit-mass satellite can be expressed according to Newton's law of motion as follows

$$\ddot{\mathbf{X}} = - \nabla V(\mathbf{X}) + \mathbf{A} \;, \tag{9}$$

where the standard sign convention is used. The vector $\mathbf{A}$ stands for the non-gravitational acceleration vector and completes formally the model. No other gravitating masses are considered in (9) but the Earth's system. No tidal effects were considered during simulations but they completed the model in case of the real data analysis. Vectors in the inertial frame, that is usually approximated by the celestial frame, are denoted by capital letters. The transformation between the two coordinate systems can be done by the well known rotation matrix operator for transformation between the terrestrial and celestial frames (McCarthy and Petit, 2004). The matrix rotating the celestial to terrestrial coordinate system is denoted as $\mathbf{M}$. At the epoch t of the observation

$$\mathbf{x}(t) = \mathbf{M}(t)\,\mathbf{X}(t)\,. \tag{10}$$

It is assumed that the geopotential V in (9) represents the gravitational potential of the rotating Earth, i.e., various direct and indirect tidal effects as well as loading effects can successfully be modelled. This concerns namely the real data analysis. We can write the inter-satellite acceleration vector of the GRACE satellites in the inertial (celestial) reference system as follows:

$$\delta\ddot{\mathbf{X}} = -\,\delta\nabla V(\mathbf{X}) + \delta\mathbf{A}\,, \tag{11}$$

with the inter-satellite non-gravitational acceleration vector $\delta\mathbf{A}$.

Combining (8) and (9), the observation equation for the first approach of type *acceleration differences* can be written in the form

$$\frac{|\,\delta\dot{\mathbf{x}}\,|^2 - \dot{\varrho}^2}{\varrho} - \ddot{\varrho} = <\delta\nabla V(\mathbf{x})\,|\,\mathbf{e}>\,, \tag{12}$$

where the geopotential gradient vectors correspond to the terrestrial geocentric Cartesian coordinate system. The velocity difference $|\,\delta\dot{\mathbf{x}}\,|$, which has to be derived by numerically differentiation from kinematic HL-SST positions, is the critical quantity for the quality of the geopotential solution in any of the proposed models. For its evaluation, see e.g. (Austen and Grafarend, 2005). Note that the non-gravitational acceleration vector difference $\delta\mathbf{A}$ is omitted starting with (12). This component was included neither in the simulated nor real data analysis.

Expanding the gradient of the geopotential at the barycenter of the two GRACE satellites leads to the second approach. Since the GRACE satellites are at any time of a similar mass, the position vector of their barycentre B can approximately be estimated as follows:

$$\mathbf{x}_B \doteq \frac{\mathbf{x}_1 + \mathbf{x}_2}{2}\,. \tag{13}$$

The geopotential gradient at the position of the first satellite can be related to the barycentre

$$\nabla_1 V(\mathbf{x}) \doteq \nabla_B V(\mathbf{x}) + \nabla_B \otimes \nabla_B V(\mathbf{x}) \; (\mathbf{x}_1 - \mathbf{x}_B) \; . \tag{14}$$

The linear term of the expansion was only considered in (14). Similarly

$$\nabla_2 V(\mathbf{x}) \doteq \nabla_B V(\mathbf{x}) + \nabla_B \otimes \nabla_B V(\mathbf{x}) \; (\mathbf{x}_2 - \mathbf{x}_B) \; . \tag{15}$$

Keeping the linear approximation in mind, subtracting (14) from (15) yields

$$\delta \nabla V(\mathbf{x}) \doteq \nabla_B \otimes \nabla_B V(\mathbf{x}) \; \delta \mathbf{x} \; . \tag{16}$$

Substituting (16) into (12) results in the *linear gradiometry equation*

$$\frac{|\, \delta \dot{\mathbf{x}} \,|^2 - \dot{\varrho}^2}{\varrho^2} - \frac{\ddot{\varrho}}{\varrho} \doteq < \nabla_B \otimes \nabla_B V(\mathbf{x}) \; \mathbf{e} \,|\, \mathbf{e} > \; . \tag{17}$$

The linearization error represents the greatest problem for its applicability (Keller and Sharifi, 2005). On the other hand, the contribution of low-degree harmonics is the most dominant one. Therefore, the smaller the low-degree coefficients, the more accurate the linear model approximates.

In this regards, we consider one of the available geopotential models and utilize the low-degree coefficients of the model as a priori information. The right-hand side of (17), for instance, can be written as

$$< \nabla_B \otimes \nabla_B V(\mathbf{x}) \; \mathbf{e} \,|\, \mathbf{e} >$$

$$= < \nabla_B \otimes \nabla_B V_l(\mathbf{x}) \; \mathbf{e} \,|\, \mathbf{e} > + < \nabla_B \otimes \nabla_B V^l(\mathbf{x}) \; \mathbf{e} \,|\, \mathbf{e} > , \tag{18}$$

where V_l and V^l stand for the spheroidal reference field and the corresponding incremental one, respectively. Accordingly, the first term on the right-hand side of (18) is split into the approximate term and the respective correction

$$< \nabla_B \otimes \nabla_B V_l(\mathbf{x}) \; \mathbf{e} \,|\, \mathbf{e} >$$

$$= \underbrace{< \nabla_B \otimes \nabla_B V_{0l}(\mathbf{x}) \; \mathbf{e} \,|\, \mathbf{e} >}_{\text{approximate value}} + \underbrace{< \nabla_B \otimes \nabla_B \Delta V_l(\mathbf{x}) \; \mathbf{e} \,|\, \mathbf{e} >}_{\text{correction}} \; . \tag{19}$$

In order to reduce the linearization error, we retain the correction term and replace the approximate one with an equation similar to (12). Consequently, (17) can be recast into

$$\frac{|\, \delta \dot{\mathbf{x}} \,|^2 - \dot{\varrho}^2}{\varrho^2} - \frac{\ddot{\varrho}}{\varrho} - \frac{1}{\varrho} < \delta \nabla V_{0l}(\mathbf{x}) \,|\, \mathbf{e} >$$

$$\doteq < \nabla_B \otimes \nabla_B \left[\; \Delta V_l(\mathbf{x}) + V^l(\mathbf{x}) \; \right] \; \mathbf{e} \,|\, \mathbf{e} > \; . \tag{20}$$

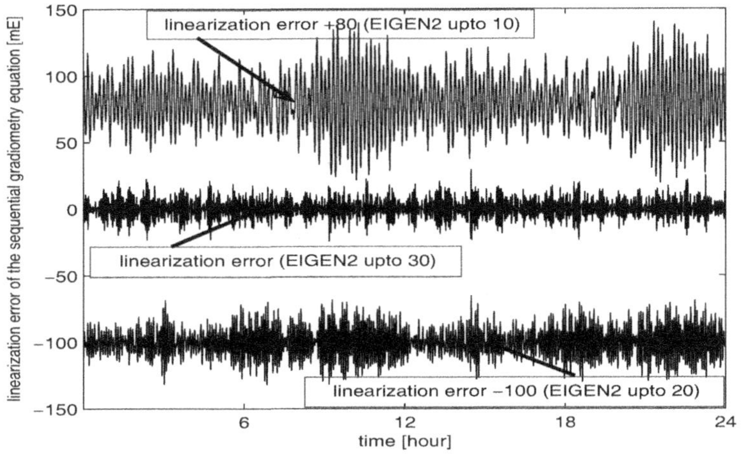

Fig. 1. Linearization error for the reference field of degree 10, 20 and 30

Equation (20) is known as the *sequential gradiometric observation equation.* It reduces the linearization error, differences between the left- and the right-hand sides of (20), dramatically. Figure 1 shows the linearization error for the reference field of degree 10, 20 and 30. The linearization error corresponding to the reference field of degree 30 is less than the observation random error (Keller and Sharifi, 2005). Note that the magnitude of the linearization errors for the full field may reach values at the level of 0.5 E.

Being one-point function and having a direct relation with the field geometry (curvature of the field at the point) are two noteworthy achievements of the alternative formulation. Besides, using an observation quantity, which is related to the second instead of the first order derivatives of the geopotential amplifies the high-frequency part of the signal. Since the transition from the first to the second order derivatives includes the application of a timely finite-differences scheme, the high-frequency part of the noise amplifies as well. Nevertheless, due to the different spectral behavior of signal and noise, in the end the second order approach leads to an improved resolution.

3.1 Spectral description of the geopotential

The geopotential satisfies everywhere outside the gravitating masses the Laplace differential equation

$$\nabla^2 V(\mathbf{x}) \;=\; 0 \,, \tag{21}$$

where ∇^2 stands for the Laplacian. Thus the geopotential can be solved for in a form of a harmonic series. Not all coordinate systems are suitable for deriving this solution, but knowing that the Earth and its equipotential surfaces are geometrically close to the sphere, geocentric spherical coordinates can be used

$$[\, r \quad \theta \quad \lambda \,]^T = \left[\sqrt{x^2 + y^2 + z^2} \quad \arctan \frac{\sqrt{x^2 + y^2}}{z} \quad \arctan \frac{y}{x} \right]^T , \qquad (22)$$

with the geocentric radius r ($0 \leq r < \infty$), reduced co-latitude θ ($0 \leq \theta \leq \pi$) and longitude λ ($0 \leq \lambda < 2\pi$), respectively. The solution of (21) can be then written for $r \geq R$

$$V(r, \theta, \lambda) = \sum_{n,m} \left(\frac{R}{r} \right)^{n+1} Y_{n,m}(\theta, \lambda) \; V_{n,m} \; , \qquad (23)$$

with radius of the reference sphere R. The spherical harmonics $Y_{n,m}$ and the Stokes coefficients $V_{n,m}$ are both degree (n) and order (m) dependent (Abramowitz and Stegun, 1972). Substituting (23) (12), (17) and (20), respectively, results in systems of linear equations with vector-valued and tensor-valued spherical harmonics, respectively. In both cases, however, their projection to the inter-satellite direction $\mathbf{e}$ is considered. Unknown values then represent the Stokes coefficients.

3.2 Boundary-value description of the geopotential

According to (21), the geopotential is a harmonic function everywhere outside the Earth's system. This property is due to the Poisson differential equation that relates the gravitational potential and the mass density distribution of the gravitating masses. The geopotential is also regular at infinity, namely

$$V(\mathbf{x}) = \mathcal{O}\left(\, |\mathbf{x}|^{-1} \, \right) , \qquad (24)$$

with the Landau symbol $\mathcal{O}$ describing in this particular case the attenuation of the geopotential with an increasing distance from the gravitating masses. Using an apparatus of Green's integrals, the solution of the Dirichlet boundary-value problem for the Laplace equation represents the well-known Abel-Poisson integral. It allows for evaluation of the geopotential at a general point P described by the geocentric radius vector $\mathbf{x}_P$ that is external to a boundary (simple smooth closed surface) on which the geopotential is known. The boundary must completely contain the gravitating masses. For the boundary S described by the geocentric radius vectors $\mathbf{y}$, this solution takes the form of the surface integral (Kellogg, 1929)

$$V(\mathbf{x}_P) = \frac{1}{S} \int_S V(\mathbf{y}) \; \mathcal{K}(\mathbf{x}_P, \mathbf{y}) \; dS(\mathbf{y}) \; . \qquad (25)$$

In geodesy, one particular geocentric geometry is being used: the reference ellipsoid. Due to its symmetricity, the integral function $\mathcal{K}$ can be constructed.

Note that not all gravitating masses are imbedded inside the surface S. The gravitational potential of all external topographic and atmospheric masses is then either neglected or corrected for by an appropriate reduction. However, this is considered to be out of the scope of this article. Applying in (25) the gradient operator in the point P yields, cf. (Ardalan and Grafarend, 2004),

$$\nabla_P \, V(\mathbf{x}) \; = \; \frac{1}{S} \int_S V(\mathbf{y}) \; \nabla_P \; \mathcal{K}(\mathbf{x},\mathbf{y}) \; dS(\mathbf{y}) \; , \tag{26}$$

with vector-valued kernel function $\nabla\mathcal{K}$. Applying the matrix direct product on the gradient operator in the point P yields (van Gelderen and Rummel, 2001) and (Martinec, 2003),

$$\nabla_P \otimes \nabla_P \; V(\mathbf{x}) \; = \; \frac{1}{S} \int_S V(\mathbf{y}) \; \nabla_P \otimes \nabla_P \; \mathcal{K}(\mathbf{x},\mathbf{y}) \; dS(\mathbf{y}) \; , \tag{27}$$

with the tensor-valued kernel function $\nabla \otimes \nabla\mathcal{K}$.

Combining Eqs.(12) and (25) then results in the observation equation

$$\frac{|\,\delta\dot{\mathbf{x}}\,|^2 \; - \; \dot{\varrho}^2}{\varrho} \; - \; \ddot{\varrho} \; = \; \frac{1}{S} \int_S V(\mathbf{y}) \; \mathcal{H}(\mathbf{x}_1,\mathbf{x}_2,\mathbf{y}) \; dS(\mathbf{y}) \; , \tag{28}$$

with the scalar three-point kernel function

$$\mathcal{H}(\mathbf{x}_1,\mathbf{x}_2,\mathbf{y}) \; = < \delta\nabla\mathcal{K}(\mathbf{x},\mathbf{y}) \mid \mathbf{e} > \; . \tag{29}$$

For a single epoch, the integral kernel $\mathcal{H}$ relates the scalar geopotential at the reference ellipsoid to the functional on the left-hand side of (28) derived from the GRACE SST data that is a function of positions of the GRACE satellites. This makes the kernel function $\mathcal{H}$ different from typical integral kernels used in geodesy that usually relate two points only such as the Stokes, Hotine or Vening-Meinesz functions. Combining Eqs.(17) and (27) then results yet in another observation equation

$$\frac{|\,\delta\dot{\mathbf{x}}\,|^2 \; - \; \dot{\varrho}^2}{\varrho^2} \; - \; \frac{\ddot{\varrho}}{\varrho} \; = \; \frac{1}{S} \int_S V(\mathbf{y}) \; \mathcal{G}(\mathbf{x}_B,\mathbf{y}) \; dS(\mathbf{y}) \; , \tag{30}$$

with the scalar-valued integral kernel

$$\mathcal{G}(\mathbf{x}_B,\mathbf{y}) \; = < \nabla_B \otimes \nabla_B \; \mathcal{K}(\mathbf{x},\mathbf{y}) \; \mathbf{e} \mid \mathbf{e} > \; . \tag{31}$$

In this case, only the barycentre of the GRACE satellites and the computation point are related.

4 Numerical analysis

The models for recovery of the Stokes coefficients were extensively tested using both simulated and actual GRACE SST data. In this section, numerical tests of the model in (12) are described including the input data. The solution in a form of the boundary integrals is still to be investigated in details, namely its application for global gravity field studies.

4.1 Description of the synthetic GRACE data

The parameters on the left-hand side of Eqs.(12) and (17) are derived by combining the GRACE SST data. To verify the applicability of the model and to test the propagation of various observation errors, a series of numerical tests was performed. In-house generated synthetic data based on the EGM96 of degree and order 120, 160 and 200 were analyzed. The IAG Section II, SC7 one-month sample data based on the EGM96 of degree and order 300 and compiled by the Universities Bonn and Delft (IAG SC7, 2001) were also used. These test datasets provide noise-free inertial position, velocity and acceleration vectors of the GRACE satellites with a 5 second sampling interval. Noise-free simulated LL-SST observations were derived from these quantities. Finally, these noise-free SST data were also contaminated with correlated errors to study an effect of various errors on the accuracy of the solution.

Unfortunately, an actual level of correlation between HL-SST data of the two low-orbiting GRACE satellites is unknown. The same applies to two adjacent positions of a single satellite. Error simulation functions for GPS data, that correlate adjacent measurements through a Gauss-Markov process, were used in our study. This approach successfully applied by (Grafarend and Vaníček, 1980) for the weight estimation in levelling is based on recursive formulas that allow for simulation of the correlated observation noise. The level of correlation is controlled by a unitless correlation factor $0 \leq \beta \leq 1$, see (Austen and Grafarend, 2005) for details. Generally, $\beta = 0$ corresponds to uncorrelated errors and $\beta = 1$ to fully-correlated errors. In order to determine the accuracy of satellites' velocity vectors computed from noisy GPS positions by numerical differentiation, various empirical tests were performed. Noise-free simulated position data contaminated with errors of a varying magnitude and correlation behaviour were numerically differentiated to obtain the velocity vectors. They were afterwards compared to the noise-free simulated velocity vectors. Based on results of these tests, we concluded that the velocity of the GRACE satellites can be determined from their GPS positions with the accuracy better than 0.3 mm s^{-1}. Correlations between kinematic orbits of the two GRACE satellites were simulated by applying the same method with an offset to the GPS positions of the satellites. Such a coloured noise was added to noise-free simulated GPS positions in order to investigate the influence of the correlated GPS data on the geopotential solution.

4.2 Propagation of observation noise

In order to assess the ability of the proposed methods for the geopotential recovery, the influence of observation noise on the coefficients' estimation process was investigated. In this contribution, namely results based on the model in (12) are presented. Figure 2 illustrates how errors in the satellites' positions determined by GPS (here a 1 cm average error was assumed for each coordinate) affect the geopotential solution. One can conclude that the overall quality of the solution strongly depends on the correlation of the position errors of both satellites. However, the analysis of synthetic data, described in this section, suggests that the Stokes coefficients up to degree and order 100 could be considered recoverable. Note that we define the maximum resolution of the model by the intersection of the signal and noise curves.

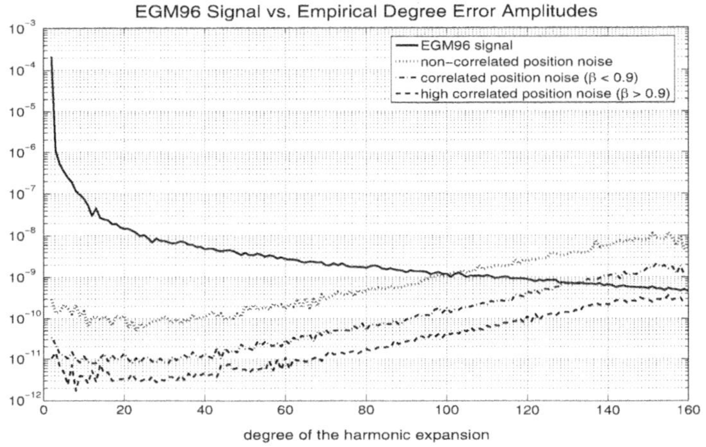

Fig. 2. Propagation of position errors ($\sigma_{x_i} = 1$ cm)

Figure 3 shows how the simulated noise in the velocity vectors of the two GRACE satellites influences the resolution of the computed model. The noise level of 0.1 mm/s is assumed for each component of the velocity vector. This noise arises from the fact that the velocity information has to be numerically derived from the noisy position vectors. As already mentioned before, noisy velocity information is the limiting factor of the overall solution quality. In absence of correlations between the two satellites' velocity vectors, i.e., for white noise, the geopotential recovery is limited to degree and order 70 given a noise level of 0.1 mm/s. To obtain the geopotential solution up to degree and order 100, the velocity errors of the GRACE satellites have to be correlated.

The impact of K-band measurement errors is displayed in Fig. 4. It can be concluded that relatively to other errors the inter-satellite position (1 cm) and inter-satellite velocity (1 μm/s) errors, that were used in our simulations, have no significant impact on the solution. Thus the GPS-based range can be used instead of the biased range from the K-band ranging system. In contrary, the effect of the inter-satellite acceleration errors is quite important. Figure 4 suggests that the accuracy of the inter-satellite acceleration at the level of 10^{-8} m/s^2 (1 μGal) is required to get the solution up to degree and order 100.

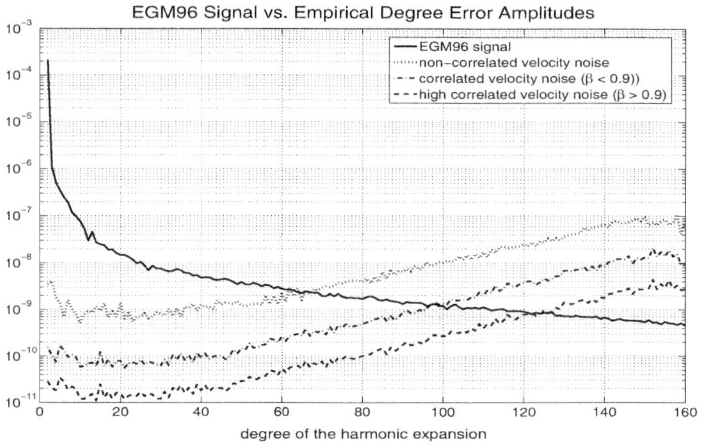

Fig. 3. Propagation of velocity errors ($\sigma_{\dot{x}_i} = 0.1$ mm/s)

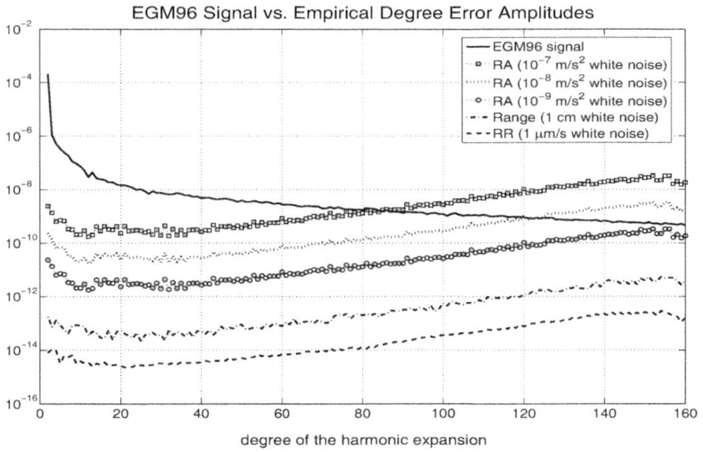

Fig. 4. Propagation of K-band measurement errors

Finally, the combined effect of all error influences (HL-SST: GPS and LL-SST: K-band) is presented in Fig. 5 for a pessimistic and optimistic mission scenario. Based on these results, the resolution of the recovered geopotential using the GRACE data and following the proposed method is expected to be in the range of degree and order 100 to 120.

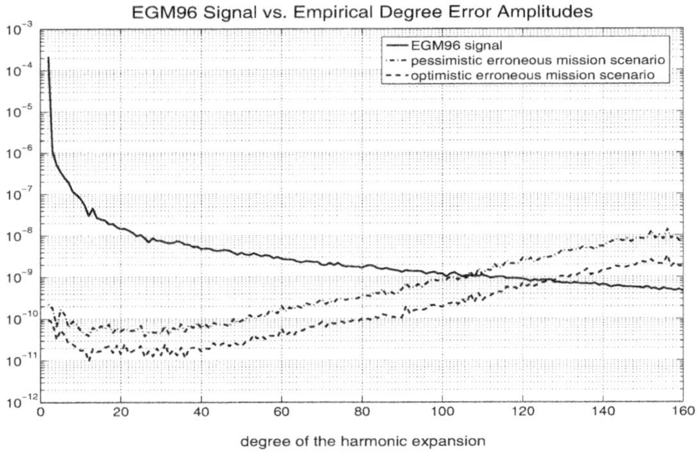

Fig. 5. Propagation of the combined measurement error influence

4.3 Processing actual GRACE data

The real data analysis was performed using approximately 88,000 observation values. The HL-SST data were kindly provided by Mr. Dražen Švehla of the Technical University Munich. Corresponding LL-SST data were then downloaded from the GRACE Data Center at the GeoForschungsZentrum (GFZ) Potsdam, Germany. Results presented in this article were extracted from 31 days of observations (03.230-03.260). Due to different sampling intervals of the data, the 30-second data were used.

Both the HL- and LL-SST data were routinely checked for outliers first. The HL-SST data were then rotated to the celestial frame and inter-satellite velocity differences were derived using numerical differentiation (Savitzky-Golay formula using the Fourier transform). This step is crucial for the approach and much attention was focused on its proper evaluation. To document the applicability of numerical differentiation schemes, Figure 6 shows an example of a comparison of intersatellite acceleration obtained by the numerical algorithm ($\varrho \rightarrow \ddot{\varrho}$) and those $\ddot{\varrho}$ provided by the GFZ Potsdam. Next, the observation vector is derived as a time series by the combination of HL- and LL-SST data according to the left-hand side of (12), see Fig. 7.

Further, the model was completed for tide-generated acceleration and reduced for a normal gravity field model. Finally, the Stokes coefficients up to degree and order 70 were recovered. The least-squares adjustment includes the computation of a full variance-covariance matrix, residuals as well as estimated a posteriori variance factor. Figure 8 shows the root mean square power spectrum of the recovered field as well as of its differences from the EIGEN-GRACE2S model of the GFZ Potsdam. Color Fig. XXIII on p. 299 then represents the synthesized differences between the recovered model and

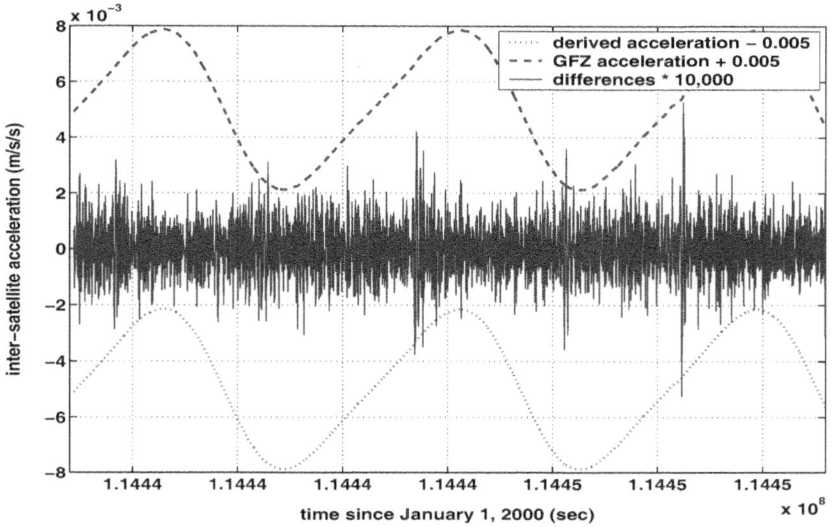

Fig. 6. Verifying the differentiation procedure using GRACE LL-SST data

the EIGEN-GRACE2S model. The maximum difference is 4.94 m and the global root mean square fit is 27 cm.

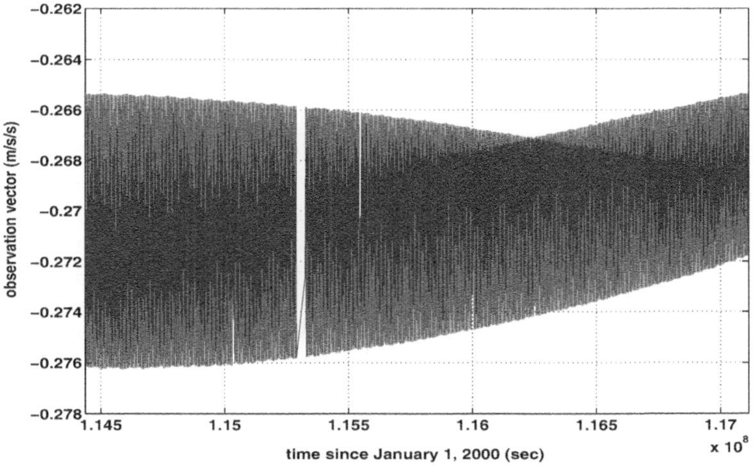

Fig. 7. Left-hand side of (12) based on actual GRACE HL- and LL-SST data

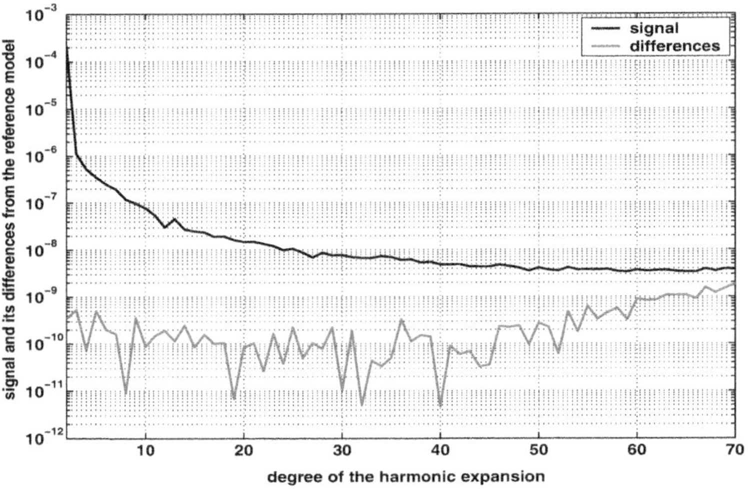

Fig. 8. Recovered coefficients and their difference to the EIGEN-GRACE02S model

5 Conclusions

Global mapping of the Earth's gravity field through the Newtonian equation of motion, see (9), combined with spectral representation, see (23), and integral representation, see (25), respectively, of the geopotential was discussed in this article. This approach represents one option for processing of the SST data of type GRACE as well as those of other satellite missions. In view of other known methods used currently in processing the SST data for geopotential recovery, the proposed approach has some significant advantages, namely simple interpretation of the relationship between observations and the unknown Stokes coefficients of the geopotential without any orbit geometry assumptions, a priori information or boundary surface, see also (Reubelt et al., 2003) and (Austen and Grafarend, 2005) for further details. On the other hand, one has to solve potentially weak points of the approach among which the most important one is handling the observation noise in the HL-SST data used for computation of the inter-satellite velocity. Due to correlated noise in the HL-SST data, the differentiation of the satellites' positions does not seem to represent any real problem to the approach.

Extensive numerical testing of the approach revealed some of its main characteristics, namely stability and sensitivity to various observation errors expected to be a part of real data. Although simulating actual errors in all their complexity is very difficult, some realistic scenarios suggest that the approach could be used for recovery of the geopotential coefficients up to degree and order of 120. Note that the time-varying component of the gravitational field was targeted with expected 30-day batches of data to be processed. From this point of view, the performance of the approach and its computer realization seems to be quite satisfactory fulfilling initial requirements of the project. Even the modified approach using the gradiometric observation equation is applicable with additional advantages of this method, see Keller and Sharifi (2005) for details.

Finally, the initial processing of the actual GRACE data provided some reasonable results proving the applicability of the proposed method for the analysis of real data. The mean difference between recovered solution and the EIGEN-GRACE2S model of the GFZ Potsdam reached the global value of 27 cm. It should be stressed that these intermediate results were obtained for a relatively small sample of data (30-day data sampled at 30 seconds), very small number of outliers removed and not all the effects included in the model. Taking into the account a large difference in the amount of input data and methodology used in their processing, this result seems to be quite promising for future investigations in this field at the Geodetic Institute of the Stuttgart University.

Acknowledgement. This is publication No. GEOTECH-151 of the programme GEOTECHNOLOGIEN of BMBF and DFG, Grant No. 03F0326A. The financial support is gratefully acknowledged as well as numerous thoughtful comments of KH. Ilk.

References

Abramowitz M, Stegun CA (1972) Legendre functions. Handbook of mathematical functions with formulas, graphs and mathematical tables. Dover, New York.

Ardalan AA, Grafarend EW (2004) High-resolution regional geoid computation without applying Stokes's formula: a case study of the Iranian geoid. *J Geod* 78: 138-156

Austen G, Grafarend EW (2005) Gravitational field recovery from GRACE data of type high-low and low-low SST. *Adv Geosc* (in press).

Blaha G (1992) Refinement of the satellite-to-satellite line-of-sight acceleration model in a residual gravity field. *Manuscr Geod* 17: 321–333

Grafarend EW, Vaníček P (1980) On the weight estimation in leveling. NOAA Technical Report NOS 86, NGS 17.

IAG SC7 (2001) Satellite gravity field missions: Simulation scenarios. [$ftp://geo@atlas.geod.uni-bonn.de/pub/SC7\_SimulationScenarios$].

Keller W, Hess D (1998) Gradiometrie mit GRACE. *Z Verm* 124: 137-144

Keller W, Sharifi MA (2005) Satellite gradiometry using a satellite pair. *J Geod*: 544-557

Kellogg OD (1929) Foundations of potential theory. Springer, Berlin.

Martinec Z (2003) Green's function solution to spherical gradiometric boundary-value problem. *J Geod* 77: 41-49

McCarthy DD, Petit G (2004) IERS Conventions (2003). IERS Technical Note 32. Verlag des Bundesamtes für Kartographie und Geodäsie, Frankfurt am Main.

Reubelt T, Austen G, Grafarend EW (2003) Harmonic analysis of the Earth's gravitational field by means of semi-continuous ephemeris of a Low Earth Orbiting GPS-tracked satellite. Case study: CHAMP, *J Geod* 77: 257-278

Rummel R, Reigber C, Ilk KH (1978) The Use of Satellite-to-Satellite Tracking for Gravity Parameter Recovery, Proceedings of the European Workshop on Space Oceanography, Navigation and Geodynamics, ESA SP-137: 153-161

Rummel R (1980) Geoid heights, geoid height differences, and mean gravity anomalies from low-low satellite-to-satellite tracking – an error analysis. Report No. 306, Department of Geodetic Science and Surveying, Ohio State University, Columbus.

Tapley BD, Reigber C (1998) GRACE: A satellite-to-satellite tracking geopotential mapping mission. Proceedings of the Second Joint Meeting of the International Geoid and Gravity Commisions, Trieste, September 7-12.

van Gelderen M, Rummel R (2001) The solution of the general geodetic boundary value problem by least squares. *J Geod* 75: 1-11

High Frequency Temporal Earth Gravity Variations Detected by GRACE Satellites

Xiaogong Hu[1], Chuang Shi, Frank Flechtner, Rolf König, Peter Schwintzer, Roland Schmidt, Ulrich Meyer, Franz H. Massmann, Christoph Reigber, and Sheng Y. Zhu

GeoForschungsZentrum Potsdam (GFZ), Department 1 'Geodesy and Remote Sensing', Potsdam, Germany
[1] On leave from Shanghai Astronomical Observatory, Shanghai, China

Summary. High frequency temporal gravity changes on sub-monthly time scales are caused by Earth's mass transport primarily originating from tidal and non-tidal atmospheric and oceanic motions. Exploitation of precise GRACE satellite-to-satellite ranging measurements now makes it possible to monitor these changes on a global scale with a moderate spatial resolution. Thus, a time series of daily 10×10 Earth gravity models has been produced for the time span from July 2 to September 30, 2003. The solid Earth, oceanic and atmospheric tidal forces are accounted for during data processing, while variations in the gravitational potential produced by non-tidal atmosphere and ocean mass transport are omitted in the a-priori force models. The recovered gravity changes are then compared to non-tidal model predictions. It is found that the agreement varies with the degree and order of the gravity harmonics. Generally, the recovered harmonics are highly correlated with the models for low order coefficients but the agreement degrades when the order becomes larger. Our results prove that GRACE is able to trace geophysical signals at short time scales, and that GRACE data can be useful to validate model predicted large-scale mass transports. Once we are able to separate tidal and non-tidal signals for a longer time span, daily gravity recovery might also prove to be useful to study gravity changes taking place at sub-monthly time scales, such as oceanic tides.

1 Introduction

The twin satellites of the US/German Gravity Recovery And Climate Experiment (GRACE) were launched in March 2002 (Tapley and Reigber, 2004). On a low altitude of about 500 km and a near-polar orbit, the GRACE satellites continuously gather high-low GPS - GRACE and low-low GRACE - GRACE satellite-to-satellite K-band ranging (KBR) data. Non-gravitational accelerations are measured onboard with a three-axis SuperSTAR accelerometer. The satellites' orientation is derived from star camera observations. Exploiting only a few months of data, GRACE has already improved the accuracy of the mean gravity field at long- and mid-wavelengths by more than one order of magnitude (Reigber et al., 2005). The time series of monthly gravity field models, GRACE's major scientific product, has shown to be consistent with geophysical predictions such as the variability of continental water storage, although the accuracy and spatial resolution has not yet reached the pre-launch expectation (Wahr et al., 2004; Schmidt et al., 2005).

Prior to the launch of GRACE (and CHAMP) gravity change monitoring was limited due to restrictions in data accuracy and, more important, limitations on the geographic distribution of the tracking data used for gravity recovery, such as ground-based satellite laser ranging or Doppler tracking (Chen and Wilson, 2003). With seasonal hydrological signals being the main focus in analyzing GRACE's monthly gravity field models, mass transports at shorter time scales must be precisely modeled and taken into account during precise orbit and gravity field partial derivatives determination to avoid aliasing effects. At time scales shorter than seasonal, the tides of the solid Earth, oceans and atmosphere as well as the non-tidal motions of the atmosphere and oceans, are important contributors to gravity changes. While the solid Earth tide models have been well established, geophysical models of the non-tidal mass transport still show differences which influence the quality of the derived monthly gravity field solutions (Flechtner et al., 2005).

It would be desirable to detect gravity changes at time scales shorter than the currently GRACE "standard" monthly resolution. However, a higher temporal resolution is usually correlated with degradation in spatial resolution. Previous work by König et al. (2004) has shown how to recover low degree spherical harmonics (zonal only, up to degree-4) from CHAMP 1.5 days time series. Our purpose in this paper is to extend that effort to daily temporal and to higher spatial resolution analyzing Level-1B GRACE data. Thus, we derived daily 10×10 degrees gravity field solutions for July 2 to September 30, 2003. The increased resolution of these gravity field solutions could help to validate short-term non-tidal model predicted mass transport, whose variability has been shown to be strong at time scales shorter than 30 days (Wahr et al., 1998), and to investigate possible errors in certain long period ocean tide components such as the Mf tide (with a period of about 13.66 days) which are impossible to ber solved by monthly GRACE solutions.

In this paper we concentrate on the first topic. The consistency between the recovered gravity changes and the geophysical model predictions yields important information on both GRACE's potential to detect high frequency gravity changes and the validity of the atmosphere and ocean de-aliasing models used in the standard monthly gravity processing of the GRACE Science Data System (Flechtner, 2003).

2 Data analysis

An integrated adjustment method as proposed in Zhu et al. (2004), where low degree spherical harmonics of the gravity field, orbits and clocks of the GRACE and GPS satellites, station coordinates of a globally distributed ground network as well as other dynamic and kinematic parameters are simultaneously estimated in one single solution, has been applied.

The geographic distribution of the GRACE ground tracks changes from day to day to provide a dense and almost uniform coverage of the globe within 30 days. For a typical day, the ground tracks consist of 15 great circles approximately passing the North and the South Pole with uncovered regions between two neighboring circles of about 24 degrees in longitude. This poor East-West sampling implies that only a low spatial resolution is possible for daily solutions. Besides, gaps in the KBR and accelerometer measurements also impose constraints on the spatial resolution. Our experiments have shown that a spatial resolution of about 2000 km half-wavelength, corresponding to degree and order 10 in terms of spherical harmonics, can be achieved for daily gravity field recovery if more than 12 hours worth of valid instrument data are available. Thus, 81 daily 10×10 GRACE gravity field solutions have been produced from 91 days out of period July 2 to September 30, 2003.

The a-priori force and measurement models used for the daily solutions are identical to those used for the routine generation of the monthly gravity models at GFZ Potsdam (Reigber et al., 2005), except that the gravitational force due to the non-tidal atmospheric and oceanic mass transport is deliberately omitted in the a priori force models. As consequence, these gravity force along with hydrological signals, unknown background model errors (e.g. ocean tide model errors) and measurement data noise should be recovered by the resulting daily solutions. At large spatial scales, which are relevant in this study, most of the contributions from continental hydrology are believed to be due to seasonal changes (Schmidt et al., 2005). Hence they will emerge only as low-frequency background signal in our comparisons.

As a-priori values for the recovered gravity harmonic coefficients we use the GFZ mean static gravity model EIGEN-GRACE02S which was obtained by analyzing 110 days of GRACE data (Reigber et al., 2005). For a typical daily solution, the root sum of squares (rss) of the least squares adjustment's formal uncertainties over all spherical harmonic coefficients from degree 1 to degree 10 is about $5 \cdot 10^{-11}$ or 0.3 mm in terms of geoid height. The formal uncertainty of an individual coefficient is a factor of 10 smaller ($5 \cdot 10^{-12}$, or 0.03 mm in terms of geoid height).

As non-tidal geophysical model, we use the GRACE Level-1B Release 01 Atmosphere and Ocean De-aliasing products (AOD1B RL01) (Flechtner (2003); also available at *http://isdc.gfz-potsdam.de/grace*). AOD1B RL01 consists of gravity harmonics complete up to degree and order 100 for every 6 hours (0, 6, 12, 18 hours UTC). For the atmospheric part, the input data are global 0.5° by 0.5° multi-layer grids of meteorological fields from the European

Center for Medium-range Weather Forecasts (ECMWF), which are integrated vertically to represent the 3D atmospheric mass distribution. The non-tidal oceanic gravity changes are derived from a barotropic ocean model provided by the Jet Propulsion Laboratory (JPL). It is forced by 6-hourly ECMWF meteorological fields and computes the component of oceanic mass transport, or barotropic sea level, due to wind stress and atmospheric pressure for an area between 65° N and 75° S on a 1.125° grid.

For comparison, an alternative de-aliasing product was provided by GRGS (Groupe de Recherche de Géodésie Spatiale) that is based on similar assumptions as AOD1B RL01, but the non-tidal ocean model is MOG2D, a barotropic, non-linear and time stepping model with global output (including the Arctic Ocean, Hudson Bay, Bering Strait and the Weddell and Ross Seas). The ice cover impact is taken into account through the diminution of the water column height, due to the submerged ice thickness. MOG2D is governed by shallow water continuity and momentum equations. The model can include tides and its main originality is a finite element space discretisation which allows to increase the resolution e.g. from 400 km in the open sea down to 20 km in areas with strong topographic gradients or in shallow waters (Carrère et al., 2003).

The differences between these two non-tidal ocean models have been investigated in Flechtner et al. (2005) and might provide an estimate of the models errors. It was found that on a global scale the monthly geoid RMS difference is in the order of about ±1.5 mm for a spatial resolution of degree and order 40. But since common ECMWF data sets are used as model input, these error estimates might be too optimistic.

3 Results and comparisons

For each of the daily solutions, a full 10×10 gravity field (except for the degree-0 harmonic representing the total mass of the Earth), consisting of 120 gravity harmonic coefficients, was recovered. It turned out, however, that geoid maps computed from the gravity field solutions bear little resemblance to those derived from the model predictions (not shown). This is due to various error sources associated with both the GRACE gravity field solutions and the models. Major error sources for the daily solutions are errors in the a-priori gravity field model, and errors in other a-priori force models, in particular the ocean tide model. For the a-priori gravity field, "true" harmonics with degree and order greater than 10 are time-varying but have been fixed instead to their mean values of EIGEN-GRACE02S. Even worse, the force model errors propagate to the recovered harmonics in a subtle way because of the non-uniform, non-global geographic distribution of the daily ground tracks. Errors in the meteorological data sets will translate into errors in the predicted atmospheric mass transport and into errors in the forcing of the non-tidal ocean model. Additionally, the complex oceanic responses to atmospheric loading may not be perfectly modeled either. For example, significant discrepancies occur for the AOD1B RL01 and MOG2D, both forced with same 6-hourly ECMWF meteorological data, degree-1 and degree-2 harmonics with negative statistical correlation for the $S_{1,1}$ harmonic. The spherical harmonic differences between the two models decrease as the spatial resolution becomes smaller, but is generally about 40% of the averaged model signal. Even when the degree-1 and degree-2 harmonics are excluded, the difference could still be as large as 0.15 mm for some harmonics, which is about five times larger than the 0.03 mm formal uncertainty associated with the daily GRACE solutions.

However, treating the 120 recovered harmonics separately, the comparison of the solution and model time series comes to encouraging agreement. Color Fig. XXIV on p. 300 shows the comparison for selected daily solutions and the MOG2D harmonics where bias, trend and "seasonal" changes (by fitting the time series with seasonal variation) are removed. The agreement is remarkably good for some coefficients, such as $C_{6,0}$ or $S_{6,1}$. Reasonable agreement can be noticed for all shown harmonics except for $C_{2,0}$, but the degree-2 harmonics are already known to contain relatively large errors for GRACE solutions (Tapley and Reigber, 2004). The correlation coefficients are about 0.9 for $C_{6,0}$ and $S_{6,1}$, about 0.7 for $S_{9,3}$ and $C_{6,3}$, about 0.6 for $C_{4,1}$, and about 0.4 for $C_{2,1}$, $S_{2,1}$ and $C_{2,0}$.

Figure 1 shows the correlation coefficients of all 120 coefficients between the MOG2D model predictions and the daily gravity solutions. The agreement depends on the harmonic degree and order and the correlation coefficients range from 0.93 for $S_{6,1}$ to -0.33 for $C_{3,3}$, where 26 out of the 120 harmonics have correlation coefficients larger than 0.7 and 39 harmonics have correlation coefficients larger than 0.6.

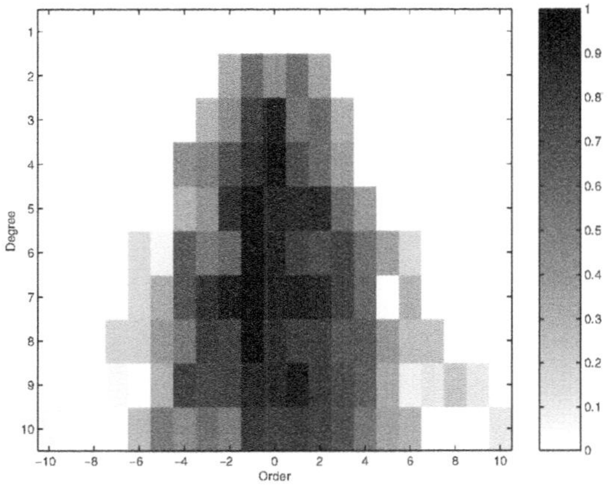

Fig. 1. Correlation coefficients between MOG2D model predictions and daily gravity solutions for S_{nm} (left) and C_{nm} (right) spherical harmonic coefficients

Since the solutions seem to agree slightly better with MOG2D than with AOD1B RL01, we focused our further comparisons on this model. Grouped by the harmonic orders, Table 1 lists the correlation coefficients along with other statistical quantities between MOG2D and the daily gravity field solutions. One statistical quantity in Table 1 is the standard deviation of different time series to characterize the variability of each harmonic. Shown are the standard deviations of the MOG2D time series (Std2), of the daily gravity field solution time series (Std3), of the daily gravity field solution minus MOG2D residuals (Std4), and after removing a trend and "seasonal" changes from the residuals (Std5). Std2 is a kind of "signal" which reduces rapidly with increasing order. The "noise" (Std3) increases with higher harmonic orders but remains more or less constant. Thus, the signal-to noise-ratio decreases with increasing order. The degree-1 and degree-2 harmonics manifest much larger variability than the other degrees (see Color Fig. XXIV on p. 118 for degree-2) and they contain also other larger geophysical information such as postglacial rebound. Thus, they are not used here in order to avoid contamination of the mean statistics in Table 1.

Although model errors cannot be completely ruled out, the systematically larger GRACE solution minus model variability of higher orders is likely due to solution errors. Based on the mean statistics (last row), the daily GRACE solutions have recovered the 10×10 harmonics (except for degree-1 and degree-2) with an accuracy of about 0.2 mm in terms of geoid height or $3 \cdot 10^{-11}$ in terms of normalized Stokes harmonics. This accuracy level is still about 6 times larger than the formal uncertainty of GRACE-EIGEN02S ($5 \cdot 10^{-12}$) but it should be emphasized that the results were obtained without any a-priori constraints.

The standard deviations of the daily solution time series are more or less reduced by removing the model predictions of MOG2D for all harmonics from order 0 to order 9, but only for order-0 and order-1 harmonics the reduction is significant: the standard deviations are reduced from 0.22 mm to 0.13 mm for order-0 and from 0.15 mm to 0.10 mm for order-1, respectively. Here, the statistical correlation is also high (greater than 0.78). The removal of a trend and "seasonal" changes from the residuals further reduces the standard deviations, but we should note again that the "seasonal" changes might not be interpreted in geophysical sense because of the short data span. Nevertheless, the daily gravity recovery looks promising and to our knowledge, represents the first effort ever to validate high frequency model predicted gravity changes.

The mean correlation coefficients between the daily GRACE gravity field solution time series and the two de-aliasing models are plotted in Fig. 2 as a function of the harmonic order. Again, the degree-1 and degree-2 terms are excluded.

Previous studies have already revealed the contribution of the atmosphere and oceans to low degree harmonics for seasonal or longer time scales with focus on the degree-2 harmonics (Chen and Wilson, 2003).

Table 1. Statistics for the comparisons between the gravity field solutions and MOG2D. [1] Number of harmonics with a fixed order excluding degree 1 and 2 harmonics; [2] Mean standard deviation of the MOG2D time series; [3] Mean standard deviation of the solution time series; [4] Mean standard deviation of the residual time series; [5] Mean standard deviation after trend and seasonal variations are removed from the residual time series; [6] Mean statistical correlation coefficients after the removal of bias, tread and "seasonal" changes. Unit for standard deviations: mm of geoid height

Order	No.[1]	Std.[2]	Std.[3]	Std.[4]	Std.[5]	Corr.[6]
0	8	0.167	0.218	0.130	0.086	0.801
1	16	0.115	0.154	0.097	0.077	0.786
2	16	0.100	0.178	0.154	0.120	0.562
3	16	0.088	0.196	0.167	0.129	0.565
4	14	0.070	0.205	0.185	0.143	0.419
5	12	0.048	0.195	0.189	0.162	0.263
6	10	0.040	0.215	0.213	0.195	0.153
7	8	0.027	0.312	0.311	0.285	0.087
8	6	0.020	0.337	0.336	0.318	0.019
9	4	0.017	0.395	0.394	0.385	0.043
10	2	0.010	0.308	0.308	0.290	-0.025
Mean		0.076	0.217	0.195	0.165	0.433

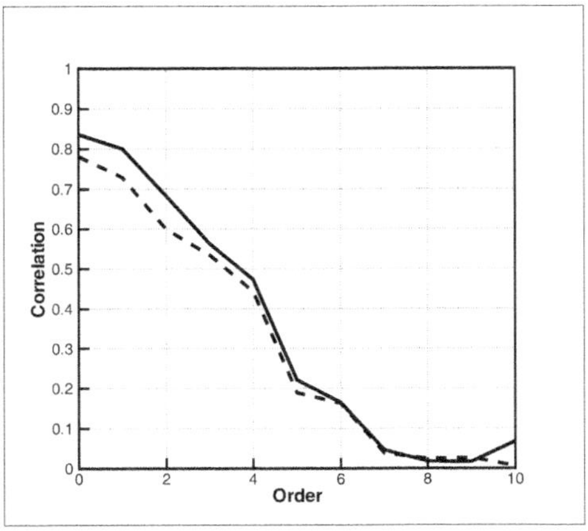

Fig. 2. Mean correlation coefficients as function of order between the daily GRACE solutions and MOG2D (solid) and AOD RL01 (dashed), respectively

The GRACE monthly solution degree-2 harmonics have relatively large errors (Tapley and Reigber, 2004), and we believe it is the same for daily solutions. In Chen et al. (2004), 18 monthly degree-2 harmonics estimated from 2 years of GRACE data are compared with both geophysical models and Earth orientation parameters. With a time span of only 3 months, the daily solution time series reported here is not long enough for a seasonal change study, but do provide more information at shorter time scales with potential to extend the validation and geophysical interpretation studies to higher temporal resolution.

The time series of $C_{2,1}$ and $S_{2,1}$ are dominated by changes at time scales shorter than 30 days, while low frequency changes are more evident in the time series of $C_{2,0}$ before the removal of a trend and "seasonal" changes. The standard deviations are 1.707 mm (geoid height) for $C_{2,0}$, 0.399 mm for $C_{2,1}$ and 0.578 mm for $S_{2,1}$ time series. After removal of the model predictions, these numbers are reduced to 1.641, 0.385 and 0.417 mm, respectively. They are further reduced to 0.750, 0.314 and 0.357 mm after removing a trend and "seasonal" changes. The 56% reduction for $C_{2,0}$ estimates suggests strong low frequency variability. Before the removal of a trend and "seasonal" changes, the statistical correlation coefficients between the solution time series and MOG2D are 0.277 for $C_{2,0}$, 0.318 for $C_{2,1}$ and 0.670 for $S_{2,1}$, what agrees with Chen et al. (2004) who found that $S_{2,1}$ is better correlated with model predictions than $C_{2,0}$, and $C_{2,1}$, despite that different models and different time scales are compared. Independent results for the degree-2 harmonics such as the EOP-derived values are not available for this study, but we'd like to note that for $S_{2,2}$ and $C_{2,2}$, the other two degree-2 harmonics not studied in Chen et al. (2004), the agreement with the models is found to be better than for $C_{2,0}$ and $C_{2,1}$. Here, the correlation coefficients are 0.545 and 0.602, respectively.

For degree-1 harmonics, i.e. the geo-center motions, large discrepancies (RMS about 5 mm) between the solutions and the models are found. But, because even the two models come to inconsistent predictions on geo-center motions, this might not be surprising.

4 Summary and discussions

Thanks to the good performance of the instruments onboard the GRACE twin satellites, time series of daily 10×10 gravity field models with a reasonable good accuracy could be produced for the first time. Daily solutions allow for validation and interpretation of high frequency gravity changes such as those produced by non-tidal atmospheric and oceanic mass transports, which possess significant variability at time scales shorter than 1 month, the present temporal resolution of GRACE gravity products. The potential of GRACE to detect high-frequency gravity changes is evaluated by comparing predictions of two different models with the daily solutions. The first is the standard GRACE Level-1B Atmosphere and Ocean De-aliasing model AOD1B used by the GRACE Science Data System to derive monthly GRACE gravity field solutions. The second model was provided by GRGS and is primarily based on an alternative ocean model (MOG2D).

Statistical correlation coefficients and standard deviations are computed to quantify the agreement between the solutions and the models. Reasonable agreement is found for only a subset of the 120 harmonics recovered; among them are the zonal and order-1 harmonics. An average uncertainty for each harmonic (except the degree-1 and degree-2 harmonics) of about $3 \cdot 10^{-11}$ or 0.2 mm (geoid height) is associated with each recovered harmonic. Although 6 times larger than the formal uncertainty, this accuracy level is obtained without any a-priori constraints and would be impossible without GRACE. $S_{2,1}$ agrees better with model predictions than $C_{2,0}$ and $C_{2,1}$, consistent with conclusions derived by other studies, although different models and time scales are looked upon. By removing a trend and "seasonal" changes presumably originating from hydrology, the agreement for $C_{2,0}$ and $C_{2,1}$ improves noticeably. Reasonable agreement is also found for $S_{2,2}$, and $C_{2,2}$.

The solution-model comparisons will be more suggestive when longer time series of daily solutions become available. This would help to better separate seasonal or other long period changes and to better understand high frequency gravity changes. From these longer time series not only the study of atmospheric and oceanic mass transport would benefit, but also weak periodic gravity changes such as those produced by oceanic tides could be investigated. It would also be interesting to obtain GRACE solutions for every two or three days to take advantage of the denser GRACE ground tracks distribution and thus to recover the gravity field with higher resolution. This might provide additional information on mass transport at smaller spatial scales.

Acknowledgement. This is publication no. GEOTECH-149 of the GEOTECHNOLOGIEN programme of BMBF and DFG, grant 03F0326A and 03F0336E. X.G. Hu is supported by a grant from the Natural Sciences Foundation of China (No. 40374002).

References

Carrére, L, and F Lyard (2003) Modeling the barotropic response of the global ocean to atmospheric wind and pressure forcing – comparisons with observations, *Geophys. Res. Lett.*, 30(6), 1275, doi: 10.1029/2002GL016473

Chen, J L, and C R Wilson (2003) Low degree gravitational changes from Earth rotation and geophysical models, *Geophys. Res. Lett.*, 30(24), 2257, doi:10.1029/2003GL018688

Chen, J L, C R Wilson, B D Tapley, and J C Ries (2004) Low degree gravitational changes from GRACE: Validation and interpretation, *Geophys. Res. Lett.*, 31, L22607, doi:10.1029/2004GL021670

Flechtner, F (2003) AOD1B Product Description Document, GRACE project document JPL 327-750, Rev. 1.0, JPL Pasadena

Flechtner, F, R Schmidt and Ul Meyer (2005) De-aliasing of short-term atmospheric and oceanic mass variations for GRACE, article in this book

König R, Ch Reigber, and S Y Zhu (2004) Dynamic Model Orbits and Earth System Parameters from Combined GPS and LEO Data, accepted by *Advances in Space Research*

Reigber, Ch, R Schmidt, F Flechtner, R König, U Meyer, K H Neumayer, P Schwintzer, and S Y Zhu (2005) An Earth gravity field model complete to degree and order 150 from GRACE: EIGEN-GRACE02S, *J. Geodynamics*, 39, 1-10, doi: 10.1016/j.jog.2004.07.001

Schmidt, R, P Schwintzer, F Flechtner, Ch Reigber, A Güntner, P Döll, G Ramillien, A Cazenave, S Petrovic, H Jochmann, and J Wünsch (2005) GRACE Observations of Changes in Continental Water Storage, accepted by *Global and Planetary Change*

Tapley B and Ch Reigber (2004) GRACE (Gravity Recovery and Climate Experiment), in *McGraw-Hill yearbook of science & technology*, pp. 135-138, McGraw-Hill, New York

Wahr, J, M Molenaar, and F Bryan (1998) Time variability of the Earth's gravity field: Hydrological and oceanic effects and their possible detection using GRACE, *J. Geophys. Res.*, 103(12), 30205-30229

Wahr, J, S Swenson, V Zlotnicki and I Velicogna (2004) Time-variable gravity from GRACE: First results, *Geophys. Res. Lett.*, 31, L11501, doi: 10.1029/2004GL019779

Zhu, S Y, Ch Reigber, and R König (2004) Integrated adjustment of CHAMP, GRACE, and GPS data, *J. Geodesy*, 78, 103-108, doi: 10.1007/s0019000403790

Part III

GOCE
The Gravity Field and Steady-State Ocean Circulation Explorer

From Kinematic Orbit Determination to Derivation of Satellite Velocity and Gravity Field

Dražen Švehla[1] and Lóránt Földváry[1,2]

[1] Institute of Astronomical and Physical Geodesy, Technical University of Munich, Arcisstrasse 21, D-80333 Munich, Germany, svehla@bv.tum.de
[2] Now at the MTA-BME Research Group for Physical Geodesy and Geodynamics, Department of Geodesy and Surveying, Budapest University of Technology and Economics, H-1521, PO Box 91, Budapest, Hungary, fl@sci.fgt.bme.hu

Summary. After an overview of approaches and results in precise orbit determination (POD) for the CHAMP satellite in the Low Earth Orbit (LEO) we focus on the relations between kinematic POD and gravity field determination. We discuss determination of kinematic velocities out of kinematic positions that enter the gravity field determination in the form of kinetic orbital energy. After testing several numerical differentiation techniques, we selected conceptually two alternative methods, the Newton-Gregory interpolation and the smoothing cubic spline function. Finally, performance of numerical differentiation techniques for the CHAMP orbit is presented based on the gravity field determination.

Key words: CHAMP, precise orbit determination, kinematic orbit, GPS, numerical differentiation, gravity

1 Introduction

Sections 1 to 3 of this paper is partially based on the paper Švehla & Rothacher (2004b).

With the pioneering satellite mission CHAMP, a new era in space geodesy and observing the planet Earth from space started. Today we are talking about gravity (CHAMP, GRACE, GOCE) and magnetic (CHAMP, ØRSTED, SWARM) fields determined from space, atmosphere sounding from space (CHAMP, GRACE, COSMIC, SWARM), monitoring oceans (TOPEX/-POSEIDON, JASON-1) and ice caps (ICESAT, CRYOSAT) from space, etc.

In all these missions, satellite orbit determination is used for geo-location of the satellite sensors on one hand and to measure the gravity field and its variations in time on the other hand, i.e., using the equation of motion to obtain information about dynamical processes in the Earth system, as e.g.

Earth tides, mass distribution, ocean circulations, etc. (Balmino et al. (1999), Rummel et al. (2003), Rothacher et al. (2004)).

In 1992, for the first time, high-precision LEO dynamic orbit determination was performed making use of GPS measurements from TOPEX/POSEIDON (Bertiger et al. (1994), Tapley et al. (1994), Yunck et al. (1994)). As a next step in this approach on board accelerometers with specific measurement bandwidth were included in order to separate gravitational from non-gravitational orbit perturbations. This was for the first time demonstrated for the CHAMP satellite, see e.g. König et al. (2001). In that case on board accelerometer is used for orbit determination as well as gravity field recovery.

Since 1992, GPS tracking has become an extremely successful method for POD and nowadays purely kinematic orbits can be determined with the same level of accuracy as orbits computed with the more common (reduced-)dynamic approach (Švehla & Rothacher (2002), Švehla & Rothacher (2004b)). Among all space geodetic techniques (SLR, DORIS, altimetry, etc.) only GPS allows purely kinematic precise orbit determination, where kinematic satellite positions are estimated epoch-by-epoch independently of orbit altitude and force models, e.g., gravity field, air-drag, solar radiation, etc. From that point of view, kinematic orbits are very well suited for the Earth observation satellites at very low altitudes, where air-drag and gravity become more difficult to model.

Using GPS data from the CHAMP satellite we showed for the first time that a LEO orbit can be estimated kinematically with the same level of accuracy (1-3 cm) as with the widely applied dynamic approaches (Švehla & Rothacher (2002), Švehla & Rothacher (2003b), Švehla & Rothacher (2004b)) and therefore kinematic POD turned out to be a new method in orbit determination. This CHAMP kinematic POD experiment initiated the computation of kinematic orbits for a longer time period, namely one year, in order to see the potential of LEO geometrical positions for the determination of, e.g., gravity field or atmosphere density parameters and for orbit comparisons and the validation of dynamic orbits and models. For the first time, several university groups estimated Earth gravity field coefficients and studied their temporal variations using these CHAMP kinematic positions together with the variance-covariance information and making use of energy balance approach or boundary value method rather than classical perturbation theory, see e.g. Gerlach et al. (2003) at TU Munich, Mayer-Gürr et al. (2005) at TU Bonn, Reubelt et al. (2004) at TU Stuttgart and Ditmar et al. (2004) at TU Delft. The validation of gravity field models computed in such a way showed that CHAMP kinematic positions contain high-resolution gravity information and that the accuracy of the derived gravity models is comparable to that of official CHAMP models, if not better. Kinematic positions with the corresponding variance-covariance information are an extremely attractive interface between the LEO GPS data and gravity field models or other interesting information that can be derived from satellite orbits, because the simultaneous adjustment of model parameters (e.g. gravity field coefficients) and a huge amount of global GPS parameters, like GPS satellite clocks and orbits, zero- or double-difference ambiguities, station coordinates, troposphere parameters, Earth rotation parameters, etc. can be avoided. However, kinematic POD provides no velocities for energy integral methods, thus they have to be derived numerically from kinematic positions.

In Švehla & Rothacher (2005) we showed for the first time that orbits of the GPS satellites can also be estimated kinematically and made comparisons with the kinematic positioning of LEO satellites and IGS stations on the ground. We demonstrated that LEO point-positioning is possible by means of GPS satellite clocks estimated solely based on phase GPS measurements and using code measurements only for a priori GPS satellite/receiver clock synchronization. In the same paper a fourth fundamental approach in precise orbit determination was introduced which we call reduced-kinematic POD, with the main characteristic, that kinematic positions between consecutive epochs are smoother. Due to small numbers of GPS satellites tracked, some kinematic epochs have worse variance-covariance properties and the reduced-kinematic approach copes with this problem in the way that kinematic position differences in time are relatively constrained to corresponding differences in a priori dynamic orbit. Therefore, reduced-dynamic POD reduces dynamics towards kinematics and in the reduced-kinematic case, kinematics is reduced towards dynamics.

2 Kinematic and Reduced-Dynamic POD

2.1 Description of Methods

Using GPS measurements the orbit of a LEO satellite can be computed using kinematic as well as (reduced)-dynamic approaches.

The kinematic approach is a purely geometrical approach without using any information on satellite dynamics (e.g. gravity field, air-drag, etc.). In our case the LEO kinematic orbit is represented by three kinematic coordinates at each epoch and estimated with a least-squares adjustment using phase measurements only. No constraints are applied to the kinematic positions.

Dynamic and reduced-dynamic POD on the other hand is based on the numerical integration of the equation of motion and the variational equations to obtain the orbit itself, as well as the partial derivatives with respect to the orbital parameters. In the reduced-dynamic case a large number of empirical (e.g. pseudo-stochastic pulses) or force field parameters are estimated in order to cope with the deficiencies in the dynamical models. In our case pseudo-stochastic pulses are set up as additional parameters in along-track, cross-track and radial direction every 6-15 min.

Kinematic or reduced-dynamic POD of a LEO may be carried out on the zero-, the double- or even the triple-difference level of GPS phase or code measurements. In this paper we will focus on zero- and double-difference procedures using phase measurements only.

In the zero-difference case, GPS satellite orbits and clocks are kept fixed and epoch-wise LEO GPS receiver clock parameters are estimated together with either epoch-wise kinematic positions in the kinematic case, or (reduced-)dynamic orbital parameters in the reduced-dynamic approach. Since phase GPS measurements are used, more than 400 zero-difference ambiguities are additional parameters for a 1-day orbit arc. POD based on zero-differences is a very fast and efficient approach, because only GPS measurements of the LEO are involved and therefore, the processing of a ground IGS network is only required in a preceding, independent step in order to obtain GPS satellite orbits and clocks.

In the double-difference approach, baselines between the LEO and GPS ground stations are formed and all clock parameters are eliminated. In our case baselines are formed between CHAMP and 40 IGS stations and all global GPS products like 3-day solutions for GPS orbits, ERPs, troposphere and weekly station coordinates are kept fixed. The main advantage of kinematic and reduced-dynamic POD based on double-differences is the possibility to resolve ambiguities to their integer values using ambiguity resolution strategies and thereby gain in accuracy. Our ambiguity resolution strategy is based on the Melbourne-Wübbena (MW) linear combination to resolve wide-lane ambiguities in a first step and on a subsequent, iterative bootstrapping to resolve narrow-lane ambiguities. 10 narrow-lane ambiguities (out of $\approx$ 5000 ambiguities) are resolved in each iteration step, i.e. before the one-day normal equation matrix is re-inverted.

More details about our zero- and double-difference approaches, ambiguity resolution and the mathematical background of kinematic and reduced-dynamic POD can be found in Švehla & Rothacher (2002), Švehla & Rothacher (2003a), Švehla & Rothacher (2003b), Švehla & Rothacher (2003c), Švehla & Rothacher (2003d), Švehla & Rothacher (2004a), Švehla & Rothacher (2004b) and Švehla & Rothacher (2005). Other approaches of kinematic POD can be found, e.g., in Colombo et al. (2002), Bock (2003) or Byun (2003).

2.2 CHAMP POD Results

Two comparisons were performed to assess the consistency and accuracy of the CHAMP orbits computed: the comparison between kinematic and reduced-dynamic orbits and the comparison of CHAMP SLR measurements to the GPS-derived orbits.

Fig. 1 shows the difference between CHAMP kinematic and reduced-dynamic orbits based on zero-difference phase measurements for GPS week 1175/2002. The consistency between these two orbits is on the level of about 2 cm over the entire week. The differences between CHAMP kinematic and reduced-dynamic orbits computed using double-differences are displayed in Fig. 2 for day 199/2002.

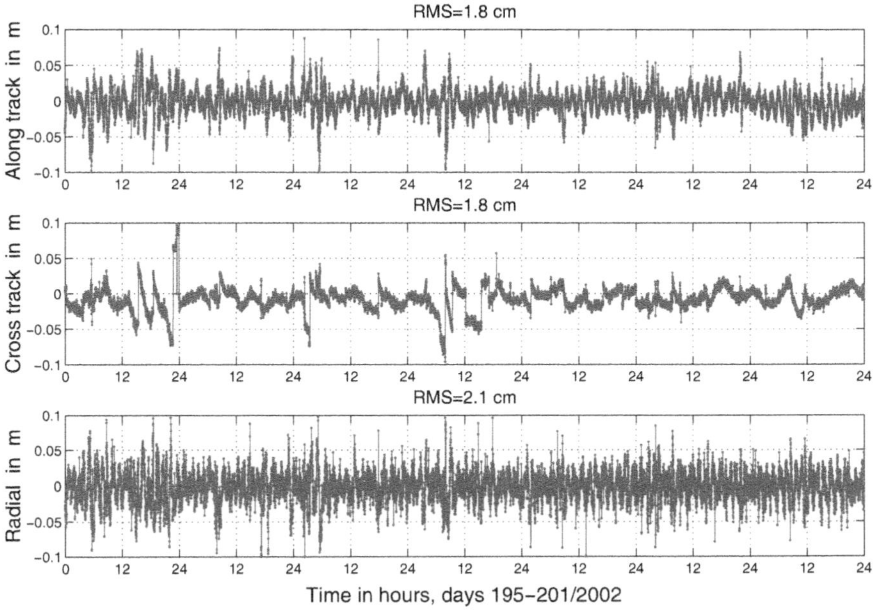

Fig. 1. Differences between CHAMP kinematic and reduced-dynamic orbit based on zero-differences, GPS week 1175/2002 (days 195-201/2002)

A more detailed look at Fig. 2 reveals that CHAMP kinematic positions sometimes exhibit large spikes, due to the small number of GPS satellites tracked and the resulting poor satellite geometry. These points can easily be recognized when looking at the variance-covariance information. Due to the nature of the GPS phase observable, kinematic positions are very smooth from epoch to epoch and as a consequence high-frequency gravity signals may be extracted from these positions. Keep in mind that in our kinematic approach no constraining is applied. Systematic deviations can be recognized in the along-track and radial component pointing at deficiencies in the gravity field and air-drag modeling of the reduced-dynamic approach.

Figure 3 shows typical correlations of LEO kinematic positions estimated using GPS measurements. One can notice a correlation length of 20-30 min and this length is very similar to the duration of continuous tracking of a single GPS satellite from a LEO orbit. Figure 3 shows that variance-covariance information between epochs has to be used when kinematic positions are used in the gravity field determination as pseudo-observations.

That we are not just talking about consistency between orbits but also about accuracy, can be seen in Fig. 4, where SLR residuals are shown for the same kinematic and reduced-dynamic CHAMP orbit as those displayed in Fig. 1. Tropospheric delays for SLR measurements were modeled using the

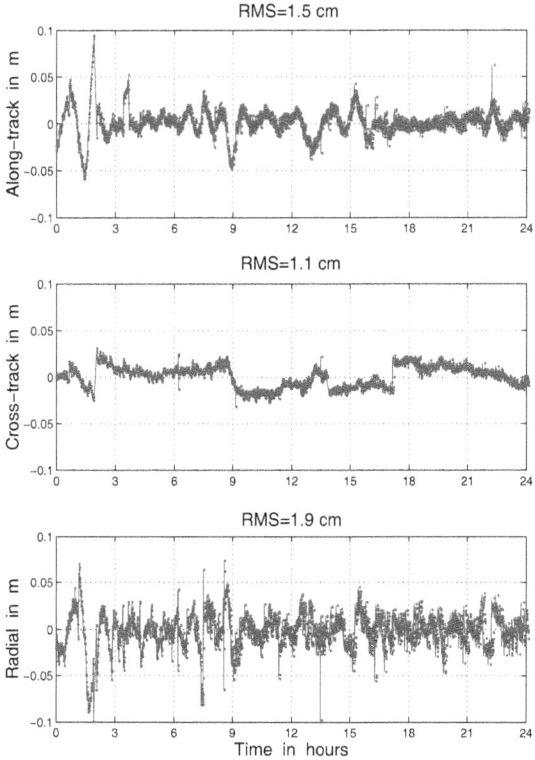

Fig. 2. Difference between CHAMP kinematic and reduced-dynamic orbits based on double-differences, day 199/2002

Marini-Murray model and standard corrections like ocean loading (GOT00.2), Shapiro relativistic effect and station velocities were applied. The analysis was performed using ITRF 2000 station coordinates, velocities and eccentricities published by ILRS at http://ilrs.gsfc.nasa.gov/. All SLR stations and SLR measurements were used in this validation (elevation cut-off 10°). Both orbit types exhibit the same quality of about 2.5 cm. It is interesting to note that the SLR residuals show a similar behaviour for kinematic and reduced-dynamic orbits and that no significant bias can be identified in the SLR residuals.

Table 1 summarizes the daily RMS of the SLR residuals for our CHAMP orbits based on four different POD approaches, namely kinematic and reduced-dynamic orbits based on zero- and double-differences.

One can see that CHAMP orbits are of similar quality for a purely kinematic and reduced-dynamic approach. This also stands for CHAMP orbits computed using either zero- or double-difference phase measurements. Slightly better orbit quality (2.56 cm) is obtained when using kinematic POD and double-differences.

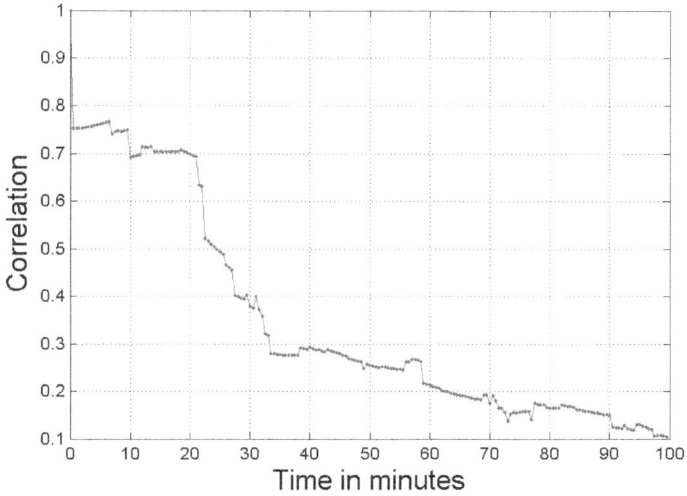

Fig. 3. Typical correlations of CHAMP kinematic positions with correlation length of 20-30 min. The x-component, day 200/2002

3 Gravity Field Determination Based on Kinematic Orbits

The use of kinematic positions together with their variance-covariance information as an interface to gravity field determination avoids the simultaneous adjustment of gravity field coefficient together with a huge amount of global GPS parameters, like GPS satellite orbits/clocks, zero- or double-difference ambiguities, station coordinates, troposphere parameters, Earth rotation parameters, etc. Comparison with reduced-dynamic orbits and external valida-

Table 1. Daily RMS of SLR residuals in cm for CHAMP kinematic and reduced-dynamic ("red.-dyn.") orbits based on zero- and double-differences (days 195-202/2002)

Day	Zero difference red.-dyn.	Zero difference kinematic	Double difference red.-dyn.	**Double difference kinematic**
195	4.02	4.17	3.22	2.66
196	2.90	2.93	3.19	3.03
197	3.40	3.11	3.29	2.90
198	2.07	2.07	1.99	1.34
199	1.94	1.66	1.91	1.70
200	1.43	1.45	1.69	1.83
201	3.59	4.65	4.32	5.00
202	2.03	2.08	1.93	2.05
Mean	2.67	2.77	2.69	**2.56**

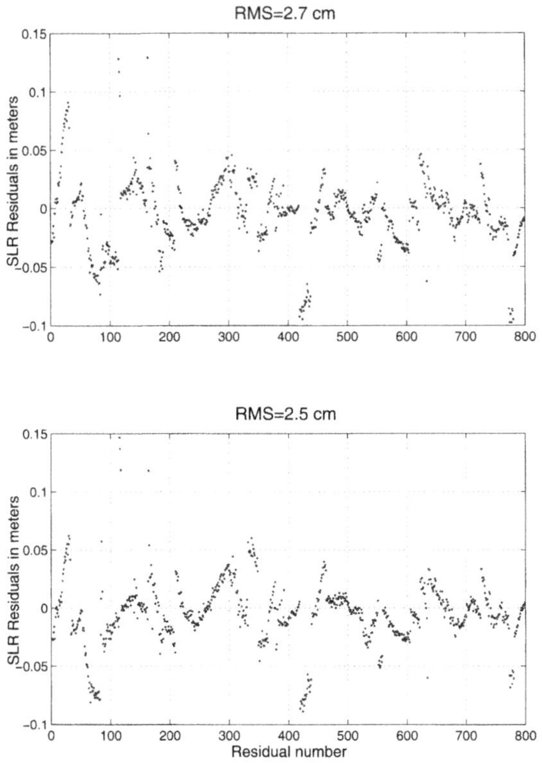

Fig. 4. SLR residuals for CHAMP kinematic (top) and reduced-dynamic orbits (bottom) for GPS week 1175/2002 (days 195-201/2002). All SLR residuals were used in the analysis, elevation cut-off 10°

tion with SLR show that, due to the nature of the phase observable, kinematic positions are very smooth from epoch to epoch (see Fig. 2 and Fig. 4) as long as there are no phase breaks. As a consequence, high-frequency gravity signals may be extracted from these positions. An elegant way to derive gravity field coefficients from kinematic positions is the use of the energy conservation law which may be written for the satellite center of mass in an inertial frame as

$$V = \frac{1}{2}(\frac{d\mathbf{x}}{dt})^2 - \int_{\mathbf{x}} \mathbf{a}_t d\mathbf{x} - \int_{\mathbf{x}} \mathbf{a}_{non} d\mathbf{x} - C. \tag{1}$$

with the gravitational potential V, the accelerations $\mathbf{a}_t$ and $\mathbf{a}_{non}$ due to the time-varying part of the gravity field (e.g. tides) and the non-gravitational forces, respectively, and the total energy constant C. A similar formulation in the earth-fixed frame can be found in Gerlach et al. (2003). An advantage of the gravity field determination based on the energy integral is that we can directly work with the gravity potential as a scalar field instead of having to integrate the equation of motion and all variational equations.

Whereas $\mathbf{a}_t$ can be obtained from models, the non-gravitational accelerations $\mathbf{a}_{non}$ are measured by accelerometers. The kinematic energy of the satellite can be computed using velocities derived from kinematic positions by numerical differentiation procedures. At the moment CHAMP kinematic positions are computed with a sampling of 30 s, which means that the spatial resolution of the estimated gravity field is limited to about 230 km and that much care has to be taken when deriving kinematic velocities. Going to a higher sampling rate, numerical differentiation will become more accurate and a higher spatial resolution will become possible.

4 Determination of Kinematic Velocity

CHAMP orbits (kinematic and reduced-dynamic) over one year have been used for determination of the TUM-1S and TUM-2Sp gravity models. Kinematic orbits are provided as satellite position and variance-covariance information, whereas reduced-dynamic orbits contain position and velocity information. The used orbits span exactly 1 year (365 days), from 11.03.2002 to 10.03.2003. The data is sampled at a rate of 30 seconds.

The raw orbit data have first been screened and pre-processed to eliminate bad data. Different thresholds have been defined and checked.

Further features become visible by screening the position differences. There can be detected clear jumps, which are actually some centimeters of the orbit change over an length of 100 km. Since the reduced-dynamic orbit is very smooth, these jumps are mainly contributed by the kinematic orbit and are caused by breaks in phase measurements and small number of tracked GPS satellites. These jumps occur between independently derived arcs of the orbit, reflecting the uncertainty of estimation of ambiguity parameters. In general, the continuous parts of the kinematic orbit contains the useful information: purely geometrical continuous positions. These are the basis for velocity determination. However, at the jumps, the edges of continuous arcs should not be employed for deriving kinematic velocities. Thus the orbit is divided into independent arcs before starting to estimate the velocity.

Thus, in the process of kinematic velocity estimation, further thresholds have been introduced. Data not used for the kinematic velocity derivation process are those, which (1) are located at jumps due to loss of phase connection between GPS satellites, which (2) are contained in arcs too short for fitting substantially an analytical function and which are (3) close to the end points of a continuous arc. This latter condition is based on screening experiences: at the end points of an arc, the fit of an analytical function is less certain than in the middle of the arc. According to these conditions another 5.28% of the data has been excluded.

Finally there is a check of the efficiency of the above criteria. The condition eliminates too large differences between kinematic and the reduced-dynamic velocities. A threshold of 1 mm/s has been defined for the velocity differences, resulting in another 0.01% of the data found to be inaccurate.

Altogether 21.57% of the total data has been eliminated from the remaining processes.

4.1 Numerical Differentiation

The kinematic velocity is derived by introducing an analytical function that fits well to the kinematic positions, followed by an analytical differentiation. Having the pre-processed kinematic orbits at hand, five mathematical methods have been tested for kinematic velocity derivation. These are cubic spline interpolation, polynomial interpolation, Newton-Gregory interpolation, smoothing cubic splines, and polynomial smoothing.

The latter two techniques are tested in order to try to smooth noise of the positions. The extent of the smoothing depends on an arbitrary parameter, which is defined directly in case of the cubic spline smoothing, while in case of the polynomial fitting it is implicitly governed by the choice of the number of points and the order of the polynomial.

One remark on kinematic velocity determination: to validate a method, a comparison for the velocity estimate is needed. Unfortunately no physically meaningful measure for comparison can be found. Due to its dependence on a gravity model the reduced-dynamic velocity is not an adequate measure, however, it indicates well the overall characteristics of kinematic velocities. With having said this, conclusions should be carefully done, and the dependence of the reference orbit on a prior gravity model should never be forgotten.

Statistics in this section refer to a test day, that is day 200 of year 2002 (equivalent to 19th July, 2002). This day was found to be representative.

Optimal Smoothing Parameter for Smoothing Splines

The extent of smoothing of a smoothing cubic spline is known to depend on the length of the time series and on the amplitude of the signal (Greville, 1967). Let us define the smoothing parameter of a smoothing spline function, S, as a linear function of the length of the arcs: $S = scale\ \ factor \cdot length(arc)$. We applied several scale factors and determined position and the first derivative (velocity) of the smoothed function. Table 2 shows the RMS of the smoothed velocities as function of the scale-factor, compared to reduced-dynamic velocities. The smoothed kinematic velocities were found to be most similar to the reduced-dynamic velocities employing 60 as a scale factor.

Velocity Estimation

Different interpolation and smoothing techniques for taking analytically the time derivative of the position are first applied on purely kinematic positions.

Table 2. Optimal scale factor for the smoothing parameter of the smoothing splines. The term 'velocity residuals' refers to reduced-dynamic minus smoothed velocities

Scale factor	RMS velocity residual [mm/s]
10	0.3336
40	0.3334
60	0.3334
100	0.3335
150	0.3335
200	0.3337

Table 3. RMS of kinematic velocity errors obtained by different mathematical techniques

RMS	velocity error [mm/s]
Spline interpolation	0.3334
Polynomial interpolation	0.3350
Newton-Gregory interpolation	0.3102
Smoothing Splines	0.3334
Polynomial Smoothing	0.2219

Table 3 shows the RMS of the velocity errors, i.e. the differences between the reduced-dynamic and the kinematic velocities.

Interpolation techniques provide similar accuracy for velocities. The Newton-Gregory interpolation performs slightly better than the other techniques. Different smoothing parameters were tested for the smoothing spline, however without any improvement in accuracy. The shown value is equivalent to $S = 0$, that is spline interpolation without smoothing. This feature is due to the peculiar filter characteristics of the smoothing spline function: in case of a signal with a characteristic large amplitude at a certain frequency (e.g. the orbital frequency), the smoothing spline minimizes the curvature of that frequency. However, in case of a nearly white distribution of the data (i.e. position differences with respect to a similar orbit), the smoothing tends to minimize amplitudes at the highest frequencies. Therefore velocity estimation was based on position differences by making use of a reference orbit in a remove-restore manner. After removing the reference orbit, we took the derivatives on residuals (having an amplitude of only few centimeters) and finally add back the residual kinematic velocity to the velocity of the reference orbit (restore).

For the reference orbit, purely dynamic EIGEN-1S orbits have been determined. Table 4 shows the RMS of the velocity errors, the differences of the kinematic and the reduced-dynamic velocities.

The results show no relevant differences among the interpolation techniques – meaning that the numerical errors of these techniques are less than

Table 4. RMS of kinematic velocity errors obtained by different mathematical approximation techniques. Reference orbit: dynamic, EIGEN-1S, max. degree: 120

RMS	velocity error [mm/s]
Spline interpolation	0.3356
Polynomial fitting (interpolation)	0.3465
Newton-Gregory interpolation	0.3185
Smoothing Splines	0.1554
Polynomial Smoothing	0.2253

10^{-8} part of the signal. It is noteworthy that the accuracy of velocities is slightly worse than that of the directly derived kinematic velocities (cf. Table 3). This feature may indicate the inconsistency of the remove-restore step, i.e. inconsistencies of reference positions and velocities. However, as expected, velocities by smoothing splines have improved a lot.

4.2 The CHAMP Velocity Results

The above mentioned tests suggest two different approaches for kinematic velocity estimation. The difference is the way of treatment of the noise of the kinematic orbit. The efficiency should depend on the spectral characteristic of the kinematic position errors.

First we tried to smooth the noise by applying smoothing cubic spline functions on position residuals. In this case we perturb the spectral characteristics of the orbit errors due to the smoothing process of the spline function. Also the reference orbit (reduced-dynamic EIGEN-2 orbit) affects the solution to some extent. Velocities of this kind have been used for the computation of the TUM-1S gravity model (Gerlach et al., 2003).

Second, we also derived kinematic velocities by fitting a 7th order polynomial to the kinematic positions with the Newton-Gregory interpolation technique. In this case all noise is included, however, its spectral characteristic remains unchanged. In case of randomly distributed kinematic position errors this method is promising. The Figure 5 shows the velocity differences derived by Newton-Gregory interpolation used for computation of TUM-2Sp (Földváry et al., 2003). The RMS of velocity differences of about 0.3 mm/s corresponds to accuracy of kinetic energy of $\approx 4.5 \cdot 10^{-8} \mathrm{m}^2/\mathrm{s}^2$.

The final conclusion on the efficiency of the two techniques can be done after the gravity field determination. Geoid heights computed from the gravity models have been compared to GPS-levelling heights along that of other gravity models. These are summed in Table 5. According to that, we can see that in case of CHAMP kinematic velocity the use of Newton-Gregory interpolation proved to be much more accurate than the smoothing splines, at least the error it produced was less than that of the smoothing splines.

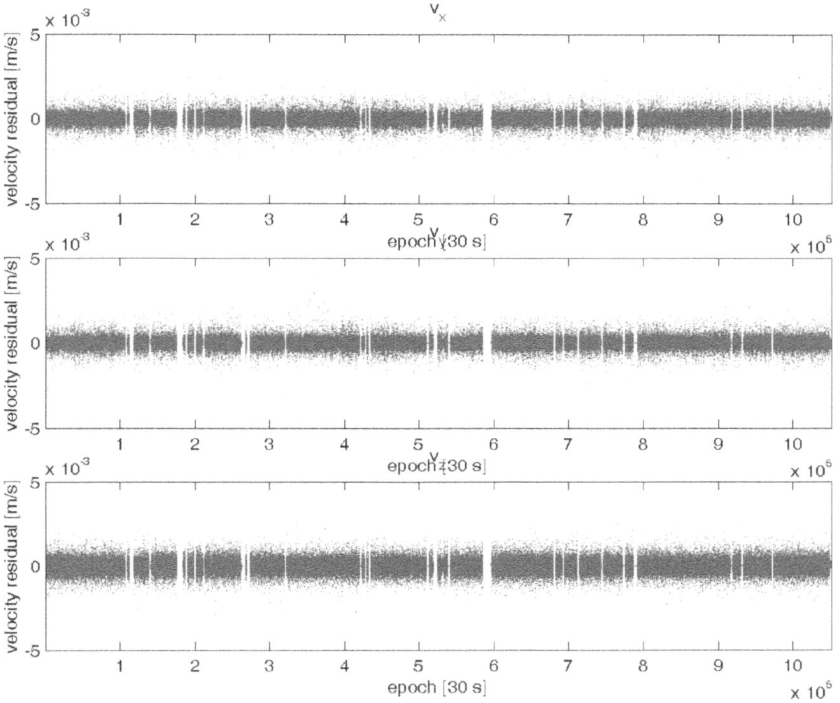

Fig. 5. Differences between the CHAMP kinematic and reduced-dynamic velocities for a period of one year (earth-fixed system). The RMS of velocity differences is about 0.3 mm/s

Table 5. Comparison between geoid heights from GPS/levelling and global potential models (low-pass filter above degree/order 60) in [cm]

data set	EIGEN-2	TUM-1S	TUM-2Sp	EIGEN-GRACE01S
USA (5168 points)	60.2	64.1	47.1	41.5
Europe (180 points)	59.3	56.4	33.1	19.4
Australia (197 points)	67.4	63.3	52.7	50.3
Japan (837 points)	69.5	65.5	54.8	51.5

5 Conclusions

Kinematic precise orbit determination (POD) is a purely geometrical approach and therefore independent of satellite dynamics (e.g. gravity field, air-drag, etc.) and orbit characteristics (e.g. orbit height, eccentricity, etc.).

Using GPS measurements the orbit of the CHAMP satellite can be determined purely kinematically with the same level of accuracy ($\approx$ 1-3 cm) as by the widely employed reduced-dynamic or dynamic approaches.

Typical correlations of LEO kinematic positions have correlation length of 20-30 min and this time is very similar to the time of continuous tracking of a single GPS satellite from a LEO orbit. Therefore the variance-covariance information between epochs has to be used when kinematic positions are used in the gravity field determination.

Conclusion on the efficiency of the several numerical differentiation techniques was drawn after the gravity field determination. Geoid heights computed from the gravity models have been compared to GPS-levelling heights along that of other gravity models. This analysis shows that in case of CHAMP kinematic velocity the use of Newton-Gregory interpolation proved to be much more accurate than the smoothing splines.

The RMS of velocity differences between derived kinematic velocities and reduced-dynamic velocities is about 0.3 mm/s corresponding to accuracy of kinetic energy of $\approx 4.5 \cdot 10^{-8}$ m^2/s^2.

Acknowledgement. This is publication no. GEOTECH-160 of the programme GEOTECHNOLOGIEN of BMBF and DFG, Grant 03F0329A.

References

Balmino G, Perosanz F, Rummel R, Sneeuw N, Sünkel H (1999) CHAMP, GRACE and GOCE: mission concepts and simulations. Bolletino di Geofisica Teorica ed Applicata. Vol. 40: 309-319

Bertiger W, Bar-Sever Y, Christensen E, Davis E, Guinn J, Haines B, Ibanez-Meier R, Jee J, Lichten S, Melbourne W, Muellerschoen R, Munson T, Vigue Y, Wu S, Yunck T, Schutz B, Abusali P, Rim H, Watkins M, Willis P (1994) GPS precise tracking of TOPEX/POSEIDON: results and implication. J Geophys Res 99, C12: 24449-24464

Byun SH (2003) Satellite orbit determination using triple-differene GPS carrier phase in pure kinematic mode. J. of Geodesy 76: 569-585

Bock H (2003) Efficient Methods for Determining Precise Orbits of Low Earth Orbiters Using the Global Positioning System. PhD thesis. Astronomical Institute University of Berne, Berne, Switzerland

Colombo OL, Luthcke SB, Rowlands DD, Chin DS, Poulouse S (2002) Filtering Errors in LEO Trajectories Obtained by Kinematic GPS with Floating Ambiguities. Presented at The ION Symposium "GPS 2002", Sept. 24-27, 2002, Portland Oregon

Ditmar P, Kuznetsov V, Van Eck van der Sluijs A, Schrama E, Klees R (2004) DEOS_CHAMP-01C_70: a new model of the Earth's gravity field derived from the CHAMP satellite data by means of the acceleration approach. Paper presented at the Joint CHAMP GRACE Science Meeting, GFZ Potsdam, Germay, 5-8 July 2004

Földváry L, Gerlach Ch, Švehla D, Frommknecht B, Gruber Th, Peters Th, Rothacher M, Rummel R, Sneeuw N, Steigenberger P (2003) Determination of the Gravity Field From CHAMP Measurements Considering the Energy Integral. In: Earth Observation with CHAMP - Results from Three Years in Orbit (eds) Reigber Ch, Lühr H, Schwintzer P, Wickert J, pp. 13-18, Springer Verlag

Gerlach Ch, Földváry L, Švehla D, Gruber Th, Wermuth M, Rothacher M, Rummel R, Sneeuw N, Frommknecht B, Peters Th, Steigenberger P (2003) A CHAMP only Gravity Field Model From Kinematic Orbits Using the Energy Integral. Geoph. Res. Letters, 30(20), 2037, doi:10.1029/2003GL018025

Greville TNE (1969) Theory and Applications of Spline Functions. Proceedings of an Advanced Seminar Conducted by the Mathematic Research Center, United States Army, at the University of Wisconsin, Madison October 7-9, 1968. Academic Press, New York, London

König R, Baustert G, Neumayer KH, Meixner H, Schweiger V, Zhu S, Reigber C (2001) Precise orbit determination for CHAMP. Paper presented at EGS XXVI General Assembly, Nice, France

Mayer-Gürr T, Ilk KH, Eicker A, Feuchtinger M (2005) ITG-CHAMP01: a CHAMP gravity field model from short kinematic arcs over a one-year observation period. J. of Geodesy, 78, (7-8): 462-480. doi 10.1007/s00190-004-0413-2

Melbourne WG (1985): The Case for Ranging in GPS Based Geodetic Systems. First International Symposium on Precise Positioning with the GPS, C. Goad (Ed), U.S. Department of Commerce, Rockville, Maryland, pp. 373-386

Reigber Ch, Schwintzer P, Neumayer K-H, Barthelmes F, König R, Förste Ch, Balmino G, Biancale R, Lemoine J-M, Loyer S, Bruinsma S, Perosanz F, Fayard T (2003) The CHAMP-only Earth Gravity Field Model EIGEN-2. Adv. in Space Research, 31(8), 1883-1888, 2003, doi: 10.1016/S0273–1177(03)00162-5

Reubelt T, Goetzelmann M, Grafarend EW (2004) A new CHAMP gravity field model based on the GIS acceleration approach and two years of kinematic CHAMP data. Paper presented at the Joint CHAMP GRACE Science Meeting, GFZ Potsdam, Germany, 5-8 July 2004

Rothacher M, Schmid R, Steigenberger P, Svehla D, Thaller D (2004) Combination of the Space Geodetic Techniques for Monitoring the Earth's System. AGU Fall Meeting, San Francisco, 13 - 17 December 2004

Rummel R (2003) How to Climb the Gravity Wall. In: Beutler G, Rummel R, Drinkwater MR, Steiger R (Eds): Earth Gravity Field From Space – From Sensors to Earth Sciences, Proceedings of an ISSI Workshop 11-15 March 2002, Bern Switzerland. Space Science Reviews, 108, (1-2): pp. 1-14

Švehla D, Rothacher M (2002) Kinematic Orbit Determination of LEOs Based on Zero- or Double-Difference Algorithms Using Simulated and Real SST Data. In: Adam J, Schwarz KP (Eds): Vistas for Geodesy in the New Millenium, Proceedings of the IAG 2001 Scientific Assembly, Budapest. Springer IAG Vol. 125, pp. 322-328. (http://tau.fesg.tu-muenchen.de/~drazen/)

Švehla D, Rothacher M (2003a) CHAMP double-difference kinematic orbit with ambiguity resolution. In: Reigber Ch, Lühr H, Schwintzer P (Eds): First CHAMP Mission Results for Gravity, Magnetic and Atmospheric Studies, Springer, pp. 70-77. (http://tau.fesg.tu-muenchen.de/~drazen/)

Švehla D, Rothacher M (2003b) Kinematic and Reduced-Dynamic Precise Orbit Determination of Low Earth Orbiters. EGSXXVII General Assembly 2002, Nice, France. Advances in Geosciences 1: 47-56 (http://www.copernicus.org/EGU/adgeo /published_papers.htm)

Švehla D, Rothacher M (2003c) Kinematic and Reduced-Dynamic Precise Orbit Determination of CHAMP satellite over one year using zero-differences. Poster presented at EGS-AGU-EUG Joint Assembly 06-11 April 2003, Nice, France (http://tau.fesg.tu-muenchen.de/~drazen/)

Švehla D, Rothacher M (2003d) Report on Validation of SRON SST Simulator. Support to End-to-End Simulations for the GOCE mission, Alenia Contract no. GO-SC-SRON-0322, Munich, Germany

Švehla D, Rothacher M, (2004b) Two Years of CHAMP Kinematic Orbits for Geosciences. Geophysical Research Abstracts, European Geophysical Society Vol. 6. ISSN:1029-7006 (http://tau.fesg.tu-muenchen.de/~drazen/)

Švehla D, Rothacher M (2004a) Kinematic Precise Orbit Determination for Gravity Field Determination. Proceedings of the International Association of Geodesy: A Window on the Future of Geodesy. Eds. F. Sanso. Springer Verlag, IAG Vol 126. pp 181-188

Švehla D, Rothacher M (2005) Kinematic positioning of LEO and GPS satellites and IGS stations on the ground. Advances in Space Research, doi:10.1016/j.asr.2005.04.066

Tapley BD, Ries JC, Davis GW, Eanes RJ, Schutz BE, Shum CK, Watkins MM, Marshall JA, Nerem RS, Putney BH, Klosko SM, Luthcke SB, Pavlis D, Williamson RG, Zelensky NP (1994) Precision orbit determination for TOPEX/-POSEIDON. J Geophys Res 99, C12: 24383-24404

Yunck T, Bertiger W, Wu S, Bar-Sever Y, Christensen E, Haines B, Lichten S, Muellerschoen R, Vigue Y, Willis P (1994) First assessment of GPS-based reduced dynamic orbit determination on TOPEX/POSEIDON. Geophys Res Letters 21: 541-544

Mission Simulation and Semi-analytical Gravity Field Analysis for GOCE SGG and SST

Martin Wermuth[1], Reiner Rummel[1], and Lóránt Földváry[1,2]

[1] Institut für Astronomische und Physikalische Geodäsie, Technische Universität München, wermuth@bv.tum.de
[2] Budapest University of Technology and Economics

Summary. GOCE will be the first satellite mission equipped with a gravity gradiometer. In order to achieve maximum precision and spatial resolution, the instrument is guided around the Earth in an extremely low orbit, employing active along track drag-free control and angular control by magnetic torquers. Furthermore, the orbit trajectory is determined very accurately by continuous and three-dimensional GPS satellite-to-satellite tracking. These mission characteristics are modelled by a system of two sequential simulators. The sensor system simulator computes the interaction of the complete sensor system and provides time series or power spectral densities of the gradiometer components. The mission simulator derives the geoid and gravity model performance. It takes as input mission and orbit parameters, expected GPS performance as well as the gradiometer error spectral densities derived from the sensor system simulator.

The computational effort of the actual data analysis can only be managed by powerful computer systems, in principle, due to the large number of observations and unknown gravity field parameters. In this article a Semi-Analytical Approach is presented; it is a simple and fast alternative to a direct solution. It is based on simplifying assumptions, which allow to use FFT-techniques. It can be divided in two approaches: the 1D-FFT approach, and the 2D-FFT approach or torus-approach. In several case studies, the basic properties of the two approaches are shown and a comparison to the direct solution is carried out. The Semi-Analytical Approach will be used as Quick-Look Tool in the official ESA GOCE gravity field processing.

Key words: GOCE, gravity gradiometry, satellite gravity gradiometer, satellite-to-satellite tracking, geodesy, satellite geodesy, Torus Approach, Simulation, gravity field analysis

1 GOCE Sensor System and Mission Analysis

Before the launch of GOCE each performance analysis and preparation of data analysis has to be based on simulations. For this reason a simulator has been developed of the GOCE sensor system. Its purpose is to obtain a realistic picture of the interaction of all sensor components and of the spectral behavior of the error budget. A second element is a mission simulator. It is built on the results of the sensor simulator. It estimates the GOCE mission performance in terms of geoid and gravity anomaly errors. Input is the orbit and mission parameters as well as the error characteristics of the gravity gradiometer system and of the GPS satellite-to-satellite tracking system.

1.1 GOCE Sensor Analysis

GOCE will be the first satellite equipped with a gravity gradiometer. It will measure gravity gradients with a precision of better than 10^{-2}E (1E = $10^{-9}s^{-2}$) in the so-called measurement bandwidth (MBW: 5mHz - 1Hz). The instrument consists of three orthogonal one-axis gradiometers, each of them composed of two three-axis accelerometers mounted at the end points of 0.5 m long rigid baselines. The gradients are derived from the acceleration differences (differential mode). Each accelerometer will be ultra-sensitive along two spatial dimensions ($10^{-12}m/s^2$ in the MBW) while the third axis will be less accurate. The sensitive axes are arranged such as to provide the gravity gradiometer tensor components Γ_{xx}, Γ_{yy}, Γ_{zz}, Γ_{xz} and Γ_{zx} with high precision. The second core instrument will be a newly developed geodetic GPS receiver. Using code and phase measurements it will deliver the satellite trajectory in all three spatial directions with cm-precision. The high precision and spatial resolution of GOCE will also be due to the extremely low orbit altitude of 240 km and due to the smooth attitude control. The orbit will be kept free of drag in along track direction by means of ion thrusters. The size of the drag is measured by the accelerometers of the gradiometer, too, in all three directions taking advantage of their common-mode - or average - signal. Attitude control was originally planned to be done by a field-electron-emission-propulsion system (FEEP). This system turned out to be not yet reliable enough, therefore one had to turn to magnetic torquing. The necessary control signals, angular velocities, are determined from a combination of star tracker signal and the Γ_{xz}- and Γ_{zx}-components of the gradiometer. These two gradiometer components allow reconstruction of the angular acceleration about the cross-track axis. The sensor simulator is modelling the entire sensor system (cf. Color Fig. XXV on p. 301).

Input to this system are the linear and angular forces along the orbit trajectory acting on the spacecraft and on the proof masses of the six accelerometers. They are, in particular, the direct and indirect gravity acceleration of Earth, Moon, Sun and planets, the non-gravitational accelerations due to air drag and solar radiation and the control accelerations determined from the common mode accelerations as well as all exterior torques (e.g. from atmosphere) and control torques. Output is time series or power spectral densities of the gravity gradient components, angular velocities and accelerations and common mode linear accelerations. The same simulator can be employed in order to model the in-orbit calibration of the gradiometer system. According to the current mission baseline this will require to incorporate cold gas thrusters. They generate random linear and angular shaking for calibration purposes. For a more detailed description of the simulator and the details of Color Fig. XXV it is referred to (Oberndorfer et al., 2000, 2002)

1.2 GOCE Mission Simulator

The sensor system simulator produces power spectral densities of all components of the gradiometer. A typical example is shown in Fig. 1.

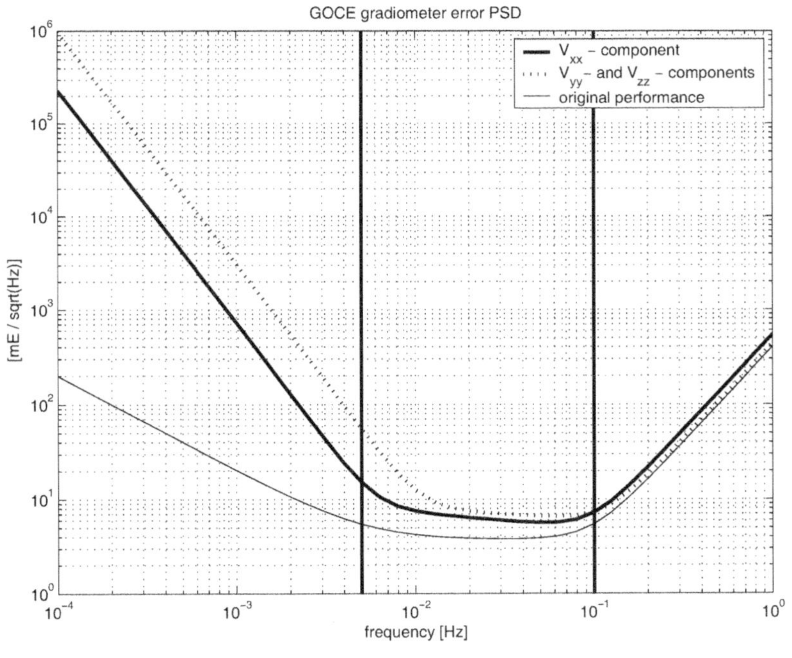

Fig. 1. Simulated power spectral densities of the GOCE gravity gradiometer

It shows that inside the measurement bandwidth a precision of below 10^{-2}E is achievable. It also shows that towards the low frequencies (below 5×10^{-3}Hz) the error behavior increases proportional with the inverse frequency. In this spectral domain support from gravitational orbit perturbations, as derived from GPS satellite-to-satellite tracking, is of high importance. The error increase at frequencies above 10^{-1}Hz would result in aliasing. It has to be eliminated by low pass filtering. The mission simulator takes the gradiometer and orbit power spectral densities as input. Input parameters are mission duration and orbit parameters (altitude, inclination and eccentricity). With this input information the expected variance-covariance matrix of the geoid and/or gravity model is derived by linear error propagation. Despite some simplifying assumptions the simulation is still very realistic. The adopted assumptions lead to a block diagonal error variance-covariance matrix. It is a semi-analytic model essentially the same as described in Sect. 2. The gravity and geoid models are expressed in a spherical harmonic expansion. The theoretical background is given in (Sneeuw, 2000). From the computed error variances and covariances of a set of spherical harmonic coefficients, error degree variances can be derived and be compared with the corresponding signal degree variances (according to e.g. Kaula's rule). An example is given in Fig. 2a. The comparison of signal and error spectrum provides insight into the relative precision of the coefficients to be deduced from the GOCE mission and into the maximum spatial resolution. From the error variances one can also determine the cumulative gravity and geoid error at each spherical harmonic degree. At degree n = 200, which corresponds to length scales of about 100 km at the Earth's surface the cumulative geoid uncertainty (commission error) will be about 2 cm; the cumulative gravity uncertainty will be 0.5 mGal, compare Fig. 2b, c, respectively.

2 Gravity Field Analysis

Due to the high spatial resolution, determination of gravity field coefficients from GOCE observations by least-squares adjustment is a very demanding computational task, and can only be managed by clusters or supercomputers. A solution up to degree $L = 250$ results in 63001 unknown coefficients, and due to the high sampling rate of 1Hz, GOCE produces several millions of observations per month. Even computer clusters with 50 nodes need several weeks to derive a gravity model in an adjustment for the whole GOCE mission. So a "Semi-Analytical Approach" has been developed to enable quicker solutions on single PCs within several hours of computation time. The approach is based on simplifying assumptions, such as constant orbit height or a repeat orbit, which allow the use of FFT-techniques for setting up the normal equations, and lead to a block-diagonal normal equation system, which can be solved much faster (Sneeuw, 2000). These assumptions introduce certain errors, which can in parts be compensated for by iteration avoiding the described simplifications during the (backward) synthesis step. The solutions become slightly less accurate than those of a direct solution, but less requirements in computation time and memory are a strong advantage.

The Semi-Analytical Approach has been accepted by ESA as an official analysis strategy, and is implemented as Quick-Look Gravity Field Analysis Tool in the WP6000 of the GOCE-HPF (high-level processing facility). The workpackage 6000 "Time-Wise Gravity Field Determination" is coordinated by the Technical University of Graz and supported by the Technical University of Munich (for the Quick-Look GFA) and the University of Bonn. It consists of a Core-Solver module, which will derive high quality solution via a direct

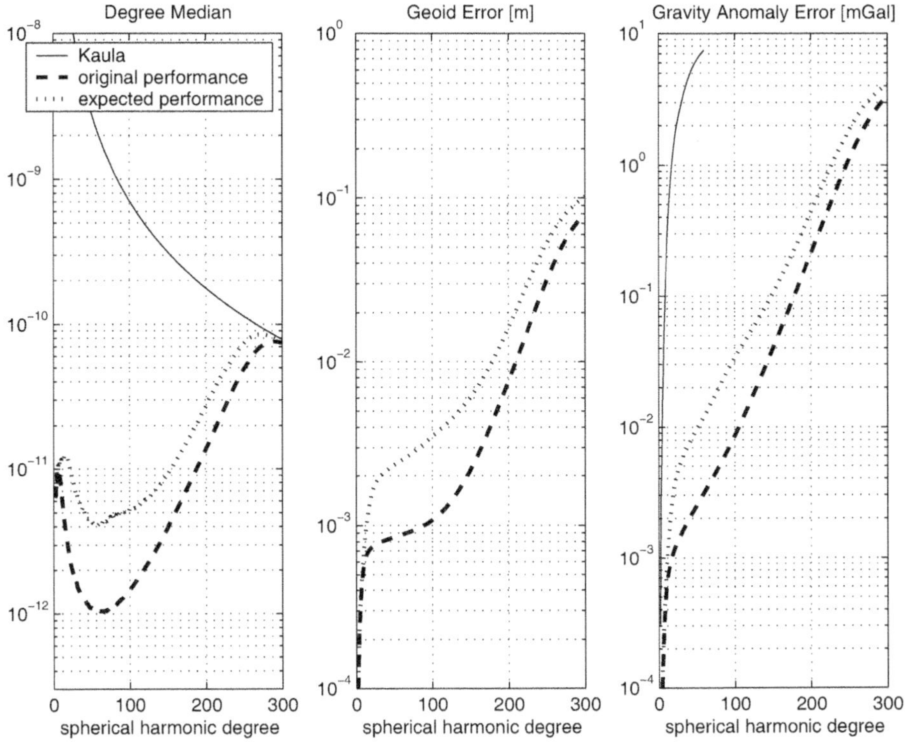

Fig. 2. Estimated signal and error degree variances per spherical harmonic degree (3a), cumulative geoid error per degree (3b), cumulative gravity error per degree (3c)

adjustment and the Quick-Look Tool for fast gravity field analysis and simulations. The purpose of the Quick-Look Tool is to derive gravity field solutions and an analysis of the sensor system with a very short latency, on the basis of rapid science orbits and partial datasets. It should detect deficiencies which are then reported to mission control.

2.1 Gravitational Potential

In this section the mathematical background for gravity field determination from SGG and SST measurements is derived. The gravitational potential in outer space $r > R$ is defined as

$$V(\theta, \lambda, r) = \frac{GM}{R} \sum_{l=0}^{\infty} \left(\frac{R}{r}\right)^{l+1} \sum_{m=0}^{l} \bar{P}_{lm}(\cos\theta) \left[\bar{C}_{lm} \cos m\lambda + \bar{S}_{lm} \sin m\lambda\right] \quad . \tag{1}$$

This can be regarded as a spectral representation on a spherical surface, with the complete and orthonormal system of basefunctions $\bar{P}_{lm}(\cos\theta)\cos m\lambda$ and $\bar{P}_{lm}(\cos\theta)\sin m\lambda$. The $\bar{C}_{lm}$ and $\bar{S}_{lm}$ are the potential coefficients, which are to be determined. As the series expansion cannot be calculated to infinity, in practice, one has to cut the computation at a maximum degree L. The error made by neglecting the higher degrees is called omission error. As $m \leq l$ and all $\bar{S}_{l0}$ - coefficients can be neglected because of $\sin(0\lambda) = 0$, the number of coefficients to be determined is $(L+1)^2$.

For SST based on the energy integral approach with kinematic orbits, a time series of disturbing potential values T enters the adjustment (cf. Svehla et al. in this volume). To obtain the observation equations, the normal potential U has to be subtracted from the potential computed by (1): $T = V - U$. As SGG measures the tensor of second derivatives of the potential, (1) has to be differentiated twice. The potential and its first and second order derivatives all can be expressed in a similar way:

$$V_{ij} = \frac{GM}{r} \sum_{l=0}^{l_{max}} \lambda_{ij} \left(\frac{R}{r}\right)^l \sum_{m=0}^{l} p\left(\alpha\cos(m\lambda) + \beta\sin(m\lambda)\right) \tag{2}$$

with λ_{ij}, p, α and β defined in Table 1. For convenience V_{ij} is written for $\frac{\partial^2 V}{\partial i \partial j}$, and the following substitutions were made: $P_{lm} = \bar{P}_{lm}(\cos\theta)$, $P'_{lm} = \frac{\partial \bar{P}_{lm}(\cos\theta)}{\partial\theta}$ and $P''_{lm} = \frac{\partial^2 \bar{P}_{lm}(\cos\theta)}{\partial\theta^2}$.

Table 1. Derivatives of the gravitational potential

differentiation w.r.t	λ_{ij}	p	α	β
-	1	$\bar{P}_{lm}$	$\bar{C}_{lm}$	$\bar{S}_{lm}$
r	$-\frac{(l+1)}{r}$	$\bar{P}_{lm}$	$\bar{C}_{lm}$	$\bar{S}_{lm}$
θ	1	$\bar{P'}_{lm}$	$\bar{C}_{lm}$	$\bar{S}_{lm}$
λ	1	$m\bar{P}_{lm}$	$\bar{S}_{lm}$	$\bar{C}_{lm}$
rr	$\frac{(l+1)(l+2)}{r^2}$	$\bar{P}_{lm}$	$\bar{C}_{lm}$	$\bar{S}_{lm}$
$r\theta$	$-\frac{(l+1)}{r}$	$\bar{P'}_{lm}$	$\bar{C}_{lm}$	$\bar{S}_{lm}$
$r\lambda$	$-\frac{(l+1)}{r}$	$m\bar{P}_{lm}$	$\bar{S}_{lm}$	$\bar{C}_{lm}$
$\theta\theta$	1	$\bar{P''}_{lm}$	$\bar{C}_{lm}$	$\bar{S}_{lm}$
$\theta\lambda$	1	$m\bar{P'}_{lm}$	$\bar{S}_{lm}$	$\bar{C}_{lm}$
$\lambda\lambda$	-1	$m^2\bar{P}_{lm}$	$\bar{C}_{lm}$	$\bar{S}_{lm}$

The gravity tensor in cartesian coordinates can be computed using the first and second derivatives of the potential. As it is necessary to obtain the tensor in a local orbit frame, it is first computed in a local, spherical, orthonormal frame, with the z-axis pointing radially away from the Earth's center, the x-axis pointing north, and the y-axis east:

$$V_{xx} = \frac{1}{r}V_r + \frac{1}{r^2}V_{\theta\theta} \ , \tag{3}$$

$$V_{xy} = \frac{\cos\theta}{r^2\sin^2\theta}V_\lambda + \frac{1}{r^2\sin\theta}V_{\lambda\theta} \ , \tag{4}$$

$$V_{xz} = \frac{1}{r^2}V_\theta - \frac{1}{r}V_{r\theta} \ , \tag{5}$$

$$V_{yy} = \frac{1}{r}V_r + \frac{1}{r^2\tan\theta}V_\theta + \frac{1}{r^2\sin^2\theta}V_{\lambda\lambda} \ , \tag{6}$$

$$V_{yz} = \frac{1}{r\sin\theta}V_{r\lambda} - \frac{1}{r^2\sin\theta}V_\lambda \ , \tag{7}$$

$$V_{xx} = V_{rr} \ . \tag{8}$$

To obtain the tensor in a local orbit frame, with the z-axis pointing radially away from the Earth's center, the x-axis orthogonal to the z axis pointing quasi along-track, and the y-axis perpendicular to the two others, the tensor has to be rotated about the z-axis by the track angle δ:

$$\delta = \arctan\frac{xrv_y - yrv_x}{sv_z - xzv_x - yzv_y}, \tag{9}$$

where $s = x^2 + y^2$. As the tensor is measured in the actual gradiometer reference frame (GRF), which will deviate from the local orbit frame by up to several degrees, an additional tensor rotation of the observation equations is necessary. As some components of the tensor are measured less accurately, the observation equations should be rotated but not the measurements, because otherwise the less accurate components would be merged with the accurate ones and the resulting gradients in the new frame would be degraded.

2.2 Direct Solution

The direct solution uses a classical least-squares adjustment to obtain the potential coefficients. Formally it can be written as

$$\begin{bmatrix}\bar{C}_{lm}\\ \bar{S}_{lm}\end{bmatrix} = \left(\mathbf{A^T P_y A} + \alpha\mathbf{R}\right)^{-1}\mathbf{A^T P_y y} \tag{10}$$

where $(\bar{C}_{lm}, \bar{S}_{lm})$ is the vector of estimated potential coefficients, $\mathbf{A}$ the matrix of partial derivatives of the observations w.r.t the potential coefficients, $\mathbf{R}$ a regularization matrix weighted by the scalar factor α, and $\mathbf{y}$ the vector containing the time series of observations along the orbit. The weight-matrix $\mathbf{P_y}$ would be by far to large for any computer system, so the information on correlations between the observations and colored noise has to be dealt with by filtering of the measurement series $\mathbf{y}$ beforehand, which is still quite time consuming. The computational effort of this least-squares adjustment can only handled by supercomputers or clusters.

2.3 Time-Wise Representation

Equation (1) is often referred to as "Space-Wise" representation. An alternative is to express it as function of Keplerian elements along the orbit. This was first introduced by Kaula (1966) and modified by Sneeuw (2000), and is referred to as "Time-Wise" representation:

$$V(r,I,u,\Lambda) = \frac{GM}{R}\sum_{l=0}^{\infty}\left(\frac{R}{r}\right)^{l+1}\sum_{m=0}^{l}\sum_{k=-l,2}^{l}\bar{F}_{lmk}(I)$$

$$[\alpha_{lm}\cos(ku+m\Lambda)+\beta_{lm}\,\sin(ku+m\Lambda)] \quad , \qquad (11)$$

with

$$\alpha_{lm} = \begin{bmatrix} \bar{C}_{lm} \\ -\bar{S}_{lm} \end{bmatrix}_{l-m=odd}^{l-m=even} \quad \text{and} \quad \beta_{lm} = \begin{bmatrix} \bar{S}_{lm} \\ \bar{C}_{lm} \end{bmatrix}_{l-m=odd}^{l-m=even} \quad . \qquad (12)$$

The $\bar{F}_{lmk}(I)$ are called *Inclination Functions.* For their computation cf. Kaula (1966) or Sneeuw (1991). Their second derivatives are derived in Sneeuw (2000):

$$V_{ij} = \frac{GM}{R}\sum_{l=0}^{\infty}\left(\frac{R}{r}\right)^{l+1}\sum_{m=0}^{l}\sum_{k}\lambda_{ij}\bar{F}_{lmk}(I)\,[\alpha_{lm}\cos(\psi_{mk})+\beta_{lm}\,\sin(\psi_{mk})] \quad , \qquad (13)$$

where λ_{ij} is given in Table 2. In the case of V_{xz}, α is replaced by β and β by $-\alpha$.

Table 2. Transfer factors for second derivatives of the gravity potential

ij:	xx	yy	xz	zz
λ_{ij}:	$\frac{-(l+1+k^2)}{r^2}$	$\frac{-((l+1)^2-k^2)}{r^2}$	$\frac{-(l+2)k^2}{r^2}$	$\frac{(l+1)(l+2)}{r^2}$

2.4 Semi-Analytical Approach

The Semi-Analytical approach is divided into two steps. The first step is the computation of the lumped-coefficients A_{mk} and B_{mk} by FFT-techniques, and the second is the adjustment of potential coefficients from the lumped-coefficients as pseudo-observables. Based on the assumption of a circular orbit - which implies, that the orbit height r and the inclination I are constant - (13) can be rewritten as two dimensional Fourier series:

$$V(u,\Lambda) = \sum_{m=0}^{L} \sum_{k=-L}^{K} A_{mk} \cos(\psi_{mk}) + B_{mk} \sin(\psi_{mk}) \quad . \tag{14}$$

The Fourier coefficients can be obtained by two different approaches: the 1D-FFT approach and the 2D-FFT approach - also known as Torus-approach.

2D-FFT Approach

The orbit coordinates and velocities can be transformed to the (u, Λ) domain via the Keplerian elements. The orbit can be imagined to be wrapped around a torus and the measurement values can be interpolated to a regular grid on the torus surface, as shown in Color Fig. XXVI on p. 302. The maximum grid size g_{max} is dependent on the maximum degree L and can be determined by the Nyquist theorem: $g_{max} = 2 \cdot \frac{360^\circ}{L}$. The lumped-coefficients A_{mk} and B_{mk} are simply obtained by a 2D-FFT of the gridded torus values.

1D-FFT Approach

In addition to the previous assumption of a circular orbit, the 1D-FFT approach assumes a repeat orbit. This means, that the satellite crosses the same point on the Earth's surface after α revolutions in β nodal days, where α and β must be integers with no common denominator. If this requirement is met, a relation between the rates of the orbital coordinates (u, Λ) can be set up:

$$-\frac{\dot{u}}{\dot{\Lambda}} = \frac{\beta}{\alpha} \quad , \tag{15}$$

which enables a spectral projection of the two dimensional Fourier series to a one-dimensional Fourier series (cf. Sneeuw (2000)):

$$\psi_{mk} \mapsto \psi_n \quad \text{with} \quad n = k\beta - m\alpha \quad . \tag{16}$$

A 1D-FFT of the observation time series produces the one-dimensional Fourier coefficients A_n and B_n, which can simply be reordered according to (16) to obtain the two-dimensional lumped coefficients A_{mk} and B_{mk}.

Block-Wise Adjustment

The spatial domain of the two dimensional Fourier series defined in (14) with the Fourier coefficients A_{mk} and B_{mk} is a torus defined in u and Λ, both being periodic. The conversion of the two dimensional Fourier coefficients A_{mk} and B_{mk} which describe the spectrum of the torus domain, to potential coefficients

$\bar{C}_{lm}$ and $\bar{S}_{lm}$ which describe the spectrum of the spherical domain, results in a least-squares adjustment:

$$\begin{bmatrix} \bar{C}_{lm} \\ \bar{S}_{lm} \end{bmatrix} = \left(\mathbf{H}_{\mathbf{lmk}}^{\mathbf{T}} \mathbf{P}_{\psi} \mathbf{H}_{\mathbf{lmk}} + \alpha \mathbf{R}\right)^{-1} \mathbf{H}_{\mathbf{lmk}}^{\mathbf{T}} \mathbf{P}_{\psi} \begin{bmatrix} A_{mk} \\ B_{mk} \end{bmatrix} \tag{17}$$

with the transfer coefficients H_{lmk}:

$$H_{lmk} = \frac{GM}{R} \left(\frac{R}{t}\right)^{l+1} \lambda_{lk} \bar{F}_{lmk}(I) \quad . \tag{18}$$

Only coefficients of the same order m are dependent on each other, which leads to a block-diagonal normal equation matrix. The adjustment can be carried out for all orders separately, which reduces enormously the requirements in terms of computation time and memory. A further advantage is, that filtering can be applied directly in the spectral domain. As each lumped coefficient can be associated with a certain frequency ψ_n, the weight matrix $\mathbf{P}_{\psi}$ works as spectral filter, which is easy to design and fast in computation.

Due to the simplifications and the neglect of correlations between coefficients of different orders, the results are less accurate that those of a direct solution. This can be compensated (at least in parts) by an iterative strategy. The residuals - the difference between the actual observation and the observations recomputed from the potential by a spherical harmonic synthesis - can be introduced to a new adjustment and corrections to the potential coefficients from the previous solution are obtained. This step can be repeated until convergence is achieved.

3 Results

3.1 Closed-Loop Simulation

In order to evaluate the method, a closed-loop simulation has been carried out. Based on the a priori gravity field OSU91a with maximum degree $L = 180$ an orbit has been integrated and the corresponding gradients have been computed. The mean altitude is about 251 km and the mean inclination about 96.6°. The orbit is a repeat orbit and closes after $\alpha = 982$ revolutions and $\beta = 61$ nodal days. The sampling rate is 10s.

The goal of this closed loop simulation is to get a gravity field solution, which is as close as possible to the a priori field. So as a first step, the "perfectly simulated" data was used without errors and the resulting differences between solution and a priori field were so small, that they can be ascribed to numerical errors. This result serves as proof of concept.

As a more realistic test, the measurements were superimposed with noise. For SST white noise velocity errors with a standard deviation of 0.1 mm/s were added to all three components. The SGG observations have been superimposed with a colored noise time series according to the specifications in Alenia (2001) (cf. Fig. 3) for V_{zz}. It was computed at the Technical University of Graz by an ARMA process (cf. Pail and Wermuth (2003)). The noise stays below $4\mathrm{mE}/\sqrt{\mathrm{Hz}}$ in the measurement band-width for the diagonal tensor components. The off-diagonal components have been neglected in this study.

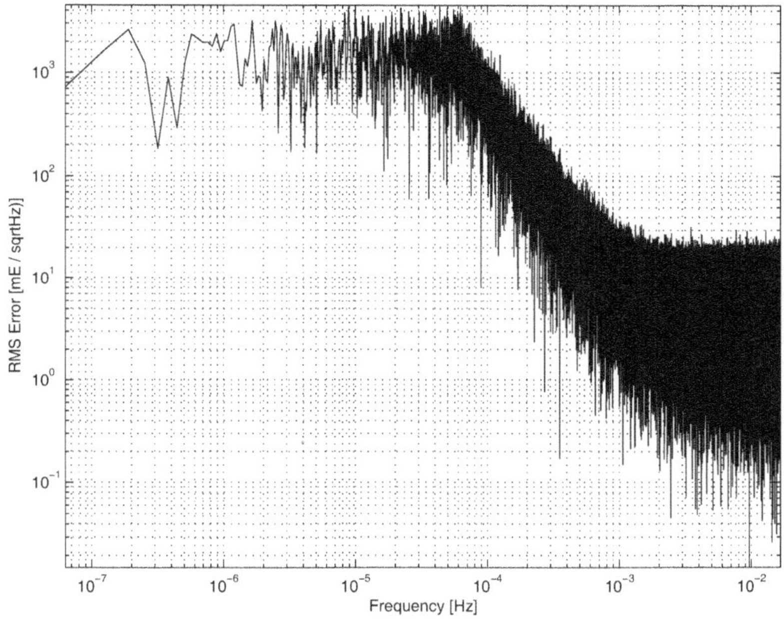

Fig. 3. PSD of colored noise for V_{zz}-component

As can be seen in Color Fig. XXVIIa on p. 302, the solution from the V_{zz} SGG-component shows large errors in coefficients with low degree l, in the zonal coefficients with low order m and in the tesseral coefficients with high degree l and high order m. The errors in the low degrees result from the assumed limited measurement bandwidth of the gradiometer, and they simply cannot be estimated well from SGG-only. The errors in the zonal coefficients result from the polar gap due to the orbit inclination of 96.6°, and the errors in the tesseral coefficients are due to an ill-posedness of the system. But the errors become smaller in the combined solution of the V_{xx}, V_{yy} andV_{zz} components, as the two additional components stabilize the system (cf. Color Fig. XXVIIb). In the combined solution of SST and SGG, which is shown in Color Fig. XXVIIc all errors became even smaller, which shows, that for an optimal result, the combination of all components is indispensable.

3.2 Spectral Leakage

FFT assumes a periodic and uninterrupted signal. In the case of the 1D-FFT approach, this implies that a strict repeat orbit without measurement gaps is required. As both conditions cannot perfectly be met in the case of a real satellite mission, the impact of deviations from the ideal case on the quality of the solution are studied in this section. The results have already been presented in Pail and Wermuth (2003), and are listed here for completeness.

Several new test scenarios have been prepared - all with similar parameters as the previous simulation in Sect. 3.1 - but with an a priori model only up to degree $L = 72$. One series of tests investigated orbits which do not close perfectly after 61 nodal days. The lateral offset at the equator between the start end the end of the orbit varies between 0 and 165 km. For the second series, data gaps of 0% to 50% of the whole data were introduced to the original orbit.

For both test series it can be concluded, that the results of a first iteration are the less accurate, the more the data deviates from the ideal case. But due to iterations with a strict synthesis step, the solution converges to a result which is only slightly worse than that of the ideal case. So the approach can deal with imperfect data - just the convergence rate (the ratio of the residual after and before an iteration step) is slower, as shown in Fig. 4.

3.3 1D-FFT vs. 2D-FFT

Both the 1D-FFT and 2D-FFT approach should theoretically produce the same set of lumped coefficients, and thus the same results. In this section a comparison of the behavior of the two methods with a realistic simulation is made. A simulated orbit based on the a priori model up to degree $L = 72$ with a lateral offset of 111 km, a data gap of 10% of the whole data, and the noise time-series from Sect. 3.1 were used. Solutions were computed for the V_{zz} SGG-component and SST only. The results in Fig. 5 show, that for low degrees up to $l = 30$ both methods are nearly equally accurate, but for higher degrees the 1D-FFT method performs better in the SGG-case and the 2D-FFT method performs better in the SST-case.

The reasons for this result are probably, that the filtering strategy of SGG-data is specially designed for the 1D-FFT approach. As the colored noise in the long-wavelength parts affects mainly the along-track direction, an isotropic interpolation on the torus seems to perform not optimal. On the other hand the assumed white noise of the SST-data (which are in any way much smoother than the SGG-data) seem to be more suitable for the torus interpolation. As a consequence, the strategy for combined solutions will be to process SGG-data with the 1D-FFT approach and the SST-data with the 2D-FFT approach. The normal equations from a mix of both approaches can be stacked without problems.

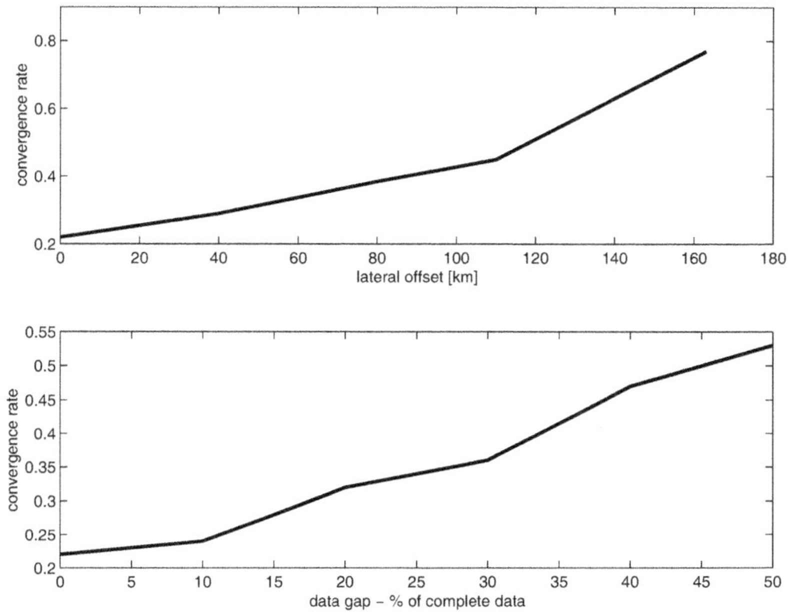

Fig. 4. Convergence rate with respect to (a) the lateral offset of a non-closing orbit, and (b the size of a data gap

3.4 Semi-Analytical Approach vs. Direct Solution

As a last case study, the semi-analytical approach is evaluated against a direct solution with real data from the CHAMP mission. CHAMP is a SST-only mission similar to the GOCE-SST part. From processing real CHAMP data, a lot of experience could be gained for the preparation of the GOCE-processing. The input data is a series of disturbing potential values derived form half a year of kinematic CHAMP orbits based on the Energy Balance Approach and derived in the context of the processing of the TUM-1S model. The processing was done similar to that described in the article of Svehla et al. in this volume. For details cf. Svehla and Rothacher (2002) and Gerlach et al. (2003). The results of this study have been published already in Wermuth et al. (2003).

The data contained about 450.000 samples, and the solution was computed up to degree $L = 100$ on one Pentium IV 3GHz processor with 1 GByte of RAM. The memory limits the application of the Direct Method to $L = 100$, while it is not a critical issue for the Semi-Analytical Approach. Due to the orbit decay of CHAMP the 1D-FFT approach could not be applied, as no constant repeat cycle could be found, therefore only the 2D-FFT approach was used. The computation time was 48 hours for the Direct Method and 10

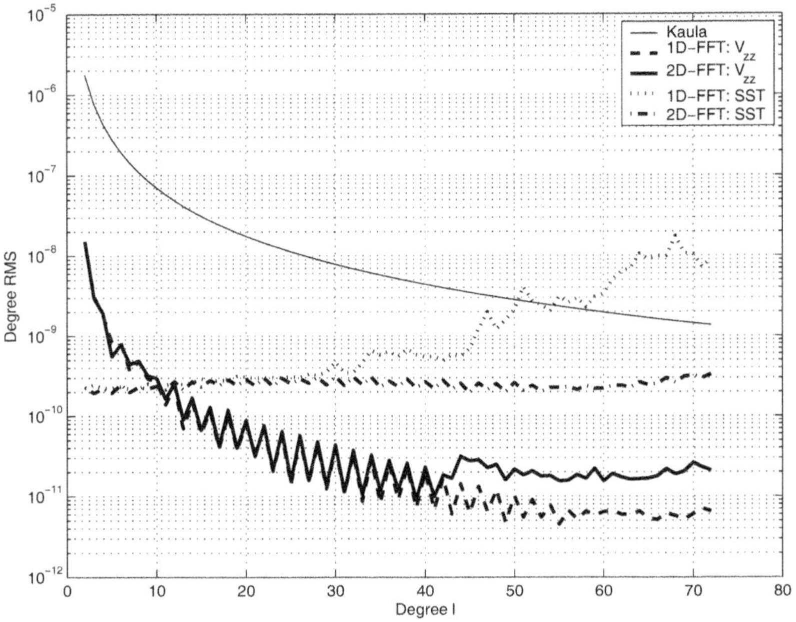

Fig. 5. Comparison of the degree RMS of 1D-FFT and 2D-FFT solutions

minutes for the Semi-Analytical Approach, where convergence was achieved after 10 iterations.

As can be seen in Fig. 6, the degree variances for both methods follow Kaula's rule of thumb until degree 60. For higher degrees the coefficients increase due to lack of signal in the data, which results in an ill-determination of the normal equations. The curve of the Torus Approach increases even stronger.

For an external evaluation the solutions were truncated at degree 60, because of the degradation of the high-frequency signal. For a set of 5168 points in the USA the geoid heights computed from the different models were compared to the geoid heights obtained from GPS/levelling. The RMS differences of both solutions range between the official EIGEN-solutions computed by GFZ (see Table 3). The semi-analytical solution is only slightly worse than the direct one.

Table 3. Geoid height RMS differences of comparison with GPS/levelling points

model	EIGEN-1S	EIGEN-2S	direct	torus
RMS [m]	0.968	0.893	0.904	0.926

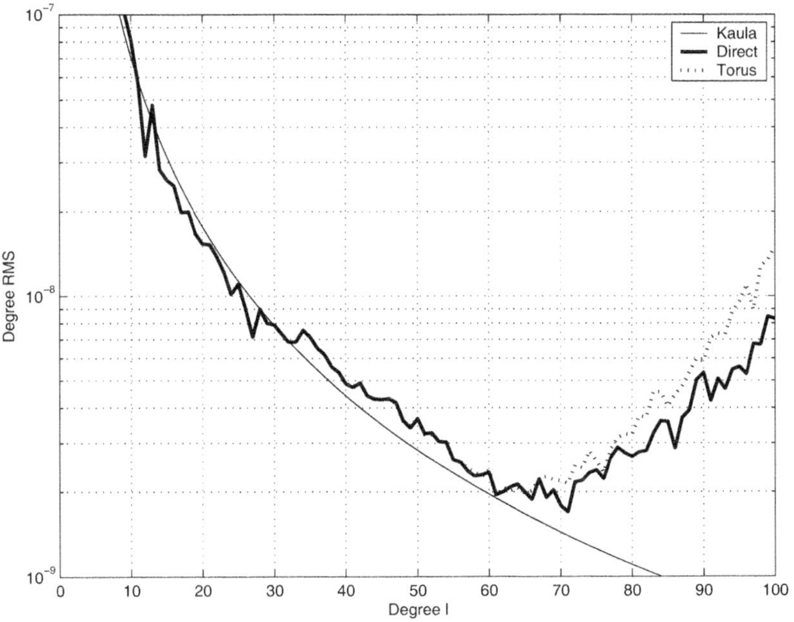

Fig. 6. Degree RMS variances of the different solutions

4 Conclusions and Outlook

The GOCE gradiometric sensor system simulator contains all essential elements of the actual gradiometric measurement system. It provides realistic insight into the interaction of the sensor components and their individual role. The gradiometer power spectral densities derived from it are input to the mission simulator, together with the essential mission parameters and the error spectra of the orbit. The mission simulator gives important feedback about the overall performance of the GOCE mission. The various test cases have shown, that the Semi-Analytical Approach is a simple and fast method for gravity field analysis. It delivers results in only a fraction of the computation time of a direct solution, at the expense of slightly lower accuracy. Thus it is an ideal tool for quick-look gravity field analysis. It is expected to perform even better for GOCE-SST than for CHAMP due to the constant orbit height and drag compensation in along track direction.

The gradiometer reference frame will deviate by several degrees from the local orbit frame. This is an issue especially for the Semi-Analytical Approach, as the observation equations should not be rotated. For SST new covariance models are provided, which contain information on the realistic noise of SST-observations. All this information has to be considered in the near future, and will be investigated in the second term of the GOCE-GRAND project.

Acknowledgement. This is publication no. GEOTECH-159 of the programme GEOTECHNOLOGIEN of BMBF and DFG, Grant 03F0329A.

The basic version of the GOCE sensor and mission simulator has been developed by H. Oberndorfer, J. Müller and N. Sneeuw in the context of a study for ESA. It has been further refined by H. Oberndorfer, B. Frommknecht and Ch. Gerlach in the context of the Geotechnologienprogramm project GOCE-GRAND.

References

Alenia (2001) Performance requirements and budgets for the gradiometric mission. Technical Note, GO-TN-AI-0027, Alenia Spazio, Turin

Gerlach C et al. (2003) A CHAMP-only gravity field model from kinematic orbits using the energy balance approach.Poster presented at EGS General Assembly, Nice

Kaula WM (1966) Theory of Satellite Geodesy. Blaisdell Publishing Company

Oberndorfer H, Müller J, Sneeuw NJ (2000) GOCE end-to-end closed loop simulation, in: GOCE End to End Performace Analysis, final report, ESTEC contract no. 12735/98/NL/GD, 186-191

Oberndorfer H, Müller J, Rummel R, Sneeuw NJ (2002) A simulation tool for the new gravity field missions, Advances Space Research,30,2,227-232

Pail R, Wermuth M (2003) GOCE SGG and SST quick-look gravity field analysis. Advances in Geosciences 1:1-5, European Geosciences Union

Svehla D, Rothacher M (2002) Kinematic orbit determination of LEOs based on zero or double difference algorithms using simulated and real SST GPS data. IAG Proceedings, Symposium No. 125, Vistas for Geodesy in the New Millenium, Springer Verlag, Heidelberg

Sneeuw NJ (1991) Inclination Functions. Delft University of Technology

Sneeuw NJ (2000) A Semi-Analytical Approach to Gravity Field Analysis from Satellite Observations. PhD. Thesis, Technical University of Munich

Wermuth M, Gerlach C, Svehla D, Földvary L (2004) Comparison of Different Gravity Field Solution Methods Applied to CHAMP Gravity Field Modelling. In: Meurers B, Pail R (eds) Proceedings of the 1st Workshop on International Gravity Field Research Graz 2003, Österreichische Beiträge zur Meteorologie und Geophysik, University of Vienna

GOCE Gravity Field Modeling: Computational Aspects – Free Kite Numbering Scheme

Christian Boxhammer and Wolf-Dieter Schuh

Institute of Theoretical Geodesy, University of Bonn, Germany
53115 Bonn, Nußallee 17, schuh@uni-bonn.de

Summary. The modelling of the Earth's gravity field by means of a high-resolving spherical harmonic analysis is a numerically demanding task, especially when realistic (non gridded) data sets are analysed. The *free kite numbering scheme*, presented in the current article, allows a flexible combination of models. It is focussed, in particular, on the combination of a model containing rotation-symmetrical, high-resolving data with a second model comprising fully correlated data, which allows the determination of the lower degrees. This kite scheme may, depending on the degree of conformance with rotation symmetry, be used both with a direct solver and to improve the convergence rate of an iterative solver.

Key words: GOCE mission, spherical harmonic analysis, preconditioner, kite numbering scheme

1 Introduction

GOCE gravity field determination constitutes a great challenge to stochastic and deterministic modelling. Obtaining optimal results from GOCE data has been the focus of research collaborations on both national and international levels. In addition to the challenge of mathematic modelling, the numerical implementation of the resulting large equation systems in particular, requires enormous efforts. As a joint venture between the Technical University of Graz, the Technical University of Munich, and the University of Bonn, a data processing chain was developed. This chain consists of three components: the *Quick-Look Tool*, which is based on the highly efficient semi-analytic approach, the *Tuning Machine* (a tailored iterative solver), and the *Final Solver*, which solves the normal equations directly. The development of this variety of methods is motivated by the complementary use of these three strategies, which all have distinct advantages as well as drawbacks.

The application of tailored numerical algorithms enabled the design of an efficient and flexible tool capable of processing the huge amount of data (approximately 58,000 parameters of a highly resolving model will be estimated from some 50 million correlated observations per measurement period of 6 month). This solution technique is based on the method of preconditioned conjugate gradients and allows a strict one-step adjustment of heterogeneous data types both in terms of observation equations and normal equations. Although this method was developed especially for GOCE data processing, it may be applied within the general context of spherical harmonic analysis. The efficiency of this iterative method, named `pcgma` (Preconditioned Conjugate Gradient Multiple Adjustment), is underlined by its ability to solve high-order spherical harmonic models (degree and order 360 and more) within a reasonable processing time. `pcgma` consists of the following main components:

- Decorrelation of the data by discrete filtering in the time domain;
- Preconditioning by means of data-adaptive sparse matrices (Kite Scheme);
- Parallelization for use on massive parallel computer.

The current article focusses on the second component and outlines the results of the latest research regarding the generation, administration, and implementation of a data-adaptive preconditioner. The proposed preconditioner constitutes an extension of the traditional *kite numbering scheme* (Schuh, 1996b) for equal efficiency with greater flexibility. Therefore, this *free kite numbering scheme* may be adjusted to various problems. Possible fields of application will be demonstrated, using as examples typical problems from satellite geodesy.

2 Spherical Harmonic Analysis

Many physical processes of our *"System Earth"* may be represented mathematically by using the solid spherical harmonics

$$r^{-(\ell+1)}\, P_{\ell m}(\sin\varphi)\, \cos m\lambda \qquad \text{and} \qquad r^{-(\ell+1)}\, P_{\ell m}(\sin\varphi)\, \sin m\lambda$$

as base functions, where r, φ, λ denote the polar coordinates of a point and $P_{\ell m}(\sin\varphi)$ the Legendre functions of degree ℓ and order m (with $\ell \leq 0$ and $m \leq \ell$). As solution of Laplace's equation outside the sphere, these base functions possess a global support. On the other hand, they build up an orthogonal system when the data coverage is global and continuous (Heiskanen and Moritz, 2000, S.29). Furthermore, these base functions constitute a complete basis, i.e. any piecewise continuous function $f(\varphi, \lambda)$ defined on the unit sphere ($r = 1$) may be represented by a linear combination of spherical harmonics as the infinite series

$$f(\varphi,\lambda) = \sum_{\ell=0}^{\infty} \sum_{m=0}^{\ell} C_{\ell m}\, P_{\ell m}(\sin\varphi)\, \cos m\lambda \;+\; S_{\ell m}\, P_{\ell m}(\sin\varphi)\, \sin m\lambda$$

with $S_{\ell 0} = 0$. The coefficients $C_{\ell m}$ and $S_{\ell m}$ may be determined independently by integration over the unit sphere, yielding

$$C_{\ell m} = \frac{1}{\iint\limits_{\sigma} \left(P_{\ell m}(\sin\varphi)\cos m\lambda\right)^2 d\sigma} \iint\limits_{\sigma} f(\varphi,\lambda)\, P_{\ell m}(\sin\varphi)\, \cos m\lambda \, d\sigma$$

and

$$S_{\ell m} = \frac{1}{\iint\limits_{\sigma} \left(P_{\ell m}(\sin\varphi)\sin m\lambda\right)^2 d\sigma} \iint\limits_{\sigma} f(\varphi,\lambda)\, P_{\ell m}(\sin\varphi)\, \sin m\lambda \, d\sigma,$$

respectively, where $d\sigma = \cos\varphi\, d\varphi d\lambda$. As the representation of a band-limited function requires only a finite number of coefficients, the infinite sum above may be terminated at a finite degree ℓ_{max}. For a graphical representation, the coefficients $C_{\ell m}$ and $S_{\ell m}$ are usually arranged as a triangle in the following manner (Fig. 1): The ordinate is defined by the degree ℓ, which increases from top to bottom, the abscissa by the order m with the cosine coefficients $C_{\ell m}$ on the left and the sine coefficients $S_{\ell m}$ on the right hand side.

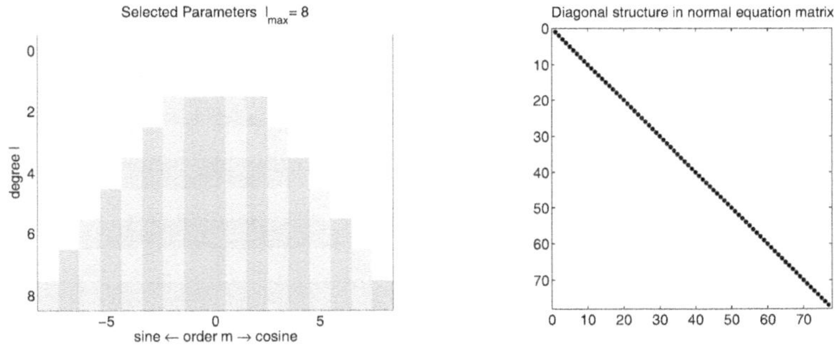

(a) Coefficient triangle (b) Normal equation matrix

Fig. 1. Representation of the coefficients $C_{\ell m}$, $S_{\ell m}$ and of their correlations

Figure 1b depicts the correlations between all of the spherical harmonic coefficients. Each non-vanishing correlation is represented by a black dot in the normal equation matrix. The fact that only the diagonal contains entries implies that all coefficients are estimable independently. Consequently, the computation of the coefficients, denoted as *spherical harmonic analysis*, constitutes, in the given case, an inverse problem which is easy to solve.

Unfortunately, the normal equation matrix will not, in general, be diagonal when discrete observations are used, because the orthogonality relations of the spherical harmonics hold only for very special discrete data distributions. In other words, the coefficients cannot, due to their correlations, be estimated independently. However, under certain assumptions regarding the local data distribution, some of the orthogonalities still hold. Especially the orthogonality relations for trigonometric functions may be exploited. As an illustration, if one full period of length 2π is sampled at $2L$ equidistant nodes, then

$$\sum_{i=0}^{2L-1} \cos m\lambda_i \, \cos k\lambda_i = (1 + \delta_{m0} + \delta_{mL}) \, L \, \delta_{mk}$$
$$\sum_{i=0}^{2L-1} \sin m\lambda_i \, \sin k\lambda_i = (1 - \delta_{m0} - \delta_{mL}) \, L \, \delta_{mk}$$
$$\sum_{i=0}^{2L-1} \cos m\lambda_i \, \sin k\lambda_i = 0.$$

where $\lambda_i = i\frac{2\pi}{2L}$ and δ_{mk} denotes the Kronecker symbol. As a consequence, all coefficients of different orders will be independent. Furthermore, the symmetries and asymmetries of Legendre's functions with respect to the equator, i.e.

$$P_{\ell m}(-\sin\varphi) = (-1)^{(\ell-m)} P_{\ell m}(\sin\varphi) \ ,$$

may be used to separate the even from the odd coefficients of a fixed order.

The standard numbering scheme, e.g. EGM96, is degree-by-degree (Rapp, 1994). Usually, the cosine coefficients $C_{\ell m}$ are enumerated first, and subsequently the sine coefficients $S_{\ell m}$ within a slightly modified loop (the coefficients $m = 0$ are omitted). The order of the coefficients arising from this numbering scheme can be demonstrated by the following loops:

Algorithm 1. Numbering scheme: degree-by-degree

```
for ℓ = 0 : ℓ_max
    for m = 0 : ℓ
        C_ℓm            % odd and even coefficients
    end
end
for ℓ = 1 : ℓ_max
    for m = 1 : ℓ
        S_ℓm            % odd and even coefficients
    end
end
```

As can be seen in Fig. 2a, this numbering scheme is in accordance with moving through the coefficient triangle along a primary horizontal track. The corresponding normal equation matrix is characterized by diagonal stripes parallel to the main diagonal (Fig. 2b), which would necessitate an elaborate mechanism for its storing.

This procedure becomes extremely simplified by an order-wise enumeration, corresponding to a vertical movement inside the coefficient triangle (Fig. 3a). This scheme, which in the following will be referred to as *block numbering scheme*, produces a block-diagonal normal equation matrix (Fig. 3b), which is now much easier to store, and for which it is easy to see that the Cholesky reduction does not produce any additional fill-in elements. The same holds also for the standard numbering but with regard to the compact storage scheme the block numbering is clearly superior.

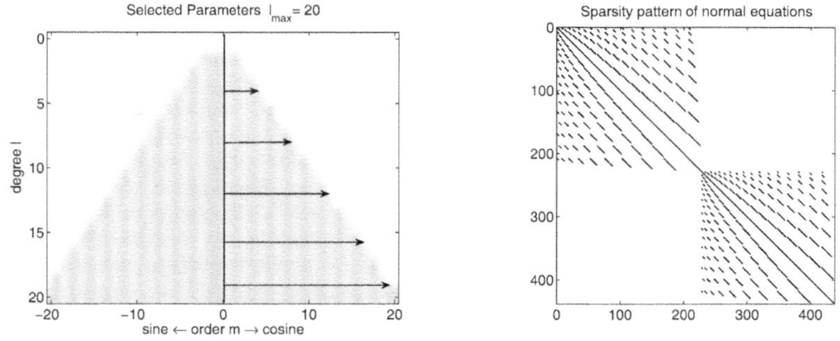

(a) Coefficient triangle (b) Normal equation matrix

Fig. 2. Standard numbering scheme

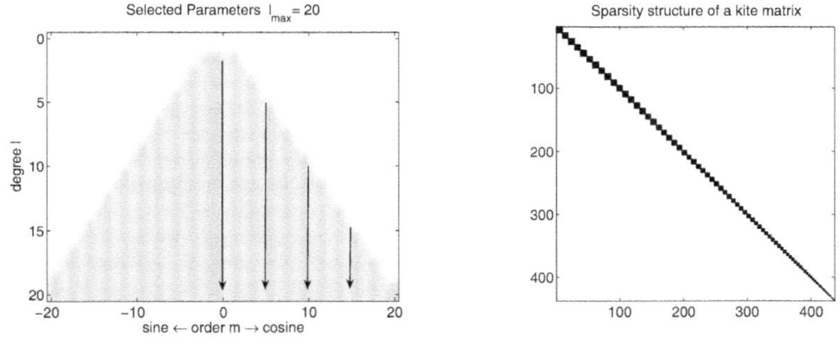

(a) Coefficient triangle (b) Normal equation matrix

Fig. 3. Block numbering scheme

The order of coefficients resulting from application of the *block numbering scheme* with additional consideration of equatorial symmetries is demonstrated in Algorithm 2. Thus, for each fixed order, coefficients of even and odd degrees are separated.

Algorithm 2. Block numbering scheme

for $m = 0 : m_{max}$
  for $\ell = m : 2 : \ell_{max}$
    $C_{\ell m}$ % *even/odd coefficients*
  end
  for $\ell = m + 1 : 2 : \ell_{max}$
    $C_{\ell m}$ % *odd/even coefficients*
  end
end

```
for m = 1 : m_max
    for ℓ = m : 2 : ℓ_max
        S_ℓm                    % odd/even coefficients
    end
    for ℓ = m + 1 : 2 : ℓ_max
        S_ℓm                    % even/odd coefficients
    end
end
```

Note that block-diagonality of the normal equation matrix holds only in the case that the nodes are, firstly, distributed rotation-symmetrically with respect to the North-South axis, secondly, aligned equidistantly along the parallels, and thirdly, located symmetrically regarding the equator. The density of the data coverage per parallel may be controlled via the grid width. Furthermore, it must be ensured that the data is of homogeneous accuracy. Although polar gaps downgrade the condition of the normal equation system, they do not destroy block-diagonality.

The orderwise independence of the coefficients consequently allows to process the data order by order. By this efficient mechanism for spherical harmonic analysis even very high-resolving models may be estimated (Colombo, 1981; Rummel *et al.*, 1993). For this reason, this block-based method is often denoted as *fast spherical harmonic analysis.*

3 Kite Numbering Scheme

Ever since satellite data became available for global gravity field determination, the idea of combining these, usually irregularly distributed measurements, with regularly distributed (gridded) data, such as gravity anomalies, has been nourished. While satellite data allows a precise determination of coefficients of lower degrees, gridded data is very suitable for estimating coefficients of higher degrees.

A simple modelling approach consists in a *"Patchwork"* technique, which, in the past, lead to a stepwise computation scheme and model refinements (confer e.g. OSU 91 (Rapp *et al.*, 1991), EGM 96 (Lemoine *et al.*, 1996)). With more complex models the correlations within the orders were strictly taken into account (Balmino, 1993). In the given context, the order-by-orderwise numbering scheme could be applied, producing the structures within the normal equation matrix as shown in Fig. 4a.

As can be seen in Fig. 4b, the Cholesky-reduced normal equation matrix contains numerous fill-in elements. The number of fill-ins increases quadratically with the maximal degree of the model. In order to simplify the structure of the reduced normal equation matrix, Bosch (1993) proposed a numbering scheme based on division of the coefficients into three zones (*three zone numbering scheme*). The first zone is built up by the fully correlated coefficients of lower degrees, the second zone by coefficients of higher degrees and

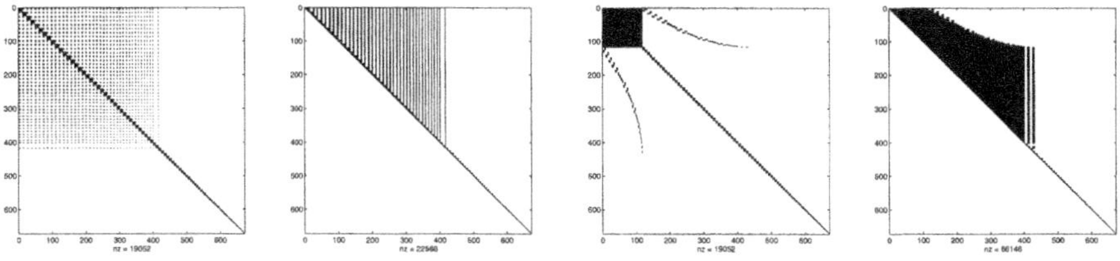

(a) normal eq. (b) red. normal eq. (c) normal eq. (d) red. normal eq.

Order-by-order numbering scheme *Three zone numbering scheme*

Fig. 4. Combined models

lower orders, and the third zone by coefficients of higher degrees and higher orders. Each zone is itself enumerated order-by-order. On the one hand, this scheme leads to a more compact occupation with non-zero elements within the normal equation matrix. On the other hand, Fig. 4d demonstrates that the Cholesky-reduction produces the same number of fill-in elements as for numbering order-by-order.

In the course of the GOCE studies a new numbering scheme was proposed by Schuh (1996a), that circumvents this fill-in effect. Reversing the order of the zones yields a scheme for which Cholesky reduction does not produce any fill-in elements. Since the structure of the normal equation matrix resembles a kite, this modified numbering scheme was termed *kite numbering scheme*. Thus, a strict combination of high-resolving, rotation-symmetrical data with arbitrarily distributed, fully correlated data is, for lower degrees, possible without fill-in elements. Therefore, this combination may be computed on a standard PC even for very high resolving models up to degree and order 720. Beside solving the equation system, partial inverses (i.e. strict inverse for selected elements) may be computed efficiently by means of this approach (Auzinger and Schuh, 1998).

For a more detailed analysis of the properties of the kite structure, the relations between single parameters and zones need to be specified more thoroughly. For this purpose, the low-resolving, fully correlated data is denoted as `DNS (dense)`, while the high-resolving, block-wise correlated data will be referred to as `BLK (block)`. The correlations between the zones may be specified as follows:

- *zone:FULL*
  All parameters are fully correlated.
- *zone:SEMI*
  The BLK parameters are correlated only within the same order, parity (odd or even), and trigonometric function (sine or cosine). Similarly, correlations between type-DNS and type-BLK data exist only within the same order, parity, and trigonometric function.
- *zone:INDEPENDENT*
  Contains all BLK parameters for whose within the same order there are

no DNS parameters. The correlations within BLK data are according to the SEMI zone. However, there are no correlations between data of types DNS and BLK.

Figure 5a illustrates the three zones arising from an exemplary combination of type-DNS observations (degree/order 2-ℓ_{DNS}) with type-BLK data (degree/order 2-ℓ_{BLK}). The "FULL" zone comprises all the parameters of degrees and orders 2 up to ℓ_{DNS}. As all these coefficients are fully correlated, the corresponding part within the normal equation matrix is fully occupied. For the "SEMI" zone, correlations between observations of types DNS and BLK exist only within the same orders and between degrees of identical parity. There are no correlations between type-DNS and type-BLK observation inside the INDEPENDENT zone.

Recall that each parameter generates one row and one column within the normal equation matrix. The kite numbering scheme produces an order-by-order sorting within each zone, which leads to their typical arrangement (*kite structure*) (Schuh, 1996b). Figure. 5b depicts the normal equation matrix for the current example. For the upper-left blocks (INDEPENDENT,INDEPENDENT) it becomes evident that Cholesky reduction does not create any additional fill-ins. The same holds also for block (SEMI,SEMI) in the middle part. At first glance, fill-ins seem to occur for (SEMI,FULL) below its wings (better: below the off-diagonal blocks). However, it can be seen that each diagonal block of (SEMI,SEMI) corresponds to exactly one block of (SEMI,FULL). As there exists only one off-diagonal block per (block-) row, the scalar products of distinct (block-) columns vanish. Consequently, no additional fill-in elements are generated within the reduction step (e.g. Cholesky).

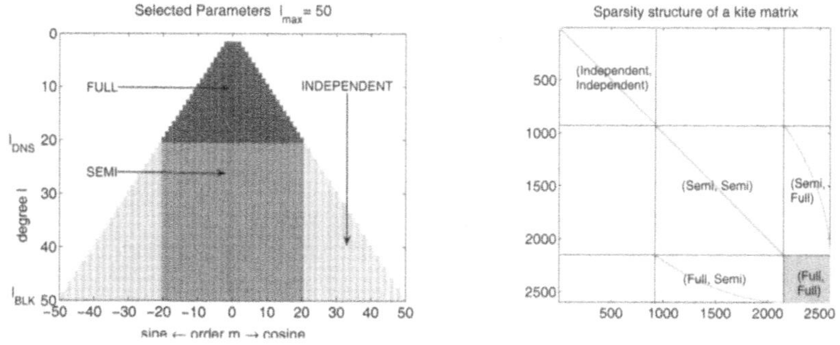

(a) correlation zones (b) normal equation matrix

Fig. 5. Combined models, *kite numbering scheme*

The *kite numbering scheme* may now be used for an efficient, yet strict solution of combined models with gridded, high-resolving data. In addition, it produces an excellent approximate solution in case the data is of almost regular spatial distribution, i.e. not necessarily rotation-symmetrical. Here, the kite structure may be used as preconditioner for a strict and efficient iterative solver. The `pcgma` algorithm (Schuh, 1996b), which was developed especially for the analysis of GOCE data, makes use of the kite scheme for the combination of SST and SGG data.

Despite the benefits described above, the latest simulations reveal some deficiencies of the kite numbering scheme. The fully correlated coefficients were, for each order m, parameterized with degrees between $\ell_{\min} = \max(2, m)$ and some fixed $\ell_{\mathrm{DNS}_{\max}} = const.$ The degree of the high-resolving model also has a prescribed limit $\ell_{\mathrm{BLK}_{\max}} = const.$ for all orders. However, for many recent Earth gravity field models(e. g. EIGEN-1S (Reigber *et al.*, 2002), EIGEN-2 (Reigber *et al.*, 2003), EIGEN-3p (Reigber *et al.*, 2004)), the limits are kept variable, for instance to take resonance frequencies into account.

In the following section a method will be presented, that allows variable limits within the *kite numbering scheme.* The *free kite numbering scheme* consists essentially of the numbering scheme and a newly developed storage scheme with respect to the normal equations. As a first step we will generalize the *block numbering scheme* to introduce flexible borders and than we will apply this concept to the *free kite numbering scheme.*

4 Free Block Numbering Scheme

To improve the flexibility of Earth gravity field modelling, it is necessary to use numbering schemes that allow selection/deletion of certain coefficients. Fixed maximal degrees throughout all orders turn out to be too inflexible as to take into account particular strengths (resonance frequencies) and weaknesses (polar gap) of a model. To circumvent this problem, the minimal degree $\ell_{\min}$ and the maximal degree $\ell_{\max}$ is fixed for each order m. These limits are stored in three vectors $\boldsymbol{m}$, $\boldsymbol{\ell}_{\min}$ und $\boldsymbol{\ell}_{\max}$, which define the set of parameterized coefficients uniquely. These vectors are of equal lengths, each with as many elements as the number o of parameterized orders (Cosine and Sine coefficients are treated symmetrically) i.e.

$$\begin{aligned} \boldsymbol{m} &= \left(m_1, m_2, \ldots, m_o\right) && \in \mathbb{R}^{o\times 1} \\ \boldsymbol{\ell}_{\min} &= \left(\ell_{\min_1}, \ell_{\min_2}, \ldots, \ell_{\min_o}\right) && \in \mathbb{R}^{o\times 1} \\ \boldsymbol{\ell}_{\max} &= \left(\ell_{\max_1}, \ell_{\max_2}, \ldots, \ell_{\max_o}\right) && \in \mathbb{R}^{o\times 1} \end{aligned}$$

As an example, let

$$\boldsymbol{m} = (0,1,2,3,4,5,6,7) \quad \boldsymbol{\ell}_{\min} = (2,2,2,3,5,5,6,7) \quad \boldsymbol{\ell}_{\max} = (4,6,7,7,7,6,7,7)\,.$$

The resulting set of parameters is illustrated in Fig. 6. The order of the involved coefficents is generated by the loops in Algorithm 3:

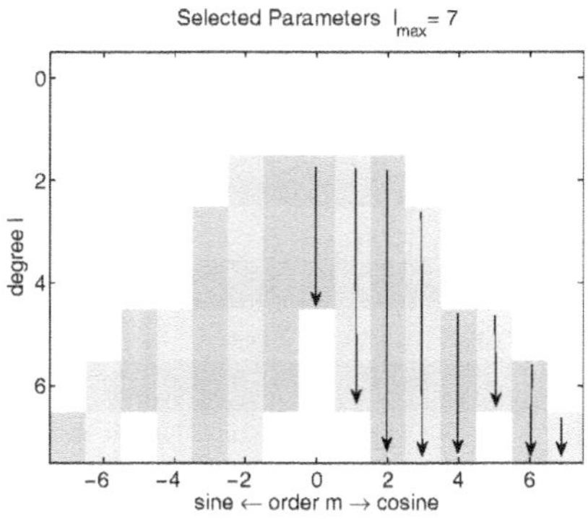

Fig. 6. Free block numbering scheme

Algorithm 3. Free block numbering scheme

for $m = 0 : o$
    for $\ell = \boldsymbol{\ell}_{\min}(m) : 2 : \boldsymbol{\ell}_{\max}(m)$
        $C_{\ell m}$ *% even/odd coefficients*
    end
    for $\ell = \boldsymbol{\ell}_{\min}(m) + 1 : 2 : \boldsymbol{\ell}_{\max}(m)$
        $C_{\ell m}$ *% odd/even coefficients*
    end
end
for $m = 1 : o$
    for $\ell = \boldsymbol{\ell}_{\min}(m) : 2 : \boldsymbol{\ell}_{\max}(m)$
        $S_{\ell m}$ *% even/odd coefficients*
    end
    for $\ell = \boldsymbol{\ell}_{\min}(m) + 1 : 2 : \boldsymbol{\ell}_{\max}(m)$
        $S_{\ell m}$ *% odd/even coefficients*
    end
end

This principle of variable block sizes will now be applied to data combination.

5 Free Kite Numbering Scheme

The newly developed *free kite numbering scheme* (FKN) allows combining two models, both having order-dependent, flexible (*"free"*) limits, as introduced in Sect. 4 for the *free block numbering scheme.* Note that it must be made sure that the normal equation matrix still is a kite matrix in order to avoid additional fill-in elements. For instance, the configuration depicted in Fig. 5 with fixed limits 2, ℓ_{DNS} and ℓ_{BLK} for *free block numbering,* is given by the following vectors:

$$m_{\mathrm{DNS}} = (0,1,2,3,4,\ldots,19,20) \qquad m_{\mathrm{BLK}} = (0,1,2,3,4,\ldots,49,50)$$
$$\ell_{\min} = (2,2,2,3,4,\ldots,19,20) \qquad \ell_{\min} = (2,2,2,3,4,\ldots,49,50)$$
$$\ell_{\max} = (20,20,\ldots,20) \qquad \ell_{\max} = (50,50,\ldots,50)\,.$$

From these vectors the symbolic parameter order and the positions of non-zero elements of the normal equation matrix can be computed. While the old algorithm for *kite numbering scheme* is based on rather complex programs with inflexible calculation rules, that do not allow any further flexibilization, the FKN scheme is built up on clearly structured, rule-based logic.

Rule-based processing

The algorithm for assembling the parameter order using the FKN scheme consists of two loops and one rule-based decision tree within the inner loop. In analogy to Algorithm 3 the outer loop is iterated over all orders and the inner loop from the minimal until the maximal degree corresponding to the current order. In contrast to the conditions of Algorithm 3, however, vectors containing the minimal and maximal degrees for both DNS and BLK data are at hand. The FKN algorithm uses, for its inner loop, order-wise the minimum of both minimal degrees and the maximum of both maximal degrees. Inside the loops three stacks are built up. Now each parameter runs through the decision tree and is, thereafter, either assigned to one of the stacks, or discarded. The resulting stacks contain, upon termination of the loop, the parameters of the zones FULL, SEMI, and INDEPENDENT. The decision tree is based on the following rules:

Algorithm 4.

Parameter $\{CS\}_{lm}$ is analysed:

a) If there are no observations of type DNS for order m, then the parameter is assigned to the INDEPEDENT zone, otherwise to either the SEMI or the FULL zone.

b) If the degree ℓ of the current parameter lies within the type−DNS observations for this order, then the parameter is assigned to the FULL zone, otherwise to the SEMI zone.

The assembling of the parameter order according to this rule goes as follows:

Algorithm 5.

Loop over all orders m
  Loop over all degrees ℓ of the current order
    Even cosine coefficients
      Assign to FULL,SEMI or INDEPENDENT zone
    Odd cosine coefficients
      Assign to FULL,SEMI or INDEPENDENT zone
    Even sine coefficients
      Assign to FULL,SEMI or INDEPENDENT zone
    Odd sine coefficients
      Assign to FULL,SEMI or INDEPENDENT zone
  end
end
String together zones

The vector of symbolic parameters is now saved as an ASCII file and may be used for data re-sorting. The elements are efficiently accessed by indirect adressing through index vectors.

The traditional graphical representing of the parameters based on the FKN scheme might generate unexpected structures if one actually exploits the full range of flexibility. Figure 7 give some unusual examples, which are, however, computable without any problems.

Rule-based computation of the correlations

Due to the definitions of the zones and their assumptions, some of the parameters are correlated, and others are not. The prescribed symbolic parameter order lead to the re-sorting of the normal equation matrix in such a way that certain operations from the domain of linear algebra are applicable to a kite matrix $\boldsymbol{K}$ without generating fill-ins. Examples are:

- Cholesky reduction;
- Partial inversion yielding $\boldsymbol{K}_{\text{part.}}^{-1}$ (strict inverse for selected elements cf. (Auzinger and Schuh, 1998));
- Solution of the equation system $\boldsymbol{K}\boldsymbol{X} = \boldsymbol{B}$ by means of Cholesky reduction $\boldsymbol{K} = \boldsymbol{R}^T\boldsymbol{R}$ ($\boldsymbol{B}$ contains multiple right hand sides).

In order to assemble a kite matrix, not only the symbolic parameter vector is required, but also the positions of its non-zero elements, which, in turn, represent the correlations between two particular parameters. The correlations are derived from the stored information regarding cosine/sine function, order, parity of degrees, and the vectors containing the minimal and maximal degrees (Algorithm 6).

Now let a kite matrix and a symbolic parameter vector in the FKN scheme be given. The symbolic structure of the kite matrix $\boldsymbol{K} \in \mathbb{R}^{n \times n}$ is stored in a symbolic parameter vector $\boldsymbol{p} \in \mathbb{R}^{n \times 1}$. $\boldsymbol{p}$ contains three entries, the degree ℓ, the order m, and the membership to one of the trigonometric functions CS

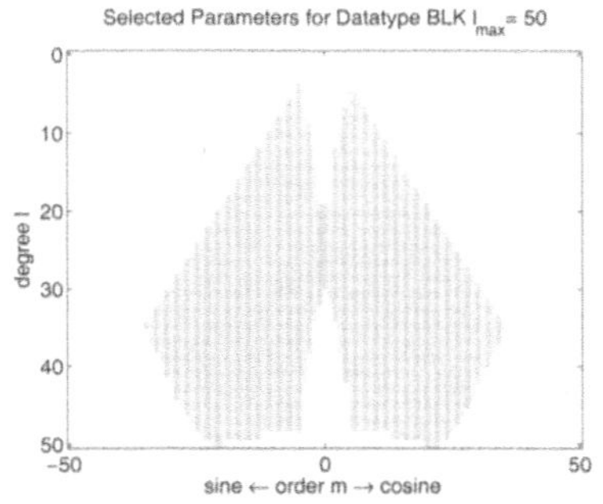

(a) Gridded Data

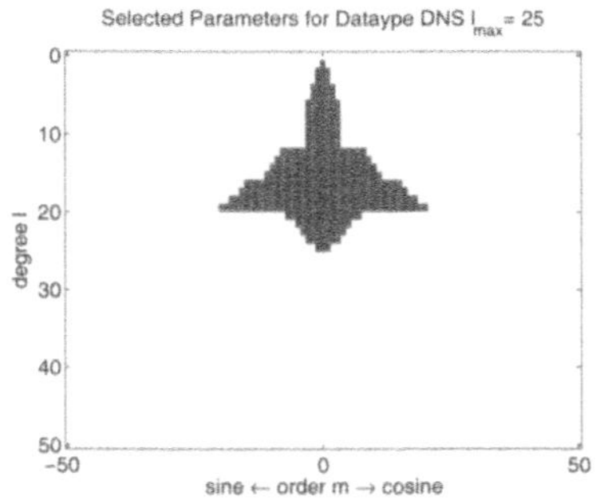

(b) Irregular Data

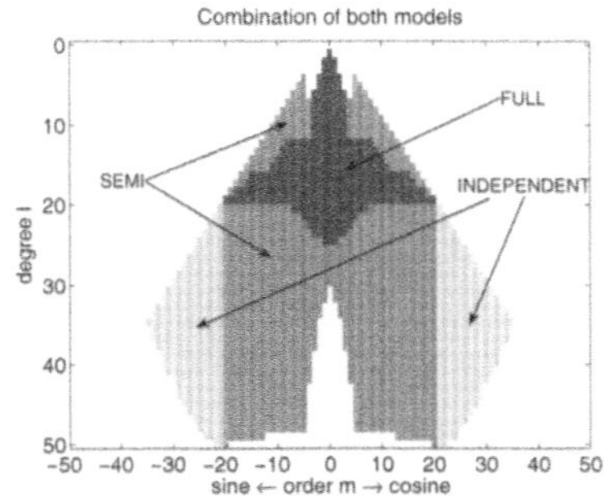

(c) Combination

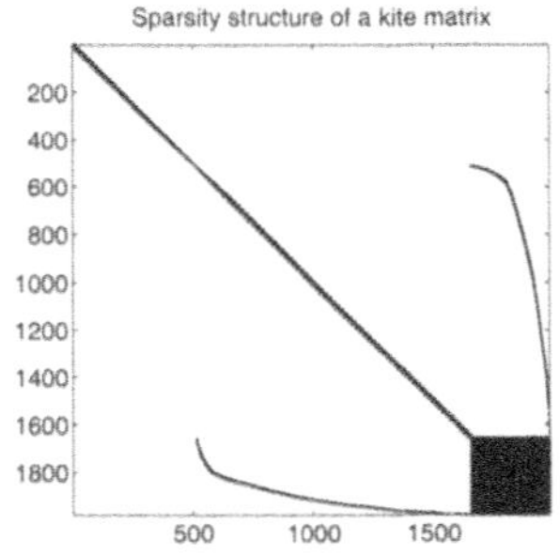

(d) Kite-Matrix

Fig. 7. Three possible correlation zones between BLK and DNS blocks

('C' for cosine and 'S' for sine). The entries of this object are accessed via the "dot" and an index variable. For instance, to test whether the fifth parameter of $\boldsymbol{p}$ has the cosine function, one would enter the statement $\boldsymbol{p}(5).CS ==' C'$. Furthermore, the kite matrix will have a non-zero element at position k_{ij} if the parameters $\boldsymbol{p}(i)$ and $\boldsymbol{p}(j)$ are correlated.

To determine the positions of the non-zero elements using the FKN scheme is computationally the most expensive step, because each of the n^2 possible positions must be evaluated. Due to the symmetry of the normal equation matrix, this number reduces to $n(n+1)/2$ positions. As can be seen from Algorithm 6, two nested loops are necessary to do the job. The outer loop is iterated over all parameters in $\boldsymbol{p}$, the inner loop over all parameters from the current position on.

Algorithm 6.

```
for i = 1 : n
    for j = i : n
        if correlated ( p(i), p(j) )
            save_indices (i, j)
        end
    end
end
```

The function correlated in line 3 of Algorithm 6 checks by means of Algorithm 7, if the two symbolic parameters currently under investigation are correlated.

Algorithm 7.

```
function correlated
  if p(i) ∈ FULL and p(j) ∈ FULL
     return true                          % p(i), p(j) correlated
  if p(i).m == p(j).m                     % same order
     and
     p(i).ℓ mod 2 == p(j).ℓ mod 2 % same parity in degree
     and
     p(i).CS == p(j).CS                   % same trig. function
     return true                          % p(i), p(j) correlated
  return false
end
```

Algorithm 6 outputs the coordinates of the non-zero elements of the given kite matrix. These coordinates are saved internally by the software. Since the non-zero elements are always clustered in blocks, this process may be considerably simplified. For each such block, the row and column index of its first element and its total number of rows and columns is stored. Figure 8a illustrates the block structure of the kite matrix. Figure 8b shows an example for block-oriented storing of the shape of the kite matrix in an ASCII file. With the information contained in this file, the kite matrix can be assembled from the design matrix using level-3 BLAS routines.

6 Conclusion and Outlook

The innovative *free kite numbering scheme* renders the way to new options both for model parameterization and preconditioning. It is now possible to adapt the parameters to the information content of the data. For instance, it might be advisable to exclude certain poorly determinable zonal coefficients from the model. In addition, any coefficients of lower degrees could be eliminated in case the data turns out to be hardly sensitive to the corresponding parameters.

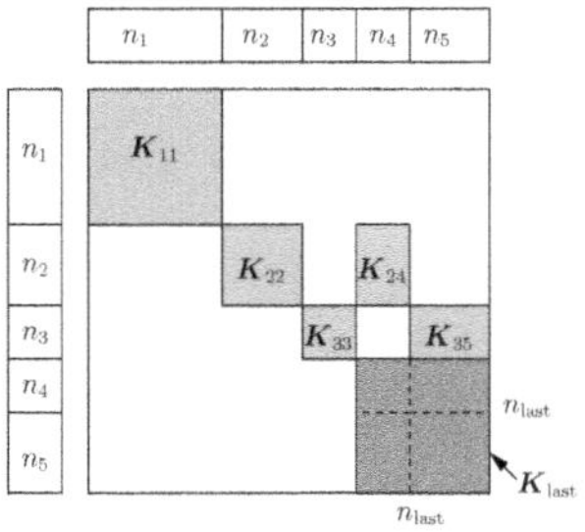

(a) Kite matrix

```
# created May 26 2005  6:37
# by class navigation (Compiled: May 20 2005,16:47:51)

# block  row   col   nrows  ncols filename
block      0     0    100    100  block0000.dat
block    100   100     99     99  block0001.dat
block    199   199     99     99  block0002.dat
block    298   298     99     99  block0003.dat
block    397   397     99     99  block0004.dat
block    496   496     99     99  block0005.dat
block    595   595    100    100  block0006.dat
block    695   695     99     99  block0007.dat
```

(b) ASCII file

Fig. 8. Storing the Kite-Matrix with block coordinates

Manifold are also the possibilities for the kite scheme to be used as a preconditioner. In that regard, one can focus on particularly correlated groups of parameters, which become included into the fully correlated part in order to improve the condition number and, therefore, the convergence rate of the iterative solver.

Consequently, kite numbering proves to be a very flexible and efficient method for a realistic modelling of the Earth's gravity field.

Acknowledgement. The research was financially supported by the BMBF GEOTECHNOLOGIEN program GOCE-GRAND (Grant 03F0329C) and the ESA Contract No. 18308/04/NL/MM. This is publication no. GEOTECH-155 of the program GEOTECHNOLOGIEN of BMBF and DFG. The authors would like to thank Boris Kargoll for translating and suggesting ways to improve the presentation of the material.

References

Auzinger, T and Schuh, W-D (1998) High-degree spherical harmonic analysis combining gridded and random distributed data sets. *Phys. Chem. Earth*, 23:19–23

Balmino, G (1993) The spectra of the topography of the earth, venus and mars. *Geophys. Res. Lett.*, 20:1063–1066

Bosch, W (1993) A rigorous least squares combination of low and high degree spherical harmonics. In *Presented Paper on "IAG General Meeting"*. Beijing 1993

Colombo, O (1981) *Numerical methods for harmonic analysis on the sphere.* Reports of the Department of Geodetic Science. Ohio State University (OSU), Ohio. No. 310

Heiskanen, W and Moritz, H (2000) *Physical Geodesy.* Institute of Physical Geodesy, Technical University Graz, Reprint

Lemoine, F, Smith, D, Kunz, L, Smith, R, Pavlis, E, Pavlis, N, Klosko, S, Chinn, D, Torrence, M, Williamson, R, Cox, C, Rachlin, K, Wang, Y, Rapp, R, and Nerem, R (1996) The development of the nasa gsfc and nima joint geopotential model. In *Proceedings of the "International Symposium on Gravity, Geoid, and Marine Geodesy"*. Tokyo, Japan

Rapp, R (1994) *The Use of Potential Coefficient Models in Computing Geoid Undulations.* Lecture Notes of the International School for the "Determination and Use of the Geoid" (Oct. 10-15, 1994), Milano

Rapp, R, Wang, Y, and Pavlis, N (1991) *The Ohio state 1991 geopotential and sea surface topography harmonic coefficient models.* Reports of the Department of Geodetic Science. Ohio State University (OSU), Ohio. No. 410

Reigber, C, Balmino, G, Schwintzer, P, Biancale, R, Bode, A, Lemoine, J-M, König, R, Loyer, S, Neumayer, H, Marty, J-C, Barthelmes, F, Perosanz, F, and Zhu, S (2002) A high quality global gravity field model from CHAMP GPS tracking data and accelerometry (EIGEN-1S). *Geophysical Research Letters*, 29(14):10.1029/2002GL015064

Reigber, C, Schwintzer, P, Neumayer, K-H, Barthelmes, F, König, R, Förste, C, Balmino, G, Biancale, R, Lemoine, J-M, Loyer, S, Bruinsma, S, Perosanz, F, and Fayard, T (2003) The CHAMP-only earth gravity field model EIGEN-2. *Advances in Space Research*, 31(8):1883–1888

Reigber, C, Jochmann, H, Wünsch, J, Petrovic, S, Schwintzer, P, Barthelmes, F, Neumayer, K-H, König, R, Förste, C, Balmino, G, Biancale, R, Lemoine, J-M, Loyer, S, and Perosanz, F (2004) Earth gravity field and seasonal variability from CHAMP. In C. Reigber, H. Lühr, P. Schwintzer, and J. Wickert, editors, *Earth Observation with CHAMP - Results from Three Years in Orbit*, pages 25–30, Berlin. Springer

Rummel, R, Sansò, F, van Gelderen, M, Brovelli, M, Koop, R, Miggliaccio, F, Schrama, E, and Scerdote, F (1993). *Spherical harmonic analysis of satellite gradiometry.* Netherlands Geodetic Commission, New Series, 39

Schuh, W-D (1996a) Least squares adjustment of high degree spherical harmonics. In m. Jacobsen, editor, *Inverse Methods - Interdisciplinary Elements of Methodology, Computation and Application*, Lecture Notes in Earth Sciences 63, pages 276–283, Heidelberg. Springer

Schuh, W-D (1996b) *Tailored numerical solution strategies for the global determination of the earth's gravity field.* Mitteilungen der Geodätischen Institute der TU, Graz. Folge 81

An Integrated Global/Regional Gravity Field Determination Approach based on GOCE Observations

Annette Eicker, Torsten Mayer-Gürr, and Karl Heinz Ilk

Institute of Theoretical Geodesy, University of Bonn, D-53115 Bonn, Germany
annette@geod.uni-bonn.de

Summary. GOCE (Gravity Field and Steady-State Ocean Circulation Explorer) is a dedicated satellite gravity field mission to be launched in the year 2006. The payload of GOCE will consist of a GPS receiver for a precise orbit determination and for recovering the long and medium spectral part of the gravity field. The high resolution spectral part of the gravity field will be derived by in-orbit gravity gradients in three spatial directions measured by a gravity gradiometer consisting of six three-axis accelerometers. In this article an integrated gravity field recovery procedure is presented that allows to determine a global gravity field solution with high long and medium wavelength accuracy and to improve this global solution in regions with characteristic gravity field features by an adapted regional recovery procedure. If necessary, several regional solutions with global coverage can be merged by means of quadrature methods to obtain an improved global solution. Simulation results are presented to demonstrate this approach. Due to the improved regionally adapted gravity field solutions this technique provides better global gravity field recovery results than calculating a spherical harmonics solution by recovering the potential coefficients directly.

Key words: GOCE, SGG, GRACE, SST, regional gravity field zoom-in, global gravity field recovery, space localizing base functions

1 Introduction

The third satellite in the sequence of dedicated gravity satellite missions after CHAMP (CHAllenging Minisatellite Payload – Reigber et al. 1999) and GRACE (Gravity Recovery And Climate Experiment – Tapley et al. 2004) will be GOCE (Gravity field and steady-state Ocean Circulation Explorer). GOCE was selected as the first Core Mission within the Living Planet Earth Observation Programme of the European Space Agency (European Space Agency, 1999)). The mission duration is planned to be 20 months beginning in the year 2006. The satellite will fly in an extremely low near polar orbit of 250 km altitude. The pure free-fall motion will be permanently maintained by ion-thrusters compensating for surface forces. The long and medium spectral part of the gravity field can be derived from a dense coverage of the GOCE orbit by GPS measurements. The gravity field determination technique for this kind of observations is based on the so-called satellite-to-satellite tracking principle in the high-low mode or, alternatively, based on the analysis of a precisely determined kinematic orbit. The high resolution spectral part of the gravity field will be derived with unprecedented accuracy by a gravity gradiometer consisting of six three-axes accelerometers to measure in-orbit gravity gradients in three spatial directions. For the first time, the short wavelength spectral part of the gravity field will be determined from space by in-situ measurements and not be based purely on the analysis of orbit perturbations.

In the past decades several gravity field analysis techniques tailored to this new kind of observables have been proposed and tested based on various more or less realistic simulation scenarios. The problem consists in processing the extremely large amount of observations and in solving the huge unstable adjustment problems. One of the key problems seems to be related to the representation of the gravity field by an appropriate set of base functions and of the associated gravity field parameters. The usual way is to model the gravity field by spherical harmonics up to a certain upper degree. This degree is limited by the significance of the gravity field signal in the observations and its ratio to the observation noise. Because of the inhomogeneous gravity field of the Earth the signal content varies in the space domain. The gravity field in regions with rough gravity field features could be recovered up to a higher degree than within regions with smooth gravity field features. This means that the degree of instability of the gravity field recovery process, which increases with increasing high frequencies in the gravity field signal, varies with the specific gravity field characteristics in different regions. A gravity field representation with base functions of global support must be regularized globally. But a global regularization causes an overall filtering of the observations leading to a mean damping of the global gravity field features with the consequence that the high frequent gravity field signal in the observations is lost again in some geographical regions.

An alternative approach, presented here, is to determine a global gravity field solution with high long and medium wavelength accuracy and improve this global solution in regions with characteristic gravity field features by an adapted regional recovery procedure. The global solution is parameterized by spherical harmonic coefficients up to a moderate degree and the regional solutions are represented by space localizing base functions, e.g., by spherical splines. This procedure provides several advantages. The regional approach allows to exploite the individual signal content in the observations and a tailored regularization for regions with different gravity field characteristics. The advantage compared to a uniform global regularization is that the regularization factor is selected for each region individually. By an individually adapted regularization it is possible to extract more information out of the given data than with a global gravity field determination process. Regions with a smooth gravity field signal for example can be regularized stronger without damping the signal. In addition, the resolution of the gravity field determination can be chosen for each region individually according to the spectral behaviour of the signal in the specific region. Furthermore, the regional approach has the advantage of dealing with regions with different data coverages more easily. If no data at all is available (e.g. the polar gap) the regional refinement can be skipped. For regions with sparse data coverage a coarser parameterization can be selected. As a first indicator of a rough gravity field the structure of the topography or geophysical a-priori information can be used as a criterium. The resolution of the regional gravity field can be further improved by a subsequent iteration step. Another aspect especially relevant for the GOCE mission with its potential to recover the high resolution gravity field is the fact that regional solutions contain less unknown parameters and therefore the computation procedure is simplified. This enables to reduce the computation costs significantly.

If necessary, several regional solutions with global coverage can be merged by means of quadrature methods to obtain a global solution, in principle, up to an arbitrary degree, only limited by the signal content of the gravity gradient observations. Simulation results are presented to demonstrate this approach. Due to the regionally adapted strategy this method provides better results than calculating a spherical harmonic solution by recovering the potential coefficients directly (see also Eicker et al. 2004).

2 Setup of the mathematical model

The GPS receiver onboard of GOCE will enable a precise orbit determination and a precise recovery of the long and medium spectral part of the gravity field, based on the satellite-to-satellite tracking principle in the high-low mode or, alternatively, based on the analysis of precisely determined kinematic orbits. The latter procedure has been described in Mayer-Gürr et al. (2005a) and applied for the determination of the dedicated CHAMP gravity field models ITG-CHAMP01 in various regularization alternatives. The procedure can be applied analogously to determine the long and medium wavelength features of the gravity field by the analysis of the kinematic free-fall orbits of GOCE. The free-fall motion of GOCE will be permanently maintained by ion-thrusters compensating for surface forces. Alternatively, the long and medium wavelength parts of the gravity field can be derived also by the analysis of short relative arcs of the GRACE twin satellites as described e.g., in Mayer-Gürr et al. (2005b).

The high resolution spectral part of the gravity field will be derived with unprecedented accuracy by a gravity gradiometer. It consists of six three-axes accelerometers to measure in-orbit gravity gradients in three spatial directions. These measurements can be transformed to second derivatives of the gravitational potential $\nabla\nabla V(t;\mathbf{r};\mathbf{x})$ located at the geocentric positions $\mathbf{r}(t)$. The potential $V(t;\mathbf{r};\mathbf{x})$ can be separated in a reference potential $V_0 = V(t;\mathbf{r};\mathbf{x_0})$ and in an anomalous potential $T(t;\mathbf{r},\Delta\mathbf{x})$,

$$V(t;\mathbf{r};\mathbf{x}) = V_0(t;\mathbf{r};\mathbf{x_0}) + T(t;\mathbf{r};\Delta\mathbf{x}). \tag{1}$$

The unknowns to be solved for are the corrections $\Delta\mathbf{x}$ to the field parameters $\mathbf{x_0}$ of the reference potential. The measurements are performed along the satellite orbit at a regular sampling rate Δt. Every observation constitutes an observation equation for the determination of the unknown field parameters, either global or regional parameters. The (global) reference potential can be formulated in the usual way as follows,

$$V_0 = \frac{GM_E}{r}\sum_{n=0}^{n_{Max}}\sum_{m=0}^{n}\left(\frac{R_E}{r}\right)^n \left(c_{nm}C_{nm}(\vartheta,\lambda) + s_{nm}S_{nm}(\vartheta,\lambda)\right), \tag{2}$$

with the surface spherical harmonics,

$$C_{nm}(\vartheta,\lambda) = P_{nm}(\cos\vartheta)\cos m\lambda, \tag{3}$$

$$S_{nm}(\vartheta,\lambda) = P_{nm}(\cos\vartheta)\sin m\lambda \quad . \tag{4}$$

The anomalous potential $T(t;\mathbf{r},\Delta\mathbf{x})$ reads either in case of a global gravity field refinement,

$$T = \frac{GM_E}{r}\sum_{n=2}^{N_{Max}}\sum_{m=0}^{n}\left(\frac{R_E}{r}\right)^n \left(\Delta c_{nm}C_{nm}(\vartheta,\lambda) + \Delta s_{nm}S_{nm}(\vartheta,\lambda)\right), \tag{5}$$

with the corrections $\Delta c_{nm}, \Delta s_{nm} \in \Delta\mathbf{x}$ to the reference potential coefficients $c_{nm},\ s_{nm} \in \mathbf{x_0}$, or in case of a regional gravity field refinement,

$$T = \sum_{i=1}^{I} a_i\ \varphi(\mathbf{r},\mathbf{r}_{Q_i}), \tag{6}$$

with the unknown local field parameters a_i arranged in a column matrix $\Delta\mathbf{x} := (a_i,\ i = 1,...,I)^T$ and the space localizing base functions,

$$\varphi(\mathbf{r},\mathbf{r}_{Q_i}) = \sum_{n=0}^{N_{Max}} k_n \left(\frac{R_E}{r}\right)^{n+1} P_n(\mathbf{r},\mathbf{r}_{Q_i}), \tag{7}$$

with the position vector $\mathbf{r}$ to an arbitrary field point P and with the position vectors $\mathbf{r}_{Q_i}$to the I nodal points Q_i. The coefficients k_n are the difference degree variances of the gravity field spectrum to be determined,

$$k_n = \sum_{m=0}^{n}\left(\Delta\bar{c}^2_{nm} + \Delta\bar{s}^2_{nm}\right) \quad , \tag{8}$$

with the (in this formula fully normalized) residual potential coefficients $\Delta\bar{c}_{nm}, \Delta\bar{s}_{nm}$. These values are certainly not known in a regional gravity field refinement *a-priori* and, therefore, have to be approximated by a variance model. As the maximum degree used for the regional recovery corresponds to a spherical harmonic expansion of, e.g., up to a maximum degree $N_{Max} = 300$, the spline kernel has been constructed so that the expected unknown gravity field features will be represented as well as possible. For the present example the expansion has been truncated at degree $N_{Max} = 300$. R_E is the mean equator radius of the Earth, r the distance of a field point from the geo-center and $P_n(\mathbf{r}, \mathbf{r}_{Q_i})$ are the Legendre polynomials depending on the spherical distance between a field point P and the nodal points Q_i of the set of base functions. With this definition the base functions $\varphi(\mathbf{r}, \mathbf{r}_{Q_i})$ can be interpreted as isotropic and homogeneous harmonic spline functions (Freeden et al., 1998). The nodal points are defined on a grid generated by a uniform subdivision of an icosahedron of twenty equal-area spherical triangles. In this way the global pattern of spline nodal points Q_i shows approximately a uniform nodal point distribution. The left graph of Fig. 1 shows the densification of the icosahedron triangles to build the system of nodal points for the harmonic spline functions. The top right graph shows an example of a space localizing base function and the bottom graph shows an example of the degree variances of a gravity field to be recovered up to degree 300.

The observation equation for gradiometer measurements is obtained by differentiating the potential twice. It reads in case of a regional gravity field refinement,

$$\nabla\nabla T(\mathbf{r}) = \sum_{i=1}^{I} a_i \, \nabla\nabla\varphi(\mathbf{r}, \mathbf{r}_{Q_i}). \tag{9}$$

The observation equations are established for short arcs over the selected regional recovery area, while the coverage with short arcs should be slightly larger than the recovery region itself to prevent the solution from geographical truncation effects. Every short arc builds a partial system of normal equations $\mathbf{N}_i$ with the right-hand sides $\mathbf{b}_i$. To consider different accuracies of the short arcs these normal equations are combined by estimating a variance factor σ_i^2 for every arc by means of variance component estimation as described by Koch and Kusche (2003). In case of a regularization with a selected regularization matrix $\mathbf{N}_x$ the regularization factor σ_x^2 can be determined by the same procedure of variance component estimation as well. The merging of the normal equations for the n short arcs within the iterative variance-covariance computation procedure including the determination of the regional regularization parameter is shown in the flow chart of Fig. 2.

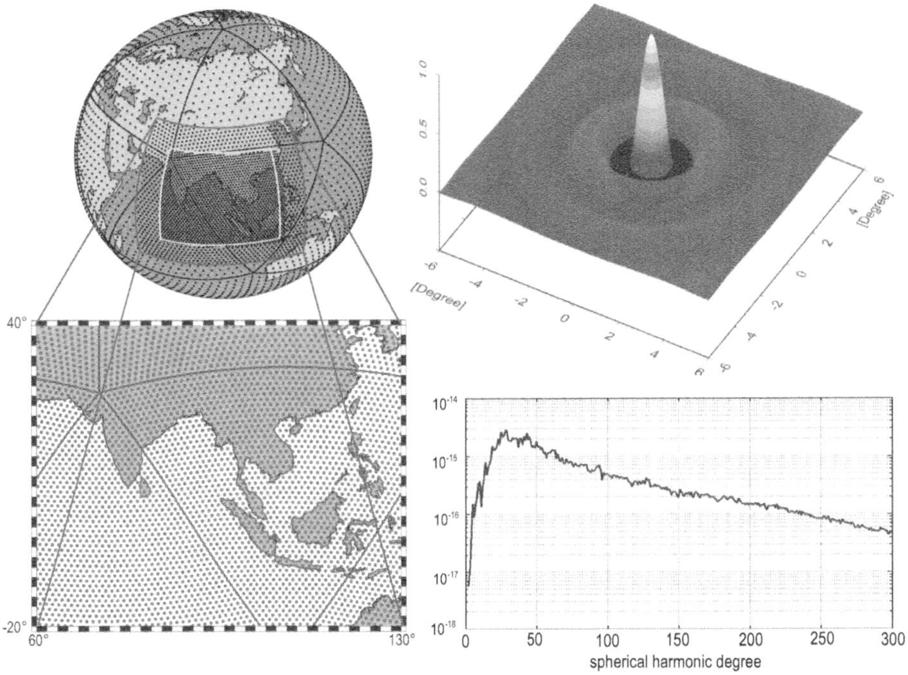

Fig. 1. Regional representation of the gravity field by space-localizing base functions. Left: densification of icosahedron triangles to build the system of nodal points for the harmonic spline functions; top right: example of a space localizing base function; bottom right: degree variances of the gravity field to be recovered (here shown up to degree 300).

3 Simulation scenario

In the present integrated gravity field recovery approach a precise global gravity field solution is refined by the high resolution regional solutions. A simulated GRACE-like solution as outlined, e.g., by Ilk et al. (2003) serves as the global solution because this mission will provide excellent results especially in the long and medium wavelength part of the gravity field spectrum. The regional refinements of this solution are then calculated on the basis of GOCE-like observations (Mayer-Gürr et al., 2003), as they cover the short periodic part of the gravity field spectrum with superior accuracy. In order to achieve a consistent data set both the GRACE and the GOCE solution were calculated from simulated observations on the basis of the gravity field model EGM96 (Lemoine et al., 1998) up to degree $n = 300$. The observations for GRACE as well as for GOCE were simulated for a period of 30 days with a sampling rate of 5sec.

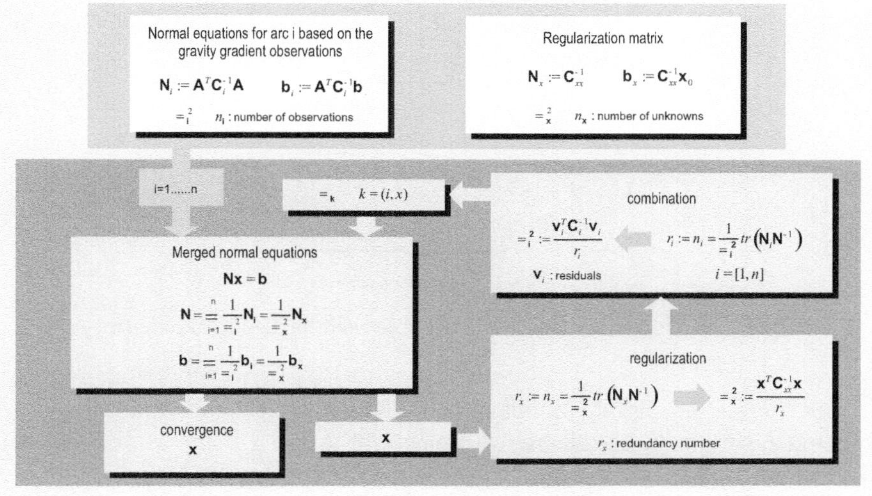

Fig. 2. Merging of normal equations for the short arcs within the iterative variance-covariance computation procedure including the determination of the regional regularization parameter.

In case of the GRACE satellites the simulated orbit positions were corrupted by white noise with a standard deviation of 3cm and the inter-satellite ranges between the GRACE twin satellites with a white noise of 10μm. The GOCE gravity gradiometer observations were corrupted by a coloured noise model with a standard deviation of 1.2mE. In this model the PSD is assumed as constant in the measurement band of the gradiometer and features increasing energy in the long wavelength part of the spectrum as the gradiometer is not able to recover these low frequencies. For the positions of GOCE again an accuracy of 3cm has been assumed. From this simulated data set a global gravity field solution based on the GRACE satellite-to-satellite low-low observations up to a spherical harmonic degree of $n = 150$ has been derived. The regional refinements to this global spherical harmonic solution were then calculated as residual fields in patches defined by a grid of geographical coordinates. Using a parameterization by splines as space localizing base functions, the spline kernels have to be adapted to the spectral range of the gravity field features to be determined. Therefore, up to degree $n = 150$, we used the error degree variances of the GRACE solution as coefficients k_n according to (8). These error degree variances represent the signal which is still in the data in addition to the GRACE solution. Above degree $n = 150$ the degree variances of the EGM96 were used for the coefficients k_n. The degree variances are shown in Fig. 3. Alternatively, the spline kernel could have been calculated according to Kaula's rule of thumb above degree 150. The latter possibility might be useful if no a-priori information at all shall be used for the regional refinements.

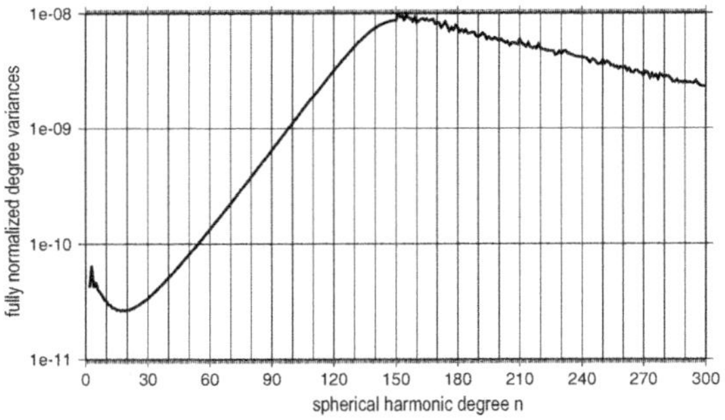

Fig. 3. Degree variances for the spherical harmonic expansion.

As there is no signal content in the GRACE field higher than degree 150 there is no jump in the degree variance graph. Even if in case of real data a jump occurred, this would have minor influence on the shape of the base function. The normalized kernel is displayed in the top right part of Fig. 1. The peak is located at the nodal points Q_i and the x- and y-axis indicate the spherical distance from these nodal points which are shown in the left part of this Fig. 1.

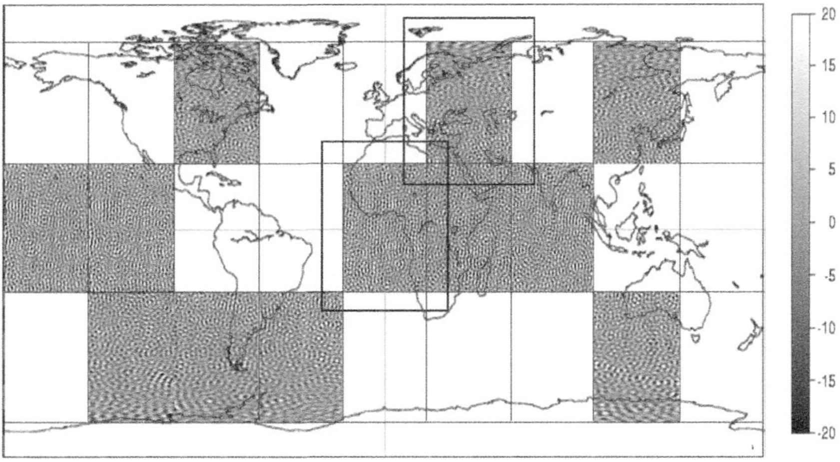

Fig. 4. Regional solutions covering the complete Earth, differences in geoid undulations to the EGM96 (cm).

4 Regional gravity field recovery

To combine the advantages of the satellite mission GOCE to cover almost the complete Earth with the advantages of regional focussing techniques, as pointed out in Sect. 1, the surface of the Earth has been divided into patches as shown in Fig. 4. To prevent the regional refinement solutions from geographical truncation effects the regions have been enlarged by an additional strip of 10° around the regions, marked by the frames drawn for two patches in Fig. 4. Then for each patch the gravity field has been determined separately. Fig. 4 shows some examples of the regional solutions and exemplary borders. The disturbing potential is represented by spline functions according to (8), located at nodal points of a grid generated as shown in Fig. 1 with a mean distance of approximately 67km. Concerning the number of unknown parameters this corresponds to a resolution of a spherical harmonic degree 300. This leads to a number of about 5000 to 9000 spline parameters to be determined for each region, the size of the patches being limited by storage restrictions. Due to the noise especially in the high frequencies a Tikhonov regularization has been applied. The regional recovery approach offers the possibility of calculating an individually adapted regularization parameter for each regional patch by means of the variance component estimation procedure according to Koch and Kusche (2003). Color Fig. XXVIII on p. 303 shows the differences of the regional solution for the Himalayan region, a region with extremely rough gravity field signals, compared to the pseudo-real field EGM96. The comparison has only been performed up to degree $n = 240$ because the higher degrees are too strongly corrupted by noise. The regional recovery procedure offers a chance to deal with the polar gap problem in a tailored way, as in regions without any data the regional refinement can either be skipped or the regularization parameter can be adjusted accordingly.

A closer look at three adjacent regional solution patches (Color Fig. XXIX on p. 303) reveals how well their residual patterns match together despite the fact that the recovery parameters such as variance factors and the regularization parameter are selected independently for each geographical region.

5 Combination of regional solutions

For many applications it seems to be useful to derive a global gravity field model by spherical harmonics without losing the details of a regional zoom-in. This can be performed by a direct stable computation step. One possibility to merge the regional solutions to a global one is to compute gravity functionals in the specific regions, in principle with arbitrary resolution and to compute the spherical harmonic coefficients by numerical quadrature. An alternative is to derive these coefficients analytically, but an adequate technique is not yet available. In our approach the coefficients of the spherical harmonic expansion are calculated by means of the Gauss-Legendre-Quadrature (see for example Stroud and Secrest (1966)). This method is also referred to as Neumann's method, as described in Sneeuw (1994) among different other quadrature methods,

$$\left\{ \begin{matrix} c_{nm} \\ s_{nm} \end{matrix} \right\} = \frac{R_E}{GM4\pi} \sum_{k=1}^{K} T_k P_{nm}(\cos\vartheta_k) \left\{ \begin{matrix} \cos(m\lambda_k) \\ \sin(m\lambda_k) \end{matrix} \right\} w_k, \tag{10}$$

with the area weights

$$w_k = \frac{2}{(1-t_k^2)\left(P'_{N+1}(\cos\vartheta_k)\right)^2} \quad , \tag{11}$$

with T_k being the gravitational potential at the K nodes of the quadrature, P_{nm} being associated Legendre functions and P'_{N+1} being the first derivatives of the Legendre polynomials of degree $N+1$, when N is the maximum degree to be determined. This method requires the data points to be located at a specific grid, called Gauss – Legendre – Grid. From the regional spline solutions the gravitational potential can be calculated at the nodes of this grid without loss of accuracy. It has equi-angular spacing along circles of latitude, along the meridian the nodes are located at the zeros of the Legendre polynomials of degree $N+1$ (Fig. 5). This quadrature method has the advantage of maintaining the orthogonality of the Legendre functions despite the discretization procedure, which allows an exact calculation of the potential coefficients. The grid used for the calculations has a spacing of $\Delta\lambda = 0.5°$ which corresponds to a number of 360 circles of latitude. It shall be pointed out that the direct computation of the spherical harmonic coefficients by solving the improperly posed downward continuation cannot provide a stable solution up to an (arbitrarily) high degree as it can be achieved here. In our application, the maximal degree should be limited only by the signal content of the gravity gradient measurements.

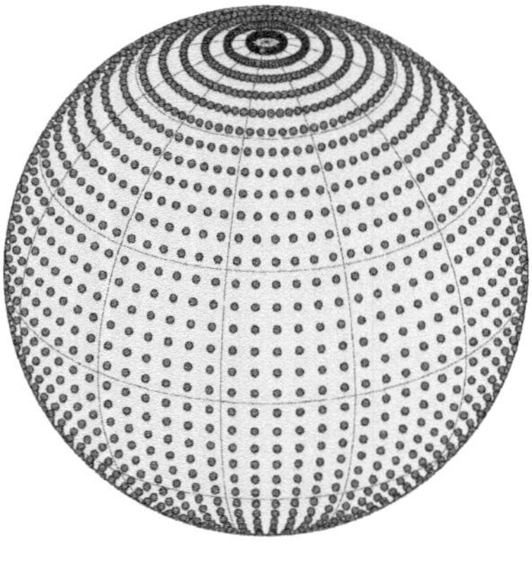

Fig. 5. Gauss-Legendre-Grid.

6 Results

The regional spline solutions have been merged to a global gravity field by predicting the gravity field functionals to the nodal points of a Gauss-Legendre-Grid. From this global field a spherical harmonics expansion has been calculated using the Gauss – Legendre - Quadrature. The differences of this combined global solution compared to the pseudo-real field EGM96 are displayed in Color Fig. XXX on p. 303. Again, the comparison has only been performed up to a spherical harmonic degree of $n = 240$.

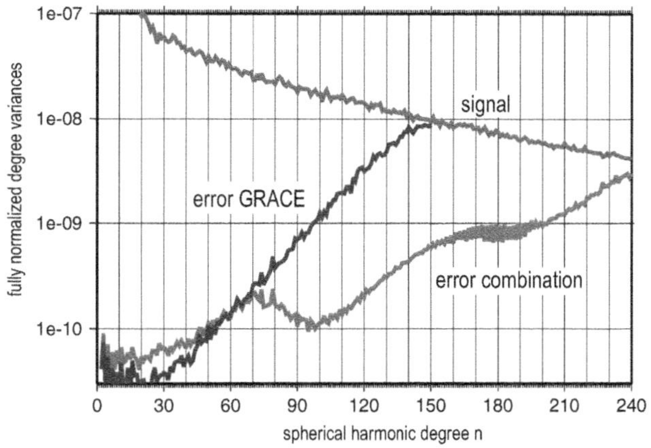

Fig. 6. Degree variances of the combined global solution.

The results show that an exact transformation of the individual regional patches to one global spherical harmonic solution is possible by this quadrature method. The differences between our solution and the EGM96 have been calculated on a $0.5° \times 0.5°$ grid of point geoid undulations, revealing a weighted global RMS of 8.6cm, which has been calculated including the poles. This accuracy corresponds to the noise described in Sect. 3 with the gradiometer observations being corrupted with a coloured noise of 1.2mE. The signal variances and the error variances of the basic global solution derived from the GRACE SST-low-low observations are shown in Fig. 6, together with the error-variances of the regionally refined solution, transformed to a global spherical harmonic solution. These results reveal the practicability of the presented procedure.

7 Conclusions and outlook

Combining regional solutions to a global gravity field solution is a reasonable alternative to deriving a global gravity field solution directly. The tailored calculation of a regularization parameter for each region allows a tailored filtering according to the individual gravity field features. The method is modest in terms of computation costs, as the complete problem is split up into much smaller problems. This procedure enables the computation of a global GOCE solution up to an arbitrary resolution on a single PC. This approach allows the calculation of a spherical harmonic expansion as well, with the advantage that the quadrature procedure does not limit the resolution to an upper degree but allows us to extract all possible information out of the given data. Adjacent regional patches fit remarkably well, even though the recovery parameters are selected individually for each region. Remaining differences are due to the different selection of the regularization parameter and to truncation errors. Further improvements are expected for example by refining the regularization strategy and tailoring it more accurately to the demand of the gravity field features in the specific regions. So far, the regions are selected according to a geographical grid, but it is planned for the future to select the region boundaries such that the global gravity field is divided into regions with homogeneous gravity field features. Furthermore, a more precise selection of the base functions and the nodal point distribution is intended in order to achieve an even better adjustment to the signal content in the regional areas. In this context multiresolution strategies might be successful as well. Other aspects that have to be taken into consideration for further investigations are for example the leakage effects originating from the patching of several regional solutions and the problem that certain global conditions (such as the conservation of the center of mass) might be violated when merging the regional solutions. Another problem that is to be solved in the future is the fact that it is necessary to take overlapping boundaries into account for the determination of the regional refinement patches in order to prevent the solution from geographical truncation errors and to ensure a smooth matching. But in this way the same data is used for the calculation of two adjacent patches and therefore the patches are not independent. This might be especially important for the derivation of a global solution by merging the patches

Acknowledgement. The support of BMBF (Bundesministerium für Bildung und Forschung) and DFG (Deutsche ForschungsGemeinschaft) of the GEOTECHNOLOGIEN programme is gratefully acknowledged. This is publication GEOTECH-157 of the programme GEOTECHNOLOGIEN of BMBF and DFG, Grant 03F0329C.

References

Eicker A, Mayer-Gürr T, Ilk KH (2004) Global Gravity Field Solutions Based on a Simulation Scenario of GRACE SST Data and Regional Refinements by GOCE SGG Observations, Proceedings of the Intern. Conference Gravity, Geoid and Space Missions – GGSM2004, 2004 August 30 – Sept. 3, Porto, Portugal

European Space Agency (1999) Gravity Field and Steady-State Ocean Circulation Explorer Mission (GOCE), Report for mission selection, in The four canditate Earth explorer core missions, SP-1233 (1), Nordwijk, The Netherlands

Freeden W, Gervens T, Schreiner M (1998) Constructive Approximatiuon on the Sphere, Oxford University Press, Oxford

Ilk KH, Feuchtinger M, Mayer-Gürr T, (2003) Gravity Field Recovery and Validation by Analysis of Short Arcs of a Satellite-to-Satellite Tracking Experiment as CHAMP and GRACE, Proceedings of the IAG Symposium G02, IUGG General Assembly 2003, Sapporo, Japan

Koch KR, Kusche J (2003) Regularization of geopotential determination from satellite data by variance components, Journal of Geodesy 76(5):259-268

Lemoine FG, Kenyon SC, Factor JK, Trimmer RG, Pavlis NK, Chinn DS, Cox CM, Klosko SM, Luthcke SB, Torrence MH, Wang YM, Williamson RG, Pavlis EC, Rapp RH, Olson TR (1998) The development of the joint NASA GSFC and the National Imagery and Mapping Agency (NIMA) geopotential model EGM96, NASA/TP-1998-206861, Goddard Space Flight Center, Grenbelt, MD

Mayer-Gürr T, Ilk KH, Eicker A (2003) Regional Gravity Field Recovery From GOCE Gradiometer Measurements and SST-High-Low Observations - A Simulation Study, Proceedings of the "1st Workshop on International Gravity Field Research", May 8-9, 2003, Graz, Austria

Mayer-Gürr T, Ilk KH, Eicker A, Feuchtinger M (2005a) ITG-CHAMP01: A CHAMP Gravity Field Model from Short Kinematical Arcs of a One-Year Observation Period, Journal of Geodesy (2005a) 78:462-480, DOI 10.1007/s00190-0004-0413-2, Springer-Verlag

Mayer-Gürr T, Eicker A, Ilk KH (2005b) Gravity field recovery from GRACE-SST data of short arcs, this volume

Reigber C, Schwintzer P, Lühr H (1999) The CHAMP geopotential mission, Boll. Geof. Teor. Appl., 40:285-289

Sneeuw N (1994) Global Spherical Harmonic Analysis by Least Squares and Numerical Quadrature Methods in Historical Perspective, Geophys. J. Int., 118: 707-716

Stroud AH, Secrest D (1966) Gaussian Quadrature Formulas, Prentice -Hall, Englewood Cliffs, N.J.

Tapley BD, Bettadpur S, Watkins M, Reigber Ch (2004) The gravity recovery and climate experiment: mission overview and early results. Geophys Res Lett 31, L09607: doi10.1029/2004GL019920

High-Performance GOCE Gravity Field Recovery from Gravity Gradient Tensor Invariants and Kinematic Orbit Information

Oliver Baur and Erik W. Grafarend

Stuttgart University, Geodetic Department, Geschwister-Scholl-Str.24D,
70174 Stuttgart, baur@gis.uni-stuttgart.de

Summary. The GOCE mission, planned to be launched in autumn 2006, will allow to determine the static Earth gravity field down to features of 100 km–70 km (half wavelength) in terms of spatial resolution. Since satellite gradiometry is restricted to the medium- to short-wavelength part of the gravitational spectrum, only its combination with satellite-to-satellite measurements in the high-low mode will meet the mission requirements as demanded by the ESA, namely a high-accurate GOCE-only terrestrial gravity field modeling. Here we apply the acceleration approach which is predominantly characterized by numerical differentiation of the kinematic GOCE orbit. Gradiometry is treated by analysis of the fundamental invariants of the gravitational tensor. These quantities neither depend on reference frame rotations nor on the orientation of the gradiometer frame in space. Linearization, computational effort and amalgamation of tensor elements provided with different levels of accuracy make this approach hard to handle. The use of high performance computing facilities, parallel programming standards and optimized numerical libraries are the key to accomplish efficient gravity field recovery.

Key words: GOCE-only Solution, Tensor Invariants, Kinematic Orbit Analysis, Numerical Differentiation, High Performance Computing

1 Introduction

Within ESA's (European Space Agency) "Living Planet" program the Earth explorer core mission GOCE (Gravity field and steady-state Ocean Circulation Explorer) planned to be launched in autumn 2006 will be the first satellite mission applying three-dimensional gradiometry in space (ESA, 1999, 2000). The observation data collected by the on-board sensors during the operational mode covering two six-month periods, interrupted by a hibernation phase of the satellite, is expected to allow for the recovery of the static terrestrial gravity field down to features of 100 km–70 km in terms of spatial resolution (half wavelength). However, satellite gravity gradiometry (SGG) is not able to recover the long-wavelength part of the Earth gravity field due to the limited measurement band width (MBW) of the gradiometer instrument ranging from 5 mHz to 0.1 Hz. Thus, a GOCE-only gravity field solution as demanded as the mission outcome by the ESA can't be provided by gradiometry alone.

Actually, the CHAMP (CHAllenging Minisatellite Payload) satellite mission established the opportunity to develop adequate algorithms for satellite-to-satellite tracking analysis in the high-low mode (hl-SST) between the high orbiting Global Positioning System (GPS) satellites and the low Earth orbiter (LEO) CHAMP (Reigber et al., 2005). Different approaches showed comparable results for the modeling of the long-wavelength part of the terrestrial gravity field such as the energy balance method (Földvary et al., 2005), short-arc analysis (Mayer-Gürr et al., 2005) and the acceleration approach (Reubelt et al., 2005). Within the scope of this contribution we focus on the acceleration approach. It is predominantly characterized by 2nd order numerical differentiation of the kinematic satellite orbit. The GPS track of the GOCE spacecraft is used to complement gradiometer observations in the dedicated frequency domain. The combined analysis of both SGG and hl-SST measurements is able to provide a GOCE-only estimate of the Earth's gravity field covering the whole gravitational spectrum.

Commonly, SGG analysis is performed on the level of gravitational gradients (GGs), namely the main diagonal elements of the gravitational tensor, being provided with highest accuracy with respect to the gradiometer frame of reference. A completely different approach is based on the rotational invariants of the observation tensor (Rummel et al., 1986). Beyond a pure methodological interest and the challenges of an efficient implementation, the main motivation for this approach is twofold. Firstly, the invariance under rotations actually means that knowledge about the gradiometer frame orientation is not required, neither with respect to the orbit frame nor to inertial space. Secondly, since this method has not been implemented before, it is independent of more conventional time-wise and space-wise approaches.

With the design and realization of the GOCE mission computational tasks gain evident importance in Satellite Geodesy, namely the access to and use of high performance computing (HPC) facilities. Due to the hardware limitations of ordinary personal computers (PCs) concerning both performance and main memory availability, only multiprocessor systems provide high-resolution gravity field estimates within a reasonable time frame.

The paper is organized as follows. The next Section deals with kinematic orbit analysis, predominantly with regard to numerical differentiation techniques to derive satellite accelerations from position information. Section 3 addresses to SGG data analysis based on the rotational invariants of the gravitational tensor. Parallel implementation of the algorithm using HPC facilities is treated in Sect. 4. Finally, the conclusions of this contribution are summarized in Sect. 5.

2 Kinematic Orbit Analysis

This Section is dedicated to the hl-SST part of the GOCE mission. First, the functional model for kinematic orbit analysis is derived followed by studies concerning numerical differentiation techniques. Finally GOCE hl-SST gravity field solutions based on the acceleration approach are presented.

2.1 Methodology

Conventionally, the parameterization of the terrestrial gravitational potential is expressed in spherical coordinates (λ, φ, r) with the unknown potential coefficients u_{lm}, cf. (1). Series truncation at a certain maximal degree $L = l_{\max}$ provides an approximation to reality. Both the geocentric constant GM and the mean Earth radius R are fixed. The normalized Legendre functions of the first kind $\bar{P}_{lm}(\sin\varphi)$ are part of the orthonormal base functions $e_{lm}(\lambda, \varphi)$, dentoted as surface spherical harmonics (2).

$$U_{\mathrm{E}}(\lambda, \varphi, r) = \frac{GM}{R} \sum_{l=0}^{\infty} \sum_{m=-l}^{l} \left(\frac{R}{r}\right)^{l+1} e_{lm}(\lambda, \varphi) u_{lm} \tag{1}$$

$$e_{lm}(\lambda, \varphi) = \begin{cases} \bar{P}_{lm}(\sin\varphi) \cos m\lambda & 0 \leq m \leq l \\ \bar{P}_{l|m|}(\sin\varphi) \sin|m|\,\lambda & -l \leq m < 0 \end{cases} \tag{2}$$

According to (3) the acceleration $\ddot{\mathbf{x}}(t) = \ddot{\mathbf{x}}(\lambda(t), \varphi(t), r(t))$ of the satellite (reduced by all disturbing effects such as tidal forces) is equal to the terrestrial attraction, namely the gradient of the Earth's gravitational potential $U_{\mathrm{E}}(\lambda, \varphi, r)$. Equation (3) is referred to as the acceleration approach and has been proven to perform well for CHAMP gravity field recovery (Reubelt et al., 2005). Kinematic orbit analysis is restricted to the determination of the long-wavelength part of the terrestrial gravity field only due to the satellite's positioning accuracy limited to some centimeters.

$$\begin{aligned} \frac{\mathrm{d}^2}{\mathrm{d}t^2}(\mathbf{e}_i x_i(t)) &= \mathbf{e}_i \frac{\mathrm{d}^2}{\mathrm{d}t^2} x_i(t) = \mathbf{e}_i \ddot{x}_i(t) = \operatorname{grad} U_{\mathrm{E}}(\lambda, \varphi, r) \\ &= \frac{GM}{R} \sum_{l=0}^{\infty} \sum_{m=-l}^{l} \left(\frac{R}{r}\right)^{l+2} \\ &\times \left\{ \mathbf{e}_\lambda \frac{\partial e_{lm}(\lambda, \varphi)}{\cos\varphi\, \partial\lambda} + \mathbf{e}_\varphi \frac{\partial e_{lm}(\lambda, \varphi)}{\partial\varphi} - \mathbf{e}_r (l+1) e_{lm}(\lambda, \varphi) \right\} u_{lm} \end{aligned} \tag{3}$$

2.2 Numerical Differentiation

The crucial point in (3) is the determination of accelerations $\mathbf{e}_i \ddot{x}_i(t)$ that serve as pseudo-observations for kinematic orbit analysis. Since hl-SST provides orbit information on the level of position coordinates, numerical differentiation techniques have to be applied. GPS positions are highly correlated, predominantly due to systematic atmospheric effects. With regard to numerical differentiation these errors can be reduced dramatically when using coordinate differences $\Delta\mathbf{x}(t_i) = \mathbf{x}(t_i) - \mathbf{x}(t_{i-1})$ instead of the absolute orbit information (Reubelt et al., 2003). Assuming the remaining systematic errors to be negligible the uncertainties of hl-SST phase measurements itself dominate the overall error budget. These random errors have to be assumed as non-correlated. However, dependent on the procedure for GOCE orbit determination based on GPS phase measurements the positions of the spacecraft will be more or less correlated. Normal distributed position errors, denoted as white noise, can be considered as the worst case. Thus, that scenario is defined as lower bound for what can be expected at least from GOCE hl-SST analysis by means of the acceleration approach. The standard deviation of random position errors is about σ_{x_i} =1-2 cm for each coordinate x_i (ESA, 1999). In most methods for numerical differentiation an interpolative polynomial is fitted through the time-series of position coordinates. A moderate number of sampling points is used to estimate the polynomial coefficients. Due to the oscillation of the polynomial at its edges, the interpolation point is specified to be in the middle of the interpolation mask which is shifted for successive satellite positions. Evaluation of polynomial derivatives finally leads to the desired pseudo-observations.

Besides the number of sampling points, the degree of the polynomial respectively, the methods for numerical differentiation predominantly differ in the kind of the interpolation polynomial, namely its smoothing behavior. Four methods have been explored in detail: (i) Gregory-Newton interpolation, (ii) spline interpolation, (iii) polynomial regression and (iv) smoothing spline interpolation. The non-smoothing methods (i) and (ii) achieve similar results as well as the smoothing methods (iii) and (iv). Thus, in the following solely Gregory-Newton (GN) interpolation (Maeß, 1988) and polynomial regression (PR) will be addressed. Actually, GN is a special case of PR, namely the least squares adjustment to estimate the polynomial coefficients is reduced to a unique estimate. For each coordinate x_i, $i = 1, 2, 3$ Table 1 contains root-mean-square (RMS) values of 500 000 acceleration residuals $\Delta\ddot{\mathbf{x}} = \ddot{\mathbf{x}}^{\text{true}} - \ddot{\mathbf{x}}^{\text{est}}$ between true accelerations $\ddot{\mathbf{x}}^{\text{true}}$ known from orbit simulation and estimated accelerations $\ddot{\mathbf{x}}^{\text{est}}$ applying GN, respectively PR (interpolation interval $\Delta t = 5\,\text{s}$). Obviously, smoothing methods seem to be by far better suited for numerical differentiation than non-smoothing ones.

Table 1. RMS values of acceleration residuals $\Delta\ddot{x}_i$ (mm), $\Delta t = 5\,\text{s}$

	$\sigma_{x_i} = 1\,\text{cm}$			$\sigma_{x_i} = 2\,\text{cm}$		
	$\Delta\ddot{x}_1$	$\Delta\ddot{x}_2$	$\Delta\ddot{x}_3$	$\Delta\ddot{x}_1$	$\Delta\ddot{x}_2$	$\Delta\ddot{x}_3$
GN	1.45	1.46	1.46	2.91	2.92	2.91
PR	0.16	0.16	0.16	0.32	0.32	0.32

The difference in RMS values between them are almost one order of magnitude. But this is not the whole truth. Figures 1 and 2 illustrate the amplitude spectra of the residuals in column 4 of Table 1. The spectra are identic in the

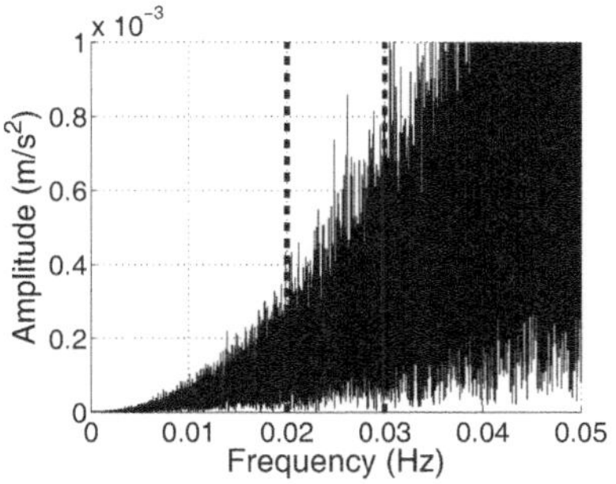

Fig. 1. Amplitude spectrum of acceleration residuals applying GN

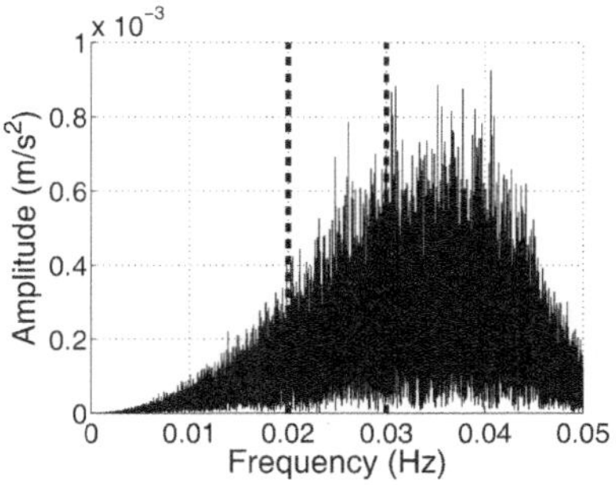

Fig. 2. Amplitude spectrum of acceleration residuals applying PR

long-wavelength part, namely up to approx. 30 mHz. The smoothing behavior dominates primarily at higher frequencies causing the differences of the RMS values in the time domain. In terms of terrestrial gravity field recovery 30 mHz corresponds to a resolution up to $L \approx 90$. This is in the range of what can be expected at most by kinematic orbit analysis, respectively what is reasonable to recover based on hl-SST observations since gradiometry covers the medium-to short-wavelength part of the gravitational spectrum. Thus, the method for numerical differentiation applying the acceleration approach is not restricted.

2.3 GOCE hl-SST Gravity Field Solutions

Numerical studies are based on a simulated GOCE data set covering one month of observation data with a sampling rate of $\Delta t = 5\,\mathrm{s}$, provided by the IAG Section II Special Commission VII (SC7). The EGM96 up to degree and order $L = 300$ is used to calculate synthetic orbit information as well as GGs with respect to the local orbit reference frame (LORF). Kinematic orbit analysis is applied to recover the terrestrial gravitational field up to degree and order $L = 100$ without any regularization. The results in terms of empirical degree error RMS are presented in Figs. 3 and 4. Due to the polar gap problem the orders $m < 5$ are not considered. Assuming non-correlated noise of 2 cm for each position coordinate a resolution up to degree $l = 70$ can be achieved whereas even $l = 80 - 85$ is reached for the more optimistic noise level of $\sigma_{x_i} = 1\,\mathrm{cm}$. Concerning the numerical differentiation method, the estimates vary for the low degrees up to approx. $l = 30$. Table 2 proves the statement that the method for numerical differentiation doesn't influence the gravity field recovery procedure significantly. The latitude weighted geoid

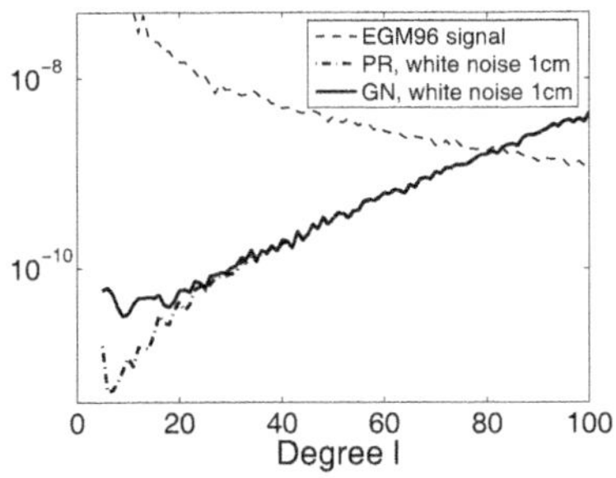

Fig. 3. Degree error RMS of hl-SST analysis with position error $\sigma_{x_i} = 1\,\mathrm{cm}$

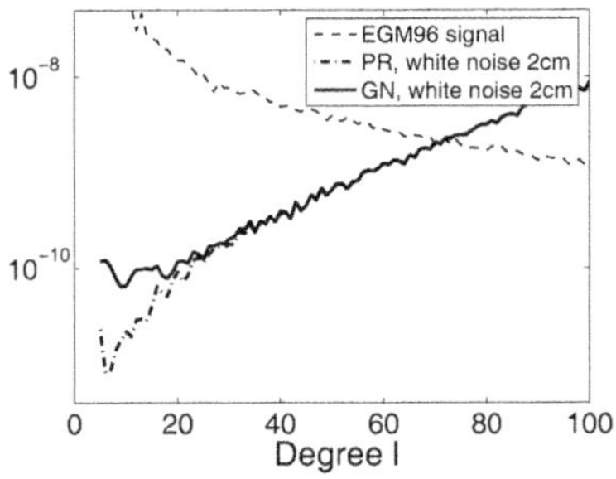

Fig. 4. Degree error RMS of hl-SST analysis with position error $\sigma_{x_i} = 2\,\mathrm{cm}$

errors (weighted with $\cos\varphi$) between the estimate and the a priori parameter set for data simulation yield almost the same results for both GN and PR.

Table 2. Latitude weighted RMS values of geoid errors (mm)

resolution L	maximal latitude $\pm 83°$				maximal latitude $\pm 90°$			
	$\sigma_{x_i} = 1\,\mathrm{cm}$		$\sigma_{x_i} = 2\,\mathrm{cm}$		$\sigma_{x_i} = 1\,\mathrm{cm}$		$\sigma_{x_i} = 2\,\mathrm{cm}$	
	GN	PR	GN	PR	GN	PR	GN	PR
70	0.89	0.88	1.78	1.77	0.85	0.85	1.70	1.69
80	1.61	1.60	–	–	1.55	1.54	–	–

3 SGG Analysis Based on Tensor Invariants

Differential mode measurements of the GOCE gradiometer lead to the gravitational tensor $\mathsf{U}_\mathrm{E} = (\mathbf{e}_i \otimes \mathbf{e}_j)\mathbf{U}_\mathrm{E}\,;\, i,j = 1,2,3$. The GGs $U_{ij} = U_{ji}$ are linear functionals of the Earth gravity field, namely the second derivatives of the terrestrial gravitational potential, compare (4). Assembling the accelerometers according to the diamant configuration (ESA, 1999), the main diagonal elements as well as the component U_{13} can be determined with an accuracy of about $6\,\mathrm{mE\,Hz}^{-1/2}$. The remaining GGs are provided three orders of magnitude worse.

$$(\mathbf{e}_i \otimes \mathbf{e}_j)\mathbf{U}_\mathrm{E} = \mathrm{grad} \otimes \mathrm{grad}\, U_\mathrm{E}(\lambda, \varphi, r) \tag{4}$$

Different methods use the observation model (4) for gravity field recovery. Dependent on the kind of data processing they are split in two groups, the space-wise and time-wise approach. For the first one, the analysis procedure is formulated in terms of a fixed boundary value problem (BVP). The time-wise approach treats the measurements in space as a time series. Solving the full normal equation system by brute-force inversion leads directly to an estimate for the unknown parameter vector. Compared to that the parameterization of (4) with inclination functions (Kaula, 1966) leads to the representation of the functional model in lumped coefficients, denoted as the time-wise approach in the frequency-domain. This method requires simplified assumptions to the orbit geometry combined with an iterative analysis process, referred to as the semi-analytical approach. The different techniques are described in Rummel et al. (1993). Further details can be found in e.g. Colombo (1981), Schuh (1996), Klees et al. (2000), Sneeuw (2000) and Pail and Plank (2002).

The accuracy of the GGs is dependent on the underlying frame of reference. This is due to the influence of the poorly known tensor elements U_{12} and U_{23}. Tensor transformation leads the transformed quantities to be a linear combination of the original components of different accuracy levels and should therefore be avoided. Actually, due to the elimination of the field emission electric propulsion (FEEP) thrusters (Saccoccia et al., 2000) from the GOCE configuration the satellite is assumed to oscillate periodically about its yaw-axis. Altogether, the orientation of the satellite in space is a delicate topic. Gravity gradient tensor invariants are free of these demands.

3.1 Methodology

The invariant properties of the gravitational tensor $\mathsf{U}_\mathrm{E} = (\mathbf{e}_i \otimes \mathbf{e}_j)\mathbf{U}_\mathrm{E}$ come along with the solution of the eigenvalue problem (5) of the coefficient matrix $\mathbf{U}_\mathrm{E}$. The cubic characteristic equation (6) is composed of the eigenvalues λ_i, $i = 1, 2, 3$ and the polynomial coefficients I_1, I_2 and I_3. They are invariant with respect to the underlying reference frame, respectively reference frame rotations, and therefore denoted as tensor invariants. For the symmetric gravitational tensor they look like shown in (7) to (9).

$$\det(U_{ij} - \lambda\delta_{ij}) = 0 \tag{5}$$

$$\lambda^3 - I_1\lambda^2 + I_2\lambda - I_3 = 0 \tag{6}$$

$$I_1 = \mathrm{tr}\,\mathbf{U}_\mathrm{E} \tag{7}$$

$$I_2 = \frac{1}{2}\left[\mathrm{tr}\,\mathbf{U}_\mathrm{E}^2 - (\mathrm{tr}\,\mathbf{U}_\mathrm{E})^2\right] \tag{8}$$

$$I_3 = \det\,\mathbf{U}_\mathrm{E} \tag{9}$$

Inserting the harmonic series expansion of the GGs in the formulae above finally leads to the functional model for invariant analysis as presented in (10) to (12).

$$I_1 = \frac{GM}{R^3} \sum_{i=0}^{\infty} \sum_{j=0}^{i} \left(\frac{R}{r}\right)^{i+3} S^1_{ij}(\varphi) \left[c_{ij} \cos(j\lambda) + s_{ij} \sin(j\lambda)\right] \tag{10}$$

$$\begin{aligned} I_2 = \left(\frac{GM}{R^3}\right)^2 \sum_{i=0}^{\infty} \sum_{j=0}^{i} \sum_{k=0}^{\infty} \sum_{l=0}^{k} \left(\frac{R}{r}\right)^{i+k+6} \{ \\ K^1_{ijkl}(\varphi) \left[-c_{ij} \sin(j\lambda) + s_{ij} \cos(j\lambda)\right] \left[-c_{kl} \sin(l\lambda) + s_{kl} \cos(l\lambda)\right] + \\ K^2_{ijkl}(\varphi) \left[c_{ij} \cos(j\lambda) + s_{ij} \sin(j\lambda)\right] \left[c_{kl} \cos(l\lambda) + s_{kl} \sin(l\lambda)\right] \} \end{aligned} \tag{11}$$

$$\begin{aligned} I_3 = \left(\frac{GM}{R^3}\right)^3 \sum_{i=0}^{\infty} \sum_{j=0}^{i} \sum_{k=0}^{\infty} \sum_{l=0}^{k} \sum_{m=0}^{\infty} \sum_{n=0}^{m} \left(\frac{R}{r}\right)^{i+k+m+9} \{ \\ D^1_{ijklmn}(\varphi) \left[c_{ij} \cos(j\lambda) + s_{ij} \sin(j\lambda)\right] \\ \times \left[-c_{kl} \sin(l\lambda) + s_{kl} \cos(l\lambda)\right] \left[-c_{mn} \sin(n\lambda) + s_{mn} \cos(n\lambda)\right] + \\ D^2_{ijklmn}(\varphi) \left[c_{ij} \cos(j\lambda) + s_{ij} \sin(j\lambda)\right] \\ \times \left[c_{kl} \cos(l\lambda) + s_{kl} \sin(l\lambda)\right] \left[c_{mn} \cos(n\lambda) + s_{mn} \sin(n\lambda)\right] \} \end{aligned} \tag{12}$$

The coefficient functions $S^1_{ij}(\varphi)$, $K^{1,2}_{ijkl}(\varphi)$ and $D^{1,2}_{ijklmn}(\varphi)$ contain the Legendre functions as well as their first and second derivatives. Analysis of the first invariant leads to the trivial solution $\mathbf{0}$ since $\operatorname{tr} \mathbf{U}_{\mathrm{E}} \equiv 0$ holds for the gravitational tensor. Thus, I_1 can't be used for potential field recovery. Non-linearity of the functional models for I_2 and I_3 makes invariant analysis to become an iterative process. Linearization of (11) and (12) with respect to the unknown coefficients c_{pq}, respectively s_{pq}, is performed by summation of the partial derivatives for each combination of $p = \text{const}$ and $q = \text{const}$, as outlined in (13) and (14).

$$\frac{\partial I_2}{\partial c, s_{p,q=\text{const}}} = \frac{\partial I_2}{\partial c, s_{i=p,j=q}} + \frac{\partial I_2}{\partial c, s_{k=p,l=q}} \tag{13}$$

$$\frac{\partial I_3}{\partial c, s_{p,q=\text{const}}} = \frac{\partial I_3}{\partial c, s_{i=p,j=q}} + \frac{\partial I_3}{\partial c, s_{k=p,l=q}} + \frac{\partial I_3}{\partial c, s_{m=p,n=q}} . \tag{14}$$

For the first iteration an approximate solution for linearization has to be introduced. Within this contribution the OSU86F model is used. Note that data simulation is based on the EGM96. Actually, the iterative procedure converges very fast. For studies using noise-free simulated data, the final solution is reached after the second or even first iteration, i.e. the linearization error is small.

3.2 Series Truncation

Unfortunately the coefficient functions $K^2_{ijkl}(\varphi)$ and $D^{1,2}_{ijklmn}(\varphi)$ are not symmetric with respect to index permutation, i.e. $K^2_{pqkl}(\varphi) \neq K^2_{ijpq}(\varphi)$ etc. holds.

Thus, for the setup of the linearized models (13) and (14) four fold, respectively six fold, nested sums have to be evaluated. Combined with an iterative procedure, especially for a high-resolution gravity field estimate, the strict method is not applicable from the computational point of view. However, the full computation can be avoided by early truncation of the inner loops (15), which virtually leads to the same results.

$$i_{\max} = L, \quad k_{\max}, m_{\max} \ll L \tag{15}$$

For noise-free SC7 data, Fig. 5 presents degree error RMS differences between SGG analysis based on the radial component U_{33} and solutions after the first iteration performing analysis of the second invariant with series truncation for the inner loops at degree 3, respectively degree 0. Due to the polar gap problem the plots don't contain low orders up to $m = 6$. Apart from the low degrees, the sum of both linearization error and truncation error is below the empirical error curve of the non-iterative estimate. Thus, early series truncation is an adequate tool to decrease runtime for invariant analysis dramatically.

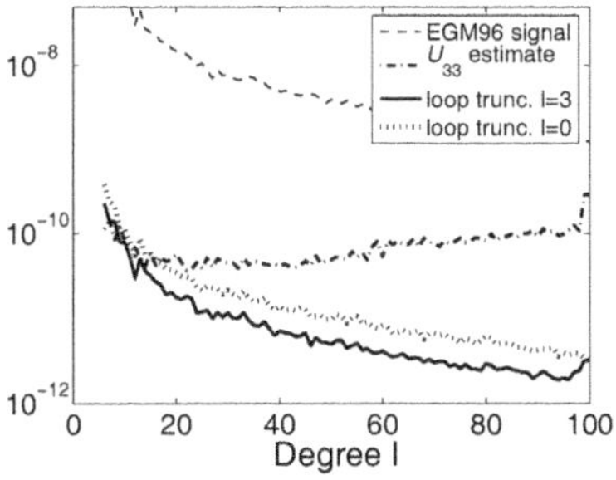

Fig. 5. Degree error RMS between U_{33} analysis estimate and I_2 solutions dependent on series truncation

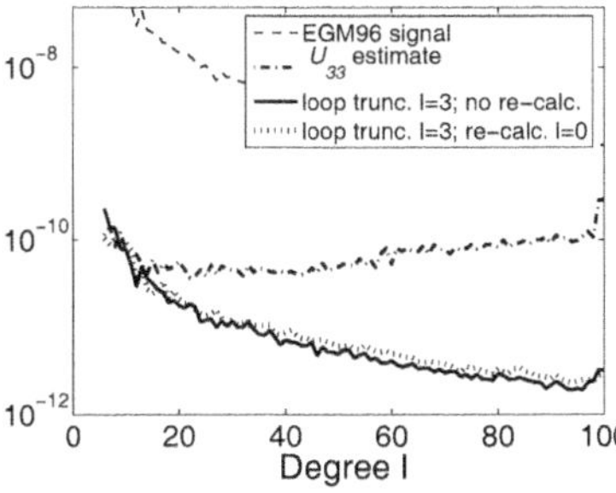

Fig. 6. Degree error RMS between U_{33} analysis estimate and I_2 solutions dependent on re-calculation of GGs

3.3 Re-calculation of Tensor Elements

Besides the computational effort for invariant analysis a further aspect has to be considered, namely the amalgamation of GGs, compare (8) and (9). The second and the third invariant are composed of products between both the tensor elements of high and reduced accuracy. Thus, the overall accuracy of invariants would be decreased as well. This problem can be circumvented by re-calculation of the elements U_{12} and U_{23} from iteration to iteration. The course of action is outlined in Fig. 7. Starting from the linearized observation equation, for the first iteration an a priori gravity field parameter set is used to calculate synthetic values for the "unknown" tensor elements U_{12} and U_{23},

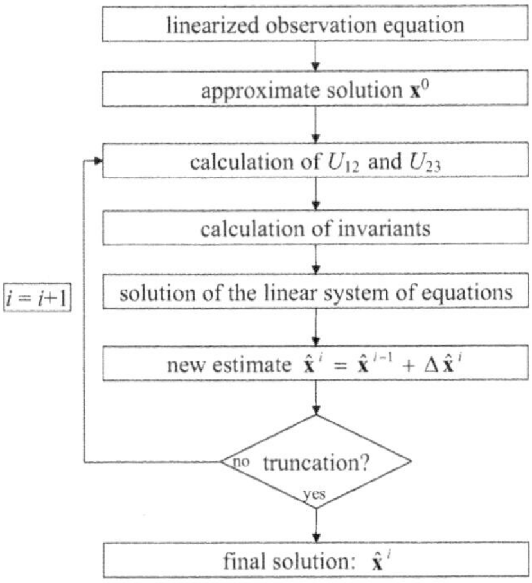

Fig. 7. Re-calculation of GGs

respectively the pseudo-observations of type invariants. For successive iterations the actual estimate is used to update the tensor component calculation. In Fig. 6 additionally to series truncation of the inner loops at degree 3, the simulated GGs U_{12} and U_{23} have been replaced by synthetic values based on the OSU86F model only considering degree 0 for evaluation. The replaced tensor elements don't influence the result significantly even when truncating their calculation at $L = 0$. That means, the analysis procedure is insensitive to the deviation of the real values from the synthetic ones. Thus, re-calculation of GGs is an adequate tool to overcome the difficulties of allocating the pseudo-observations for invariant analysis.

4 High Performance Computing in Gravity Field Research

From the computational point of view the problem dimension for GOCE data analysis has to be treated according to both memory requirement and runtime. Parallelization of the least squares adjustment procedure using multi-processor architectures is indispensable for solving for the unknown gravity field parameters. During the operational mode of the GOCE mission observations amount to several millions of gradiometer measurements as well as position coordinates of the spacecraft. Whereas the setup of the design matrix, respectively the normal matrix, can be done blockwise, for direct inversion of the normal equation system (NES), referred to as the brute-force approach, at least one triangle of the symmetric normal matrix has to be kept in the memory. For a resolution of the terrestrial gravity field up to degree and order $L = 300$ this equals a memory availability of 33 GB. Ordinary PCs don't come along with these requirements. Remedy can be found by using iterative solvers such as conjugate gradient (CG) methods (Hestenes and Stiefel, 1952) or the LSQR algorithm (Paige and Saunders, 1982a,b) which have been applied successfully for potential field recovery (Ditmar and Klees, 2002; Baur and Austen, 2005). Due to their design, matrix-matrix operations are replaced by repeated vector-vector operations. Thus, memory requirements are small. Despite of that, for both the brute-force approach and iterative solvers runtime of the analysis process makes parallelization of the implementation necessary.

4.1 Parallelization

With regard to the brute-force approach the setup of the design matrix $\mathbf{A}$ and computation of the normal matrix $\mathbf{N}$ is done blockwise according to (18).

$$\mathbf{A}\mathbf{x} = \mathbf{y} \tag{16}$$

$$\mathbf{A}^T\mathbf{A}\hat{\mathbf{x}} = \mathbf{A}^T\mathbf{y} \leftrightarrow \mathbf{N}\hat{\mathbf{x}} = \mathbf{b} \tag{17}$$

$$\mathbf{N} = \mathbf{A}_1^T\mathbf{A}_1 + ... + \mathbf{A}_j^T\mathbf{A}_j = \sum_{i=1}^{j}\mathbf{A}_i^T\mathbf{A}_i = \sum_{i=1}^{j}\mathbf{N}_i \tag{18}$$

This approach is indispensable, given that for a large amount of observations and many unknowns the design matrix can not be kept in the main memory. Indeed, a blockwise procedure even allows for the use of parallel matrix-matrix and matrix-vector routines provided by the numerical libraries such as e.g. Lapack and Blas (Anderson et al., 1999) to repeatedly compute $\mathbf{N}_i = \mathbf{A}_i^T\mathbf{A}_i$. Further parallelization is achieved for the setup of each design matrix block $\mathbf{A}_i$ by distributing the observations contributing to each block to several central processing units (CPUs).

Due to the character of iterative methods, matrix-matrix and matrix-vector multiplications are avoided by means of repeated vector-vector operations. Since neither the design matrix nor the normal matrix must be kept in the main memory, storage requirements are by far smaller as compared to direct solvers at the expense of an increased amount of operations. Fortunately, these multiplications can be done separately for each observation. Additionally, the major computational costs occur within the calculation of the design matrix. It is therefore reasonable to distribute the amount of observations on several CPUs of a multiprocessor computation platform to build up the design matrix line by line, each line referring to a separate observation. Within the scope of this contribution it is exclusively concentrated on the brute-force approach.

4.2 High Performance Computing

Different platform architectures for parallel implementation with OpenMP and MPI have been considered to investigate their benefit for the setup and solution of the NES by means of the brute-force approach. All the systems are supported by the High Performance Computing Center Stuttgart (HLRS), namely (i) NEC TX-7 (ii) NEC SX-6 (iii) Cray Strider. The architecture of platform (i) is ccNUMA, i.e. similar to shared-memory systems. Platform (ii) is, considering only one node, a shared-memory array processor system. Platform (iii) is in principle (disregarding that each single node is in fact a SMP node of 2 CPUs) a distributed-memory cluster. Table 3 lists some specific values of the super computers. To evaluate the performance of the least squares procedure regarding the different platforms a comparable test peak performance according to the last column of Table 3 has been chosen. Table 4 summarizes the performance, efficiency respectively, of a test scenario with maximal resolution $L = 50$ and half a million of observations of type hl-SST. The SX-6 is more than three minutes slower than the TX-7 and only achieves an efficiency of 37% as compared to 53% for the ccNUMA architecture. This is surprising since vectorization of the algorithm should increase the performance significantly. Actually, the efficiency of the platform is decreased by comparatively time-consuming data reading. The time for real computations is about seven minutes which corresponds to a performance of 24.6 GFlops, respectively an efficiency of 55%. This is still disappointing but due to the moderate problem dimension. To proof that, an additional calculation has been performed on the SX-6 with a resolution up to $L = 100$ for the terrestrial gravitational potential. The impact of vectorization is obvious since the efficiency amounts to 88% which is near to the theoretical peak performance. Thus, the array processor system SX-6 is suited very well for the brute-force approach. The result for the Opteron cluster Cray Strider turns out to be

Table 3. Specific values of computation platforms (PP–peak performance)

platform	architecture	number of CPUs	main memory (GB)	theoretical PP (GFlops)	test PP (GFlops)
NEC TX-7	ccNUMA	16	240	$16 \cdot 6$	$8 \cdot 6 = 48$
NEC SX-6	cluster	$6 \cdot 8$	$6 \cdot 64$	$6 \cdot 8 \cdot 9$	$5 \cdot 9 = 45$
Cray Strider	cluster	$125 \cdot 2$	$125 \cdot 4$	$125 \cdot 2 \cdot 4$	$12 \cdot 4 = 48$

Table 4. Achieved performance for hl-SST analysis

platform	wall time	user time	performance (GFlops)	efficiency (%)
NEC TX-7	6 m 58 s	54 m 7 s	25.4	53
NEC SX-6	10 m 22 s	38 m 32 s	16.6	37
Cray Strider	9 m 16 s	111 m 12 s	18.6	39

worse. This has been expected in advance since compared to the TX-7 the cache of the Opteron CPUs is considerably smaller. Additionally, runtime costs for the communication between the cluster nodes by using MPI has to be accepted.

The impact of HPC for GOCE gravity field recovery is summarized in Tables 5 and 6. The calculations are performed on the TX-7 dependent on the number of observations and the number of threads used. The runtimes in

Table 5. Impact of HPC for hl-SST analysis ($L = 100$)

number of observations (mill.)	number of CPUs	time setup $\mathbf{A}$ (min)	time setup $\mathbf{N}, \mathbf{b}$ (min)	time NES inversion (min)
1.5	1	8	560	4
1.5	4	4	140	1
1.5	8	2	70	0.5

Table 6 correspond to SGG analysis on the level of gravity gradients, respectively second invariant analysis with truncation of the inner loops at degree 0. Decorrelation is not applied regarding the time to set up the design matrix $\mathbf{A}$. For both hl-SST and SGG analysis optimal scaling is achieved with respect to the number of observations as well as the number of CPUs. The main computational effort is within the calculation of the NES, namely the algebraic operations $\mathbf{N} = \mathbf{A}^T\mathbf{A}$ and $\mathbf{b} = \mathbf{A}^T\mathbf{y}$. The time for NES inversion is comparatively short.

Table 6. Impact of HPC for SGG analysis ($L = 200$)

number of observations (mill.)	number of CPUs	time setup $\mathbf{A}$ (min)	time setup $\mathbf{N}, \mathbf{b}$ (min)	time NES inversion (min)
0.5	1	320	2700	160
0.5	4	80	680	40
0.5	8	40	340	20
1.5	8	120	1050	20

5 Conclusions

For purposes of estimating a GOCE-only gravity field we have analyzed hl-SST data using the acceleration approach. Based on one month of kinematic orbit data, dependent on the error budget at least a resolution up to degree $L = 70$ in terms of spherical harmonics in the long-wavelength part can be achieved. The method for numerical differentiation of the spacecraft's position information is not restricted.

Invariant analysis has been proven to be applicable in three-dimensional gradiometry. It is associated, though, with a more complicated processing strategy as compared to conventional approaches, in which the main diagonal tensor elements are analyzed directly. Both the enormous numerical effort and the mixture of high and low accuracy tensor components can be circumvented by early series truncation, respectively re-calculation of the poorly known gravitational gradients from iteration to iteration based on the previous estimate of the unknown parameter vector. Stochastic properties of the invariant approach have not been treated yet.

Parallel implementation of the least squares adjustment procedure on different HPC platforms has been achieved successfully. Linear scaling in runtime with respect to both the number of observations and the number of CPUs together with the use of optimized numerical libraries ensure highly efficient data processing.

Acknowledgement. This is publication No. 155 of the program GEOTECHNOLOGIEN of BMBF and DFG, Grant 03F0329B. The authors thank the High Performance Computing Center Stuttgart (HLRS) for the opportunity to use their computing facilities.

References

Anderson E, Bai Z, Bischof C, Blackford S, Demmel J, Dongarra J, Du Croz J, Greenbaum A, Hammarling S, McKenney A, Sorensen D (1999) LAPACK Users' Guide (third edition). SIAM Publications, Philadelphia

Baur O, Austen G (2005) A parallel iterative algorithm for large-scale problems of type potential field recovery from satellite data. Manuscript submitted to Advances in Geosciences

Colombo O (1981) Numerical methods for harmonic analysis on the sphere. Department of Geodetic Science, Report No. 310, Ohio State University, Columbus, Ohio

Ditmar P, Klees R (2002) A method to compute the Earth's gravity field from SGG/SST data to be acquired by the GOCE satellite. Delft University Press

ESA (1999) Gravity Field and steady-state ocean circulation - The four candidate Earth explorer core missions. ESA Publications Division, ESA SP-1233(1), ESTEC, Noordwijk, The Netherlands

ESA (2000) From Eötvös to milligal. Final report ESA/ESTEC, Contract No. 13392/NL/GD

Földváry L, Švehla D, Gerlach C, Wermuth M, Gruber T, Rummel R, Rothacher M, Frommknecht B, Peters T, Steigenberger P (2005) Gravity Model TUM-2Sp Based on the Energy Balance Approach and Kinematic CHAMP Orbits. In: Reigber C, Lühr H, Schwintzer P, Wickert J (eds.) Earth Observation with CHAMP - Results from Three Years in Orbit, 13–16, Springer, Berlin

Hestenes MR, Stiefel E (1952) Methods of conjugate gradients for solving linear systems. Journal of Research of the National Bureau of Standards, 49, 409–436

Kaula WM (1966) Theory of satellite geodesy. Blaisdell, Waltham, MA

Klees R, Koop R, Visser P, van den IJssel J (2000) Efficient gravity field recovery from GOCE gravity gradient observations. JoG, 74, 561–571

Maeß G (1988) Vorlesungen über numerische Mathematik II. Akademie Verlag, Berlin

Mayer-Gürr T, Ilk KH, Eicker A, Feuchtinger M (2005) ITG-CHAMP01: A CHAMP Gravity Field Model from Short Kinematic Arcs over a One-Year Observation Period. JoG, 78, 462–480

Paige CC, Saunders MA (1982a) LSQR: An algorithm for sparse linear equations and sparse least squares. ACM Transactions on Mathematical Software, 8, 43–71

Paige CC, Saunders MA (1982b) LSQR: Sparse linear equations and least squares problems. ACM Transactions on Mathematical Software, 8, 195–209

Pail R, Plank G (2002) Assessment of three numerical solution strategies for gravity field recovery from GOCE satellite gravity gradiometry implemented on a parallel platform. JoG, 76, 462–474

Reigber C, Jochmann H, Wnsch J, Petrovic S, Schwintzer P, Barthelmes F, Neumayer KH, König R, Förste C, Balmino G, Biancale R, Lemoine JM, Loyer S, Perosanz F (2005) Earth Gravity Field and Seasonal Variability from CHAMP. In: Reigber C, Lühr H, Schwintzer P, Wickert J (eds.) Earth Observation with CHAMP - Results from Three Years in Orbit, 25–30, Springer, Berlin

Reubelt T, Austen G, Grafarend EW (2003) Harmonic analysis of the Earth's gravitational field by means of semi-continuous ephemerides of a low Earth orbiting GPS-tracked satellite. Case study: CHAMP. JoG, 77, 257–278

Reubelt T, Götzelmann M, Grafarend EW (2005) A new CHAMP gravitational field model based on the GIS acceleration approach and two years of kinematic CHAMP data. Manuscript submitted to Advances in Geosciences

Rummel R (1986) Satellite Gradiometry. In: Sünkel H (ed.) Mathematical and Numerical Techniques in Physical Geodesy, Lecture Notes in Earth Sciences 7, Springer

Rummel R, Sansò F, van Gelderen M, Brovelli M, Koop R, Miggliaccio F, Schrama E, Scerdote F (1993) Spherical harmonic analysis of satellite gradiometry. Netherlands Geodetic Commission, New Series, 39

Saccoccia G, Gonzales del Amo J, Estublier D (2000) Electric Propulsion: A Key Technology for Space Missions in the New Millenium. ESA Bulletin 101, ESTEC, Noordwijk, The Netherlands

Schuh WD (1996) Tailored Numerical Solution Strategies for the Global Determination of the Earth's Gravity Field. Mitteilungen der Geodätischen Institute der TU Graz, 81, Graz

Sneeuw N (2000) A semi-analytical approach to gravity field analysis from satellite observations. DGK, Series C, No. 527, Munich

The Impact of Temporal Gravity Variations on GOCE Gravity Field Recovery

Oleg Abrikosov[1], Focke Jarecki[2], Jürgen Müller[2], Svetozar Petrovic[1], and Peter Schwintzer[1]

[1] GeoForschungsZentrum Potsdam, Department 1: "Geodesy and Remote Sensing", Telegraphenberg A17, D-14473 Potsdam, Germany, abrik@gfz-potsdam.de, sp@gfz-potsdam.de
[2] Institut für Erdmessung, Universität Hannover, Schneiderberg 50, D-30167 Hannover, Germany, jarecki@ife.uni-hannover.de, mueller@ife.uni-hannover.de

Summary. Since the main goal of the GOCE mission is the derivation of a static gravity field, significant temporal gravity changes from mass redistributions in the System Earth have to be removed from the measurement data in a dealiasing step. Furthermore, a method for gravity field recovery has to be developed, which is capable to process different kinds of data simultaneously. The effects of different mass redistribution systems, like atmosphere, oceans or hydrology, are investigated in terms of geoid and gravity gradients. Main focus is laid on hydrology effects, since global models of the continental water storage turned out to be rather inconsistent, compared to models of the other systems. However, they may benefit from the newly available GRACE gravity field models. It is shown that all time variable gravity effects are small compared with the gradiometer performance; nevertheless it is recommended to use the data from geophysical models and from monthly GRACE gravity field solutions to diminish aliasing effects in the GOCE measurements. In order to simplify the assimilation of gradiometric and satellite-to-satellite-tracking data (e.g. also from GRACE), a method for gravity field recovery has been developed, which is capable to handle the gradiometric data directly in the gradiometer reference frame. It benefits from a filter algorithm based on colored noise for the decorrelation of the gradients and applies powerful parallelization techniques. A high degree gravity field is recovered from simulated SGG data by this approach.

Key words: gravity satellite mission, GOCE, GRACE, gravity gradiometry, global hydrological model, temporal gravity field variations, gravity field recovery, colored noise filtering

1 Variations of the Earth's Gravity Field with Time

The main goal of the GOCE satellite mission (ESA, 1999) is to achieve a high-resolution model of the Earth's static gravity field, which can be defined as a mean state over some considered time period. If the satellite observations were available for a sequence of epochs covering this time period with a dense regular time step and for each of these epochs in a rather regular distribution in space domain, there would be no problem to recover a static gravity field model just by simple averaging. However, a satellite (GOCE or any other) can only observe the gravity field at one instant only at one place along the orbit. Therefore, there is no regular coverage of the space domain and for every epoch there is an observation only at one place. The consequence is that averaging cannot yield a mean field in the above sense. The fact that for GOCE two separate observation phases in corresponding periods of two consecutive years are planned and no full seasonal cycle will be covered makes the problem even worse. Hence, it is necessary to eliminate time variable effects from satellite observations in an independent way.

1.1 Eliminating Temporal Variations from Satellite Observations

The variations of the Earth's gravity field in time are primarily caused by redistributions of masses in the System Earth. The largest impact comes from solid Earth and ocean tides, from redistributions of masses in the atmosphere and ocean, and, last but not least, from changes in the continental water storage and cryosphere. These mass changes can be represented using available physical models, like:

- solid Earth tides: IERS CONVENTIONS,
- ocean tides: FES2004 (Le Provost, 2002) or CSR 4.0 (Eanes and Bettadpur, 1995),
- ocean models: barotropic (e.g. JPL), baroclinic,
- atmosphere: ECMWF, NCEP,
- continental water storage: global hydrological models like WGHM (Döll et al., 2003), H96 (Huang et al., 1996; Fan and van den Dool, 2004), LaD (Milly and Shmakin, 2002) or GLDAS (Rodell et al., 2004),
- models for ice shields (e.g. Sasgen et al., 2005).

These models are given in various representations and data formats. In order to use them for the elimination of temporal variations from satellite observations they all have to be transformed into the same form which is convenient for further processing. Hence, individual physical effects are represented as time series of epochs, every epoch being represented as a spherical harmonic expansion (e.g. Wahr et al., 1998).

The order of magnitude of seasonal changes of the Earth's gravity field induced by the atmosphere, oceans, hydrology and Antarctic ice was estimated by taking spherical harmonic expansions of monthly means of these effects for 2002 and computing degree variances of respective geoid height variations (Fig. 1). The atmospheric and oceanic influences seem to be significantly stronger than the hydrological ones, at least up to degree and order $n, m = 10$. However, the joint impact of the atmosphere and oceans is weaker

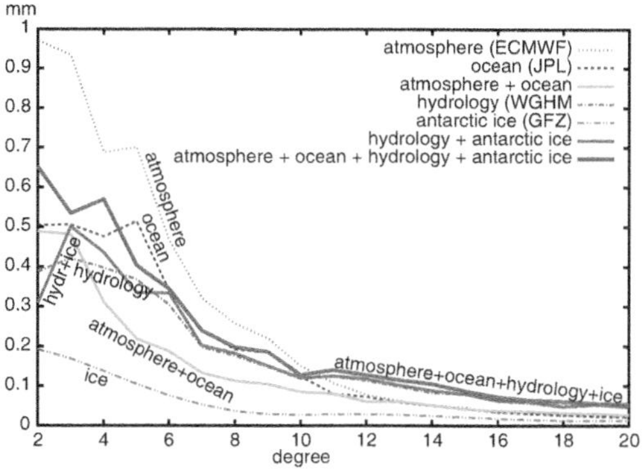

Fig. 1. Impact of some physical effects on the geoid height variations (degree variances for the year 2002 computed from monthly mean values)

than the impact of the continental water storage (cf. Fig. 1), obviously due to the inverse barometer effect. It should also be noted that a direct comparison of the presented power curves is not quite realistic, since the data coverage of different physical quantities extends over different parts of the Earth's surface. This is the main reason why completing the hydrological signal by the influence of the Antarctic ice shield does not bring any significant change in the degree variances.

Both the high-frequency and the seasonal effects of the atmosphere and the ocean can be removed for the most part from satellite observations using the above models. For instance, at GFZ Potsdam the GRACE dealiasing products (AOD = Atmosphere Ocean Dealiasing Product) are routinely being produced using a barotropic ocean circulation model and ECMWF atmospheric data, see (Flechtner, 2003), which can be adapted for GOCE, see Sect. 2.3.

However, the use of different atmospheric or ocean models leads to differences which cannot be neglected without further investigations. Hence, the remaining modeling errors can produce time-variable aliasing effects, see (Han et al., 2004; Thompson et al., 2004). Even the modeling of ocean tides can not be regarded as sufficiently good for all purposes of satellite geodesy. Taking the differences of two ocean tide models as a measure for the uncertainty of ocean tide modeling it can be shown that the errors of this modeling might affect GRACE-derived monthly gravity field solutions, see (Knudsen, 2003; Wünsch et al., 2005). For GOCE gradiometry the errors of ocean tide modeling do not represent a big concern, see (Ray et al., 2003) and Sect. 2.

Compared to the relatively small modeling uncertainties of atmospheric and oceanic effects, global hydrological modeling is still essentially less reliable and requires special attention.

1.2 Uncertainties of Global Hydrological Modeling

In order to check the reliability of different global hydrological models, the gravity variations induced by them were compared. Color Fig. XXXI on p. 304 shows a comparison of variations of the Earth's gravity field deduced from global hydrological models WGHM and H96. Global patterns in the two left columns are rather coherent and the well pronounced features, especially in large tropical river basins, are clearly visible in both representations. However, regional differences (cf. Color Fig. XXXI, third column) achieve almost the same order of magnitude as the effects themselves at some places. The differences are especially, but not exclusively, visible in regions with ice or snow cover.

The global comparison of the gravity field variations deduced from WGHM and H96 is quantified in Table 1. The differences are considerable and reach almost the same order of magnitude as the geoid variations. However, the global correlation coefficients listed in the last column are rather high.

The right column of Color Fig. XXXI shows correlation coefficients between the WGHM–deduced and H96–deduced geoid variations computed using a moving window. The major part of the continents is covered with high correlation coefficients (above +0.8). Although low or negative correlation coefficients mostly appear in regions where the effects are rather small, such areas are in no case negligible and it should be noted that the patterns for individual epoch differences vary considerably.

Comparing LaD with H96 or WGHM results in similar conclusions. This is no wonder, since the existing global hydrological models were originally designed for different specific applications and the modeling of total continental water mass displacements, which is necessary for the elimination of the associated gravity field variations from the GOCE observations, was not the primary goal. Consequently, neither the same effects are included in different models nor the modeling of the same effect is done in a compatible way. In particular, the treatment of ice and snow masses varies, but to some extent also that of all other components.

A comparison of gravity field variations resulting from the three considered global hydrological models with the variations recovered from GFZ's monthly GRACE gravity field solutions (which are already dealiased for the atmosphere and oceans, see Flechtner, 2003) shows a basic qualitative agreement, especially in the regions where an acceptable hydrological modeling can be expected. However, the variations observed by GRACE are significantly larger than hydrological predictions. The differences are considerable and lie in the

Table 1. Comparison of geoid changes in mm deduced from WGHM and H96

	WGHM			H96			WGHM-H96			correl.
months	min	max	wrms	min	max	wrms	min	max	wrms	coeff.
05/02–08/02	-7.19	8.37	1.70	-7.42	8.88	1.75	-4.92	4.29	1.04	0.82
08/02–11/02	-3.68	5.47	1.05	-2.85	3.42	0.66	-1.89	2.67	0.56	0.88
11/02–03/03	-13.17	4.89	1.74	-11.55	7.26	1.96	-3.82	6.00	1.10	0.83
03/03–05/03	-3.89	3.41	0.73	-2.35	2.30	0.47	-1.95	2.63	0.57	0.63
05/03–08/03	-7.28	7.86	1.71	-8.75	8.41	1.87	-4.44	4.16	1.09	0.82

same order of magnitude as the variations of the hydrological part of the geoid and the correlation coefficients are not especially high. More details on comparisons of gravity field variations resulting from global hydrological models with the variations recovered from GRACE observations can be found e.g. in (Wahr et al., 2004; Schmidt et al., 2005).

Current global hydrological models are not suitable for the elimination of time variations from the GOCE observations. However, an assimilation of GRACE-derived variations of continental water budget into hydrological modeling and its further development might soon make their application for this purpose possible.

2 Mass Transport and GOCE Gradients

This section concentrates on investigations, whether models of geophysical effects like those discussed in Sect. 1.1 and monthly GRACE gravity fields, as proposed in Sect. 1.2, are useful to reduce corresponding variations in the new measurement type, gravity gradients, observed within the upcoming GOCE satellite mission and described in Sect. 3.1.

2.1 Amplitudes of Temporal Gravity Changes at GOCE Altitude

Before addressing geophysical models, one should take a look at the amplitudes of the gravity variations between monthly GRACE solutions, dealiased for the atmospheric and oceanic signals, at GOCE altitude. To achieve global amplitude information, appropriate signal degree variances for the differences between two sets of GOCE-like gradients from monthly GRACE models are used, see (Jarecki et al., 2005). Exemplarily, the earth-pointing radial gradient V_{rr} as defined e.g. in (Ditmar and Klees, 2002) is chosen, because it shows good global comparability even in different reference frames, see (Müller, 2003). The mean of all these degree variances, representing a mean gravity change, is shown as solid line in both parts of Fig. 2. It reaches only small values, remaining after consecutive subtraction of the monthly models and upward continuation. Comparing the cumulative degree variances (gray lines), the signal (solid) does not exceed the error curve (dashed) at the lower degrees due to the reduced accuracy of the very low coefficients. In contrast, the mean degree-wise signal (black) exceeds the mean error curve (dashed) between $n = 4$ and 11 and has some peaks at higher degrees. Because of missing temporal equidistance of the GRACE models (the time span between available consecutive monthly solutions varies between one and four months), this averaged view is not really satisfying. The results from Sect. 1 indicate at least seasonal geoid changes, which should cause significant changes in the GOCE gradients. Consequently, a closer look onto the single differences in Table 2 shows much better results for longer time spans. Hence, the extrema of the error degree variances, representing the best and the worst scenario (smallest

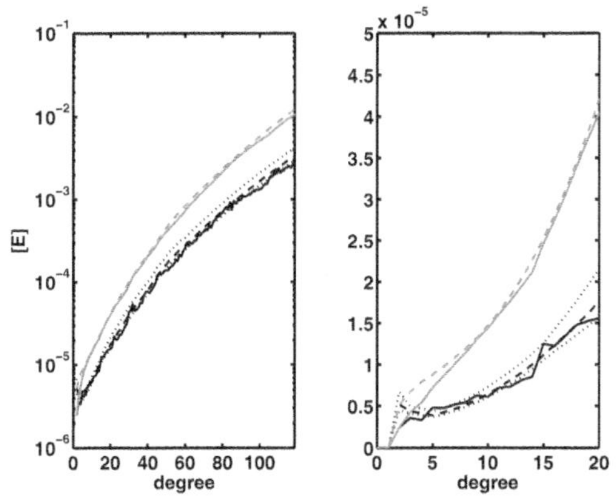

Fig. 2. Mean signal (solid) and error (dashed) degree variances of V_{rr} in GOCE altitude from monthly GRACE field differences. Grey lines represent appropriate cumulative degree variances. The right plot focuses on lower degrees

and largest error of all monthly differences), are given for each degree, too (dotted lines in Fig. 2).

Generally, the mean signal slightly exceeds the smallest available error of monthly differences (lower dotted line), but it drops below this best-case scenario several times and it is always below the curve representing the worst case (upper dotted line). At the level of $n = 12 \dots 15$ the signal curve clearly approaches and even intersects the best-case error curve (cf. right plot of Fig. 2). The maximum degree for the application of GRACE temporal varying gravity to GOCE should be searched in this part of the spectrum.

2.2 Temporal Variations of Gravity Gradients at GOCE Altitude

The models from Sect. 1.1, especially the combination of WGHM with the ice shield model of (Sasgen et al., 2005), here denoted with WGHM+ice, are used in their spherical harmonic expansions to compute V_{rr}, radial gravity gradients in an satellite–fixed Earth–pointing system. Color Fig. XXXII on p. 304 shows the minimum and maximum monthly changes and their rms for the whole WGHM+ice period (1992 to 2003) on a global $1° \times 1°$ grid of V_{rr} in 260 km altitude, which is representative for the GOCE orbit, see (Alenia, 2001). The spherical harmonics are derived from the models (considering their spatial resolution) up to a maximum degree of $n = 100$ and used completely. The amplitude reached by these gravity changes from modeled hydrology and ice is not critical for GOCE gradiometry at all, as it does not reach the mE-level even in extreme constellations. Furthermore, the time scale represented by these models is not fitting the gradiometer's measurement bandwidth.

Although it does not seem to affect the gradiometric measurements itself, studies were carried out to find a threshold for the maximum degree for application of GRACE monthly solutions in the gradients to reduce possible systematic influences. The complete results can be found in (Jarecki et al., 2005). For the WGHM+ice model, they show a reasonable comparability up to $n_{\max} = 13$. So, the application of this low degree part of the GRACE

Table 2. Gradient changes in V_{rr} from GRACE and WGHM+ice, $n_{\max}$=13, (mE)

	GRACE			WGHM+ice			difference			correl.
months	min	max	rms	min	max	rms	min	max	rms	coeff.
08/02 - 11/02	-0.152	0.089	0.022	-0.057	0.045	0.012	-0.106	0.075	0.020	0.403
11/02 - 03/03	-0.097	0.221	0.029	-0.066	0.153	0.015	-0.074	0.091	0.021	0.722
03/03 - 04/03	-0.039	0.061	0.011	-0.018	0.015	0.004	-0.041	0.055	0.011	0.163
04/03 - 05/03	-0.052	0.058	0.012	-0.032	0.035	0.006	-0.042	0.040	0.012	0.288

models might improve the error estimates as well as the whole analysis of the gradiometric data. Table 2 supports this assumption, showing very small influences on the gradients itself, but reasonable correlations between the GRACE solutions and the models, especially on longer time spans.

Referring to Fig. 2 again, there is no difference in the significance of the GRACE derived gradient signal, but the signal strength increases several orders of magnitude at the higher degrees. Therefore, just cutting off the GRACE models at low degrees, like suggested above, produces a large formal omission error (about 10 mE for $n_{\max} = 13$, i.e. 500 times the used signal). To include this signal, which is contaminated by a rising noise level, the models were evaluated completely up to $n_{\max} = 120$ (resp. $n_{\max} = 100$ for WGHM+ice) and filtered in the spatial domain to remove the noise. A moving average window and a $\cos^2$–based filter algorithm have been applied. Furthermore, a subset of gradients from the latitudes between 80° S and 80° N (denoted as ±80° dataset) has been selected to consider possible GRACE shortcomings in the polar regions.

The results are shown in Table 3, which presents the approximated maximum degree of a spherical harmonic expansion compared with the averag-

Table 3. Global rms of the V_{rr} changes from filtered high degree models, (mE)

$n_{\max}$	filter window	dataset coverage	GRACE	WGHM +ice	difference GR-(W+i)	correl. coeff. mvg.avg.	$\cos^2$
27	no filter	global	0.062	0.019	0.061	0.125	
		±80°	0.048	0.014	0.046	0.025	
27	37°x 37°	global	0.012	0.007	0.013	0.195	0.207
	($n \approx 10$)	±80°	0.008	0.005	0.006	0.055	0.057
27	27°x 27°	global	0.018	0.009	0.018	0.199	0.203
	($n \approx 13$)	±80°	0.011	0.006	0.010	0.060	0.060
120/100	37°x 37°	global	0.177	0.007	0.177	0.007	0.008
	($n \approx 10$)	±80°	0.100	0.005	0.100	0.031	0.035
120/100	27°x 27°	global	0.276	0.009	0.276	0.004	0.005
	($n \approx 13$)	±80°	0.174	0.007	0.174	0.028	0.031

ing window filter applied and global rms for the derived gradient grids. The last two columns contain the mean correlation coefficients of the datasets derived from GRACE and WGHM+ice for all consecutive monthly differences, when filtered with either the moving average or the $\cos^2$–based filter. Each filtered high resolution GRACE/WGHM+ice model combination shows large differences. Correlation is below the 1% level and rises to about 3% when neglecting the polar regions. The global rms differences seem to be unaffected by the hydrological model. Nevertheless, some differences show very high correlation, remember Table 2, which even rises significantly focusing on the $\pm 80°$ dataset. Filtering does not enhance the (limited) usefulness of a complete $n_{\text{max}} = 120/100$ model, although the $\cos^2$–based filter, which considers more of the gradient's typical local sensitivity, works slightly better. Anyway, spatial filtering does not seem to be an adequate approach for the highly location-sensitive gradient data. A combination of spectral (cut-off) and spatial (averaging) filtering might be another way to go. For instance, the medium resolution cut-off model with $n_{\text{max}} = 27$ looks slightly enhanced (correlation grows from 12.5% to around 20%) by the filtering. Nevertheless, most of the monthly GRACE signal beyond $n_{\text{max}} = 13$ does not seem to be explained by hydrological models. Therefore, these residuals represent a significant non-hydrological signal, a hydrological signal, which is for some reason not contained in the models, or just measurement noise.

2.3 Time Variable Gravity Signals in the GOCE SGG Time Series

Section 2.2 shows, that hydrology seems to be the main systematic signal in the monthly GRACE solutions. Hence, the GRACE dealiasing products for oceanic and atmospheric mass changes can be assumed to be good. Therefore they have been used to calculate gradients along simulated GOCE tracks, modeling a realistic gradiometer attitude. The power spectra of this time series led to signals in the $\text{mE}/\sqrt{\text{Hz}}$-range, which is below the actual gradiometer performance.

Figure 3 shows combined atmosphere and ocean influence on the test orbit as a PSD plot. The superposition of these signals with the easy to model influ-

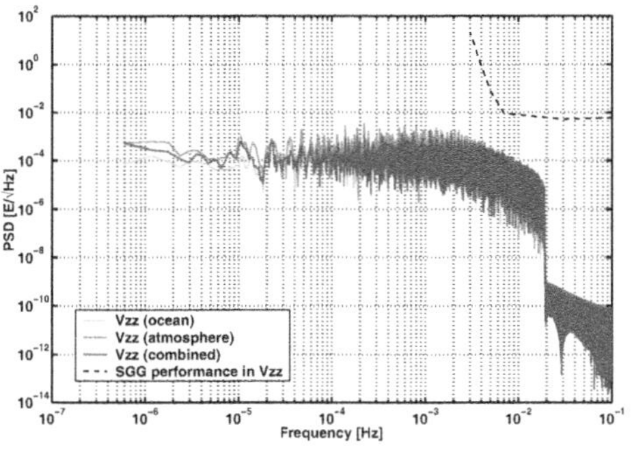

Fig. 3. Signal PSD of dealiasing V_{rr} along GOCE test orbit, based on GRACE dealiasing product for ocean and atmosphere (AOD1B)

ences of ocean and solid Earth tides (producing mainly a 1 cpr peak) and third body potential, which is at the same order of magnitude, results in the time variable gravity components proposed for the dealiasing of GOCE gradients, see (Abrikosov and Schwintzer, 2004). Same as in the case of hydrology, they do not reach the actual GOCE performance, but the dealiasing with validated models should improve the stochastic behavior of the gradiometric data.

3 Recovery of the Earth's Gravity Field

In numerical simulations for the GOCE mission, two concepts to invert SGG data into a model of the Earth's gravity field are to be distinguished. These are the time-wise approach and the space-wise approach (Rummel et al., 1993). The direct approach, i.e. creating observation equations for gravity field recovery in a least squares sense directly at the point of the measurement, is a third approach in treating SGG data. Being a classical concept, it considers measurements as belonging to the space-time domain and estimates parameters of the Earth potential in a rigorous manner but it demands considerable computer resources.

This section deals with the functional and stochastic model to formulate the observation equations and with the simulation of gravity field recovery using the direct approach.

3.1 Computation of Gravity Gradients in the Gradiometer Reference Frame

Because a frame transformation of the measured gravity gradients largely degrades the accuracy, we have developed a technique for a direct computation of observation equations of GOCE SGG data in the gradiometer reference frame. We suppose the direction vectors $\mathbf{e}_j = \left(e_{j1}\; e_{j2}\; e_{j3} \right)^T$ of the gradiometer axes ξ_j ($j = 1, 2, 3$) to be known in the Earth-fixed system $Ox_1x_2x_3$ of Cartesian coordinates, to which the used gravity field model is referred. Then the second-order derivatives of the gravitational potential $V = V(x_1, x_2, x_3)$ with respect to directions of the gradiometer axes are

$$\frac{\partial^2 V}{\partial \xi_j \partial \xi_k} = \sum_{i=1}^{3} \sum_{l=1}^{3} e_{ji} e_{kl} \frac{\partial^2 V}{\partial x_i \partial x_l} \,. \tag{1}$$

By inserting the well-known expressions (Cunningham, 1970) for $\frac{\partial^2 V}{\partial x_i \partial x_l}$ into the right-hand side of (1) we get the explicit formula

$$\frac{\partial^2 V}{\partial \xi_j \partial \xi_k} = GM \sum_{n=0}^{n_{\max}} R^n \sum_{m=0}^{n} \bar{\mathbf{h}}_{nm}^T \frac{\partial^2}{\partial \xi_j \partial \xi_k} \bar{\mathbf{u}}_{nm} \,, \tag{2}$$

where GM is the product of the universal gravitational constant G by the Earth mass M, R is the mean radius of the Earth, $\bar{\mathbf{h}}_{nm} = \left(\bar{C}_{nm}\ \bar{S}_{nm}\right)^T$ are the real-valued vectors of fully normalized harmonic coefficients of the potential. The second-order derivatives of fully normalized spherical harmonics $\bar{\mathbf{u}}_{nm}$ in (2) are real-valued vectors and

$$\begin{gathered}\frac{\partial^2}{\partial\xi_j\partial\xi_k}\bar{\mathbf{u}}_{nm} = b_{nm}^{(-2)}\mathbf{D}_{jk}\bar{\bar{\mathbf{u}}}_{n+2,m-2} - b_{nm}^{(-1)}\mathbf{F}_{jk}\bar{\bar{\mathbf{u}}}_{n+2,m-1} \\ +b_{nm}^{(0)}\frac{3e_{j3}e_{k3}-\delta_{jk}}{2}\bar{\bar{\mathbf{u}}}_{n+2,m} + b_{nm}^{(1)}\mathbf{F}_{jk}^T\bar{\bar{\mathbf{u}}}_{n+2,m+1} + b_{nm}^{(2)}\mathbf{D}_{jk}^T\bar{\bar{\mathbf{u}}}_{n+2,m+2}\ ,\end{gathered} \tag{3}$$

δ_{jk} is Kronecker's delta, the coefficients $b_{nm}^{(l)}$ are

$$b_{nm}^{(l)} = \sqrt{(2-\delta_{0m})(2n+1)\frac{(n-m+2-l)!(n+m+2+l)!}{(n-m)!(n+m)!}},\ l = -2\ldots 2 \tag{4}$$

and matrices $\mathbf{D}_{jk}$, $\mathbf{F}_{jk}$ are functions of components of direction vectors of the gradiometer axes:

$$\mathbf{D}_{jk} = \mathbf{B}_j\mathbf{B}_k,\ \mathbf{F}_{jk} = e_{k3}\mathbf{B}_j + e_{j3}\mathbf{B}_k,\ \mathbf{B}_j = \tfrac{1}{2}\begin{pmatrix} e_{j1} & -e_{j2} \\ e_{j2} & e_{j1}\end{pmatrix}\ . \tag{5}$$

Values of Schmidt-normalized spherical harmonics $\bar{\bar{\mathbf{u}}}_{nm}$ in the right-hand side of (3) can be computed as functions of the Earth-fixed coordinates x_1, x_2, x_3 of the gradiometer center of mass by standard recursive formulas (Chapman and Bartels, 1940).

Now we have a basis for the computation of observation equations coefficients from SGG data. Obviously, one can perform all computations for each component of the gradient tensor separately (e.g. for the three diagonal components only) and directly in the gradiometer reference frame without a transformation of measured gradients to any other frame. This possibility is extremely important in view of the degradation of the tensor components' accuracy when re-orienting them by attitude angles.

3.2 Decorrelating of Gravity Gradients Affected by Colored Noise

Colored noise $\mathbf{n}$ contaminating GOCE SGG data can be handled by taking its inverted covariance matrix $\mathbf{C}_{nn}^{-1}$ into account. The approach used in this paper implicitly includes $\mathbf{C}_{nn}^{-1}$ into the procedure of SGG data processing. One can define a matrix $\mathbf{F}$ such that $\mathbf{F}^T\mathbf{F} = \mathbf{C}_{nn}^{-1}$. Then, similar to (Klees et al., 2003), we can apply it to the right-hand side $\mathbf{L}+\mathbf{n}$ of the system of observation equations as well as to the design matrix $\mathbf{A}$: $\mathbf{FAX} = \mathbf{FL}+\mathbf{Fn}$, and after that, solve the equivalent system: $\tilde{\mathbf{A}}\mathbf{X} = \tilde{\mathbf{L}} + \tilde{\mathbf{n}}$, containing uncorrelated noise $\tilde{\mathbf{n}}$. We follow the assumption that $\mathbf{C}_{nn}$ is a circulant Toeplitz matrix of order N equal to the number of observation points. If we additionally require $\mathbf{F}$ to be a matrix of the same type, we can define it completely, in analogy to $\mathbf{C}_{nn}^{-1}$ in (Ditmar and Klees, 2002; Ditmar et al., 2003), by its first row

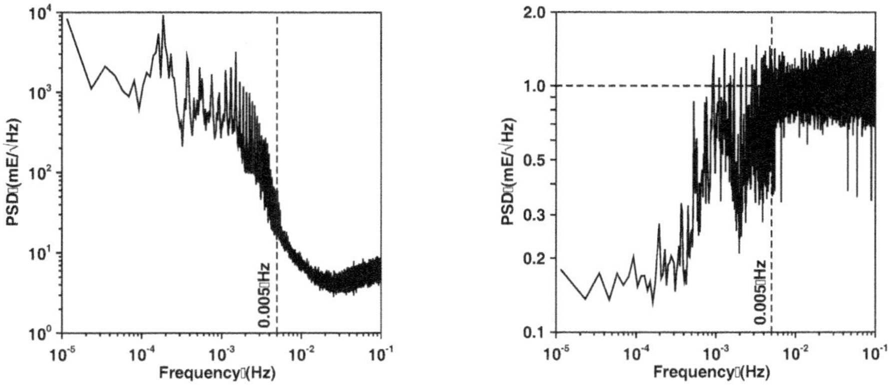

Fig. 4. PSD of noise in GOCE SGG data (left) and of noise after filtering (right)

$$\mathbf{f}_{jk} = \mathcal{F}^{-1}\left(\frac{1}{N\left|\mathcal{F}\left(\mathbf{n}_{jk}\right)\right|}\right), \tag{6}$$

where $\mathbf{n}_{jk}$ is the colored noise in the jk-th component of the gradient tensor. In (6), $\mathcal{F}(\bullet)$ and $\mathcal{F}^{-1}(\bullet)$ denote direct and inverse Fourier transforms, respectively. Now, application of the matrix $\mathbf{F}$ to the observation equations is a cyclic convolution of columns of the design matrix $\mathbf{A}_{jk}$ and the right-hand side $\mathbf{L}_{jk}+\mathbf{n}_{jk}$ with the vector $\mathbf{f}_{jk}$ from (6). Because the first row of the matrix $\mathbf{C}_{nn}^{-1}$ is a convolution of the vector $\mathbf{f}_{jk}$ with itself, one can treat equation (6) as a result of deconvolution. The latter problem, however, has usually no unique solution. Thus, we should note that (6) is just one among possible filters for colored noise in GOCE SGG data.

For practical reasons, the filter should be truncated to a relatively short length comparable with the measurement bandwidth (MBW) of GOCE gradiometry. We have tested such a truncation for two practical cases: (a) amplitudes $\left|\mathcal{F}\left(\mathbf{n}_{jk}\right)\right|$ in (6) come from a-priori given (instrument characteristics) power spectral densities of the gradiometer noise and (b) these are estimated as discrete Fourier transforms of the sum of the measured diagonal components of the gradient tensor. Both cases lead to results of similar quality. Figure 4 (left) shows typical colored noise spectrum, and Fig. 4 (right) the results obtained with a filter covering a time interval of 450 s. The filter significantly decreases the power of filtered noise at low frequencies. Because the filter is truncated, the resulting noise is not fully uncorrelated. However, for filters longer than 400 s, remaining correlations are already practically negligible.

3.3 Simulation of GOCE SGG Data

A GOCE-like orbit has been simulated for a 28-day interval by using GFZ's EPOS-OC 5.4 software. All conservative and non-conservative orbit perturbations were taken into account during the computation of the simulated orbit.

Initial values of semimajor axis $a = 6621\,\text{km}$ and inclination $I = 96.5°$ were chosen in accordance with the nominal parameters of the GOCE orbit (Alenia, 2001). Because GOCE orbital eccentricity can reach $e = 0.0045$ (Alenia, 2001), we choose just this value for the initial eccentricity. Due to the upper limit (0.1 Hz) of the MBW proposed for the GOCE gradiometry, the state vectors of the satellite were stored with a sampling rate $\tau = 5\,\text{s}$ at 483840 points regularly spaced along the 28-day orbital arc. In such a manner, we got a quasi-regular coverage of the Earth's surface by sub-satellite points with along-track distances of about 40 km between consecutive points (within 1 revolution of the satellite) and distances from 50 km to 170 km between two ground-tracks.

Unlike the orbit computations, we took into account only the set of coefficients of the GPM98CR model up to degree 720 for the simulation of the gravity gradients along the modeled GOCE orbit. In this simulation, we supposed the gradiometer axes to be directed along the axes of "radial - quasi-transversal - normal" (RTN) local orbital reference frame.

3.4 Results of Earth's Gravity Field Recovery

To estimate the Earth gravity model, we used the diagonal components of the gradient tensor. Colored noise with PSD shown in Fig. 4 (left) was generated for each component separately and then added to these data. The filter with the length of 450 s was constructed on the basis of the sum of the simulated diagonal components of the gradient tensor and then applied to the system of observation equations. The system of normal equations has been computed for 40397 unknown coefficients up to degree 200. The normal matrix has a diagonally dominant structure. Being arranged for harmonics in sequence "order-by-order", the matrix demonstrates that many off-diagonal coefficients are very small in comparison with the diagonal ones. Figure 5 shows maximal magnitudes of coefficients of the normal matrix in percentage of magnitudes of corresponding diagonal coefficients in the block for harmonics up to order 70. Such distribution is valid for the whole normal matrix and can be explained (Ditmar and Klees, 2002) by peculiarities of the GOCE orbit. Thus, the system of normal equations is well-conditioned and can be solved without any stabilization by means of the Cholesky decomposition.

Degree differences of the geoid heights (δN_n)

$$\delta N_n = R\sqrt{\Delta_n}, \quad \Delta_n = \sum_{m=0}^{n} \left(\left(\bar{C}_{nm} - \bar{C}_{nm}^{\text{true}}\right)^2 + \left(\bar{S}_{nm} - \bar{S}_{nm}^{\text{true}}\right)^2 \right) \tag{7}$$

are shown in Fig. 6 to demonstrate the errors of the obtained results. Because of the huge errors in low degree coefficients caused by the effect of colored noise at low frequencies, we used reverse cumulative differences

$$\delta N_n^{\text{cumulative}} = \sqrt{\sum_{k=n}^{n_{\max}} \delta N_k^2} \tag{8}$$

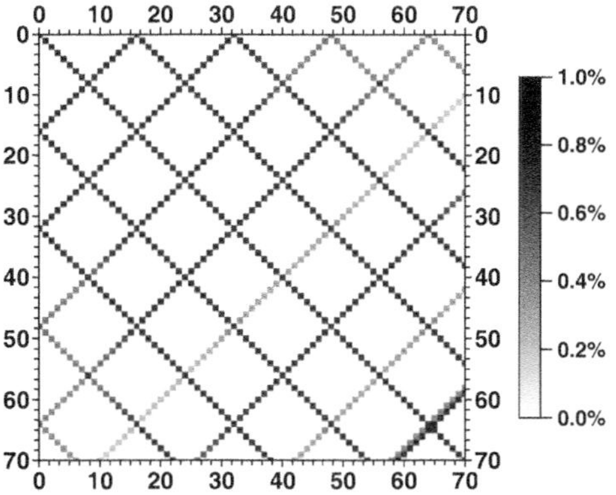

Fig. 5. Magnitudes of coefficients of the normal matrix

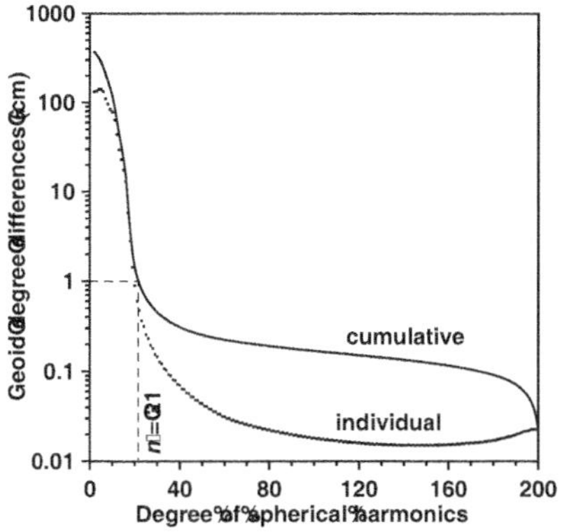

Fig. 6. Degree differences of geoid heights

instead of usual ones to get a picture of error accumulation in the solution. Now one can conclude that the GOCE mission baseline accuracy for the geoid (1 cm) is fulfilled for harmonic coefficients from degree 22 to at least 200.

4 Conclusion

In order to achieve a high-resolution model of the Earth's static gravity field based on the GOCE mission it is necessary to eliminate time variable effects from satellite observations. These can be modeled using available physical models for redistributions of masses in the atmosphere, ocean, cryosphere and continental water storage. The last mentioned global hydrological models are most critical and in their present form do not seem to be suitable for the elimination of time variations from satellite observations.

The investigation based on available physical models indicates that a reduction of time variable effects for GOCE gradiometric measurements should not be necessary. Neither of the effects delivers simulated gradient influences in the amplitude of the GOCE gradiometer specifications. However, for the detection of the spatial and temporal long wavelength part of the gravity field, the satellite-to-satellite-tracking, i.e. the GOCE orbit determination, will be used as well. In this part the large scale physical effects play an important role and have to be considered. A possible solution would be to use, besides physical models for atmospheric and oceanic effects, also the series of GRACE monthly gravity field solutions for modeling both the seasonal and interannual hydrological variations and the remaining effects. However, this is possible only up to a degree of about 15 in the spherical harmonic expansion.

Furthermore, various aspects of the direct approach to process GOCE SGG data were investigated. Observation equations were computed directly in the gradiometer reference frame in order to avoid frame transformations degrading the exploitation of GOCE SGG data. The observations were to a large extent de-correlated by filtering, taking into account a realistic colored noise model. A solution for the Earth's gravity field has been obtained complete to degree and order 200 meeting the baseline accuracy requirements, without taking into account the long wavelength part ($\lambda > 1000$ km), which will be recovered by GPS-GOCE satellite-to-satellite tracking.

Acknowledgement. This is publication no. GEOTECH-154 of the programme GEOTECHNOLOGIEN of BMBF and DFG, Grant 03F0329D.
The presented work also largely benefited from the activities of the European GOCE Gravity Consortium (EGG-C). The authors thank Chris Milly, Yun Fan and Huug van den Dool, as well as Petra Döll and Andreas Güntner for providing the LaD, H96 and WGHM data, respectively.
The authors dedicate this paper to the memory of their dear coauthor and colleague Dr. Peter Schwintzer, who sadly passed away unexpectedly on December 24, 2004.

References

Abrikosov O, Schwintzer P (2004) Recovery of the Earth's gravity field from GOCE satellite gravity gradiometry: a case study. In: GOCE, the geoid and oceanography. ESA-SP569, ESA Publications Division, ESTEC, Nordwijk

Alenia Spazio S.p.A. (2001) Performance requirement and budgets for the gradiometric mission. Technical note GO-TN-AI-0027, Turin

Cunningham LE (1970) On the computation of the spherical harmonic terms needed during numerical integration of the orbital motion of an artificial satellite. Celestial Mechanics, 2:207–216

Chapman S, Bartels J (1940) Geomagnetism. Vol. II, Oxford University Press, New York

Ditmar P, Klees R (2002) A method to compute the Earth's gravity field from SGG/SST data to be acquired by the GOCE satellite. Delft University Press

Ditmar P, Klees R, Kostenko F (2003) Fast and accurate computation of spherical harmonic coefficients from satellite gravity gradiometry data. J Geod, 76:690-705

Döll P, Kaspar F, Lehner B (2003) A global hydrological model for deriving water availability indicators: model tuning and validation. J Hydrol 270:105–134

Eanes R, Bettadpur S (1995) The CSR 3.0 global ocean tide model. Tech Memo CSR–TM–95–06, Center for Space Research, University of Texas, Austin

ESA (1999) Gravity field and steady-state ocean circulation Mission, Reports for mission selection, ESA SP-1233(1)

Fan Y, van den Dool H (2004) The CPC global monthly soil moisture data set at 1/2 degree resolution for 1948–present. J Geophys Res 109, D10102, doi:1029/2003JD004345

Flechtner F (2003) AOD1B product description document. GRACE project documentation, JPL 327–750, Rev. 1.0, JPL, Pasadena, Ca.

Han SC, Jekeli C, Shum CK (2004) Time–variable aliasing effects of ocean tides, atmosphere, and continental water mass on monthly mean GRACE gravity field. J Geophys Res 109, B 04403, doi:10.1029/2003/JB002501

Huang J, van den Dool HM, Georgakakos KP (1996) Analysis of model-calculated soil moisture over the United States (1931–1993) and applications to long-range temperature forecasts. J Climate 9:1350–1362

Jarecki F, Müller J, Petrovic S, Schwintzer P (2005) Temporal gravity variations in GOCE gradiometric data. In: Jekeli C, Bastos L, Fernandes J (eds) Gravity, geoid and space missions GGSM04 IAG International Symposium Porto, Portugal August 30 – September 3, 2004. International Association of Geodesy symposia series, vol. 129, Springer, Berlin Heidelberg New York (in print)

Klees R, Ditmar P, Broersen P (2003) How to handle colored observation noise in large least-squares problems. J Geod, 76:629-640

Knudsen P (2003) Ocean tides in GRACE monthly averaged gravity fields. Space Science Reviews 108(1–2):261–270

Le Provost C (2002) FES2002 – A new version of the FES tidal solution series. Abstract Volume, Jason–1 Science Working Team Meeting, Biarritz, France

Milly PCD, Shmakin AB (2002) Global modeling of land water and energy balances. Part I: The Land Dynamics (LaD) model. J Hydrometeorology 3(3):283–299

Müller J (2003) GOCE gradients in various reference frames and their accuracies. Adv in Geosciences 1:33–38

Ray RD, Rowlands DD, Egbert GD (2003) Tidal models in a new era of satellite gravimetry. Space Science Reviews 108(1–2):271–282

Rodell M, Houser PR, Jambor U, Gottschalck J, Mitchell K, Meng CJ, Arsenault K, Cosgrove B, Radakovich J, Bosilovich M, Entin JK, Walker JP, Lohmann D, Toll D (2004) The Global Land Data Assimilation System. Bull Amer Meteor Soc 85(3):381–394

Rummel R, van Geldern M, Koop R, Scharma E, Sansò F, Brovelli M, Miggliaccio F, Sacerdote F (1993) Spherical harmonic analysis of satellite gravity gradiometry. Publ Geodesy, New series, No 39, Nederlands Geodetic Commission, Delft

Sasgen I, Wolf D, Martinec Z, Klemann V, Hagedoorn J (2005) Geodetic signatures of glacial changes in Antarctica: rates of geoid-height change and radial displacement due to present and past ice-mass variations. Scientific Technical Report STR05/01, GFZ Potsdam

Schmidt R, Schwintzer P, Flechtner F, Reigber Ch, Güntner A, Döll P, Ramillien G, Cazenave A, Petrovic S, Jochmann H, Wünsch J (2005) GRACE observations of changes in continental water storage. Global and Planetary Change 48(4):259–273 (scheduled)

Thompson PF, Bettadpur SV, Tapley BD (2004) Impact of short period, non-tidal, temporal mass variability on GRACE gravity anomalies. Geophys Res Lett 31, L06619, doi:10.1029/2003GL019285

Wahr J, Molenaar M, Bryan F (1998) Time variability of the Earth's gravity field: Hydrological and oceanic effects and their possible detection using GRACE. J Geophys Res, 103(B12):30205–30229

Wahr J, Swenson S, Zlotnicki V, Velicogna I (2004) Time-variable gravity from GRACE: First results. Geophys Res Lett 31(11), L11501, doi:10.1029/2004GL019779

Wünsch J, Schwintzer P, Petrovic S (2005) Comparison of two different ocean tide models especially with respect to the GRACE satellite mission. Scientific Technical Report STR05/08, GFZ Potsdam

Quality Assessment of GOCE Gradients

Focke Jarecki, Karen Insa Wolf, Heiner Denker, and Jürgen Müller

Institut für Erdmessung, University of Hannover, Schneiderberg 50,
30167 Hannover, Germany, `jarecki/wolf/denker/mueller@ife.uni-hannover.de`

Summary. To meet the accuracy requirements of the GOCE mission, the gradiometer has to be calibrated and validated internally as well as externally. An internal quality assessment of the observed GOCE data is possible by comparisons of observations at the same satellite position, i.e. at satellite track cross-overs. Due to the orbit characteristics of the mission, satellite ground track cross-overs have to be used instead of identical repeat positions. Therefore, an appropriate reduction concept has to be applied to consider the differences caused by different satellite altitudes and orientations. It is shown here, that present global gravity field models meet the accuracy and resolution requirements for the reduction concept, and hence for the relative validation of GOCE gradients.

For an external calibration or validation based on regional data sets, terrestrial gravity anomalies are upward continued to gravitational gradients at GOCE altitude. The computations are done with synthetic data in a closed-loop simulation. Two upward continuation methods are considered, namely least-squares collocation and integral formulas based on the spectral combination technique. Both methods are described and the results are compared numerically with the ground-truth data. Finally, the results of a regional calibration experiment with simulated noisy GOCE gradients are described.

Key words: calibration, validation, cross-overs, upward continuation

1 Introduction

GOCE is a challenging project from the instrumental and conceptual point of view. To achieve the planned accuracy of the gravitational gradients $V_{ij} = \frac{\partial^2 V}{\partial i \partial j}$ $(i, j = X, Y, Z)$ at the mEötvös (mE) level, several calibration steps (in orbit and in post-processing) are required. In addition, independent validation procedures are needed to assure the accuracy standards of the resulting gravity field quantities (e.g., gradients, spherical harmonic coefficients or geoid heights). In this paper, the use of cross-overs for in-orbit validation is addressed as one possible validation strategy. In the second part of the paper, the upward continuation of terrestrial gravity anomalies for the computation of reference gradients for a regional external calibration is discussed.

2 In-Orbit Validation of GOCE Gradiometric Data

Validation understood as monitoring the temporal behaviour of the gradiometer should be performed in near real-time. Therefore, it can not be based upon a GOCE-derived geopotential model (GPM) which would not be available before some time of data assimilation and processing. Here, in contradiction to the validation of data products defined by Koop et al. (2001), an in-orbit method is needed. Such a method is provided by the analysis of the gradiometer measurements in the same geographical position, which is known as cross-over validation from satellite altimetry, see, e.g., Shum (1990), and was recommended for GOCE, too, e.g., by Albertella et al. (2000) or Schrama (2001).

2.1 Orbit Characteristics

Several studies, such as (Albertella et al., 2000) or (Schrama, 2001), show the concept of cross-over or repeat orbit validation for satellite gravity gradiometry (SGG) data in principle. In addition to those investigations, which used simplified assumptions of the orbit geometry, in this study a "more realistic" GOCE test data set is applied, which was derived from the IAG-SC7 simulation, see IAG-SC7 (2003), within the GOCE-GRAND project of the GEOTECHNOLOGIEN research programme. For this GOCE-GRAND test data set, the IAG-SC7 data was improved mainly in terms of a higher sampling rate, 1 Hz instead of 0.2 Hz, in agreement with the mission design. The 30 days orbit of these test data sets, however, shows an increase of the orbit eccentricity and a rotation of the orbital ellipse. This leads to different satellite altitudes at the same geographical position reached after varying time intervals and inhibits the formulation of constant conditions for the data processing. Therefore, on one hand, a concept for the determination of ground-track cross-overs on a well defined projection surface has to be developed. In contradicition to satellite altimetry, where short-term repeat orbits are realised leading to repeated cross-overs, in this scenario, every single cross-over is unique and, therefore, has to be processed individually. On the other hand, reductions have to be calculated to consider the measurement differences caused by the different altitudes and orientations of the satellite. The orientation of the measurement has not to be considered in the satellite altimetry concepts, as the altimetric range signal is rather invariant against the satellite attitude.

2.2 Cross-Over Determination and Reduction Concept

The first step of the cross-over determination is the conversion of the orbit positions, given in a cartesian inertial system, to geographical coordinates by

Table 1. Error budget in (mE) for the cross-over consistency check of the radial gradient V_{ZZ} applying different interpolation techniques for satellite position, orientation and measurements

interpolation method	0.2 Hz test data mean	rms	max	1 Hz test data mean	rms	max
nearest neighbour	0.458	92.562	639.825	0.077	18.572	131.582
linear	0.006	0.910	14.615	0.000	0.037	1.352
cubic/quadratic	0.000	0.006	1.011	0.000	0.005	1.475

well known formulas, according to the definitions in IAG-SC7 (2003). Here, a spherical earth model is used for the determination of the projection direction and for the determination of the ground-track cross-over positions. The cross-overs are found by the intersection of the difference vectors between each two pairs of consecutive sample positions at the projected ascending and descending track, respectively.

Because of the sparse sample points in the investigated data set (sample point distance is 8 km for the realistic sampling rate of 1 Hz, see ESA (1999), growing to 40 km for the 0.2 Hz data set, for which noise models are available from IAG-SC7, see Sect. 2.4), a linear interpolation for the cross-over determination is not accurate enough. Especially at higher latitudes, where the ground tracks look more curved, the cross-over position obtained from the linear intersection differs up to 300 m from the correct cross-over point in case of the 0.2 Hz test data. This would lead to gradient differences up to 2.5 mE in the radial gradient component V_{ZZ}. The way of interpolation of the satellite orientation angles has a large effect, too. Table 1 presents the statistics of different interpolation approaches (for the cross-over position and the corresponding satellite orientation and gradiometer measurements) applied to the test data sets mentioned above in the latitude range between 80°S and 80°N in a consistency check. Whereas the pure statistics look quite promising for linear interpolation of 1 Hz data, it has to be mentioned, that the used data covers over 100 000 cross-overs, causing a reasonable rms value, although a significant number of cross-overs show inconsistencies in the range of 1 mE. Therefore a more sophisticated interpolation algorithm is recommended for the computation of the exact position of the ground-track cross-over (φ, λ) as well as for the interpolation of the satellite altitude (radii r_1, r_2 on the ascending and descending track, respectively), the orientation of the gradiometer $(\vartheta_{XYZ1}, \vartheta_{XYZ2})$ and the measured gradients (to be validated), even in the case of 1 Hz sampling. The use of polynomials in the projected sample point positions, cubic ones for the positions and bivariate quadratic ones for the interpolation of the observed gradients, delivers sufficient position and gradient accuracy (better than 1 m resp. 1 mE except a small number of outliers from numerical insufficencies).

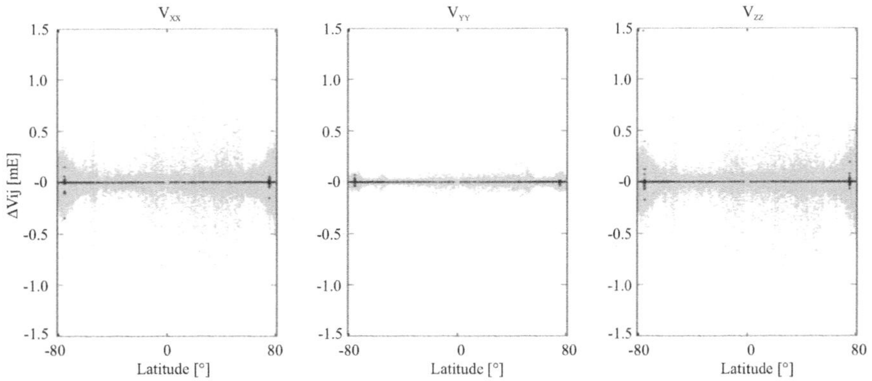

Fig. 1. Comparison of V_{XX}, V_{YY}, V_{ZZ} in cross-over points of the 0.2 Hz (grey) and 1 Hz (black) test data set with height reduction applied.

The differences in the gradients in two satellite positions can be calculated from an existing geopotential model using (1). Here, these differences are designated as *height reduction*, although orientation differences are considered, too. The reductions are assumed to be much more accurate than the gradients in the original satellite positions, as many errors drop out due to differentiation. Corresponding calculations are presented, e.g., in Jarecki and Müller (2003) and are summarised below. The height reduction is given by

$$\Delta V_{ij}(\varphi,\lambda) = V_{ij}(\varphi,\lambda,r_1,\vartheta_{xyz1}) - V_{ij}(\varphi,\lambda,r_2,\vartheta_{xyz2})\,, \tag{1}$$

where the V_{ij} are gradients in the local orbital frame ($i,j = X,Y,Z$), which is used as reference system for the gradiometer here.

Figure 1 shows the differences of the three diagonal tensor elements in the cross-overs of the data sets with the described height reduction applied. In this case, the height reduction is computed from EGM96 up to degree and order 300, just like the simulated measurements. Therefore, the differences should disappear in this consistency check. The remaining differences, which are strongly correlated with the latitude, are much larger for the 0.2 Hz data. They are mainly caused by interpolation errors for the orientation as described above. Another - much smaller - class of non-zero differences is spread over the whole latitude band in a somewhat periodic manner. On a global map (see Color Fig. XXXIII on p. 305) these differences show up in areas with an outstanding rough gravity field. In these areas, the interpolation is not able to model the correct measurements in the cross-over points from the used sparse sample points. Both kinds of inconsistencies are caused by interpolation errors. The results from the 1 Hz test data set are shown in Fig. 1, too. Here, interpolation errors vanish besides some numerical artifacts due to the crossing angle of the tracks at a certain latitude around 75°. But even the errors in the 0.2 Hz results do not affect the validation procedure, as they are much smaller than the gradiometer error limits aimed for, which are in the range of a few mE, see ESA (1999).

2.3 Accuracy of the Reduction Method

The accuracy of the reductions derived from (1) depends strongly on the errors of the utilised geopotential model, which can be split into three error parts, see, e.g., Wenzel (1985): 1) The omission error representing those high frequency parts of the gravity field which are not covered by the maximum degree of the spherical harmonic expansion, 2) the commission error reflecting the uncertainties of the coefficients itself, and 3) the errors caused by the inaccurate input values, in this case the cross-over positions and orientations.

All three error parts have been checked with respect to the height reduction calculation. The omission error for the height reductions will not exceed 0.02 mE utilising EGM96 up to degree and order 300, which is a negligable order of magnitude. The commission error from the EGM96 coefficient variances (again up to degree and order 300) reaches a critical amount of about 3 mE, but it is strongly correlated with the satellite height difference in the cross-overs and, as most of the cross-overs show relative small height differences, a huge number of validation points with adequate accuracy ($>$ 80% of the tested cross-overs are better than 1.5 mE) can be selected. The contribution of the coordinate errors as third error part has to be split in the φ, λ and the height component. As mentioned in Sect. 2.2 the geographical position does not play an important role: even hundreds of meters only cause some mE of error, so orbit accuracies in the range of one to some meters should be accurate enough. The vertical component is much more sensitive: even height reductions for small ($<$ 1 km) height differences turn out to be as inaccurate as 1.8 mE for a height error of 1 m. Hence one has to pay attention to the height determination accuracy in both, the orbit determination itself and the cross-over interpolation to achieve a value less than 1 m. Keeping this margin, the height reduction derived from EGM96 is accurate engough for a huge number of validation points, which can easily be selected with a condition depending on the height differences.

2.4 Results of Cross-Over Validation with Simulated Noisy Data

As the method has been proven in the consistency check and accuracy studies above, it has also been applied to simulated noisy GOCE data. This test data set is based on the IAG-SC7 files again, but now together with an error model from the SRON GOCE E2E simulator considering the Onera gradiometer specifications, see Smit et al. (2000). Figure 2 shows the noise added on the diagonal tensor elements of the 0.2 Hz IAG test data set. Besides strong biases, it shows a realistic rms of e.g. 12.0 mE for the V_{XX} component over the whole 30 day time series.

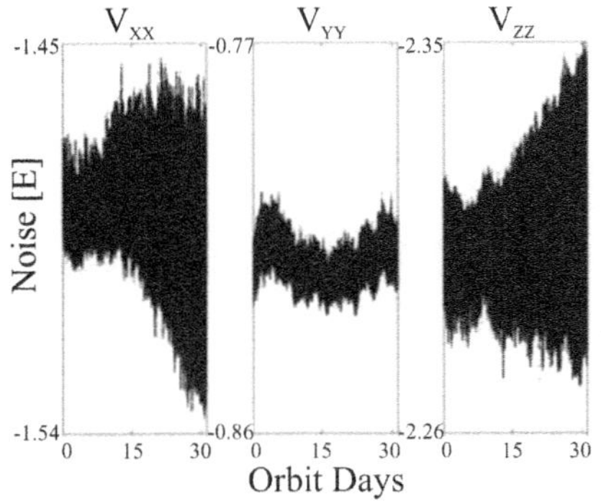

Fig. 2. Noise model applied to the diagonal tensor elements of the IAG-SC7 orbit.

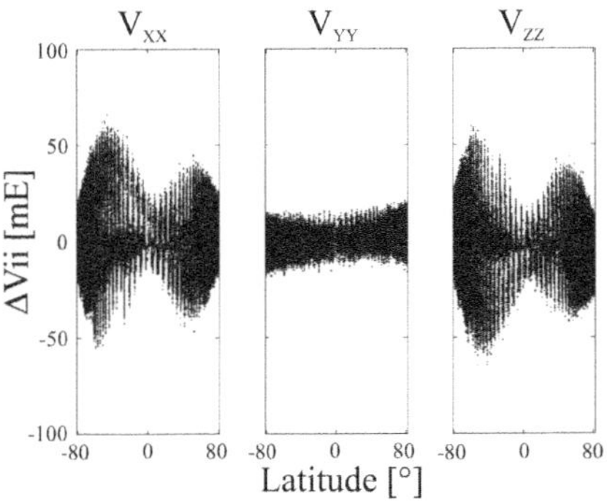

Fig. 3. Gradient differences recovered from noisy diagonal tensor elements in cross-overs, height reduction applied.

As a relative validation procedure utilises gradient differences, the rms of the differences, which should be $\sqrt{2}$ times the original noise rms (16.8 mE in the case of V_{XX}), and their bias, which should vanish, are adequate statistical quantities. In the cross-overs selected in Sect. 2.2, the rms of the recovered noise was 16.6 mE for V_{XX}, which fits well with the theoretical value. The mean bias determined was -3.4 mE. These global characteristics of the noise model were detected with sub-mE accuracy by the cross-over validation approach as introduced in Sect. 2.2, again applying the appropriate reductions from EGM96 up to degree and order 300. The recovered gradient differences themselves are shown in Fig. 3 with respect to the cross-over latitude. Here, a strong latitude-dependence of the gradient differences can be seen, which is caused by the 1-(and 2-)cpr-peaks in the GOCE gradient error curves, see Müller (2001).

Summarising, all investigations show that the cross-over method is able to monitor the gradiometer performance.

3 Upward Continuation for External Calibration

In this study, the external calibration procedure for GOCE gradients is subdivided into two steps. At first, reference gradients are computed from terrestrial gravity data in combination with a global geopotential model (GPM). Then calibration parameters are derived in an adjustment approach by comparing the reference gradients with the (simulated) GOCE gradient observations. The calibration by regionally upward continued gravity data over well surveyed areas was already proposed in Arabelos and Tscherning (1998), Pail (2002) and Bouman et al. (2004). Here, the accuracy of the upward continuation is investigated in a closed-loop computation using synthetic data. Two methods for the upward continuation are described and their results are compared. All six components of the symmetric gradient tensor of the disturbing potential T,

$$T_{ij} = \frac{\partial^2 T}{\partial i \partial j} = \begin{bmatrix} \frac{\partial^2 T}{\partial x^2} & \frac{\partial^2 T}{\partial x \partial y} & \frac{\partial^2 T}{\partial x \partial z} \\ \frac{\partial^2 T}{\partial y \partial x} & \frac{\partial^2 T}{\partial y^2} & \frac{\partial^2 T}{\partial y \partial z} \\ \frac{\partial^2 T}{\partial z \partial x} & \frac{\partial^2 T}{\partial z \partial y} & \frac{\partial^2 T}{\partial z^2} \end{bmatrix} , \quad (2)$$

are computed, with x pointing North, y East and z Radial.

For the upward continuation of regional data, a remove-restore procedure is applied based on a reference geopotential model (GPM_0) up to spherical harmonic degree l_{max}. Residual gravity anomalies $\Delta g'$ are generated by subtracting the long-wavelength GPM_0 part, Δg_0, from the observations Δg, yielding

$$\Delta g' = \Delta g - \Delta g_0 \, . \quad (3)$$

After the upward continuation of the residual anomalies, the predicted residual gradients T'_{ij} have to be complemented by $T_{ij,0}$ to get finally the complete signal T_{ij}. As the computations in this study are based on synthetic data with a limited resolution, no terrain reductions are considered for the modelling of high frequency effects.

For the calibration, the upward continued gradients can be rotated from the local geographic (x, y, z) to the gradiometer (X, Y, Z) reference system to avoid a rotation of the GOCE gradients, thus preserving the high accuracy of the gradiometer diagonal tensor elements.

3.1 Synthetic Data

Synthetic data sets are produced for a closed-loop computation. A blended geopotential model according to Wolf and Denker (2005) is introduced as ground-truth model GPM^{true}. It is combined from an actual GRACE GPM ($l = 0...89$, (JPL, 2003)), EGM96 ($l = 90...360$, (Lemoine et al., 1998)), and GPM98C ($l = 361...1300$, (Wenzel, 1999)). From this model different data sets are derived:

1. (clone) model, GPM_M, up to degree $l_{max} = 360$, where noise is added to the GPM^{true} coefficients according to the standard deviations of the coefficients; from this model, both, gravity anomalies on the ellipsoid (Δg_0) and gradients at GOCE altitude ($T_{ij,0}$) are derived and utilized in the remove-restore procedure.
2. Δg^{true} up to degree $l = 1300$ in a geographical grid on the ellipsoid; from this data set two simulated observation data sets are computed by adding white noise (Δg^{WN}) and coloured noise (Δg^{CN}), respectively.
3. T_{ij}^{true} up to degree $l = 1300$ in a geographical grid at GOCE altitude, serving as ground-truth in the closed-loop computations.

White noise is computed with a MATLAB routine based on independent normal distributed numbers. Coloured noise is computed using the Cholesky factor of the error covariance matrix of the terrestrial gravity data, cf. Wolf (2005). As the simulated data are on a geographical grid, the error covariance matrix has a block Toeplitz structure. This can be exploited for memory reduction and efficient factorisation routines (e.g., SLICOT package, cf. Brenner et al. (1999)).

3.2 Computation Methods

Two methods are used for the upward continuation of terrestrial gravity data to gravitational gradients at GOCE altitude, namely the spectral combination method with integral formulas (IF) and least-squares collocation (LSC). In the next two paragraphs both methods are described briefly. The computation of all six components is already outlined in Kern and Haagmans (2005) and Wolf and Denker (2005). A more detailed discussion and theoretical comparison of the IF and LSC approach is given in Moritz (1976) and the effect of limited regional data areas is investigated in de Min (1995) for both methods.

Spectral Combination Method with Integral Formulas

For the spectral combination technique (Wenzel, 1982), the basic integration formula for the residual disturbing potential T' is

$$T'(r_P, \theta_P, \lambda_P) = \frac{R}{4\pi} \iint_{\sigma} K(\psi_{PQ}, r_P, R) \Delta g'(R, \theta_Q, \lambda_Q) d\sigma \,, \tag{4}$$

with the kernel function

$$K(\psi_{PQ}, r_P, R) = \sum_{l=2}^{\infty} \left(\frac{R}{r_P}\right)^{l+1} \frac{2l+1}{l-1} w_l P_l(\cos(\psi_{PQ})) \,, \tag{5}$$

where r_P, θ_P, λ_P, R, θ_Q and λ_Q are the spherical coordinates of points P and Q, w_l are spectral weights and $P_l(\cos(\psi_{PQ}))$ are Legendre's polynomials depending on the spherical distance ψ_{PQ} between points P and Q.

The spectral weights are derived by a least-squares adjustment approach:

$$w_l = \frac{\sigma_l^2(\epsilon_{\Delta g_0})}{\sigma_l^2(\epsilon_{\Delta g_0}) + \sigma_l^2(\epsilon_{\Delta g})} \,. \tag{6}$$

The $\sigma_l^2(\epsilon_{\Delta g_0})$ and $\sigma_l^2(\epsilon_{\Delta g})$ are the error degree variances of the geopotential model GPM_0 and of the terrestrial gravity data, respectively.

The error degree variances of the terrestrial gravity anomalies are derived for an uncorrelated (white noise) and correlated (coloured noise) scenario from a given error covariance function, as described in Wolf (2005). For the white noise case, the error degree variances are derived by equally distributing the noise over a certain spectral range. The width of the spectral range is chosen depending on the resolution of the gridded terrestrial data which is common practice. However, the error covariance function based on these degree variances differs from a white noise covariance function (cf. Color Fig. XXXIV on p. 305), i.e., significant correlations remain for short distances. This has to be considered in the discussion of the results. For the coloured noise scenario, the error degree variances are derived by a Legendre transform of the covariance function, cf. Wenzel (1981). For the error covariance function, the following model based on Weber and Wenzel (1983) is used:

$$E_{\Delta g,\Delta g} = \sigma^2_{\Delta g} e^{-4.0\cdot\psi[^\circ]}\,, \tag{7}$$

with $\sigma_{\Delta g}$ as noise level. The resulting spectral weights for white noise (depending on the assumed data resolution) and coloured noise scenarios are shown in Color Fig. XXXV on p. 305.

For the derivation of the integral formulas for the gradients T_{ij} from (4), the partial derivatives according to (2) are needed (for details see, e.g., Moritz (1971)). The derivations with respect to spherical coordinates θ, λ, r can be found, e.g., in Wolf and Denker (2005). For this purpose, the kernel function has to be differentiated with respect to the radius r and the spherical distance ψ, yielding $K_\psi, K_r, K_{\psi\psi}, K_{rr}$ and $K_{\psi r}$. Combining (2) and (4) and considering the above abbreviations gives:

$$T'_{xx} = \frac{R}{4\pi}\iint_\sigma \Bigg(\frac{1}{r}K_r + \frac{1}{2r^2}\Big((K_{\psi\psi} + K_\psi \cot\psi) + (K_{\psi\psi} - K_\psi \cot\psi)\cos(2\alpha)\Big)\Bigg)\Delta g'(R,\theta_Q,\lambda_Q)d\sigma\,,$$

$$T'_{xy} = \frac{R}{4\pi}\iint_\sigma \frac{1}{2r^2}(K_{\psi\psi} - K_\psi \cot\psi)\sin(2\alpha)\Delta g'(R,\theta_Q,\lambda_Q)d\sigma\,,$$

$$T'_{xz} = \frac{R}{4\pi}\iint_\sigma \frac{1}{r}\left(\frac{1}{r}K_\psi - K_{\psi r}\right)\cos(\alpha)\Delta g'(R,\theta_Q,\lambda_Q)d\sigma\,,$$

$$T'_{yy} = \frac{R}{4\pi}\iint_\sigma \Bigg(\frac{1}{r}K_r + \frac{1}{2r^2}\Big((K_{\psi\psi} + K_\psi \cot\psi) - (K_{\psi\psi} - K_\psi \cot\psi)\cos(2\alpha)\Big)\Bigg)\Delta g'(R,\theta_Q,\lambda_Q)d\sigma\,,$$

$$T'_{yz} = \frac{R}{4\pi} \iint\limits_{\sigma} \frac{1}{r} \left(\frac{1}{r} K_{\psi} - K_{\psi r} \right) \sin(\alpha) \Delta g'(R, \theta_Q, \lambda_Q) d\sigma ,$$

$$T'_{zz} = \frac{R}{4\pi} \iint\limits_{\sigma} K_{rr} \Delta g'(R, \theta_Q, \lambda_Q) d\sigma . \quad (8)$$

The practical evaluation of the above equations is done by 1D FFT according to Haagmans et al. (1993), as used already by Denker (2003) for the radial component T_{zz}.

Least-squares Collocation

The theoretical background of LSC is described in detail in Moritz (1980). The basic formula of LSC is

$$\hat{T}'_{ij} = C_{T'_{ij}\Delta g'} (C_{\Delta g' \Delta g'} + E_{\Delta g \Delta g})^{-1} \Delta g' , \quad (9)$$

where $C_{T'_{ij}\Delta g'}$ and $C_{\Delta g' \Delta g'}$ are the signal covariance matrices, $E_{\Delta g \Delta g}$ is the error covariance matrix, and $\hat{T}'_{ij}$ and $\Delta g'$ are the residual signals (gradients, gravity anomalies).

The complete formalism for the derivation of the signal covariances is described in Tscherning (1976). All covariances are based on the covariance function C_{TT} of the disturbing potential T

$$C_{TT} = \sum_{l=2}^{\infty} \sigma_l^2(T) \left(\frac{R_B^2}{r_P r_Q} \right)^{l+1} P_l(\cos(\psi_{PQ})) , \quad (10)$$

with the degree variances $\sigma_l^2(T)$ of T and the radius R_B of the Bjerhammer sphere.

The degree variances $\sigma_l^2(T)$ are taken from the error degree variances of the GPM_0 up to degree l_{max}, and above degree l_{max} from the Tscherning and Rapp (1974) degree variance model. The error covariance matrix is assumed as a diagonal matrix for the white noise scenario, and as a full matrix according to the exponential noise model (7) for the coloured noise scenario. For the computation of the signal covariance values, subroutines implemented by C. C. Tscherning (Tscherning, 1974, 1976) are used in an updated version. As the numerical experiments in this study are based on data in geographical grids, again the block Toeplitz structure of the covariance matrices can be exploited for memory reduction and faster equation solver routines (e.g., SLICOT package, cf. Brenner et al. (1999)).

3.3 Numerical Experiments

The IF and LSC are applied in a closed-loop computation. The first set of input gravity anomalies (Δg^{WN} and Δg^{CN}, noise level $\sigma_{\Delta g} = 5\,\text{mgal}$) has a

Table 2. Statistics of the differences between ground-truth and predicted gradients from IF for the *white* noise scenario with a noise level of 5 mgal. Unit mE.

WN	mean	std	min	max
$T_{xx}^{true} - T_{xx}^{IF}$	0.0	0.8	−3.5	2.8
$T_{xy}^{true} - T_{xy}^{IF}$	0.0	0.7	−2.5	2.4
$T_{xz}^{true} - T_{xz}^{IF}$	0.0	1.1	−4.0	3.6
$T_{yy}^{true} - T_{yy}^{IF}$	0.0	1.9	−6.9	6.9
$T_{yz}^{true} - T_{yz}^{IF}$	0.0	2.0	−7.1	7.7
$T_{zz}^{true} - T_{zz}^{IF}$	0.0	2.4	−8.6	7.9

Table 3. Statistics of the differences between ground-truth and predicted gradients from IF for the *coloured* noise scenario with a noise level of 5 mgal. Unit mE.

CN	mean	std	min	max
$T_{xx}^{true} - T_{xx}^{IF}$	0.0	2.3	−7.6	8.8
$T_{xy}^{true} - T_{xy}^{IF}$	0.0	1.5	−5.6	5.9
$T_{xz}^{true} - T_{xz}^{IF}$	0.0	2.8	−8.8	10.9
$T_{yy}^{true} - T_{yy}^{IF}$	0.0	3.4	−14.0	12.7
$T_{yz}^{true} - T_{yz}^{IF}$	0.0	3.7	−14.2	15.0
$T_{zz}^{true} - T_{zz}^{IF}$	0.0	4.6	−15.8	18.7

resolution of 0.05° and covers an area of 37°...50°N and 0°...13°E. The model GPM_M is used as the reference model GPM_0. The predictions from both methods are compared in the central 2° by 2° area. The maximum differences between IF and LSC are less than 0.5 mE for the coloured noise scenario. The corresponding differences for the white noise scenario go up to 2.4 mE due to the modelling of the white noise in the IF approach (cf. Color Fig. XXXIV on p. 305).

In view of the calibration procedure, a second larger data set is introduced in the closed-loop procedure applying IF. An area of 42.5°S...3.5°N and 35°W...15°E is covered with a resolution of 0.1°. The statistics of the differences between predicted and ground-truth gradients in the central 40° by 40° area are shown in Table 2 for the white noise scenario and in Table 3 for the coloured noise scenario, both with a noise level of 5 mgal. Even with a relatively high noise level of the input data, the standard deviations of the differences are in the range of a few mE. The component T_{zz} has the largest signal variance, and correspondingly, it shows the largest error variance. The standard deviation of the differences in T_{zz} agrees with the estimation in Wolf (2005), where 4.8 mE and 3.6 mE were derived by error propagation based on error degree variances, and 5.0 mE and 2.5 mE by LSC for coloured and white noise (input noise level 5 mgal and data resolution 0.1°), respectively. Assuming 1 mgal noise for the terrestrial gravity data, errors of 1...2 mE can be expected for the gradients.

4 External Calibration

Previous investigations for the external gradiometer calibration done in, e.g., Bouman and Koop (2003) and Bouman et al. (2004), are based on global GPM data and regional data sets. In this study, calibration parameters are

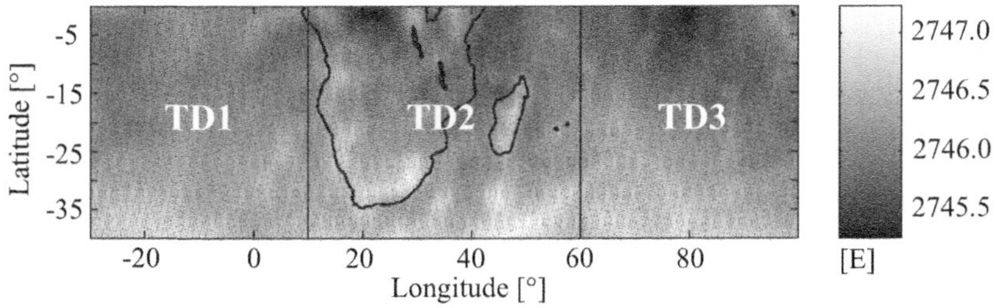

Fig. 4. Regional areas TD1, TD3 (used for the derivation of the calibration parameters) and TD2 (used for testing the derived calibration parameters)

computed for the radial component V_{ZZ} over two regional areas (TD1, TD3 in Fig. 4) and then applied to a third area (TD2), followed by an evaluation of the results.

4.1 Calibration Model

The functional calibration model in a least-squares adjustment approach is chosen according to Bouman and Koop (2003):

$$V^G_{ZZ}(t) = \lambda \left[V^R_{ZZ}(t) + \Delta V_{ZZ} + V'_{ZZ} t + \sum_{k=1}^{4} a_k \cos(k\omega(t)) + b_k \sin(k\omega(t)) \right] .$$

The observations V^G_{ZZ} are simulated GOCE gradients (see below) and V^R_{ZZ} are the reference gradients. The unknown calibration parameters are the scale factor λ, the bias ΔV_{ZZ}, the trend V'_{ZZ}, and the Fourier coefficients a_k and b_k modelling revolution dependent errors (in cycles/rev.).

4.2 Numerical Experiments

In this study, simulated 0.2 Hz V_{ZZ} time series are used as observations. This data set from the IAG-SC7 (cf. IAG-SC7 (2003)) also includes gradiometer noise. Gradients over the two regional areas (TD1, TD2) can be upward continued as shown in Sect. 3.3 for the area TD1. However, due to time restrictions, the actual upward continued gradients were not used so far. Instead, the calibration procedure is tested here with the ground-truth reference gradients with 1 mE white noise added.

Gradients over the areas TD1 and TD3 (Fig. 4) were selected for the external calibration. The derived calibration parameters were then applied to the time series over area TD2. The standard deviation of the differences between the calibrated gradients and ground-truth gradients decreases by a factor of two. The statistics of the errors of the gradients over this area before and after the calibration are shown in Table 4.

Table 4. Statistics of the errors of simulated V_{ZZ} time series over regional area TD2 before and after calibration. Unit E.

Status	mean	std	min	max
Before Calibration	−2.3159	0.0115	−2.3497	2.2869
After Calibration	0.0010	0.0056	−0.0232	0.0181

5 Summary and Conclusions

Two procedures for the quality assessment of the GOCE gradients have been investigated in this study. First, the use of cross-overs to relatively compare gradients in orbit has been discussed, where all computation steps (cross-over detection, measurement interpolation and height and orientation reduction) were addressed and possible error sources (e.g., from orbit determination) were considered. It has been shown, that this procedure is suited very well to monitor the gradiometer behaviour.

Secondly, reference gradients for a regional gradiometer calibration or validation were upward continued from terrestrial gravity data using two methods, spectral combination with integral formulas (IF) and least-squares collocation (LSC). Differences of a few mE in a closed-loop procedure based on synthetic data were obtained. The application of upward continued gradients over regional areas in an external calibration procedure has been described. The error standard deviation of a time series of the radial gradient component over a neighbouring area decreases by a factor of two by applying the derived calibration parameters.

Acknowledgement. Thanks go to K.-H. Ilk and IAG-SC7 for making the GOCE test data available and to C.C. Tscherning for providing his covariance computation routines. This is publication no. GEOTECH-158 of the programme GEOTECHNOLOGIEN of BMBF and DFG, Grant 03F0329D.

References

Albertella A, Migliaccio F, Sansó F (2000) The space-wise approach - Overall scientific data strategy. In: Sünkel H (ed.), ESA From Eötvös to Milligal Final Report. ESA/ESTEC Contract No. 13392/98/NL/GD, pp. 267-298, Graz

Arabelos D and Tscherning CC (1998) Calibration of Satellite Gradiometer Data Aided by Ground Gravity Data. J. Geod. 72:617–625

Bouman J and Koop R (2003) Calibration of GOCE SGG Data Combining Terrestrial Gravity Data and Global Gravity Field Models. In: Tziavos IN (ed.), Gravity and Geoid 2002, Proc. 3rd Meeting of the International Gravity and Geoid Commission, pp. 275–280, Zeta Publ., Thessaloniki.

Bouman J, Koop R, Tscherning CC and Visser P (2004) Calibration of GOCE SGG Data using High-low SST, Terrestrial Gravity Data and Global Gravity Field Models. J. Geod. 78:124–137

Brenner P, Mehrmann V, Sima V, Van Huffel S and Varga A (1999) SLICOT – A Subroutine Library in Systems and Control Theory. NICONET Technical Report 97-3

Denker H (2003) Computation of Gravity Gradients Over Europe For Calibration/-Validation of GOCE Data. In: Tziavos IN (ed.), Gravity and Geoid 2002, Proc. 3rd Meeting of the International Gravity and Geoid Commission, pp. 287–292, Zeta Publ., Thessaloniki

ESA (1999) Gravity Field and Steady-State Oean Circulation Mission, Reports for mission selection, The Four Candidate Earth Explorer Core Missions, SP-1233(1), ESA Publications Division, Noordwijk

Haagmans R, de Min E and van Gelderen M (1993) Fast Evaluation of Convolution Integrals on the Sphere Using 1D FFT, and a Comparison with Existing Methods for Stokes' Integral. Man. Geod. 18:227–241

IAG-SC7 (2003) Satellite Gravity Field Missions: Simulation Scenarios, IAG - Special Commission VII (Section II), Satellite Gravity Field Missions, `http://www.geod.uni-bonn.de/SC7/index.html`, 13.10.2003

Jarecki F and Müller J (2003) Validation of GOCE Gradients Using Cross-Overs, In: GEOTECHNOLOGIEN; Observation of the System Earth from Space, GEOTECHNOLOGIEN Science Report No. 3, pp. 79-84, Potsdam

JPL (2003) GRACE Homepage, Jet Propulsion Laboratory, `http://podaac.jpl.nasa.gov/grace/`, 15.12.2003

Kern M and Haagmans R (2005) Determination of Gravity Gradients from the Terrestrial Gravity Data for Calibration and Validation of Gradiometric GOCE Data. In: Jekeli C, Bastos L and Fernandes J (eds.), Proc. IAG International Symposium Gravity, Geoid and Space Missions, Porto, Aug. 30 - Sept. 3, 2004, International Association of Geodesy Symposia, vol. 129. Springer, Berlin Heidelberg New York (in print)

Koop R, Visser P and Tscherning CC (2001) Aspects of GOCE Calibration. In: Drinkwater MR (ed.), Proc. International GOCE User Workshop. European and national user group activities (Noordwijk), pp. 51–56. ESA/ESTEC, Noordwijk

Lemoine FG, Kenyon SC, Factor JK, Trimmer RG, Pavlis NK, Chinn DS, Cox CM, Klosko SM, Luthcke SB, Torrence MH, Wang YM, Williamson RG, Pavlis EC, Rapp RH, and Olson TR (1998) The Development of the Joint NASA GSFC and NIMA Geopotential Model EGM96. Technical Paper NASA/TP-1998-206861, NASA, Greenbelt, Maryland

Min E de (1995) A Comparison of Stokes' Numerical Integration and Collocation, and a New Combination Technique. Bull. Géod. 69:223–232

Moritz H (1971) Kinematical Geodesy II. Reports of the Department of Geodetic Science No 165, Department of Geodetic Science, Ohio State University, Columbus, Ohio

Moritz H (1976) Integral Formulas and Collocation. Man. Geod. 1:1–40

Moritz H (1980) Advanced Physical Geodesy. Herbert Wichmann Verlag, Karlsruhe

Müller J (2001) Die Satellitengradiometriemission GOCE, Theorie, technische Realisierung und wissenschaftliche Nutzung, Veröffentlichungen der Deutschen Geodätischen Kommission, Reihe C, Nr. 541, München

Pail R (2002) In-orbit Calibration and Local Gravity Field Continuation Problem, In: Sünkel H (ed.), ESA From Eötvös to Milligal+ Final Report, ESA/ESTEC Contract No. 14287/00/NL/GD, pp. 9-112, Graz

Schrama EJO (2001) External geophysical validation and calibration of an orbiting gravity gradiometer. In: Drinkwater MR (ed.), Proc. International GOCE User Workshop. European and national user group activities (Noordwijk), pp. 57–62. ESA/ESTEC, Noordwijk

Shum CK, Zhang BH, Schutz BE, Tapley BD (1990) Altimeter Crossover Methods for Precision Orbit Determination and the Mapping of Geophysical Parameters, J. Astronaut. Sc. 38:355-368

Smit M, Koop R, Visser P, vd IJssel J, Sneeuw N, Müller J, Oberndorfer H (2000) SID 2000: GOCE End to End Performance Analysis, ESA/ESTEC Contract No. 12735/98/NL/GD, Noordwijk

Tscherning CC (1974) A FORTRAN IV Program for the Determination of the Anomalous Potential Using Stepwise Least-squares Collocation. Reports of the Department of Geodetic Science No 212, Department of Geodetic Science, Ohio State University, Columbus, Ohio

Tscherning CC (1976) Computation of the Second-Order Derivatives of the Normal Potential Based on the Representation by a Legendre Series. Man. Geod. 1:71–92

Tscherning CC and Rapp RH (1974) Closed Covariance Expressions for Gravity Anomalies, Geoid Undulations, and Deflections of the Vertical Implied by Anomaly Degree Variance Models. Reports of the Department of Geodetic Science No 208, Department of Geodetic Science, Ohio State University, Columbus, Ohio

Weber G and Wenzel H-G (1983) Error Covariance Functions of Sea Gravity Data and Implications for Geoid Determination. Marine Geodesy 75(1-4):199–226

Wenzel H-G (1981) Zur Schätzung von Anomalie-Gradvarianzen aus lokalen Kovarianzfunktionen. ZfV 106(5):234–243

Wenzel H-G (1982) Geoid Computation by Least-Squares Spectral Combination Using Integral Kernels. In: Proc. General Meeting of the IAG, pp. 438–453, Tokyo. Springer, Berlin Heidelberg New York

Wenzel H-G (1985) Hochauflösende Kugelfunktionsmodelle für das Gravitationspotential der Erde, Wissenschaftliche Arbeiten der Fachrichtung Vermessungswesen der Universität Hannover, Nr. 137, Hannover

Wenzel H-G (1999) Schwerefeldmodellierung durch ultra hochauflösende Kugelfunktionsmodelle. ZfV 124(5):144–154

Wolf KI (2005) Considering Coloured Noise of Ground Data in an Error Study for External GOCE Calibration / Validation. In: Proc. GOCINA Workshop, April, 13-15, 2005, Cahiers du Centre Europeen de Geodynamics et de Seismologie. Luxembourg. (submitted)

Wolf KI and Denker H (2005) Upward Continuation of Ground Data for GOCE Calibration / Validation Purposes. In: Jekeli C, Bastos L and Fernandes J (eds.), Proc. IAG International Symposium Gravity, Geoid and Space Missions, Porto, Aug. 30 - Sept. 3, 2004, International Association of Geodesy Symposia, vol. 129. Springer, Berlin Heidelberg New York (in print)

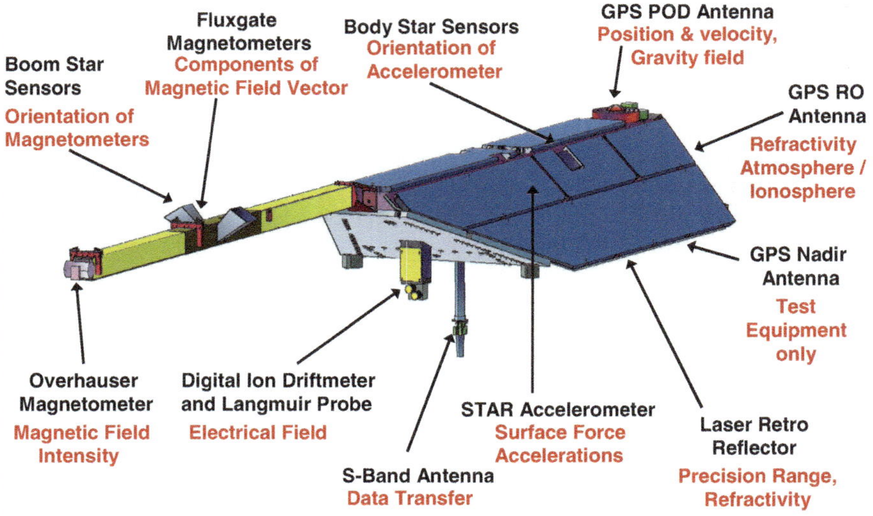

Color Fig. I. CHAMP spacecraft and science instruments

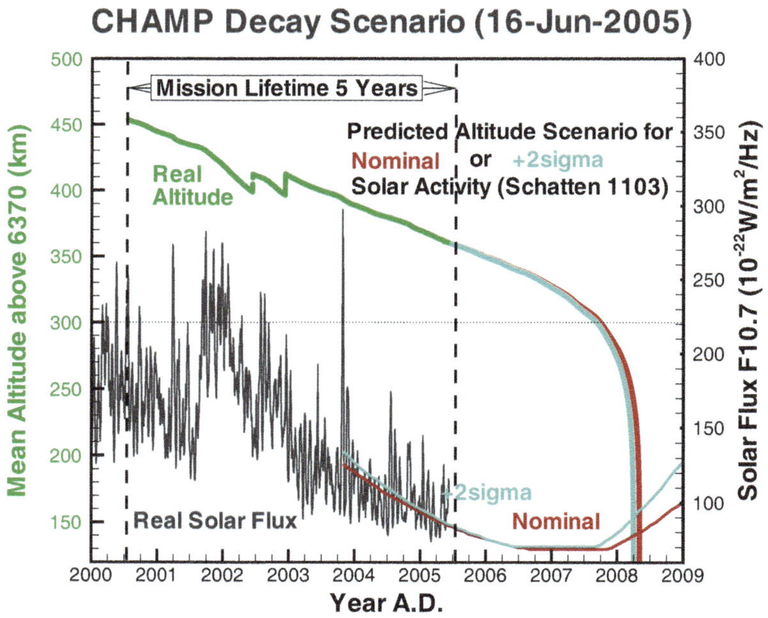

Color Fig. II. CHAMP spacecraft and science instruments

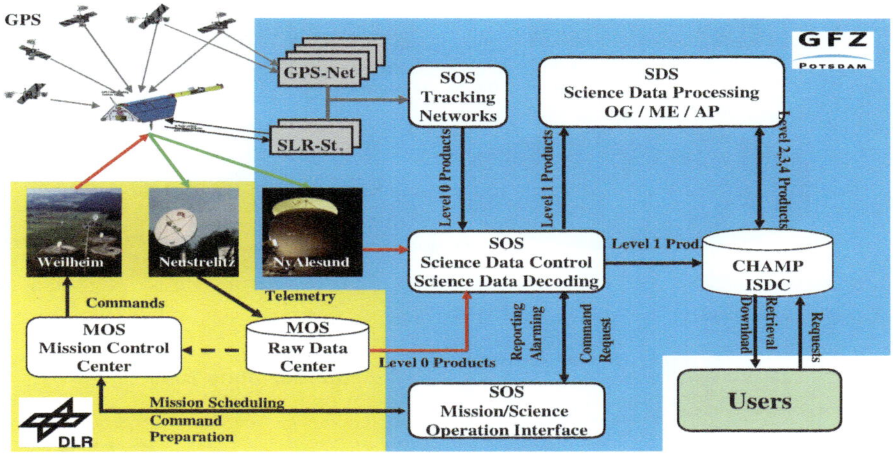

Color Fig. III. CHAMP orbit evolution (real and predicted part)

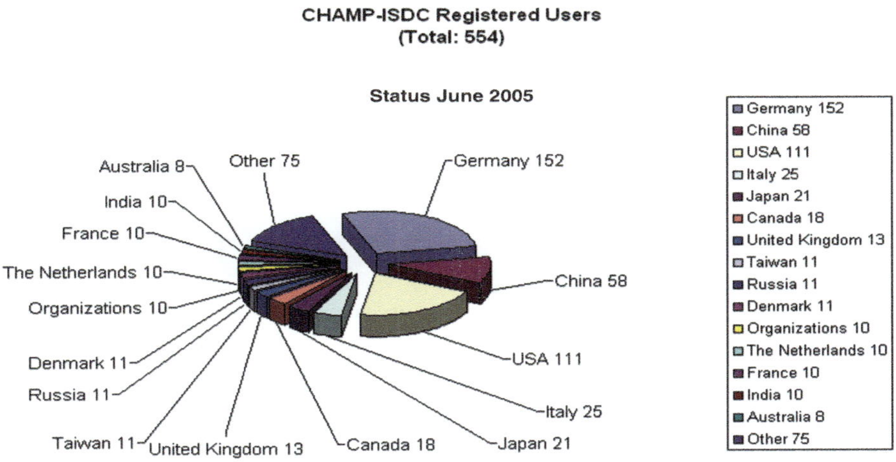

Color Fig. IV. CHAMP ground system elements and responsibilities

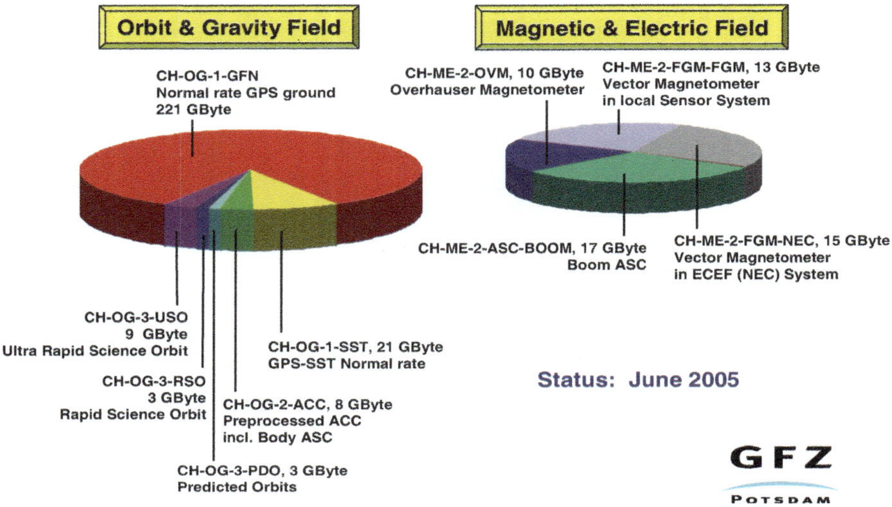

Color Fig. V. CHAMP registered users per country

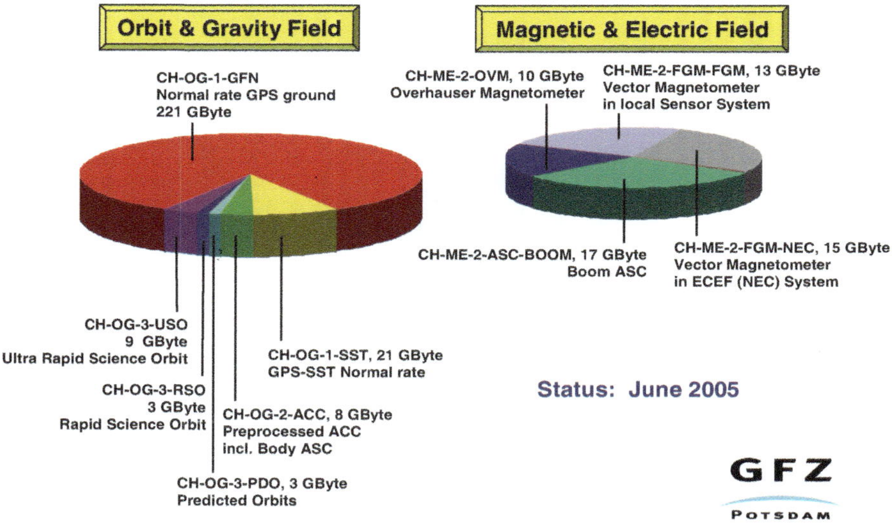

Color Fig. VI. Amount of accessible CHAMP atmosphere and inosphere dta since launch

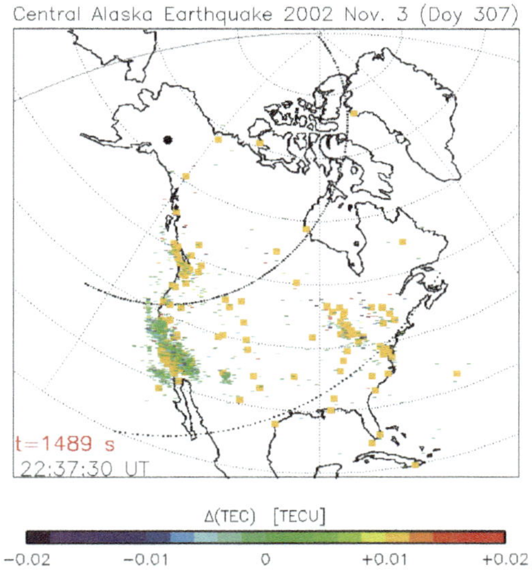

Color Fig. VII. Propagation and observation geometry of the Alaska earthquake on November 3, 2002. The color coded pixels represent the amplitudes of the ionospheric wave whereas the yellow square symbols mark the GPS sites used for observation

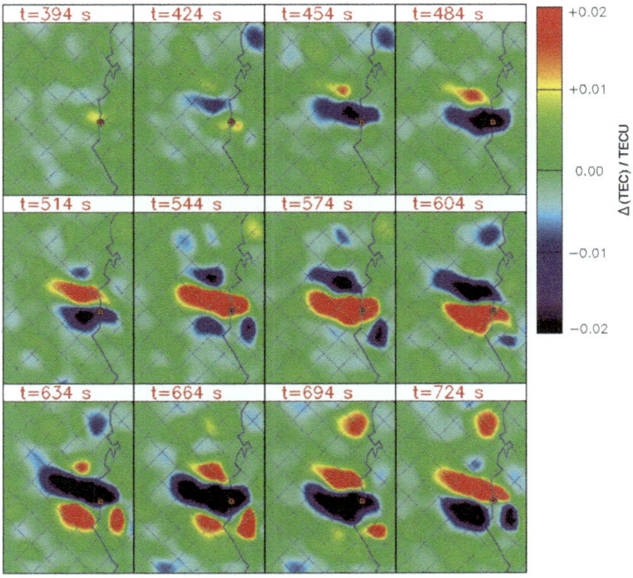

Color Fig. VIII. Time series of amplitude maps of band pass filtered vertical TEC data reconstructed by TEC data assimilation into a background model (zero amplitude values). The epicenter is marked by a red point

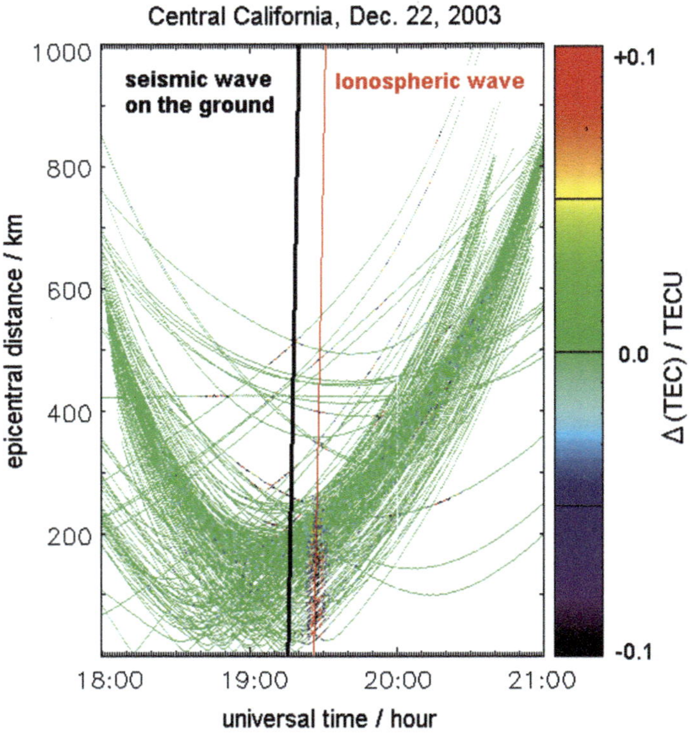

Color Fig. IX. Subsequent band pass filtered vertical TEC data as a function of their distance from the epicenter. The full black line indicates an assumed Rayleigh wave propagating at the Earth surface with a velocity of 3.5 km/s

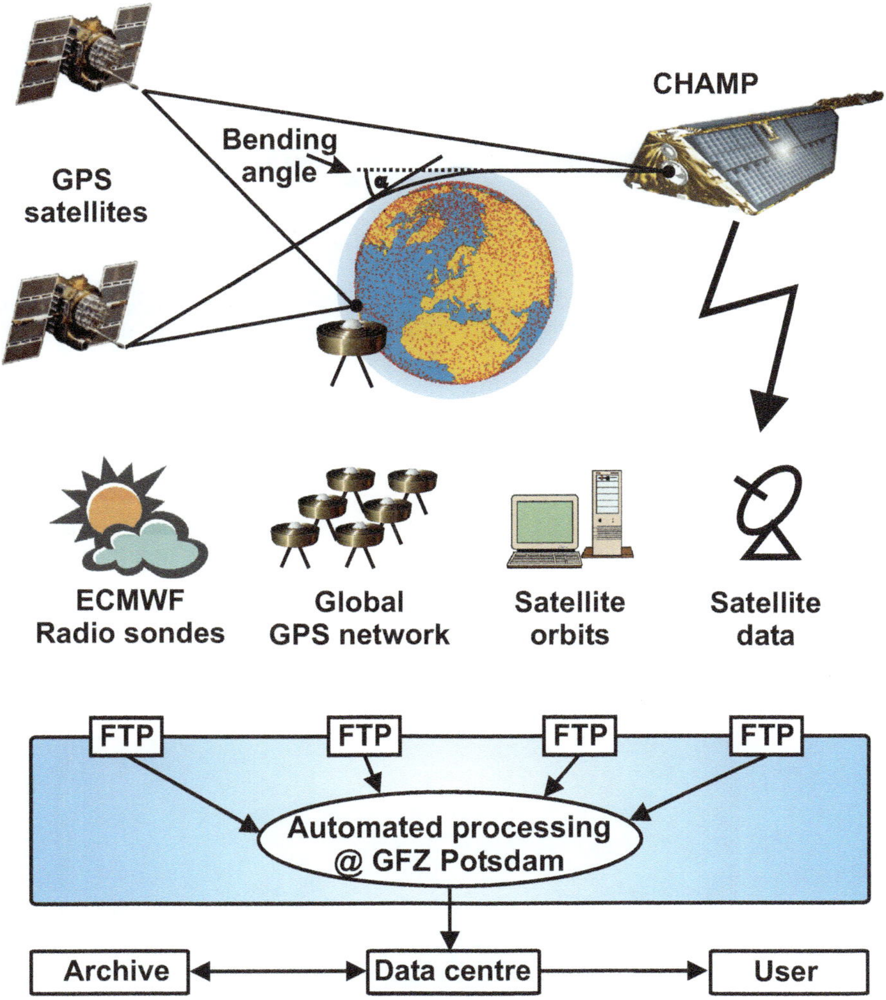

Color Fig. X. Schematic overview CHAMP GPS radio occultation experiment and the related infrastructure (modified from Wickert (2002))

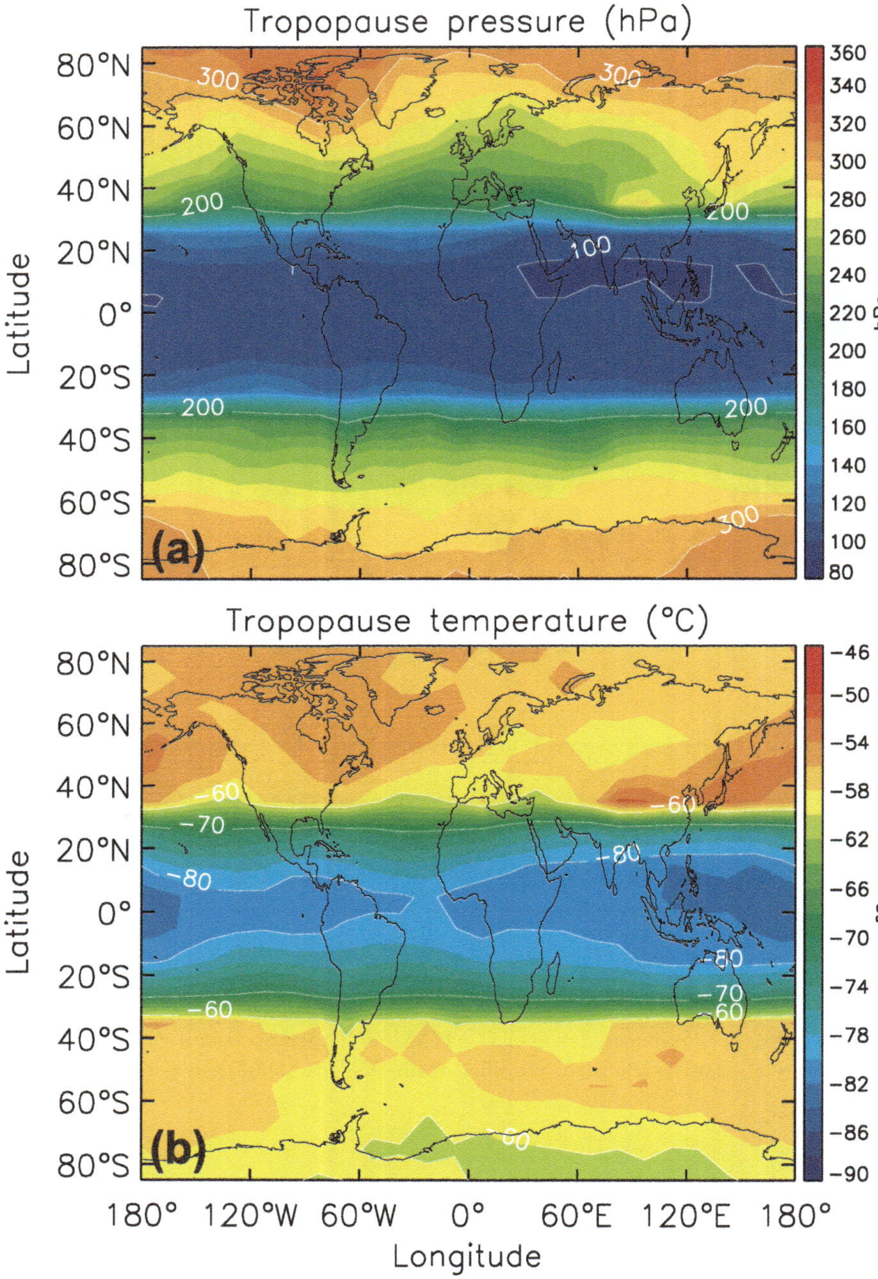

Color Fig. XI. Lapse-rate tropopause pressure (upper panel) and temperature (lower panel). Contour interval: 10 hPa (a) and 2 K (b) respectively. The plots are derived using CHAMP and SAC-C occultation data between May 2001 and April 2005 (update from (Schmidt et al., 2005a))

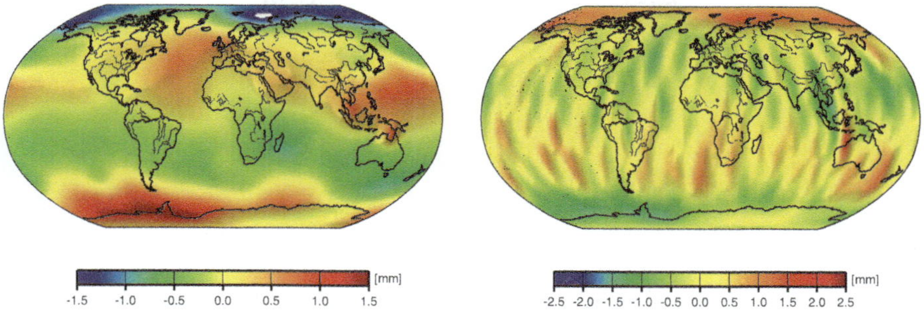

Color Fig. XII. Mean geoid height differences [mm] between PPHA and MOG2D (left) and corresponding gravity field products (right) for February 2004. Figures have been smoothed by a Gaussian filter of 500 km radius, C20 variations have been ignored

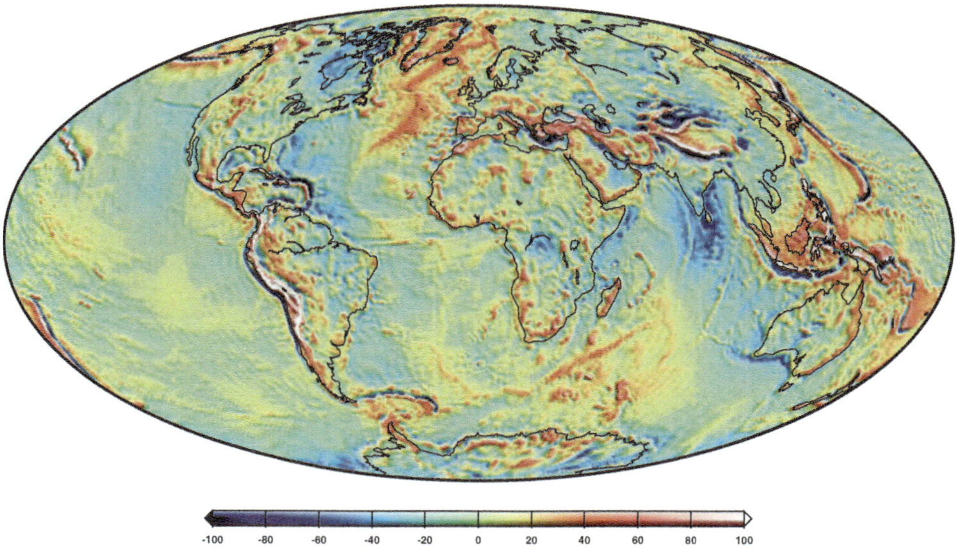

Color Fig. XIII. EIGEN-CG03C free air gravity anomalies (mgal; $0.5° \times 0.5°$ resolution)

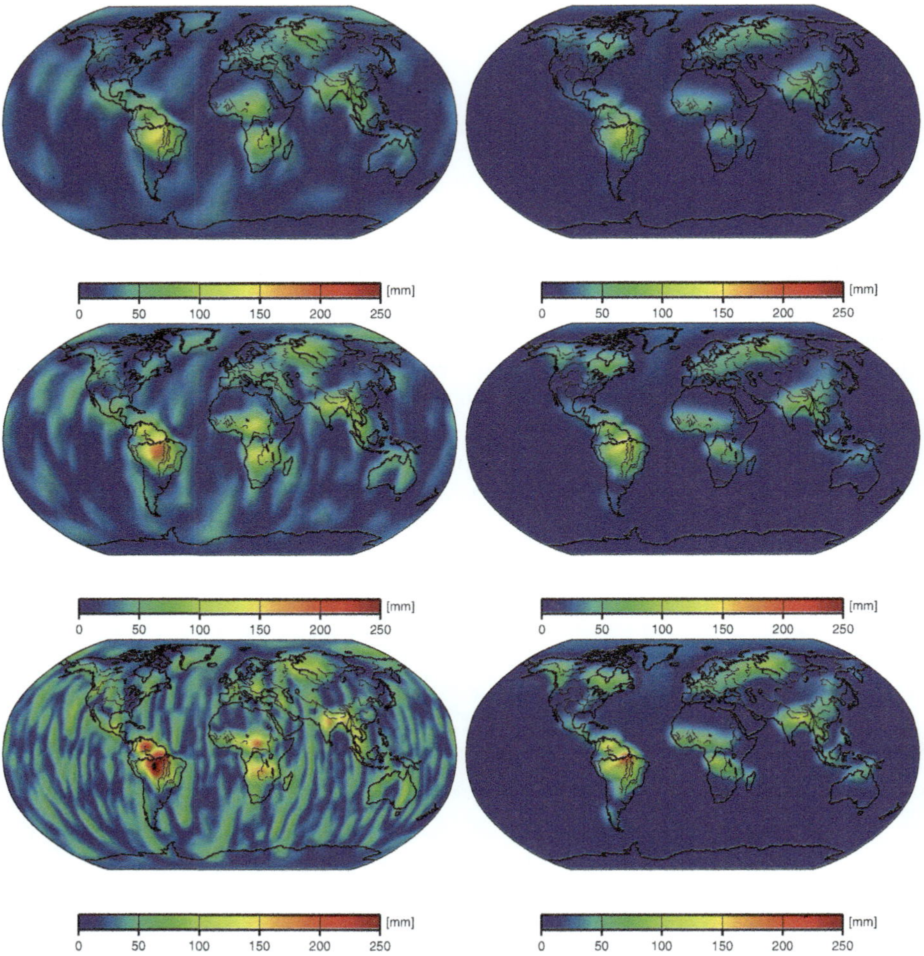

Color Fig. XIV. Amplitudes of best fitting annual signal of surface mass variability derived from GRACE (left) and WGHM (right) filtered by Gaussian averaging radii of 1000, 750 and 500 km (from top to bottom) in terms of thickness of equivalent mass of water

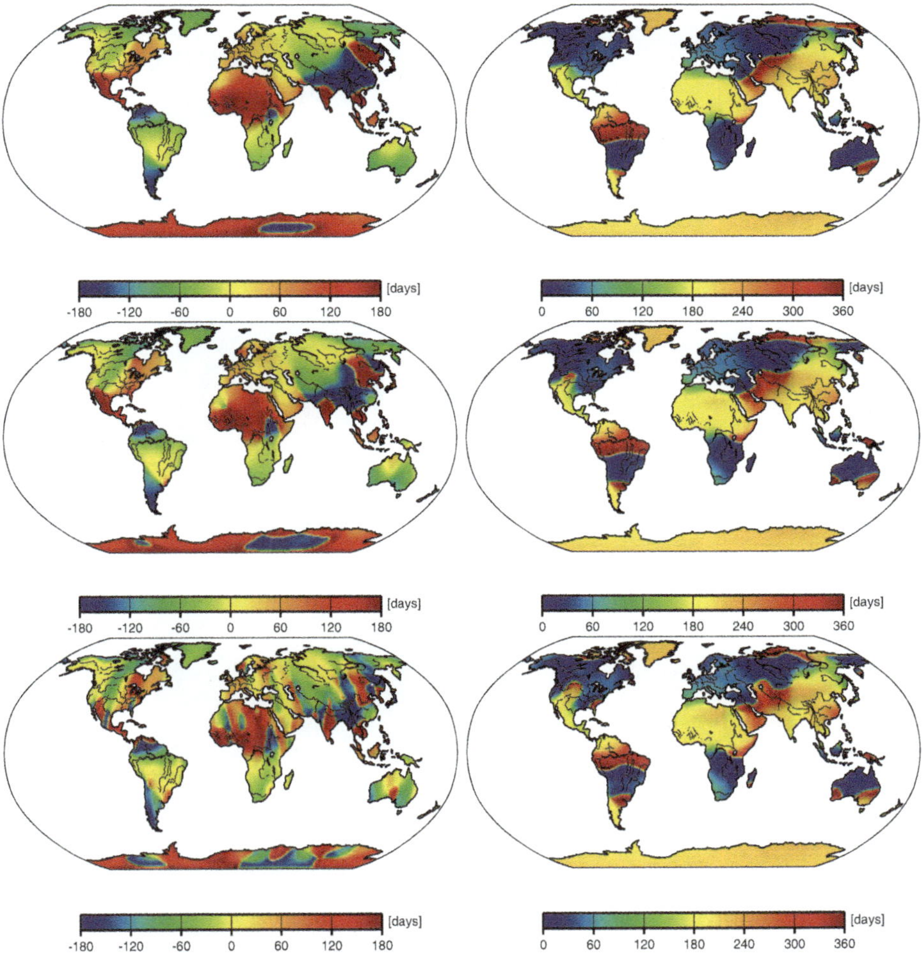

Color Fig. XV. Phases of best fitting annual signal of surface mass variability derived from GRACE (left) and absolute values of the GRACE minus WGHM differences (right) filtered by Gaussian averaging radii of 1000, 750 and 500 km (from top to bottom) in terms of days (Note: Greenland and Antarctica not modeled in WGHM)

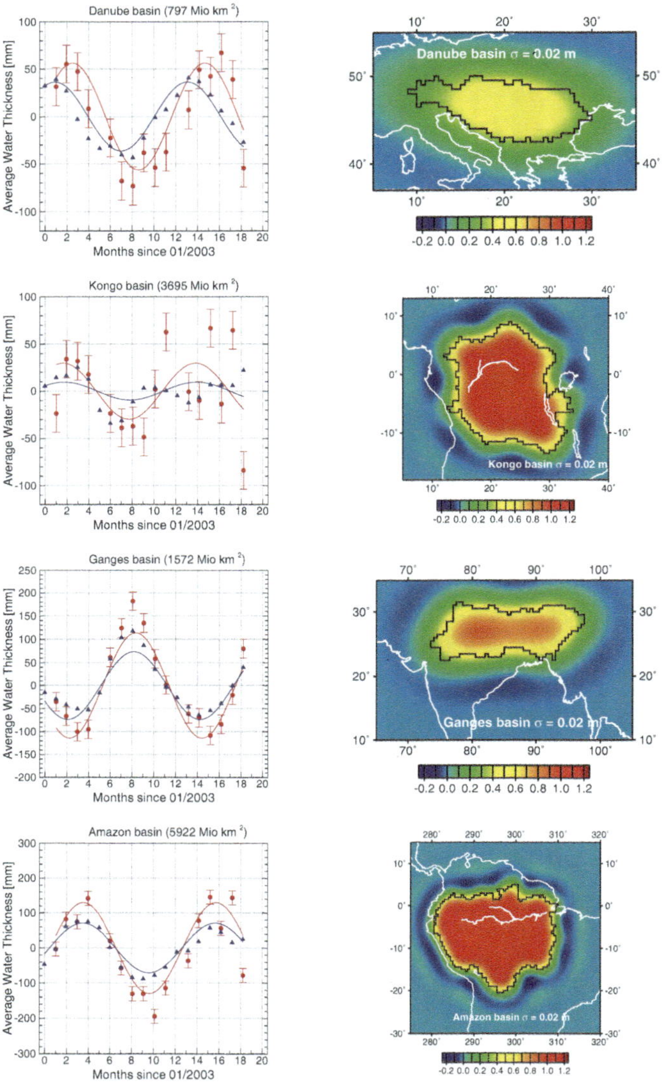

Color Fig. XVI. Left column: Regional surface mass variability (dots) and estimated annual signal (solid lines) derived from GRACE (in red) and WGHM (in blue) in mm of the thickness of an equivalent mass of water for selected river basins (Amazon, Ganges, Congo, Danube from bottom to top) using approximate averaging kernels from Swenson and Wahr (2003) constrained to an uncertainty of 2 cm of the surface mass variability from GRACE. Right column: Corresponding approximate basin functions for the selected river basins (Amazon, Ganges, Congo, Danube from bottom to top)

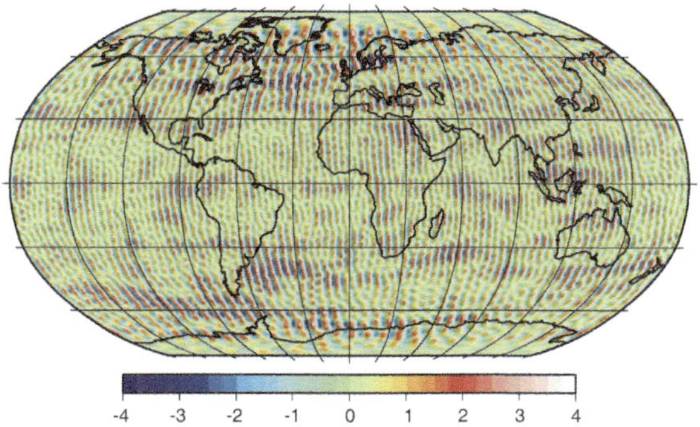

Color Fig. XVII. Simulation case 1: geoid height differences in cm (spherical harmonic expansion up to degree n=110: RMS 1.23cm, avg 0.97cm, min/max -4.76/5.07cm)

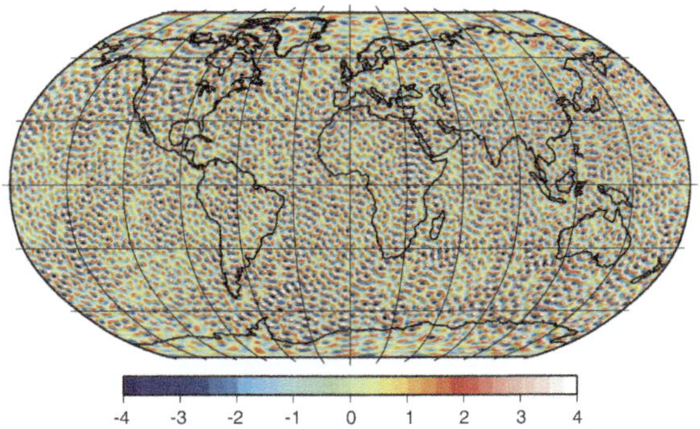

Color Fig. XVIII. Simulation case 2: geoid height differences in cm (spherical harmonic expansion up to degree n=110: RMS 2.11cm, avg 1.69cm, min/max -8.61/8.82cm)

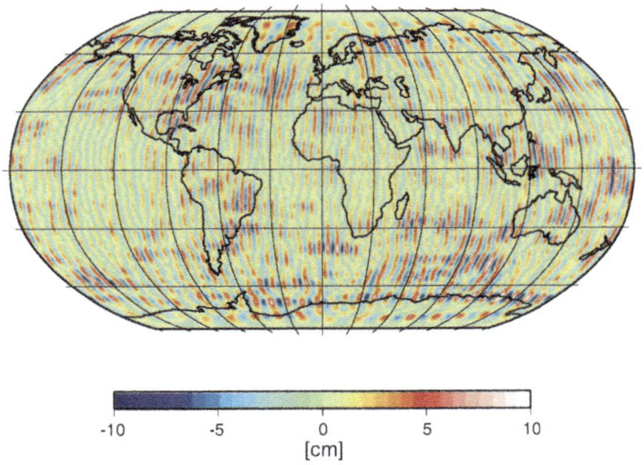

Color Fig. XIX. Geoid height differences: preliminary global solution ITG-GRACE-2003-08 – GGM02C (spherical harmonic expansion up to degree n=90: RMS: 2.6cm, avg: 2.0cm,min/max: -12.7/14.7cm)

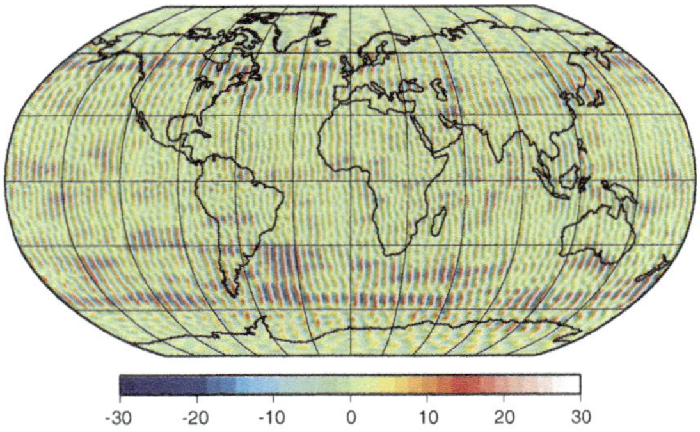

Color Fig. XX. Geoid height differences: global solution with regional refinements ITG- GRACE-2003-08 – GGM02C (spherical harmonic expansion up to degree n=110: RMS 7.27cm, avg 5.61cm,min/max -28.7/30.6cm)

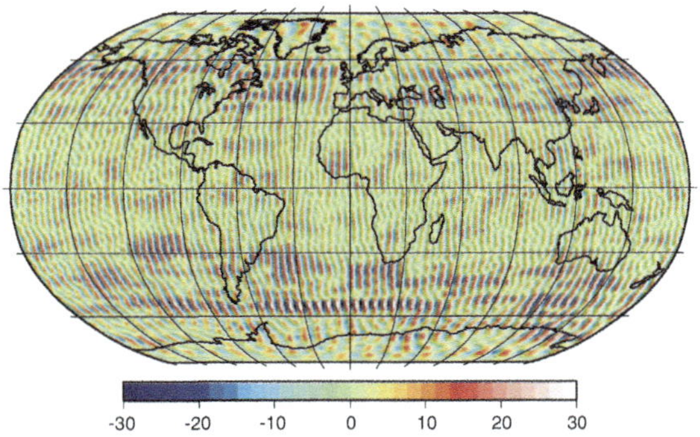

Color Fig. XXI. Geoid height differences: global solution with regional refinements ITG- GRACE-2003-08 – EIGEN-CG03C (spherical harmonic expansion up to degree n=110: RMS 8.91cm, avg 6.88cm, min/max -44.2/42.7cm)

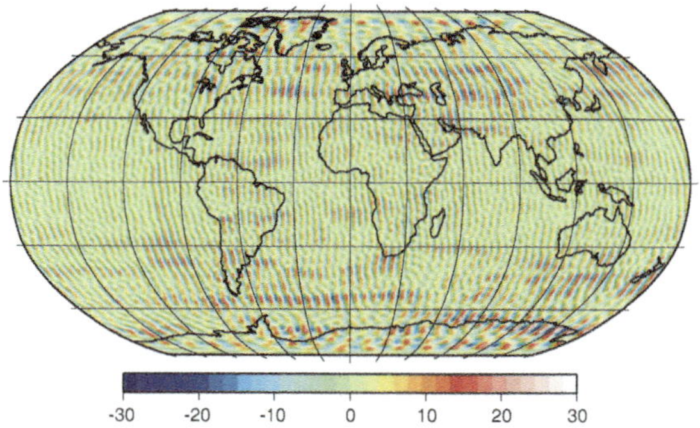

Color Fig. XXII. Geoid height differences GGM02C - EIGEN-CG03C (spherical harmonic expansion up to degree n=110: RMS 6.20cm, avg 4.81cm, min/max -35.9/36.1cm)

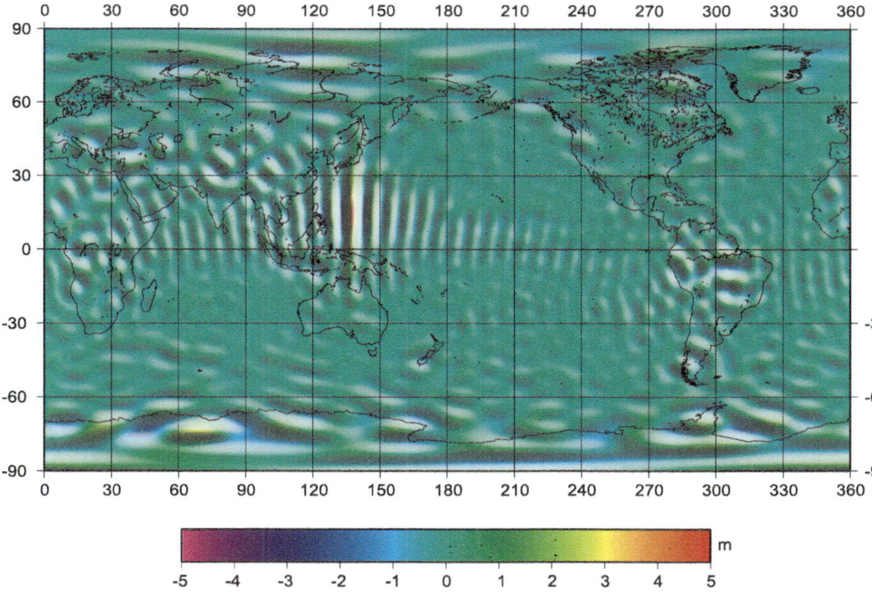

Color Fig. XXIII. Synthesized differences of the solution and the EIGEN-GRACE02S model

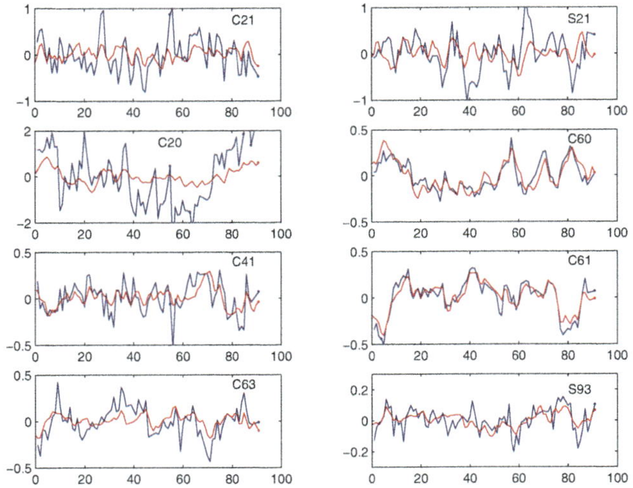

Color Fig. XXIV. Comparison of selected harmonics (unit: mm of geoid height) between the daily GRACE solutions (blue) and the MOG2D model (red). Bias, trend and "seasonal" changes have been removed. Harmonics with degree and order larger than 10 are fixed to their mean values of the 150 × 150 EIGEN-GRACE02S gravity field model, which also provides a-priori values for the estimated harmonics. No a-priori constraints have been imposed on the estimated harmonics

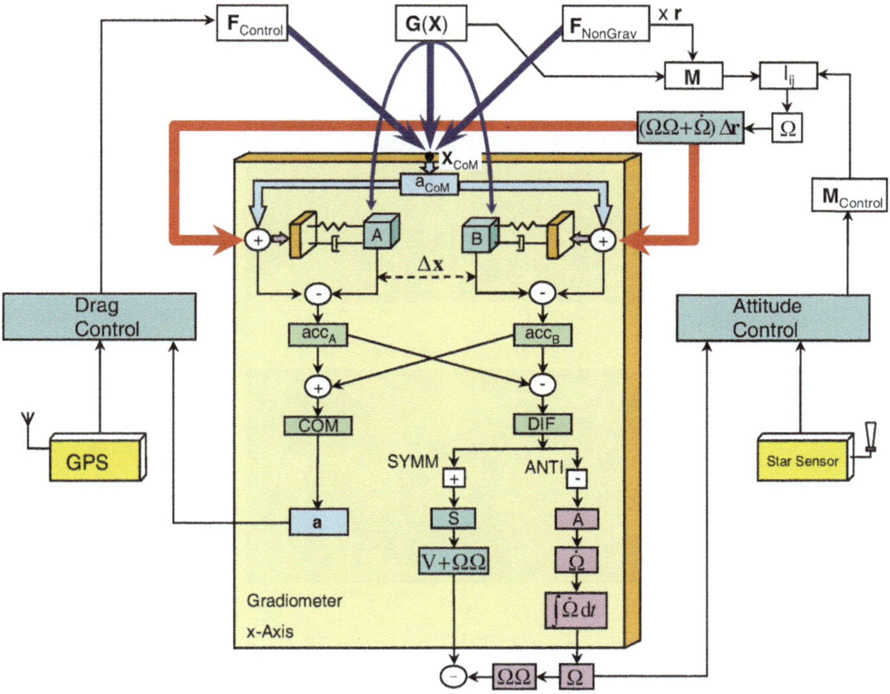

Color Fig. XXV. Elements of the GOCE sensor system simulator

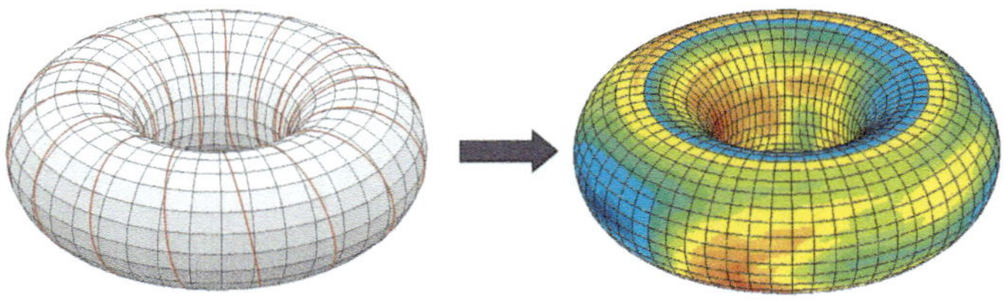

Color Fig. XXVI. Projection of the Time-Series along the Orbit on a Torus

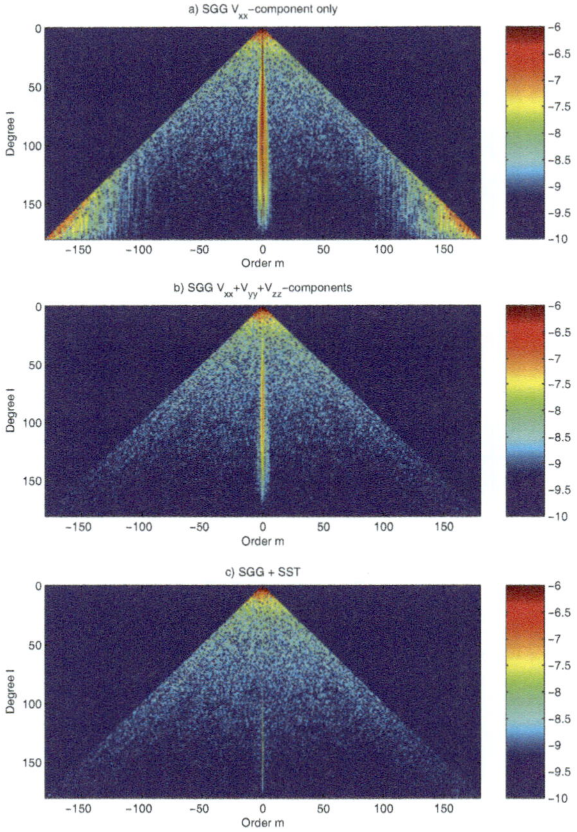

Color Fig. XXVII. Results of the Closed-Loop Simulation. Coefficient errors of a) the V_{zz}-component only, b) the three diagonal SGG-Components and c) SGG+SST

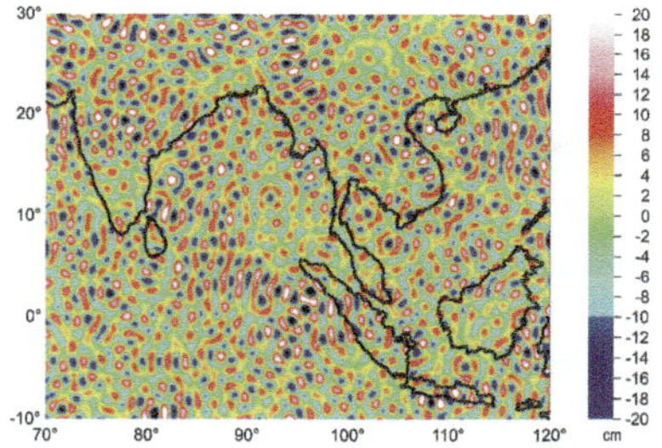

Color Fig. XXVIII. Differences in geoid undulations between a regional solution and the global gravity field model EGM96 up to degree 240 (cm), RMS: 10.1cm

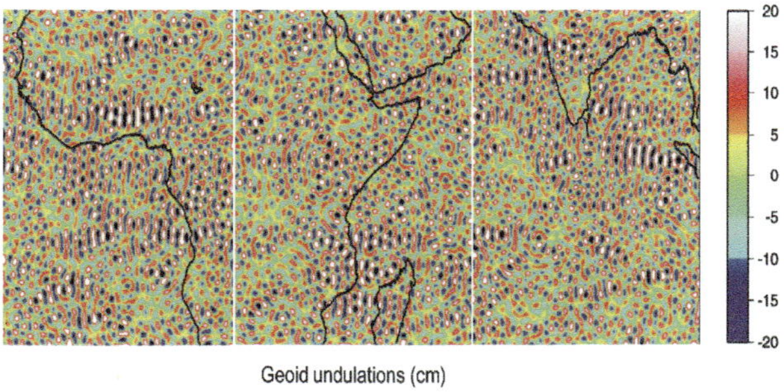

Color Fig. XXIX. Differences of adjacent regional solutions to the pseudo-real gravity field EGM96 in terms of geoid undulations to demonstrate the excellent matching of the residual patterns

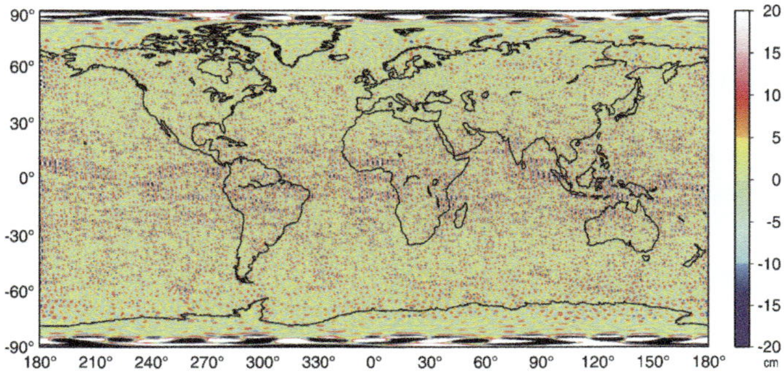

Color Fig. XXX. Differences in geoid undulations between the merged regional solutions and EGM96 compared up to degree n=240 (cm), RMS: 8.6cm

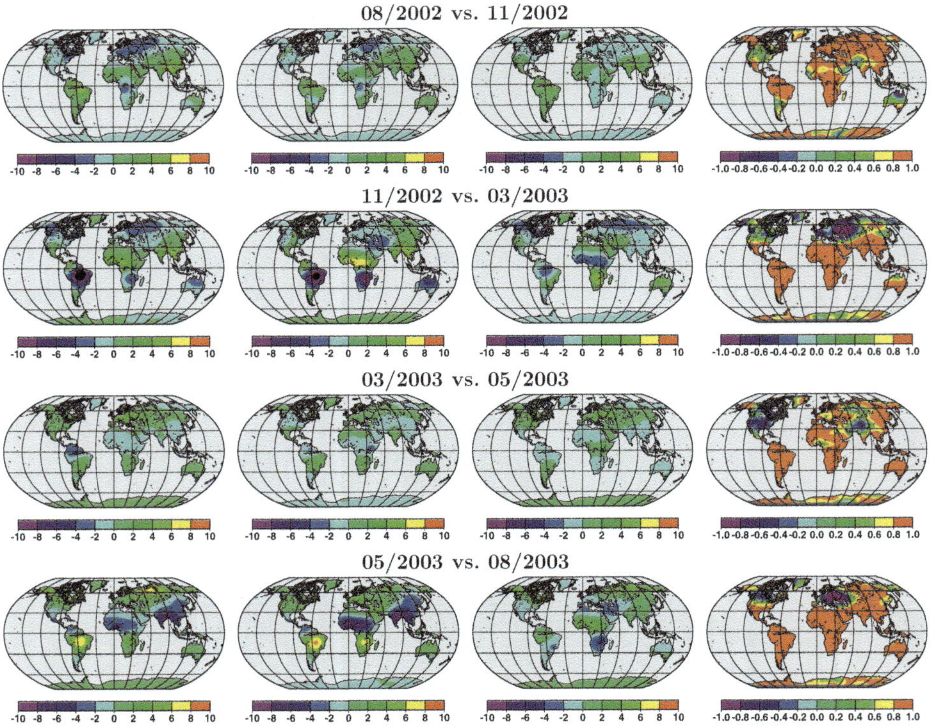

Color Fig. XXXI. Geoid variations in mm deduced from WGHM (left) and H96 (second column), differences between the both (third column) and correlation coefficients between the WGHM–deduced and H96–deduced geoid variations (right)

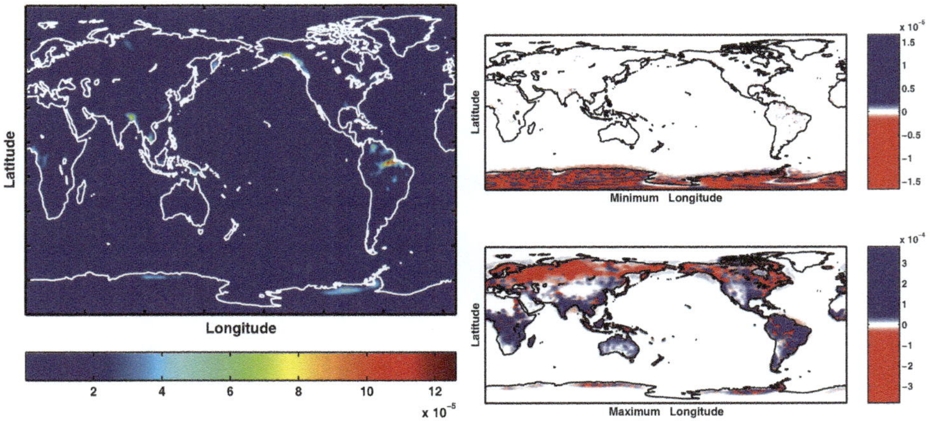

Color Fig. XXXII. Rms, absolute min and max of monthly V_{rr} changes from WGHM+ice, (E)

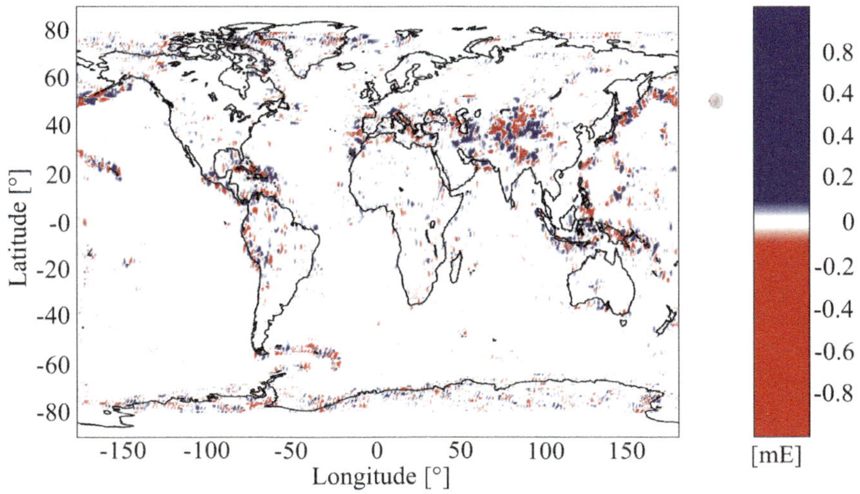

Color Fig. XXXIII. Comparison of tensor element V_{XX} in cross-over points of the 0.2 Hz test data set with height reduction applied, showing inconsistencies mainly in regions with rough gravity field

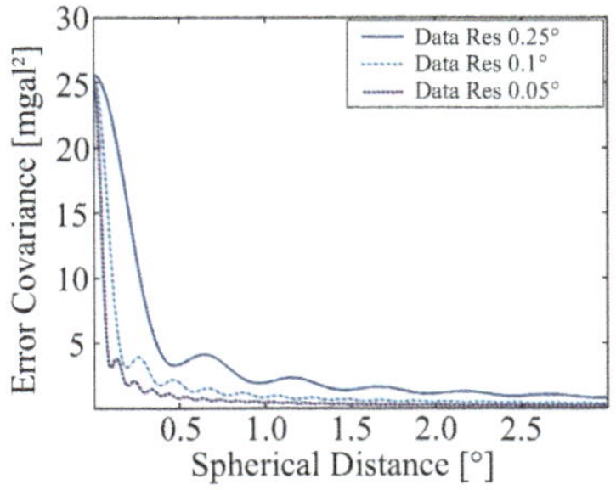

Color Fig. XXXIV. Error covariance functions from accumulated error degree variances for white noise scenario with noise level of 5 mgal

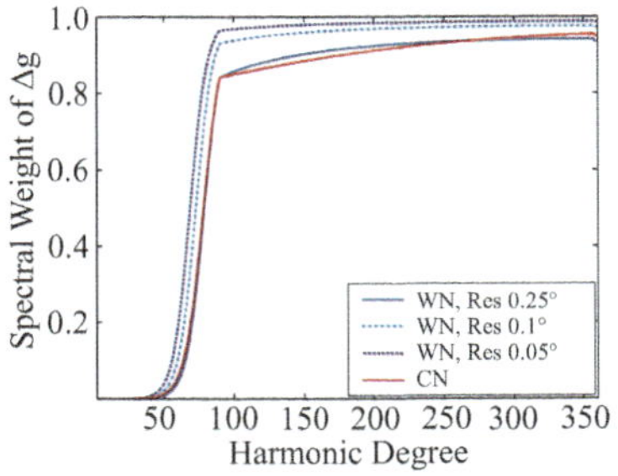

Color Fig. XXXV. Spectral weights of terrestrial gravity anomalies derived for white noise (WN) and coloured noise (CN) scenarios with noise level of 5 mgal. Spectral weights for white noise scenarios depend on assumed data resolution

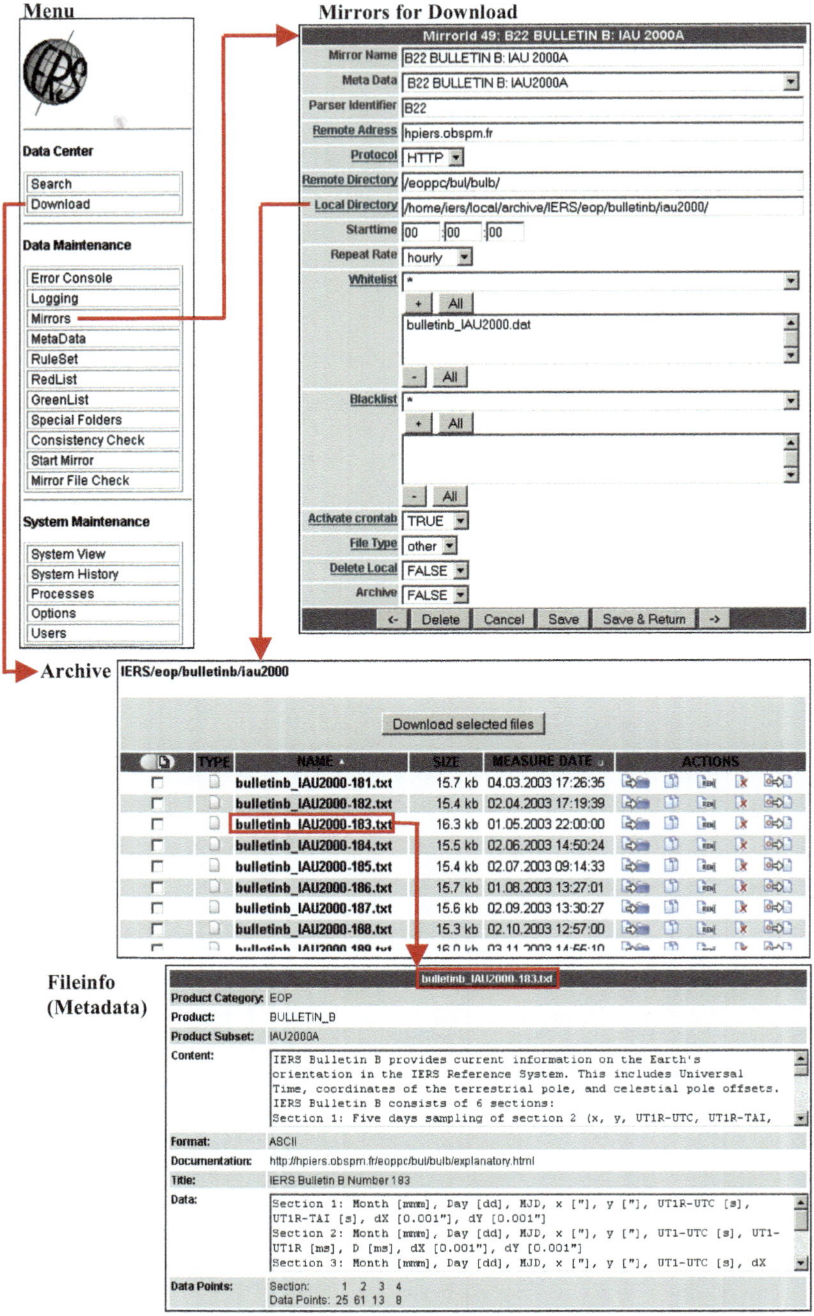

Color Fig. XXXVI. The Data Management System used to automatically collect and archive all IERS products from the various product centres

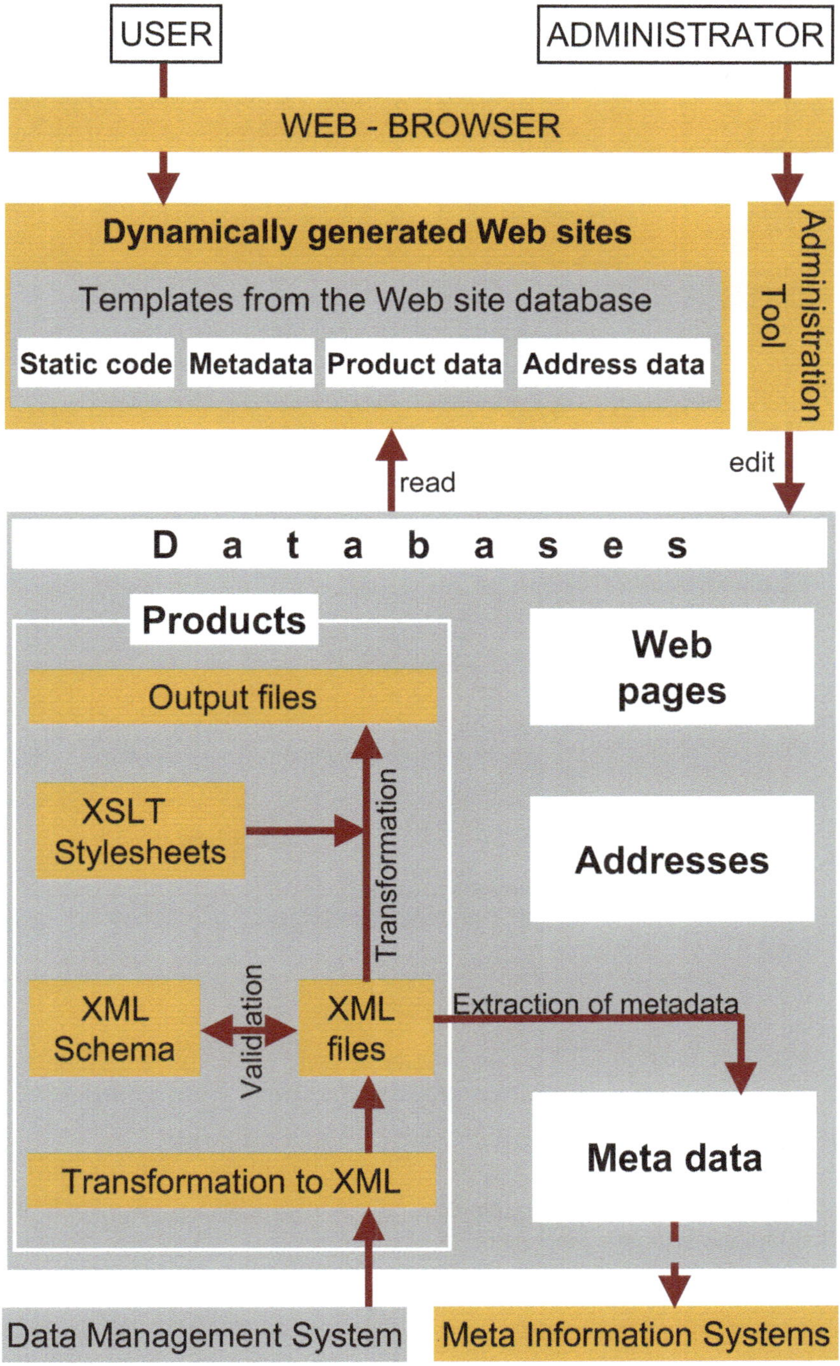

Color Fig. XXXVII. The concept of the IERS Data and Information System

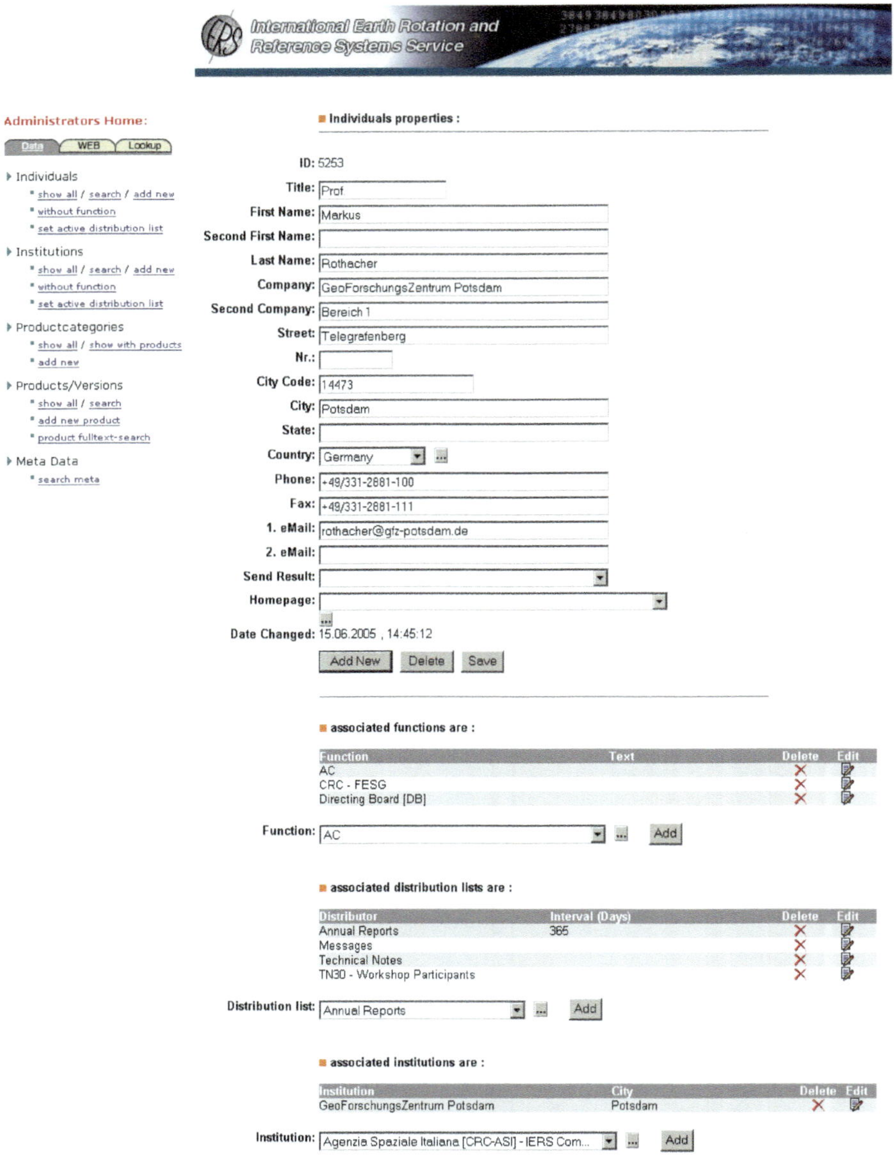

Color Fig. XXXVIII. The Administration Tool to manage the content of the IERS Data and Information System

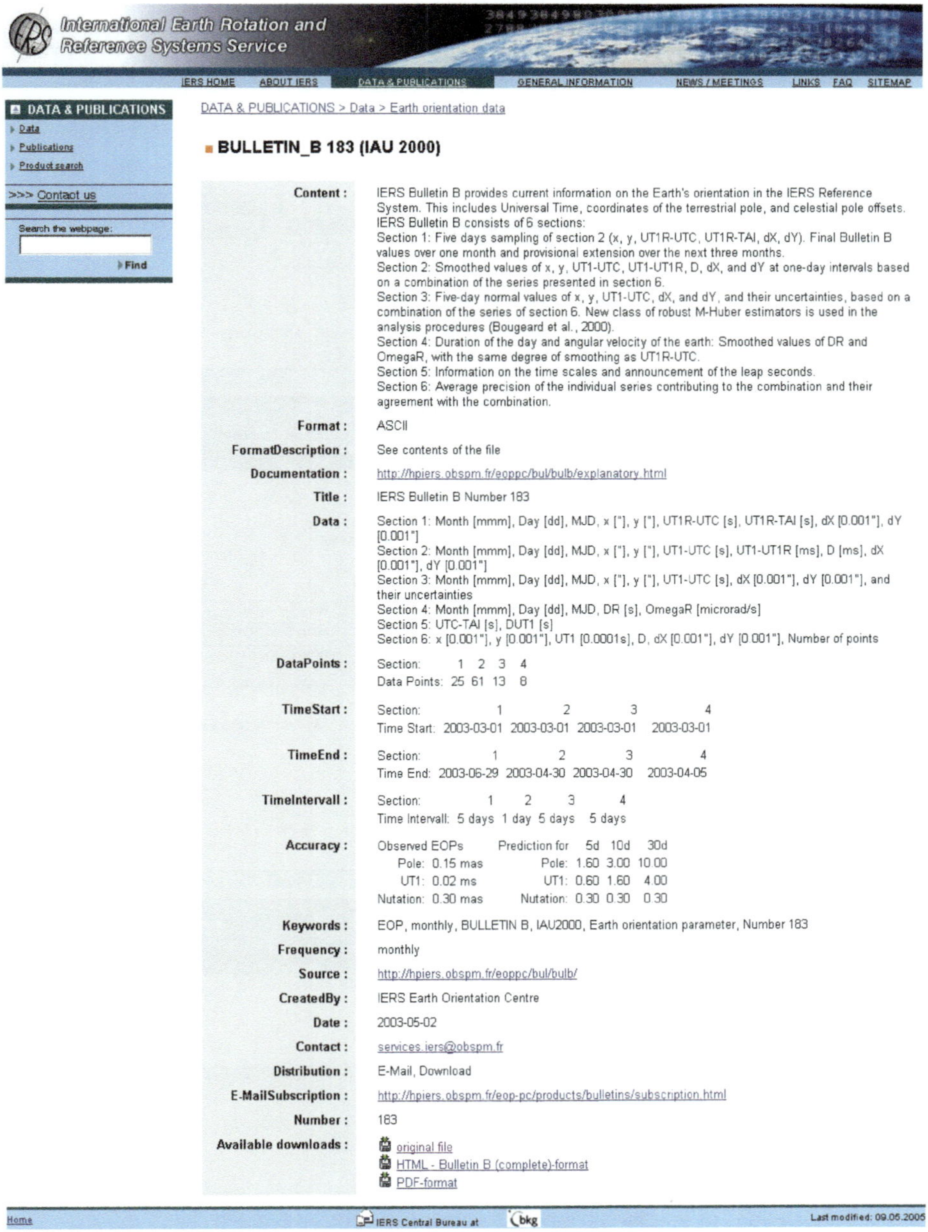

International Earth Rotation and Reference Systems Service

IERS HOME | ABOUT IERS | DATA & PUBLICATIONS | GENERAL INFORMATION | NEWS / MEETINGS | LINKS | FAQ | SITEMAP

DATA & PUBLICATIONS
- Data
- Publications
- Product search

>>> Contact us

Search the webpage:

Find

DATA & PUBLICATIONS > Data > Earth orientation data

BULLETIN_B 183 (IAU 2000)

Content :	IERS Bulletin B provides current information on the Earth's orientation in the IERS Reference System. This includes Universal Time, coordinates of the terrestrial pole, and celestial pole offsets. IERS Bulletin B consists of 6 sections: Section 1: Five days sampling of section 2 (x, y, UT1R-UTC, UT1R-TAI, dX, dY). Final Bulletin B values over one month and provisional extension over the next three months. Section 2: Smoothed values of x, y, UT1-UTC, UT1-UT1R, D, dX, and dY at one-day intervals based on a combination of the series presented in section 6. Section 3: Five-day normal values of x, y, UT1-UTC, dX, and dY, and their uncertainties, based on a combination of the series of section 6. New class of robust M-Huber estimators is used in the analysis procedures (Bougeard et al., 2000). Section 4: Duration of the day and angular velocity of the earth: Smoothed values of DR and OmegaR, with the same degree of smoothing as UT1R-UTC. Section 5: Information on the time scales and announcement of the leap seconds. Section 6: Average precision of the individual series contributing to the combination and their agreement with the combination.
Format :	ASCII
FormatDescription :	See contents of the file
Documentation :	http://hpiers.obspm.fr/eoppc/bul/bulb/explanatory.html
Title :	IERS Bulletin B Number 183
Data :	Section 1: Month [mmm], Day [dd], MJD, x ["], y ["], UT1R-UTC [s], UT1R-TAI [s], dX [0.001"], dY [0.001"] Section 2: Month [mmm], Day [dd], MJD, x ["], y ["], UT1-UTC [s], UT1-UT1R [ms], D [ms], dX [0.001"], dY [0.001"] Section 3: Month [mmm], Day [dd], MJD, x ["], y ["], UT1-UTC [s], dX [0.001"], dY [0.001"], and their uncertainties Section 4: Month [mmm], Day [dd], MJD, DR [s], OmegaR [microrad/s] Section 5: UTC-TAI [s], DUT1 [s] Section 6: x [0.001"], y [0.001"], UT1 [0.0001s], D, dX [0.001"], dY [0.001"], Number of points
DataPoints :	Section: 1 2 3 4 Data Points: 25 61 13 8
TimeStart :	Section: 1 2 3 4 Time Start: 2003-03-01 2003-03-01 2003-03-01 2003-03-01
TimeEnd :	Section: 1 2 3 4 Time End: 2003-06-29 2003-04-30 2003-04-30 2003-04-05
TimeIntervall :	Section: 1 2 3 4 Time Intervall: 5 days 1 day 5 days 5 days
Accuracy :	Observed EOPs / Prediction for 5d 10d 30d Pole: 0.15 mas / Pole: 1.60 3.00 10.00 UT1: 0.02 ms / UT1: 0.60 1.60 4.00 Nutation: 0.30 mas / Nutation: 0.30 0.30 0.30
Keywords :	EOP, monthly, BULLETIN B, IAU2000, Earth orientation parameter, Number 183
Frequency :	monthly
Source :	http://hpiers.obspm.fr/eoppc/bul/bulb/
CreatedBy :	IERS Earth Orientation Centre
Date :	2003-05-02
Contact :	services.iers@obspm.fr
Distribution :	E-Mail, Download
E-MailSubscription :	http://hpiers.obspm.fr/eop-pc/products/bulletins/subscription.html
Number :	183
Available downloads :	original file HTML - Bulletin B (complete)-format PDF-format

Home | IERS Central Bureau at bkg | Last modified: 09.05.2005

Color Fig. XXXIX. Example for the presentation of the IERS products within the IERS Data and Information System (IERS Bulletin B No. 183, IAU 2000)

Color Fig. XL. The search interface to search for specific products

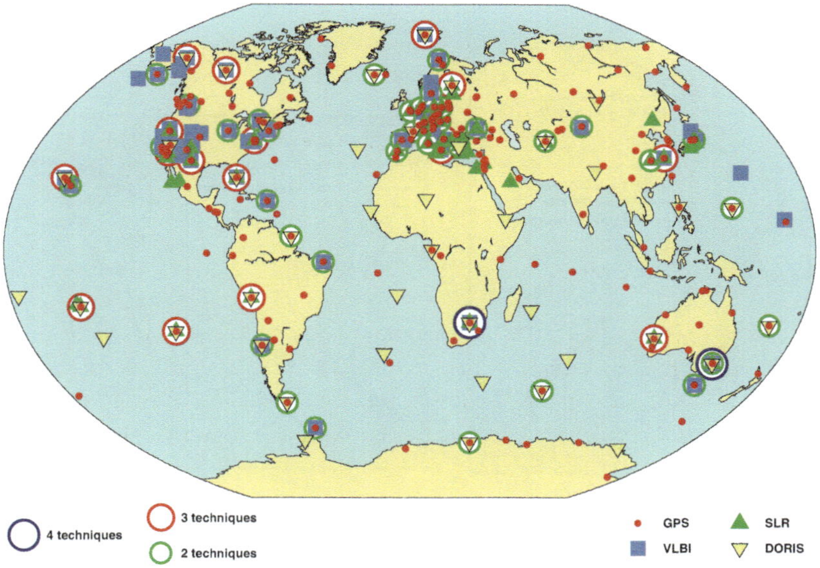

Color Fig. XLI. Different techniques co-location sites for inter-technique combination

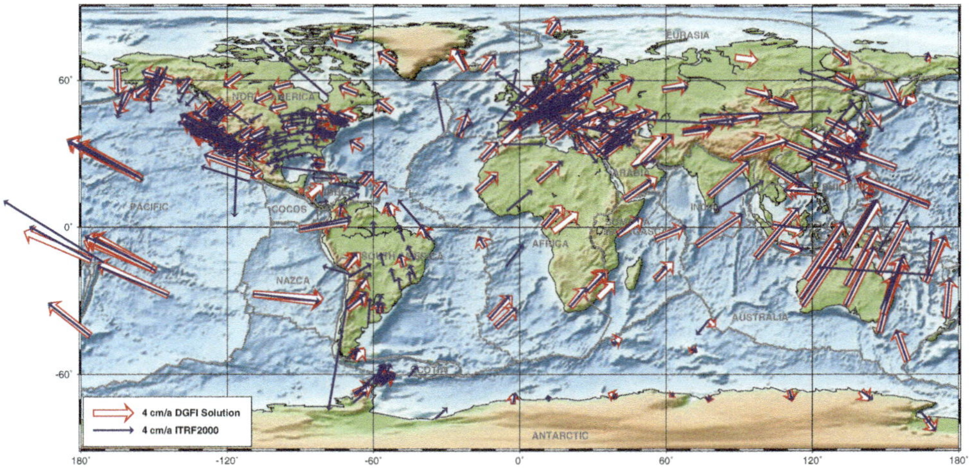

Color Fig. XLII. Comparison of DGFI TRF03 with ITRF2000

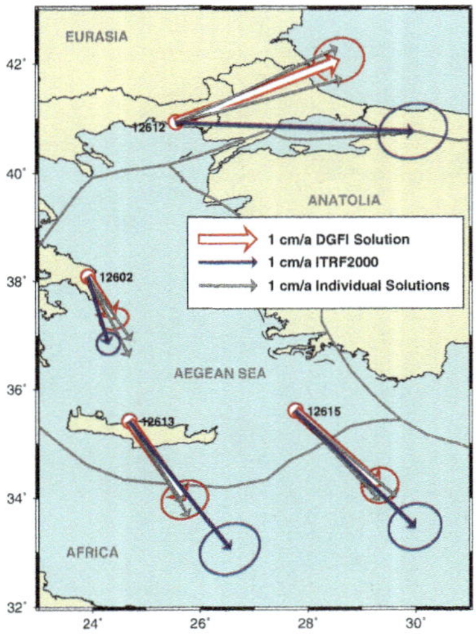

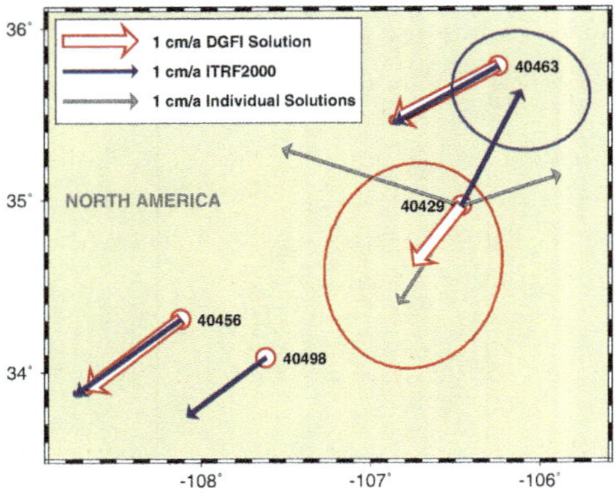

Color Fig. XLIII. Examples of blunders in the velocities of stations with short observation periods in the Aegean (left) and New Mexico (right)

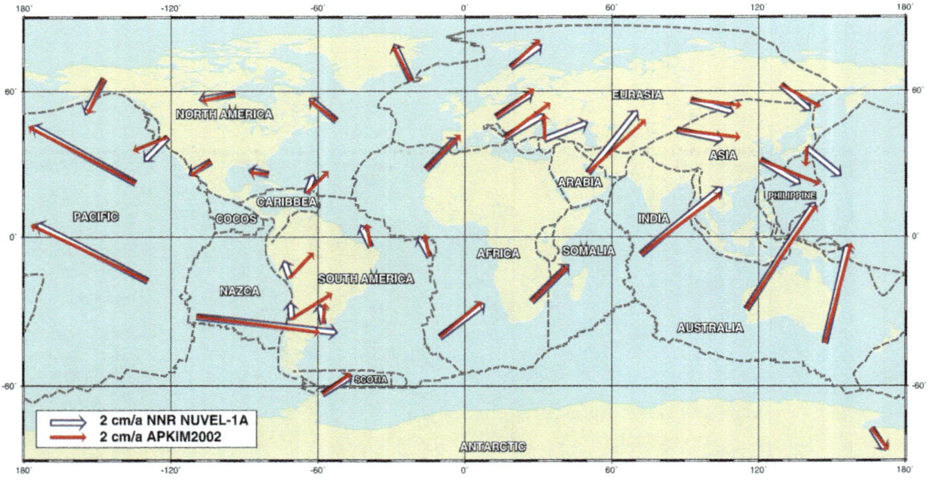

Color Fig. XLIV. Comparison of the geologic-geophysical model NNR NUVEL-1A and the Actual Plate KInematic and deformation Model APKIM2002

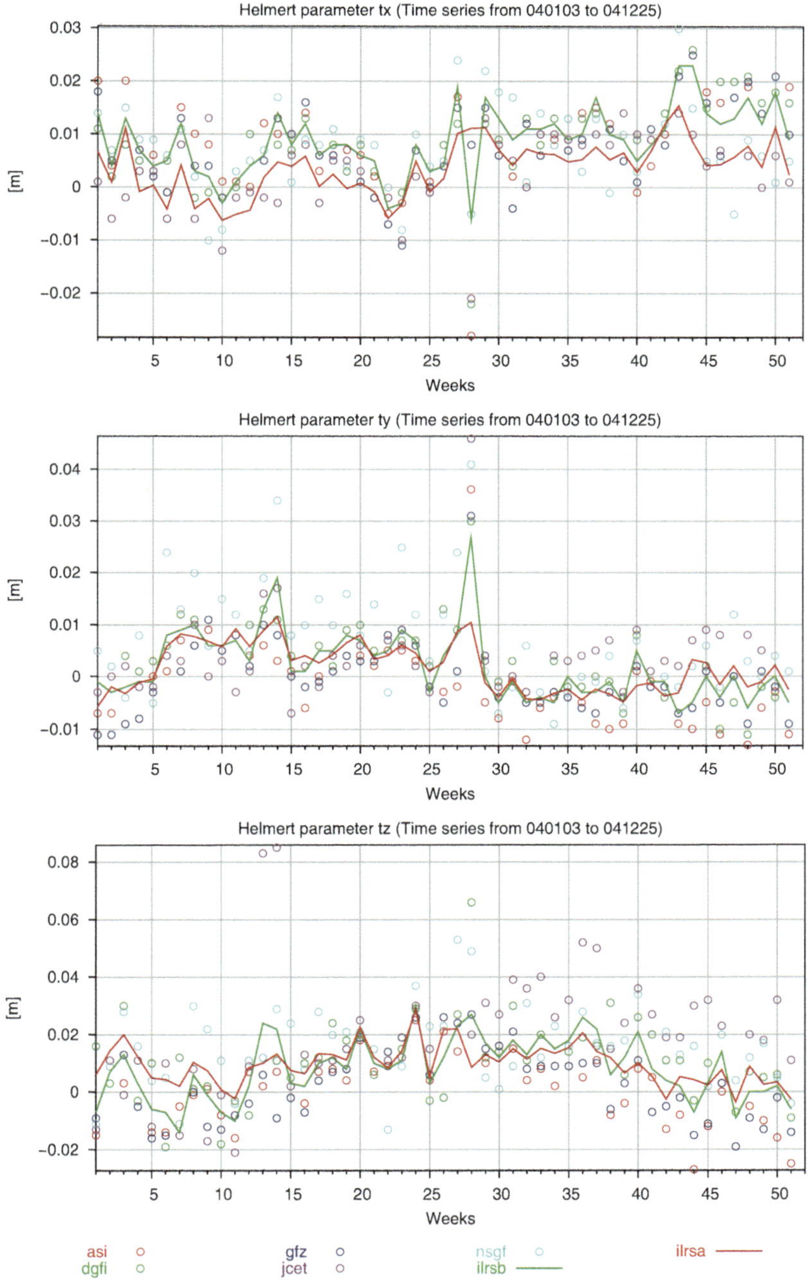

Color Fig. XLV. Time series of weekly translation parameters for the individual SLR solutions and the combined solutions of ASI (ilrsa, red) and DGFI (ilrsb, green) for 2004

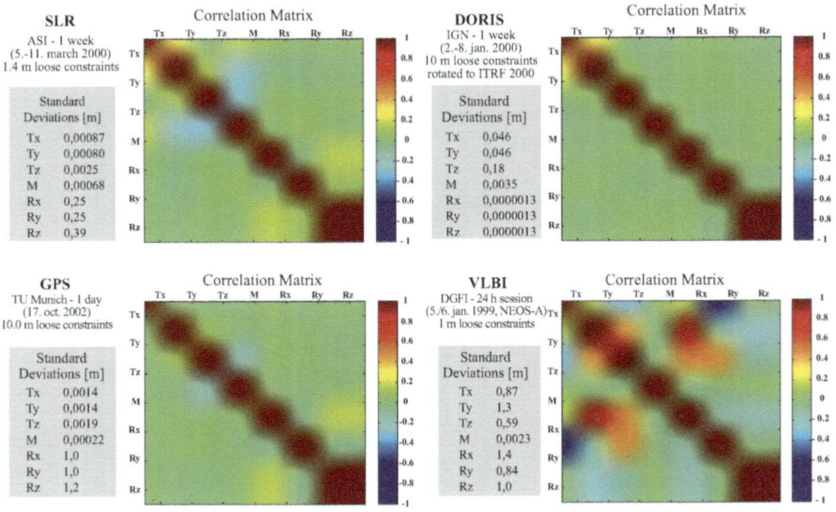

Color Fig. XLVI. Standard deviations and correlation matrix of the datum information of different types of solutions. The standard deviations are in [m] on the Earth surface

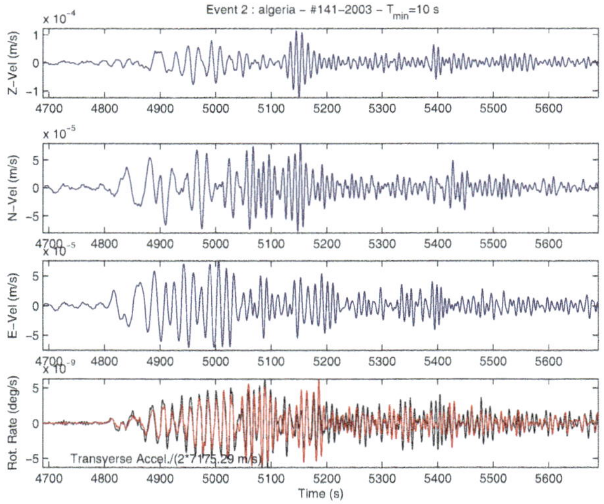

Color Fig. XLVII. Three components of translations (blue) and comparisons of transverse acceleration (black) and rotation rate (red) for one example in the event data base (superimposed traces at the bottom). Note the consistent match for most of the seismogram except towards the end where scattering may lead to phases arriving from out-of-plane directions. Conversion phase velocities are given in the lower left corner

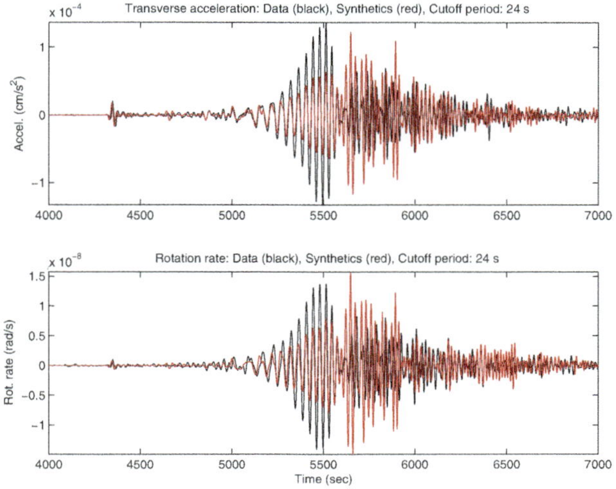

Color Fig. XLVIII. Modelling of translational and rotational ground motions of the M8.1 Tokachi-oki event, September 25, 2003. Top: Observed (black) and modeled (red) transverse acceleration. Bottom: Observed (black) and modeled (red) rotation rate. All traces are shown with absolute amplitude. Note the excellent fit in waveform and amplitude of the direct S-wave (first onset) and the phase match of the fundamental mode Love waves. After 5500 seconds, not unexpectedly, the match is far from excellent, caused by the arrival of multipathing, scattered and/or non-plane waves

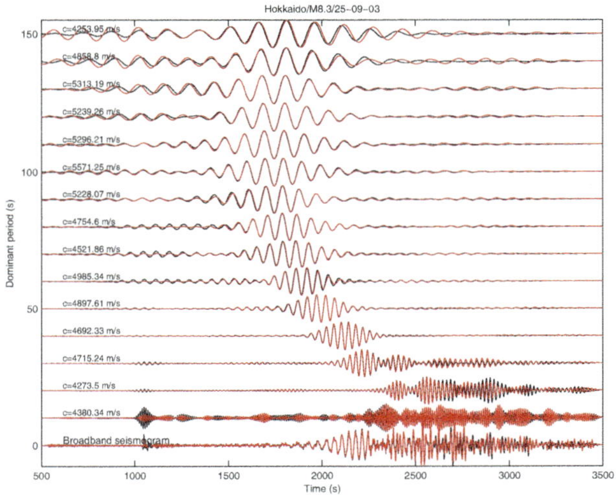

Color Fig. XLIX. Frequency dependent comparison of rotations (red) and accelerations (black). The vertical axis is central period of a very narrow band-pass filter. Note the excellent match at periods down to 150 s. Note also the general increase in best-matching phase velocities as expected from theoretical Love-wave dispersion

Color Fig. L. System components of gravitymeter (UniBwM)

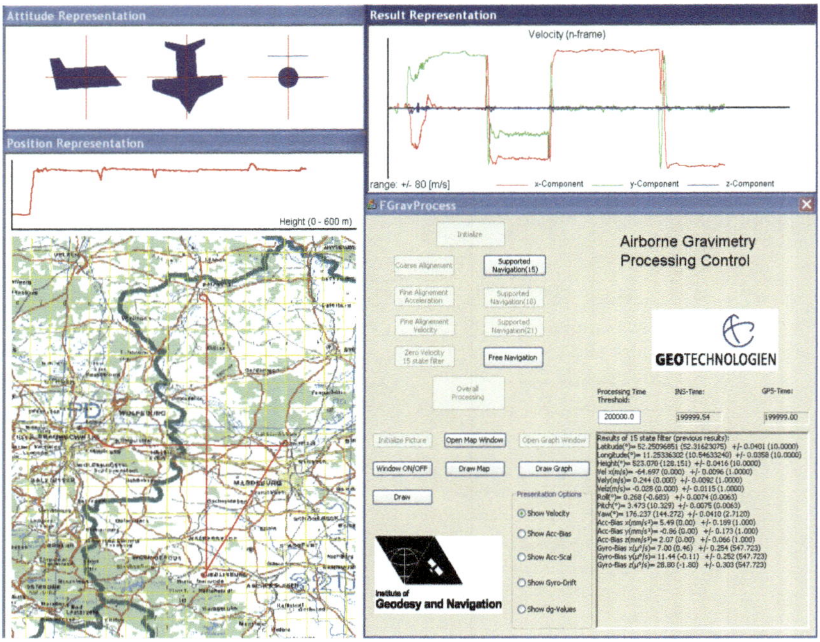

Color Fig. LI. Graphical user interface of processing software

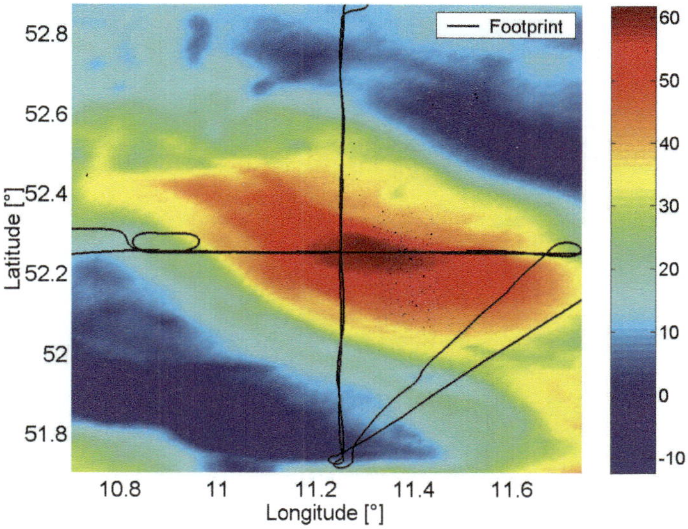

Color Fig. LII. Flight track and reference gravity field [mGal]

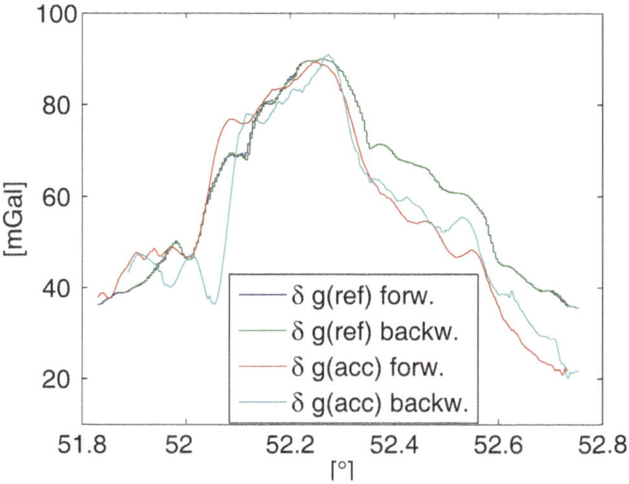

Color Fig. LIII. Calculated gravity profiles (forward/backward direction, reference)

Part IV

IERS
The International Earth Rotation and Reference Systems Service

IERS Data and Information System

Wolfgang Schwegmann and Bernd Richter

Bundesamt für Kartographie und Geodäsie, Richard-Strauss-Allee 11,
D-60598 Frankfurt am Main, Germany, schwegmann@iers.org, richter@iers.org

Summary. In the framework of the *Geotechnologien* project "Integration of space geodetic techniques and development of a user centre for the International Earth rotation and Reference systems Service (IERS)" the IERS Data and Information System[1] has been developed at BKG. The system allows the proper maintenance of all data and products of the IERS and provides user-friendly interfaces to browse the data, to search for specific data and to download the data. The system is database-driven to guarantee the timeliness and consistency of the contents of the information system. Additionally, meta data of all products and publications are modelled in a database to allow the users to search for specific data or topics with respect to space, time and content. In order to be able to link related data the heterogeneous formats of the products are transformed into a common format. The eXtensible Markup Language (XML) will be used, to perform this ambitious task. The usage of XML not only links related data but also supports the exchange of data and the output in different formats like html, pdf, etc. The system is being completed by an administration tool to manage and coordinate the tasks of the Central Bureau and by a general information system with respect to IERS-related topics. The IERS Data and Information System is the basis for the development and implementation of a German contribution to a "Global Geodetic Observing System (GGOS)[2]" realising a central interface to transfer information between the highly complicated system of measurement and analysis procedures and the users. The standardised database tools will promote an easy exchange of information with and links to other databases within the GGOS project to realise a powerful instrument to serve the Earth observing system.

Key words: IERS, Earth rotation, reference systems, GGOS, Information System, Data Centre, XML

[1] http://www.iers.org
[2] http://www.ggos.org

1 Introduction

The primary objectives of the International Earth rotation and Reference systems Service (IERS) are to serve the astronomical, geodetic and geophysical communities by providing terrestrial and celestial reference systems (and their realisations), Earth orientation parameters, global geophysical fluids data as well as standards, constants and models. Starting January 1, 2001, BKG has taken responsibility for the Central Bureau (CB) of the IERS (Dick and Richter, 2004). According to the IERS Terms of Reference (Dick and Richter, 2002, pp. 103–107) the CB is in charge of the general management and the coordination of the IERS. Furthermore one important task of the CB is the development of a data and information system based on modern Internet technologies, providing all data and products as well as general information about the IERS and the topic of Earth rotation to the users of the IERS (Richter et al., 2005).

The main goals of the new IERS Data and Information System are to archive all data and products, to provide them through the World Wide Web, ftp or electronic mail to professional customers as well as to a broad scientific and non-scientific community. Moreover, the users should be able to search for specific data and tools should be offered for the integrated usage of the data sets. Providing these facilities the system will represent a fundamental basis for the development and implementation of a German contribution to a GGOS (Drewes and Reigber, 2005) realising a central interface to transfer information between the highly complicated system of measurement and analysis procedures and the users. The system is being developed in the framework of the *Geotechnologien* project "Integration of space geodetic techniques and development of a user centre for the IERS" in which all kinds of components (Technique Centres, Product Centres, Combination Centres, Analysis Coordinator, Central Bureau) of the IERS are involved. The close connection of all IERS components within the research project facilitates the formulation and testing of the structure, requirements, efficiency and reliability of the new system. The requirements for the new system, the concept and its implementation will be described in more detail in the next sections.

2 Requirements for the IERS Data and Information System

All products within the IERS are generated by an enormous investigation of human and financial resources. They are used by professional customers and should be available also to a broad scientific and non-scientific community. However, the usage of these data is very much limited to professional users because of the format of the data and the way information on as well as access to the data are accommodated.

Presently, the IERS information system provides a unified schematic description of all IERS products (ITRF, ICRF, EOP, Global Geophysical Fluids, etc.) while the respective data are available for download only at the individual product centres. To collect the data the user has to navigate through various Web sites where he is able to download the data in the provided (fixed) format that is not unique for all the data. To link related data to each other the user has to take care about the various formats. Moreover, the integrated usage of the data sets is limited to some specialists at a few research centres. The main reason is the lack of standardised data structures, formats and specifications as well as the out-dated information technology used to provide the data. However, the integrated usage of data from geodetic, geodynamic or other Earth sciences is one of the key tasks of a GGOS.

In addition to the product information the system contains information about IERS publications, IERS's structure and components as well as news and general information on Earth rotation and reference frames. Currently, all this information is presented by static Web pages, so that rather often the same information is stored at different html pages. Thus, it is tedious to keep the information up-to-date and consistent without redundancy. To avoid these disadvantages the contents of the information system will be managed dynamically supported by a database.

Another deficiency in the way the products are presented to the users is the lack of interfaces to search for specific data. Therefore, it is difficult to discover specific data of interest, especially for new or less experienced users. A powerful search interface has to provide all necessary information to retrieve easily specific data with respect to space, time and content allowing extensive and comfortable inquiries with respect to all IERS products and related data as well as to download the data.

The new IERS Data and Information System will provide an easy, uniform and central access to the distributed IERS products, their associated information and the related data via the Internet. In addition, it will allow to extract information directly from the data and to download the data in appropriate formats. Thereto, an easy-to-use and self-explaining Web interface to browse through the products with respect to special user requests of the broad scientific and non-scientific community will be developed. Only server-side technologies will be used, so that the user needs no special programs or browser plug-ins. The user will be able to navigate through the data using simple or complex search functions and will be able to evaluate, visualise, connect and download the data. Therefore, clearly defined standard formats are necessary in order to be able to model the meta data about all products stored in the database. The standardised description of the datasets will also allow their comparison and the exchange of information with and links to other databases within the GGOS project to realise a powerful instrument to serve the Earth observing system. To complete the provided information other related information systems like the Scripps Orbit and Permanent Array Center (SOPAC)[3] or the new NASA geodetic database which will be developed within the "INDIGO" project (Borgen et al., 2002) have to be included at least by a link.

According to the requirements described above four main goals have been derived for the new IERS Data and Information System:

1. Archive all IERS products and all the data necessary to re-compute these products.
2. Support the research activities within the IERS components, e.g. the combination of IERS products.
3. Provide user-friendly interfaces to search the data with respect to time, space and content and to download the data.
4. Facilitate the maintenance of a consistent, non-redundant and up-to-date information system.

3 Concept for the IERS Data and Information System

In order to consider all tasks of the data and information system the concept consists of three major components:

1. The **Data Management System** used to
   a) search all IERS product centres for new data and
   b) automatically download new data,
   c) extract meta data and
   d) import the products into the IERS Data and Information System.
2. The **Information System** includes programs to
   a) transform the imported products into XML,
   b) create various output formats as requested by the user, and
   c) present the meta data as well as information on the Web pages and on address data, all of these stored in databases, to the users via a Web browser.
3. The **Administration Tool** to easily manage all the contents for the information system.

These components will be described in Sects. 3.2, 3.3 and 3.4.

3.1 Key Technologies

To realise the IERS Data and Information System the concept considers three key technologies:

1. The usage of database technologies.
2. The extensive usage of standardised meta data.
3. The application of XML and related technologies.

[3] http://sopac.ucsd.edu

As recommended by a concept paper for the efficient geodata management of the German federal government (Arbeitsgruppe "Konzeption Geodatenmanagement", 2000) all components of the new IERS Data and Information System are based on open source software. The information system will be run on an Apache Web Server, while the open source relational database management system MySQL will be used to store the data. Access to the database for searching, viewing and downloading is best provided via the Internet. To be independent of computer platforms and locations the Web interfaces to access and browse this information are realised using the scripting language PHP.

Databases

To build a modern dynamic system with an efficient, consistent and non-redundant content base that can be kept up-to-date more easily several databases will be built. Meta data about the products are stored in a meta database whereas information on the available products is kept in a product database. In addition all information to manage the Web site, including the Web site structure, the Web pages themselves and the information on the structure of the IERS and its components will be maintained in a Web site database and an address database. The databases will be the core of the new data and information system and will be populated automatically or manually with data (cf. Sect. 3.3).

Standardised Meta Data

Meta data are information about the available data sets that allow the associated data to be independently understandable by the user community. Moreover, the meta data can be read by the search interface of the IERS Data and Information System to provide an extensive and comfortable product search. Therefore, meta data have to be extracted from the archived data and to be stored in the meta database.

The meta data have to be created according to the *ISO 19115 Metadata* standard. The standard facilitates the exchange and integration of data and information by giving a systematised description of the identification, extent, quality, spatial and temporal scheme, spatial reference and distribution specifics of geospatial data. Thus, the meta data can be easily integrated into *Meta Information Systems (MIS)*, as for example *GeoMIS.Bund*[4] developed at BKG. Such MIS allow the directed search for integrated data sets. Moreover, datasets from neighbouring geo-sciences can easily be found and new users of the IERS products may be gained.

---

[4] http://puppis.geomis.bund.de/geoportal/index.jsp

The eXtensible Markup Language (XML)

The most ambitious task is the development of format descriptions for the various heterogeneous IERS products by using only one technique in order to be able to exchange and combine the data, but representing the data as flexible as possible. Therefore, the eXtensible Markup Language (XML) (Harold and Means, 2002) and related technologies are used to represent all IERS products, to provide them in various formats and to facilitate the transfer of the data.

XML is a recommendation of the World Wide Web Consortium (W3C) to describe data and document structures as well as the behaviour of programs used to process such documents (Bray et al., 2004). Originally designed to meet the challenges of large-scale electronic publishing, XML is playing an increasingly important role in the field of Web publishing and in the description, representation and exchange of data on the Web and elsewhere.

The most important benefits of XML are:

- Simplicity: Information coded in XML is plain ASCII text and thus easy to read and understand.
- Openness: XML is a platform independent and non proprietary W3C standard, endorsed by software industry market leaders.
- Extensibility: XML is, by definition, extensible and can be used as a type of grammar to define a markup language for specific purposes and schemas for all kinds of data models. XML documents can be validated against the respective XML Schema (Fallside, 2001) definition.
- Separates content and structure from presentation: XML tags describe meaning not presentation. Thus, it is possible to use associated techniques like the eXtensible Stylesheet Language Transformations (XSLT) (Kay, 2005) to generate several output formats of the same input XML document.
- Can embed existing data: Mapping existing data structures like file systems or relational databases to XML is simple. XML supports multiple data formats and can cover all existing data structures.
- Can embed multiple data types: XML documents can contain any possible data type – from multimedia data to active components.

During the last years XML has been applied in more and more Earth sciences applications for format definitions (cf. Cox et al. (2004), Neumann et al. (2004), GalileoTech News). By using XML as standardised language to describe the format of all products of the IERS the validation and integrated usage of the data will be possible and the data can be provided in individual formats as requested by the user.

3.2 The Data Management System

Some screenshots from the Web interface of the Data Management System which is used to collect automatically all IERS products from the various product centres are shown in Color Fig. XXXVI on p. 306. The system maintains all processes to download and archive the data, to extract meta data, to browse the file archive, to search for specific data and to check the system maintenance. The upper left picture displays the menu of the system to perform all these tasks.

The download of the data from the product centres is controlled by so-called mirrors specifying which files are expected from which source. The most important specifications within such a mirror definition are the remote address of the server storing the data, the remote directory, a whitelist listing rules for filenames to be downloaded, a blacklist for filenames not to be downloaded, a start time when to start the download and a repeat rate specifying whether the check for new data files should be done hourly, daily, weekly or monthly. Another entry in the mirror definition is called *Meta Data* giving the name of a meta data table containing the meta data that will be associated to the data files downloaded by this mirror. These product specific meta data are completed by product version specific meta data which are parsed when importing a new file into the system.

The files are archived in a file system that can be browsed via the Web interface as shown in the middle of Color Fig. XXXVI. For each available file the listing of each directory shows a box to mark the file for download, a column to modify the associated meta data, the filename, size and date of creation, as well as symbols to move, copy, rename or delete the file or to inherit meta data from another file or product. When clicking on the file name the meta data stored for this file will be shown in a new window (cf. lower part of Color Fig. XXXVI).

3.3 Data Flow Within the IERS Data and Information System

The general concept of the IERS Data and Information System is shown in Color Fig. XXXVII on p. 307. Each time the Data Management System downloads a new data file, a processing chain is being started automatically to import the file into the system and to update the database tables of the product database and the meta database from which the Web sites are generated automatically.

At first, the new file is being transformed into XML by JAVA programs and the meta data for the file are being extracted and stored in the meta database automatically. The meta data valid for the associated product (and all its dedicated versions) have been entered manually into the database before. The structure of the meta data is based on a proposal compiled by the group of the IERS Analysis Coordinator to describe all IERS products by the same keywords. This structure can easily be transformed into an ISO 19115 consistent form which will allow the export of the meta data into Meta Information Systems like *GeoMIS.Bund*. This has already been done for a few selected IERS products.

The XML files will be validated with respect to the respective XML Schema definition. Format violations are automatically discovered and reported to the administrator who can get into contact with the originator of the file in order to clarify the format violations. In that case the import of the file will be stopped. Otherwise the XML file will be processed together with the associated XSLT stylesheets to create standard output files in various formats like html, pdf or even plots of the data. Afterwards, the information on the imported product version and the output files are stored in the product database where also the associated XML Schema and XSLT stylesheet files are stored.

The Web site database includes information on the structure and organisation of the IERS Web site, i.e. which pages are available under which topic, which content is generated dynamically, etc. The Web pages are based on XML and each page has an associated XSLT stylesheet to render the XML page into HTML for viewing in a Web browser. In connection with the administration tool (cf. Sect. 3.4) the Web site database realises a Content Management System. The data for the Web site database have been entered manually.

The address database comprehends information on all registered IERS users regularly receiving products from the IERS and will be used to distribute the products. Currently the address database consists of nearly 3000 entries.

The Web pages of the information system are created dynamically accessing all database contents from the product, meta data, Web site and address databases. The users can view the pages, download product files or search for specific data via a standard Web browser (cf. Sect. 3.5). The administrator is able to manage all databases using the administration tool via a Web browser, too (cf. Sect. 3.4).

One of the first applications of the XML technology was the development of an XML Schema definition for the SINEX V2.0 format. The SINEX format primarily a GPS product related format was adopted for the TRF and EOP products also by the other technique services. The SINEX V2.0 is an extended version developed under the leadership of the IERS Analysis Coordinator in cooperation with the space geodetic technique services. Using the new XML Schema definition all SINEX related data can be transformed into XML. As one of the advantages the XML documents are tested automatically against the respective XML Schema definition to validate the input data.

For example SINEX files can be checked by simply using a freely available XML parser instead of a user-specific "SINEX checker". In that way general inconsistencies of technique-specific SINEX files could be discovered as well as individual format violations.

In contrast to the SINEX file format, the so-called geophysical fluids data, another important product of the IERS, are not based on a common file format. The development of an XML Schema definition is much more complicated and will be based on a thorough investigation of the various geophysical fluids data which has not been done yet.

3.4 The Administration Tool

The administration tool is used to control the import of new data files into the system, to manage the Web site structure and the contents of the Web sites, to specify or modify the meta data associated to the products and product versions as well as to maintain the address database. Using this tool the Web pages can be kept consistent, without redundancy and up-to-date with a minimum effort compared to the old information system.

Color Fig. XXXVIII on p. 308 shows an example where the administration tool is used to manage the contents of the address database. As soon as information is changed all Web pages containing this information will automatically display the changes without further work to be done.

In more detail the administration tool allows the following operations:

- Managing the address database:
  - Specify or modify address information with respect to individuals and institutions.
  - Managing functions of individuals and institutions within the IERS and associate it to specific individuals or institutions.
  - Managing distribution lists for the disposition of the IERS products.
- Managing the product database:
  - Organise a hierarchy of product categories, products and product versions and assign XML Schema files for the validation and XSLT stylesheets for the generation of output files.
  - Specify and modify meta data and associate it to product categories, products and product versions.
- Managing the Web site database:
  - Organise a hierarchy of Web topics and associated Web pages.
  - Assign XSLT stylesheets to the XML Web pages in order to render the HTML pages.
- Monitor the import of new products into the database.

3.5 Example Web pages from the IERS Data and Information System

Although the new Web pages cover almost the same information as the old ones they are much easier to maintain and the consistency and actuality can be guaranteed because of the dynamic content. For example the listing of the IERS Directing Board Members can be created dynamically from the Web site and address database. New functionalities have been added to the new Web site compared to the old one. The most important are search capabilities and the availability of all products of the IERS for download.

Color Fig. XXXIX on p. 309 presents an example of an IERS product within the IERS Data and Information System (IERS Bulletin B No. 183, IAU 2000). All meta data belonging to the product are displayed to describe this specific product version. Based on this description the user can decide whether these data are useful for further investigations. The download can be initiated from this page, too.

Apart from a fulltext Web page search a product search based on the meta data stored in the database is available. Color Fig. XL on p. 310 exhibits the search interface to go for specific product versions. The upper part shows the start page for the product search. Here a fulltext search can be initiated, for example to search within documents, as well as a metasearch, e.g. with respect to a specific product. The middle part of Fig. XL shows the search interface for a metasearch for the product *IERS_MESSAGE*. The user can select specific values for some *Meta fields*. The bottom part of the figure shows the results for the selections made for the search.

4 Summary

In the framework of the *Geotechnologien* project "Integration of space geodetic techniques and development of a user centre for the International Earth rotation and Reference systems Service (IERS)" of the BMBF a user and data centre has been realised at BKG. It allows the proper administration of all IERS products and related data.

Using the Data Management System the data and products of the IERS are collected, administrated and archived in their original version. The general concept with respect to the further processing of the data is based on the usage of the eXtensible Markup Language (XML) as central data format for the exchange and the work with the data. From the XML data different output formats can easily be generated and processed by the user by applying the eXtensible Stylesheet Language for Transformations (XSLT). Due to its openness, platform independence and free availability XML has become a world-wide standard with respect to the exchange of data over the Web. Without these advantages the integrated use and combination of heterogeneous data would hardly be possible.

Standardised meta data according to ISO 19115 are produced for all data and products and stored in a database. These meta data provide all necessary information to the users to easily retrieve specific data within the data centre. The search interface of the new IERS Data and Information System directly accesses the meta database and thus allows extensive and comfortable inquiries with respect to all IERS products and related data as well as to download the data from the data centre. When integrating the meta data in *Meta Information Systems (MIS)*, as for example *GeoMIS.Bund* developed at BKG, new users of the IERS products may be gained.

Within the IERS Combination Pilot Project[5] the IERS data centre is used as the central platform to exchange the necessary input and output data sets

[5] http://www.iers.org/iers/about/wg/wg3/cpp.html

of the participating institutions and to guarantee the information flow as well as the integration, documentation and availability of the data.

The new system will support the Central Bureau in its routine work (e.g. distributing information and data electronically). It will provide a modern Web service to answer requests on information and data with respect to the fields of Earth rotation and reference systems to support scientific work as well as to bring these topics into a broad public. Within the *Geotechnologien* project the IERS Data and Information System serves as central platform for the exchange of data and information.

Acknowledgement. This is publication no. GEOTECH-167 of the programme GEOTECHNOLOGIEN of BMBF and DFG, Grant 03F0336D.

References

Arbeitsgruppe "Konzeption Geodatenmanagement" (2000) Konzeption eines effizienten Geodatenmanagements des Bundes (19. September 2000). Geschäftsstelle des Interministeriellen Ausschusses für Geoinformationswesen (IMAGI), Frankfurt am Main, http://www.imagi.de.

Borgen R, Moore A, Neilan R, Noll C, Pearlman M, Stowers D, Vandenberg N, Webb F, Yunck T (2002) Inter-Service Data Integration for Geodetic Operations - Indigo. Proposal submitted to NASA Cooperative Agreement (CAN) CAN-02-OES-01.

Bray T, Paoli J, Sperberg-McQueen CM, Maler E, Yergeau F (eds) (2002) Extensible Markup Language (XML) 1.1, W3C Recommendation 04 February 2004, edited in place 15 April 2004. http://www.w3.org/TR/2004/REC-xml11-20040204/.

Cox S, Daisey P, Lake R, Portele C, Whiteside A (eds) (2004) OpenGIS Geography Markup Language (GML) Implementation Specification Version 3.1.0. Date: 2004-02-07. Category: OpenGIS Recommendation Paper. Copyright (c) 2004 Open GIS Consortium, Inc. and ISO. Reference: OGC 03-105r1. 601 p.

Dick WR, Richter B (eds) (2002) IERS Annual Report 2001. Verlag des Bundesamtes für Kartographie und Geodäsie, Frankfurt am Main, 123 p.

Dick WR, Richter B (2004) The International Earth rotation and Reference systems Service (IERS). In: André Heck (ed) Organizations and Strategies in Astronomy, Vol. 5. (Astrophysics and Space Science Library; vol. 310) Kluwer Academic Publishers, Dordrecht Boston London, p. 159-168.

Drewes H, Reigber C (2005) The Global Geodetic Observing System (GGOS) of the International Association of Geodesy – Objectives and Status. In proceedings: From Pharaos to Geoinformatics, FIG Working Week 2005 and GSDI-8, Cairo, Egypt April 16–21, 2005.

Fallside DC (2001) XML Schema Part 0: Primer, W3C Recommendation. http://www.w3.org/TR/xmlschema-0/.

GalileoTech News: Ground Segment Data Model & Data Standard (GXML). ESA-ESTEC: Galileo project office.
http://esamultimedia.esa.int/docs/galileo/ESA_Galileo_GXML.pdf.

Harold ER, Means WS (2002) XML in a Nutshell. O'Reilly, Beijing.

Kay M (ed) (2005) XSL Transformations (XSLT) Version 2.0 W3C Working Draft 04 April 2005, http://www.w3.org/TR/xslt20/.

Neumann K, Mathiak B, Kuper A (2004) Der Einsatz von GML, XSLT und SVG am Beispiel von ATKIS-DLM-Daten. In: Arbeitsgruppe Automation in der Kartographie – Tagung 2003, Mitteilungen des Bundesamtes für Kartographie und Geodäsie, Band 31, Frankfurt am Main, 230 p.

Richter B, Schwegmann W, Dick WR (2005) Development of an Information and Database System for the IERS: status and outlook. J Geodyn, in print.

IERS Analysis Coordination

Markus Rothacher, Robert Dill, and Daniela Thaller

Forschungseinrichtung Satellitengeodäsie, Technische Universität München, Germany, rothacher@bv.tum.de

Summary. Until today the products of the International Earth Rotation and Reference Systems Service (IERS), like International Terrestrial Reference Frame (ITRF), International Celestial Reference Frame (ICRF) and Earth Orientation Parameters (EOP), are combined independently, neither intra-technique nor inter-technique combinations, including the full variance-covariance information are performed. To overcome this deficiencies in the present IERS product generation the IERS implemented a new structure in January 2001. This includes the new IERS Combination Research Centres (CRC) and the IERS Analysis Coordinator (AC). He is responsible for the long-term and internal consistency of the IERS products. To achieve the highest accuracy and consistency, it is crucial to proceed towards a fully rigorous combination of all the parameters common to more than one space geodetic technique. Since 2001 the IERS AC initiated and coordinated many different projects and campaigns towards this overall goal. Now the results of the last three years build the theoretical and practical base for the latest project, the Combination Pilot Project (CPP). This project will prepare the generation of a combined IERS product on a routine basis.

Key words: International Earth Rotation and Reference Systems Service, IERS, Analysis Coordination, EOP Alignment Campaign, SINEX Combination Campaign, Combination Pilot Project

1 Introduction

The Analysis Coordinator (AC) of the International Earth Rotation and Reference Systems Service (IERS) is responsible for the long-term and internal consistency of the IERS reference frames and other products. He is responsible for ensuring the appropriate combination of the IERS Technique Centres (TC) products into the single set of official IERS products and the archiving of the products at the Central Bureau (CB) or elsewhere. The AC serves for a four-year term, renewable once by the IERS Directing Board (DB). The responsibility of the AC is to monitor the TC's (International VLBI Service

VLBI, International GNSS Service IGS, International Laser Ranging Service ILRS, International DORIS Service IDS) and Product Centre's (PC) activities to ensure that the IERS objectives are carried out. This is accomplished through direct contact with the independent TC Analysis Coordinators or equivalent. Specific expectations include quality control, performance evaluation, and continued development of appropriate analysis methods and standards. The AC interacts fully with the Central Bureau, the Product Centers and the Combination Research Centers (CRC). Although this new structure of the IERS was already implemented on January 1, 2001, the work of the current IERS Analysis Coordinator officially started only at the IERS Directing Board Meeting in September 2001 in Brussels. This report will cover the activities since September 2001.

2 Goals and Time Schedule

The major goals of the IERS Analysis Coordination, as presented in Brussels in September 2001, may be grouped into five major tasks as follows (see also Fig. 1):

1. Coordination of CRC activities: an ongoing process.
2. Status / list of present IERS products and their methods of generation.
3. Plan for the optimization of the consistency and accuracy of the products. A plan has to be developed with the different IERS components on how to achieve this goal.
4. Implementation of new combination strategies. A procedure in two steps is proposed:
   - "Weekly" solutions
   - "Multiyear" solutions
5. Routine quality control of all IERS products.

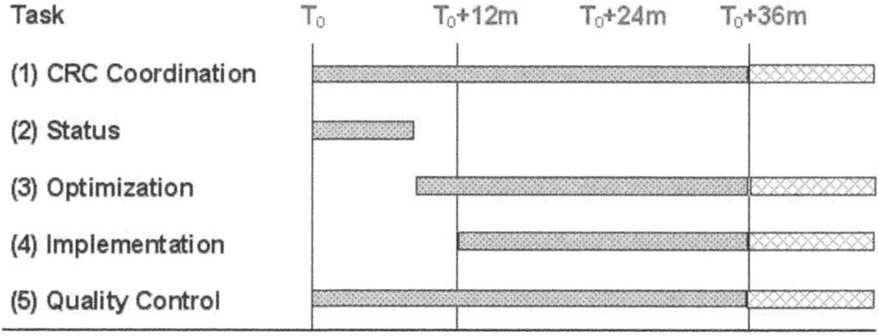

Fig. 1. Time schedule of the IERS Analysis Coordination activities. T0 = September 2001

Today the various products of the IERS, especially the International Terrestrial Reference Frame (ITRF), the International Celestial Reference Frame (ICRF) and the Earth Orientation Parameter (EOP) series, are still combined independently. With the exception of the ITRF and the IGS products neither intra-technique nor inter-technique combinations are correctly performed including the full variance-covariance information. In this way neither the consistency of the products can be guaranteed nor can the different strengths of the individual space geodetic techniques be exploited to improve the products. This means that there are clear deficiencies in the present IERS product generation. To achieve the highest accuracy and consistency, it is crucial to proceed towards a fully rigorous combination of all the parameters common to more than one space geodetic technique, especially in respect to the challenges the IERS is facing with the new satellite missions (gravity, altimetry, astrometry), with the Global Geodetic Observing System (GGOS) project of the International Association of Geodesy (IAG), and with many other applications.

In order to make progress towards the goals mentioned above, the following primary activities were initiated:

- Web page for coordination, communication and scientific transfer between CRCs.
- Compilation of a complete list of IERS products. The present status of the list has fully been implemented in the IERS web pages (see `http://www.iers.org/iers/products/`).
- IERS Analysis Campaign to align EOPs to ITRF2000/ICRF.
- IERS SINEX format unification
- IERS SINEX Combination Campaign
- IERS Combination Pilot Project
- IERS Long Time Series
- IERS Working Groups

3 Activities and Projects of the IERS Analysis Coordination

This section will give only a short overview of the past and ongoing activities of the IERS AC. More detailed information on the results of each project will be presented in the following chapters.

3.1 Web Pages

According to the proposed procedure, presented at the IERS Directing Board Meeting No. 34 in September 2001 in Brussels, a new web site was planned and installed in November 2001. This site provides a communication and data exchange platform to the combination research community. It should also stimulate and support their ongoing activities. The web site http://tau.fesg.tu-muenchen.de/~iers/ presents the current activities and research intentions of the Combination Research Centres (CRC), provides a communication platform with mail forum and e-mail exploder and builds the main coordination of the IERS SINEX Campaign and the IERS Analysis Campaign to align EOPs to ITRF2000/ICRF (EOP Alignment Campaign).

3.2 List of IERS Products

A complete list of the IERS products was compiled by Daniela Thaller at the end of 2001 and finalized in the first few months of 2002. In mid 2002 the detailed information for all products was added in a uniform way to the IERS Web pages at the IERS Central Bureau by Wolfgang Schwegmann (see http://www.iers.org/iers/products/). This list serves as the "status quo", i.e., as the starting point for improving the IERS product consistency and accuracy. In view of the fact that only a rigorous combination of station coordinates, EOP and quasar coordinates can guarantee a consistent IERS product palette, the emphasis of the IERS AC work was put on stimulating the combination efforts and the development of appropriate software algorithms. In the long run, this strategy will be more efficient than trying to improve individual products (e.g., improving EOP series independent of the reference frame).

3.3 EOP Alignment Campaign

In September 2001 the IERS Analysis Coordinator presented an IERS Analysis Coordination Campaign to align EOPs to ITRF2000/ICRF (EOP Alignment Campaign) as originally proposed by Jim Ray. The intention of the EOP Alignment Campaign was to create EOP series with the highest possible consistency with ICRF and ITRF2000. The aim of this project was to analyze and understand the origin of systematic errors belonging to the reference frames. The EOP Alignment Campaign was started at the end of September 2001 with an initial call for participation. The campaign was subdivided into two parts. In a first step, the Technique Centres were asked to produce EOP series with a reference frame fixed to the ITRF2000 / ICRF at the level of uncertainty. In addition, they were asked to produce solutions with different constraints on ITRF2000 (ICRF). The second step consisted of the analysis of the submitted EOP series and comparisons with the official solutions (C04, Bulletin A) and by studying the consistency between the various series. Until May 2002, 21 proposals were received. Twelve of them contributed to the first step and produced more than 40 different EOP series from all four techniques (VLBI, SLR, GPS, DORIS) with various constraints (fixed site coordinates, significant or minimum constraints on the site coordinates) to realize the ITRF2000 reference frame. A first overview of the EOP series was given by Robert Dill at the EGS General Assembly 2002 in Nice. As a second step of the campaign, twelve groups analyzed the submitted EOP series. At the IERS Combination Workshop in Munich in November 2002 it was decided to continue the EOP Alignment Campaign until the time of the EGS General Assembly, April 2003, where the final results were presented by a few different groups (see, e.g., IERS Annual Report 2003 (Dick and Richter, 2004), section 3.6.2.4). It was clearly demonstrated that considerable biases do exist between official IERS products (e.g., between ITRF2000 and the C04 polar motion series). There was a relevant inconsistency of about 0.2 mas in the y-pole coordinate between the official IERS C04 series and the newly computed series correctly aligned to ITRF2000. It was concluded that follow-up studies of these biases should be based on the SINEX files (Solution (Software/technique) INdependent EXchange format) of the IERS SINEX Combination Campaign instead of the EOP time series alone, to allow for more detailed analysis and for rigorous combination approaches including EOP and station coordinates.

3.4 SINEX Format

The web pages and the communication platform was used to develop a new SINEX format (Version 2.00). The objective was to create a single, shareable and uniform SINEX format. This was a very basic requirement for the combination of SINEX files from all different technique centres. The most common SINEX format descriptions were analyzed and a new uniform consistent version was proposed. The latest version of the SINEX format is available at `http://tau.fesg.tu-muenchen.de/~iers/web/sinex/format.php` (see as well IERS Message No. 26).

3.5 IERS SINEX Combination Campaign and SINEX Data Pool

It was the intention of this campaign to combine "weekly" solutions from SINEX files of different techniques with station coordinates and EOP (and ICRF) and to assess systematic biases between the individual space geodetic techniques. The web pages were prepared and the campaign was initiated at the beginning of the year 2002. The goals, the procedure and the participants can be found at `http://tau.fesg.tu-muenchen.de/~iers/web/sinex/campaign.php`. Eleven groups sent back a proposal. The campaign was divided into two parts:

- as a first step, solution series of the space geodetic techniques were produced by several analysis centres for the whole year of 1999, including at least station coordinates and EOP as parameter types in the SINEX files;
- as a second step, these solution series in SINEX format were then combined by several groups with the goal to develop appropriate combination software and to assess systematic biases between the individual space geodetic techniques.

The first step was finished by the end of the year 2002 and all solution series were added to the SINEX data pool, which was prepared already in 2001. First results of the combination step were presented at the IERS Workshop in November 2002 in Munich looking very promising. Further results were presented by several groups at the EGS General Assembly 2003 in Nice in April 2003. At the IERS Retreat in Paris (see below) it was decided that the IERS SINEX Combination Campaign should evolve into a pilot phase of routine product generation, i.e., into the IERS Combination Pilot Project (CPP).

Parallel to the SINEX Combination Campaign a SINEX data pool was created. Various sets of SINEX files, especially the contributions to the SINEX Combination Campaign, were collected and archived to provide the CRC community a user friendly data base for their research. An up-to-date list and links to a selected subset of SINEX files suitable for combination research and software testing is available in the SINEX file archive, `http://tau.fesg.tu-muenchen.de/~iers/web/sinex/datapool.php`.

3.6 IERS Retreat 2003

Together with the IERS Central Bureau considerable effort was put into the organization of the IERS Retreat 2003 in Paris. At this retreat, among others, the following important decisions were taken:

- the IERS Combination Pilot Project (CPP) should be initiated,
- an IERS Working Group on "Site Survey and Co-location",
- an IERS Working Group on "Combination" to coordinate the IERS CPP and
- an IERS Working Group on "ITRF Datum" should be established.

For more information and details concerning the outcome of the IERS Retreat 2003 we refer to `http://www.iers.org/iers/meetings/IERSRetreat-2003/`

3.7 IERS Combination Pilot Project (CPP) and IERS Working Group on Combination

At the IERS Retreat in Paris in April 2003 (see previous section above) it had been decided, that

- an IERS Working Group (WG) on Combination was set up and that
- the IERS SINEX Combination Campaign was converted into a pilot project, namely the IERS Combination Pilot Project(CPP), to be started in spring 2004, and to prepare the IERS product generation on a routine basis.

To establish the IERS WG on Combination (IERS WG3) according to the IERS Terms of Reference a charter and a preliminary list of participants were drafted and discussed during the December 2003 IERS Directing Board meeting. The working group was set up in the beginning of 2004 and, as a first action, the Call for Participation (CfP) for the IERS Combination Pilot Project was launched. The IERS CPP consists of three steps:

1. The computation of one single combined SINEX file per week by each of the four Technique Centres (IGS, IVS, ILRS, IDS) starting from SINEX files of the individual analysis centres of each technique service (intra-technique combination).
2. The computation of combined weekly SINEX files by so-called IERS Combination Centres based on the "weekly" intra-technique SINEX files and local tie information.
3. The validation of the combined inter-technique solutions through comparisons, repeatability studies, and use of external information (geophysical fluids, models, ...).

The sequence of the different steps for a "weekly" routine product generation is visualized in Fig. 2.

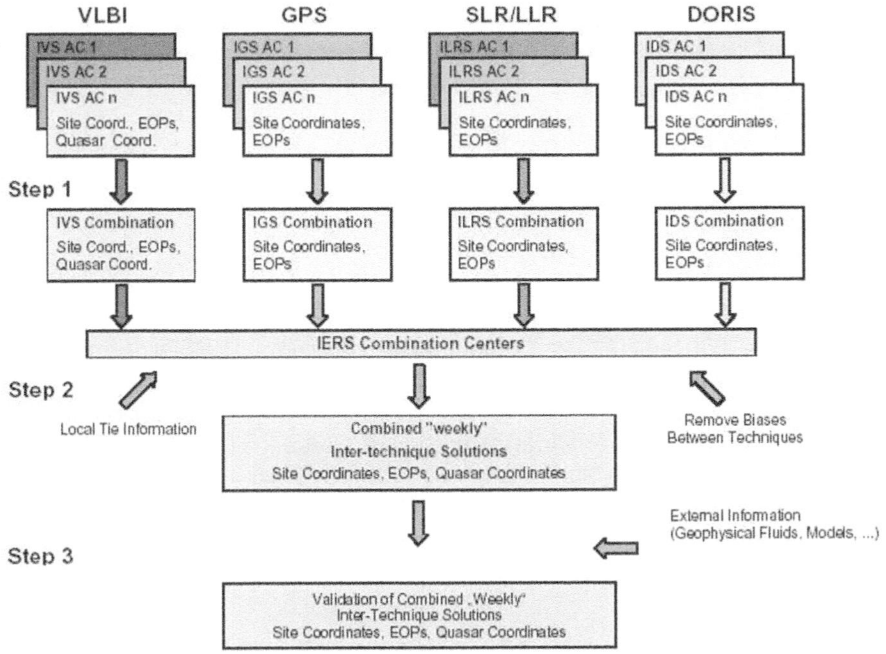

Fig. 2. Combination and validation of "weekly" SINEX solutions in the framework of the IERS Combination Pilot Project (CPP)

More information about the CPP and the present status may be found at `http://www.iers.org/iers/about/wg/wg3/cpp.html`.

3.8 IERS Long Time Series

During the IERS CPP it was recognized that the availability of long time series is essential for a weekly rigorous combination of space geodetic solutions. The weekly combined solutions cannot be generated consistently over longer time spans as long as there are no time series available as a reference for datum definition. Especially the derivation of weighting schemes and the application of local ties is only possible in a reasonable way if a longer series of weekly solutions or a long-term solution is available. Due to the call for IERS Long Time Series in December 2004, combined intra-technique solutions from all techniques are now available for long time spans. It must be emphasized that this is the first time that the technique services IGS, IVS and ILRS place a combined solution of the respective technique including station coordinates and EOP at the disposal of all users. Additionally, all the solutions were reprocessed, except for IGS, hence they form a homogeneous set of time series. The available solutions are archived by the IERS Central Bureau and are integral part of the IERS Central Bureau information and database system: `http://iers1.bkg.bund.de/info/listFileITRF2004.php`

Furthermore, the long time series, together with the routine CPP solutions, are the basis for the future generation of a consistent set of IERS products. The next main step along this line is the generation of the ITRF2004 (station coordinates, velocities and EOP) by the ITRF Combination Centres, Institut Géographique National (IGN), Deutsches Geodätisches Forschungsinstitut (DGFI) and Natural Resources, Canada (NRCan).

3.9 IERS Working Groups

At the IERS Retreat 2003, as mentioned above, it was also decided that, besides the IERS WG3 on Combination, two more IERS working groups should be established, namely the IERS WG1 on Datum Definition of Global Terrestrial Reference Frames, jointly with IAG Sub Commission 1.2 WG 1 (SC1.2-WG1) and IAG Inter-Commission Committee on Theory (ICCT), and the IERS WG2 on Site Survey and Co-locations, jointly with IAG Sub-Commission 1.2 WG 2 (SC1.2-WG2). In the meantime, two of the three working groups (IERS WG2, IERS WG3) have been fully established (see `http://www.iers.org/iers/about/wg/`).

3.10 IERS Workshop on "Site Co-location" in Matera

For the first time in IERS history a workshop was organized solely concentrating on the topic of site surveying, site co-location and all issues related to the co-location of instruments at fundamental geodetic observatories. It was one of the major goals of this workshop to demonstrate that the topic of site surveying etc. should play a much more prominent role in space geodesy and especially in the combination of the space geodetic results into a consistent set of IERS products (see IERS CPP). The two most important recommendations resulting from the workshop are listed here:

- All local ties between co-located instruments should be determined with an accuracy of 1 mm or better in the ITRF (global, cartesian) and the full variance / covariance information should be made available in SINEX format (Recommendation 1).
- Local survey measurements should have the same importance as and should be treated like any of the space geodetic techniques (Recommendation 2).

The establishment of highly accurate local ties for all fundamental geodetic observatories should be considered a "conditio sine qua non" for a successful combination of the space geodetic techniques and a consistent set of IERS products (i.e., consistency between techniques as well as consistency between parameter groups like site coordinates, EOP and quasar coordinates). A major effort should therefore be devoted to this task. For additional information and the presentations given at the workshop see `http://www.iers.org/workshop_2003_matera/`.

4 Outlook

In the future the IERS will most certainly generate at least the following four principle types of products (besides such products as the IERS Conventions etc.) that have different characteristics concerning latency, generation cycle, accuracy and, possibly, computation strategy (see Table 1):

1. Multi-year solutions for consistent ITRF, EOP and ICRF realizations: IERS200x=ITRF200x+EOP200x+ICRF200x, based on long-time series from reprocessing efforts and recombination by all techniques, using the most up-to-date standards for modelling and parameterization.
2. Weekly final solutions to generate combined EOP series on a routine basis: generation from the weekly intra-technique combined solutions with a strategy consistent with the most recent IERS200x realization, monitoring of station coordinates (Earthquakes, station problems, etc.).
3. Daily rapid solutions for EOP: based on a combination of VLBI Intensive sessions (e-VLBI) and GPS rapid solutions (and SLR/DORIS rapid solutions).
4. Extrapolated EOP solutions for real-time users: based on the rapid daily products.

Acknowledgement. This is publication no. GEOTECH-162 of the programme GEOTECHNOLOGIEN of BMBF and DFG, Grant 03F0336A.

Table 1. Four different types of IERS products

Product Type	Main Parameters	Generation Cycle	Maximum Latency	Based on
Multi-Year	ITRF EOP ICRF	Yearly	Maximum 1 year	Reprocessed high-accuracy combined long term intra-technique series
Weekly "Final"	EOP Station Coordinates	Weekly	2 weeks	Based on the "final" routine intra-technique combined products
Daily "Rapid"	EOP Station Coordinates Troposphere Parameters	Daily	1 day	VLBI Intensives and IGS Rapid Products and ...
Daily "Predicted"	EOP	Daily	1 day	Combined daily rapid IERS products

References

Dick W., B. Richter (2004) IERS Annual Report 2003. Verlag des Bundesamts für Kartographie und Geodäsie, Frankfurt am Main.

Analysis and Refined Computations of the International Terrestrial Reference Frame

Hermann Drewes, Detlef Angermann, Michael Gerstl, Manuela Krügel, Barbara Meisel, and Wolfgang Seemüller

Deutsches Geodätisches Forschungsinstitut, München, drewes@dgfi.badw.de

Summary. The International Terrestrial Reference Frame (ITRF) is realized by epoch positions and linear velocities of a set of geodetic points on the Earth's surface. Up to the present, i.e. the ITRF2000, the computation is done by a 14 parameter similarity transformation (7 for the stationary and 7 for the kinematic coordinates) of individual solutions from the different space geodetic observations (VLBI, SLR, GPS, DORIS) and the simultaneous adjustment of the position and velocity coordinates. The analysis of the ITRF2000 shows some problems resulting from this transformation procedure. The refined TRF computations done by DGFI use unconstrained normal equations from the solutions of the individual techniques. After a thorough analysis and editing they are combined in a first step internally by accumulation per technique. In the second step the normal equations of the unique techniques are accumulated for an inter-technique combination. The datum of the final TRF solution is attained by no net rotation w.r.t. ITRF2000. Detailed comparisons show a generally good agreement between the DGFI TRF and the ITRF2000. Some outliers are discussed.

Key words: Precise positioning, Terrestrial Reference Frame computation, VLBI, SLR, GPS, DORIS

1 Introduction

The computations of the International Terrestrial Reference Frame (ITRF) go back to the first solution ITRF88 (Boucher and Altamimi, 1990) where position coordinates of 120 stations are derived from the combination of five solutions from Very Long Baseline Interferometry (VLBI) and six solutions from Satellite Laser Ranging (SLR), all provided by individual analysis centres. A total of ten ITRF realizations have been computed since then until the latest one, the ITRF2000 (Altamimi et al., 2002; Boucher et al., 2004) with positions and velocities of 764 points at 477 sites obtained from the combination of various VLBI, SLR/LLR, GPS and DORIS solutions (see below). The methodology of the combination was a similarity transformation

(Helmert-transformation) of all individual solutions to a common geocentric datum along with the adjustment of the position and velocity coordinates.

A problem occurs in this procedure, if the individual solutions are not free of geometric deformations, i.e., if there are more constraints in the individual solutions than allowed by the degree of freedom of the datum definition. In a stationary three-dimensional reference frame (or station network) we have to define the datum by 7 parameters (3 for the translation of the origin, 3 for the rotations around the three axes, 1 for the scale). In kinematic reference frames we need the same set of parameters for the linear derivatives (velocities), i.e., 14 in total. If in the data processing and parameter adjustment more datum relevant parameters are constrained, e.g., by station positions or velocities, Earth rotation or gravity field parameters fixing the origin (C_{10}, C_{11}, S_{11}) or the orientation (C_{21}, S_{21}) of the satellite orbits, the solution can numerically be deformed (Drewes and Angermann, 2003). In this case the similarity condition is no longer fulfilled and the Helmert transformation leads to wrong results.

In the frame of the programme "Geotechnologien" financed by the German Bundesministerium für Bildung und Forschung (BMBF) and the Deutsche Forschungsgemeinschaft (DFG), the German Geodetic Research Institute (Deutsches Geodätisches Forschungsinstitut, DGFI) analysed the methodology and results of the present ITRF realizations and computed a terrestrial reference frame (TRF) with a different approach using nearly the same input data. The main objective was to avoid as much as possible the deformations of the station network by over-constrained individual technique solutions.

2 General characteristics of the ITRF2000

The ITRF2000 contains 764 occupations of individual techniques located at 477 sites. A total of 3 VLBI, 7 SLR, 1 LLR, 6 global and 9 regional GPS, 2 DORIS and 2 multi-technique solutions were used for the simultaneous computation of transformation parametes, station positions and linear velocities (Altamimi et al., 2002). The result are 3-D epoch position coordinates for each occupation. The velocities are in general equated for all occupations at the same site (exceptions are velocities before and after earthquakes). The resulting standard deviations for positions and velocities are shown as a statistics in Table 1.

We see that more than half of the number of stations have got precise positions ($\leq \pm$ 1 cm) and velocities ($\leq \pm$ 5 mm/a). Approximately 10 % of the stations, however, have got position errors greater than $\pm$ 10 cm and velocity errors greater $\pm$ 5 cm/a. Bearing in mind that real-time positioning, e.g. with GPS, is possible today with a few centimetres precision, and that the fastest motions of tectonic plates are 15 cm/a only, one may state that the above numbers do not fulfil the criteria for an accurate reference frame. The reason is in most cases that the observation period for a particular station was very

Table 1. Statistics of 3-D position and velocity standard deviations in ITRF2000

Interval	No. of positions	Interval	No. of velocities
$\leq \pm$ 5 mm	229 = 30%	$\leq \pm$ 1 mm/a	62 = 13%
$\pm$ 5 mm ... $\pm$ 1 cm	216 = 28%	$\pm$ 1 mm/a ... $\pm$ 5 mm/a	192 = 40%
$\pm$ 1 cm ... $\pm$ 5 cm	195 = 26%	$\pm$ 5 mm/a ... $\pm$ 1 cm/a	106 = 22%
$\pm$ 5 cm ... $\pm$ 1 dm	42 = 5%	$\pm$ 1 cm/a ... $\pm$ 5 cm/a	76 = 16%
$\pm$ 1 dm ... $\pm$ 5 dm	52 = 7%	$\pm$ 5 cm/a ... $\pm$ 1 dm/a	13 = 3%
$\geq \pm$ 5 dm	30 = 4%	$\geq \pm$ 1 dm/a	28 = 6%

short, sometimes less than one year, e.g., by occupations with mobile systems. The velocities of these stations are determined with large uncertainties and the connection to the long-term reference frame is very weak.

The consequence for the present TRF computation is that stations with a continuous observation period less than one year and "bad" observations, e.g. old systems, were excluded. The principal criterion is that reliability and quality of the results is more important than the quantity of stations.

3 Methodology of combination of position and velocity solutions

The best way of computing a TRF is to start from the original observations and to use identical constants, models and parameters according to the latest conventions (McCarthy and Petit, 2004) for the estimation of the unknowns in a common adjustment. This requires a sophisticated software package for the generation of observation equations of all geodetic techniques and does not allow a distributed computation at different analysis centres. It therefore has not been attempted in operational solutions up to date. If the constants, models and parameters in the individual processing approaches per technique are the same, then the use of normal equations is identical to the use of observation equations.

The advantage with respect to a similarity transformation of adjusted solutions is, that all the parameters included in the normal equations may individually be changed in the adjustment procedure, while in a similarity transformation only the transformation parameters (e.g., 7) are estimated. Another advantage of using normal equations is, that they can easily be checked whether and how they are constrained, e.g., by rank defect analyses. This could become problematic due to numerical problems if the large variance-covariance matrices of the solutions get large.

Although complete identity in the individual processing is not guaranteed, we see the use of normal equations as an essential benefit in combining geodetic observations from different techniques.

A problem in combining various individual solutions from different analysis centres is that they all use nearly the same observation data. Combining n of those solutions implies that identical data enter n times into the adjustment procedure. The weighting of techniques' data depends thus on the number of individual solutions provided for the combination.

As a consequence we use in the present TRF computation only one solution per technique by combining the individual techniques' solutions before in an "intra-technique" combination. The combination methodology includes thus the following steps:

- Analysis of individual solutions and generation of normal equations,
- Realization of the common datum,
- Intra-technique combination of individual solutions,
- Inter-technique combination for the final solution.

The procedure is shown in principle in Fig. 1. Details may be found in (Angermann et al., 2004).

4 Analysis of individual solutions as input data

The data used for the DGFI TRF computation were in a first step taken from the original data sets provided for the ITRF2000 (see above). The datum constraints of the solutions should be specified by the providers as removable constraints, minimum constraints and loose constraints. If the solutions are removable or minimum constrained, the reduction of the reported constraints must result in unconstrained, i.e. singular, normal equations. Loosely constrained solutions must result in large r.m.s. errors of positions and velocities.

In a first step the input data sets were analysed with respect to these criteria. It was found that for some solutions the a priori constraints were not or not clearly reported in the SINEX files so that they could not be removed. In other solutions the given constraints were removed but the normal equations did not become singular. Remaining datum constraints may deform the network of stations, therefore it should be avoided to use these solutions for the TRF computation as far as possible. The final input data include the following sets provided for ITRF2000:

- VLBI solution of Geodetic Institute University Bonn, Germany (GIUB)
- VLBI solution of NASA Goddard Space Flight Center, USA (GSFC)
- VLBI solution of Shanghai Astronomical Observatory, China (SHA)
- SLR solution of Communications Research Laboratory, Japan (CRL)
- SLR solution of Center of Space Research, Austin, USA (CSR)
- SLR solution of Joint Center for Earth System Technology, USA (JCET)
- DORIS solution of Groupe de Recherche de Géodésie Spatiale (GRGS)

Besides these we used for this TRF realization also some later multi-year solutions available to us containing more recent observations, which were added or exchanged with those used in ITRF2000, respectively:

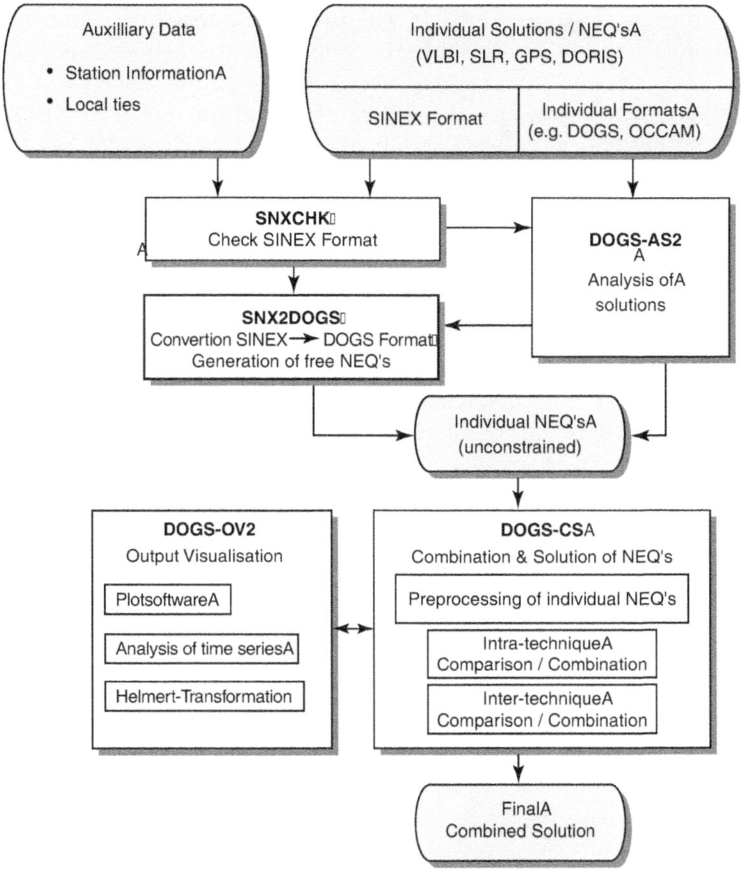

Fig. 1. Combination methodology of individual normal equations (DOGS = DGFI Orbit and Geodetic parameter estimation Software)

- VLBI solution of German Geodetic Research Institute (DGFI02R02)
- SLR solution of German Geodetic Research Institute (DGFI01L01)
- GPS combined solution of the International GPS Service (IGS03P01)
- DORIS solution of Institut Géographique National, France (IGN02D04)

The data processing started with the generation of normal equations by removing the datum constraints from the individual solutions when possible (GIUB, CRL). The loosely constrained solutions (GSFC, SHA, JCET) were taken as they are. The CSR solution is constrained by a rotation datum which could not be removed. It was rotated onto the ITRF2000 datum. The datum of DORIS solutions (GRGS, IGN) could not be removed. They were transformed to the ITRF2000 datum. Both solutions of DGFI (VLBI and SLR) were unconstrained, i.e., the singular normal equations were directly introduced. The cumulatively combined GPS solution (IGS) is over-constrained with non-removable constraints. It was taken into the processing by adding 14 parameters of a kinematic similarity transformation (positions and velocities).

The position coordinates of some of the solutions (JCET, GRGS, IGS) refer to different epochs. They were all transformed to the epoch of ITRF2000 (1997.0). The a priori parameters were different in most of the solutions. They were all transformed to the ITRF2000 datum. Furthermore, stations with an observation period of less than one year were excluded from the processing.

As mentioned above, the reduction of constraints was not possible for some of the selected solutions, and consequently these solutions are not compatible with the combination strategy on the level of unconstrained normal equations. We decided to include them in order to get redundancy for VLBI and SLR, and to have any GPS and DORIS data at all. The selected loosely constrained solution may not bias the combination results. The problems of reducing contraints could have been avoided, if the analysis centers would have provided SINEX files with unconstrained normal equations, which was not the case for ITRF2000.

5 Realization of the common datum

The terrestrial reference frame shall be a geocentric system with orientation in a mean Earth rotation axis and a long-term stable scale. Space geodetic observations are not directly given in such a system but they may be referred to it by the data processing procedure. The origin of the coordinate system is related to the Earth's centre of mass (geocentre) by means of the one degree and order terms of the spherical harmonics of the gravity field (C_{10}, C_{11}, S_{11}) used by satellite techniques for the orbit computation. If gravity field models with $C_{10} = C_{11} = S_{11} = 0$ are used, the satellite positions refer to the geocentre. In principle, all satellite techniques can realize a geocentric origin. SLR gives directly precise distances between satellite and terrestrial positions and thereby relates the position coordinates of the reference frame to the geocentre. The differential approach (range differences) used in GPS because of the clock corrections necessary in one-way ranging systems eliminates most of the reference to the geocentre. VLBI is not related to the gravity field at all. The analysis of time series of translation parameters obtained from weekly solutions has shown that GPS and DORIS results are obviously affected by systematic effects, especially for the z-component of the origin (Angermann et al., 2005b). As a consequence, the origin of the TRF is fixed by the SLR solutions only.

The orientation of the reference frame is a conventional definition given by a mean Earth rotation axis. It is realized in the ITRF computations by aligning it always to the previous realizations (Altamimi et al., 2002). Satellite orbits refer to the axis of maximum inertia by means of the gravity field models (C_{21}, S_{21}). As the relation between the axis of maximum inertia and rotation cannot be described analytically with the necessary precision, satellite methods (SLR, GPS, DORIS) cannot contribute directly to the determination of the orientation parameters. VLBI is the only technique providing the orientation of the Earth in space. To use this capacity one would have to derive the precession/nutation parameters, orientation of the terrestrial reference frame (pole position) and UT1 in a common adjustment relative to the Celestial Reference Frame (CRF). This has not yet been done at present but will be a future procedure. For the present computation the orientation of the TRF is aligned to the ITRF2000.

The scale of the reference frame is in general given by the speed of light transforming the measured travel times into metric distances. It is well-known that microwave satellite techniques (here GPS and DORIS) suffer from uncertainties of the antenna phase centre variations depending on the elevation angle of the received satellite signals, instrumental environment and other effects. These affect the scale of the observed station network (Zhu et al., 2003). Therefore the scale of the TRF is only determined by the SLR and VLBI solutions.

The kinematic reference system provides the orientation of the station velocities on the Earth's surface. It includes also the translation, rotation and scale parameters. The translation parameters are constrained to the geocentre (no net translation, NNT). As any instantaneous motion parallel to a sphere can be described by a geocentric rotation vector it is directly related to Earth rotation. The TRF is defined to have no net rotation (NNR) with respect to the mean global Earth rotation. The NNR condition may be derived from the geodetic observations (see Sect.9). To be consistent with previous ITRF realizations we do not use this approach at present and align the kinematic parameters (station velocities) to the ITRF2000 velocities which refer to the geologic-geophysical model NNR NUVEL-1A (DeMets et al., 1994). The scale parameter is implicitely constrained to be time invariant.

The geodetic datum constraints described above are introduced into the data processing procedure by minimum constraints on a set of selected core stations (sum conditions as pseudo observations). The selection was done with regard to quality and spatial distribution of the stations.

6 Intra-technique combination

The individual solutions of each technique shall be combined to one unique solution. Before doing this, outliers have to be identified and rejected. This was done in an iterative procedure. For each station of each solution the position and velocity differences with respect to the weighted average of the other solutions were computed. Discrepancies of 3 cm (VLBI), 5 cm (SLR) or 7 cm (DORIS), respectively, were considered as a priori outliers and rejected. The normalized differences (absolute difference divided by the standard deviation) served as a test value for subsequent rejections. 32 VLBI, 91 SLR and 18

DORIS station occupations were eliminated from the individual solutions by this procedure. There is only one GPS solution which was compared with another IGS cumulative solution (IGS02P32) and with ITRF2000. Two stations were rejected as outliers because the differences were greater than 3 cm.

In principle the individual solutions should have the same level of accuracy because they use the same observation data. However, the standard deviations of the estimated parameters may differ a little because of small differences in used models, weighting, parameterisation, included observations (e.g. outliers), etc. Therefore the generated normal equations were scaled against each other. The scaling factor was computed from the mean standard deviation of positions of a selected set of core stations.

Various sites were occupied several times by the same or different instruments or with changed equipment (e.g. antennae). The velocities of the site may be identical or may be different due to local deformations. To decide whether the estimated velocity values can be equated in the combination procedure all different occupations were first adjusted separately. The normalized differences were then compared. The velocities were equated when the differences were not significant.

After the procedures described above and principally valid for all techniques the normal equations of the individual data sets were accumulated, provided with the necessary datum constraints and solved. The individual solutions were then compared with the final combined solution. The r.m.s. errors of positions and velocities are shown in Table 2.

Table 2. r.m.s. differences between the individual solutions and the combined technique solutions, [mm] and [mm/a], respectively

	VLBI				SLR				DORIS	
	GIUB	*GSFC*	*SHA*	*DGFI*	*CRC*	*CSR*	*JCET*	*DGFI*	*GRGS*	*IGN*
Pos.	±2.7	±2.2	±3.1	±3.4	±4.1	±4.9	±3.7	±2.3	±6.4	±7.4
Vel.	±0.7	±0.2	±0.8	±0.4	±1.6	±1.4	±1.2	±0.7	±1.8	±1.8

7 Inter-technique combination

The input data for the inter-technique combination are the normal equations of the techniques' combined solutions. These include 81 VLBI, 65 SLR, 202 GPS and 53 DORIS stations. As described above the corresponding normal equations are unconstrained or loosely constrained or over-constrained according to the input data. The scaling of the normal equations is quite different due to the standards, models and parameterisations used. The first step is thus the weighting of the combined technique solutions, i.e., the scaling of the normal equations.

For the weighting two approaches were applied, the comparison of average standard deviations obtained from the intra-technique combination (cf. Sect. 6, Table 2) and the comparison with the ITRF2000 station positions and velocities. The results are given in Table 3. They show in general the same behaviour. Therefore the weighting was done with respect to these numbers.

Table 3. Average standard deviations of combined techniques' solutions from intra-technique combination and with respect to ITRF2000

	VLBI	*SLR*	*GPS*	*DORIS*
Intra-techniques				
Positions [mm]	±2.9	±3.8	--	±6.9
Velocities [mm/a]	±0.6	±0.9	--	±1.8
w.r.t. ITRF2000				
Positions [mm]	±3.9	±4.7	±1.6	±14.6
Velocities [mm/a]	±0.5	±1.2	±0.9	±2.7

The techniques' reference frames are realized by the observation systems at the stations. To combine the different techniques a connection has to be given by local ties between co-located instruments at TRF sites. There are nearly 100 of those sites. The local tie vectors play therefore an important role. However, the quality of the local surveys, reported from about 80 sites, is quite different. To validate the local tie vectors given by the ITRF Product Centre (ftp://lareg.ensg.ign.fr/pub/itrf/itrf2000) they were compared with the station positions obtained from the intra-technique solutions. The differences range from less than 5 mm (about 7 % of the sites) to more than 2 cm (more than 50 %) in position and from less than 1 mm/a (about 10 %) to more than 5 mm/a (about 40 %). The best agreement is between GPS and VLBI or SLR, respectively, largest discrepancies appear in the comparisons with DORIS.

The selection of the used local ties in the combination procedure was done by an iterative process (Krügel and Angermann, 2005). In the first step only the best local ties (six, each three between GPS and VLBI or SLR, respectively) were used to perform a combined adjustment of the complete network. The resulting coordinate differences between the techniques' locations were then again compared with the given local ties. For the second and following steps all the co-locations with differences less than a given limit (± 34 mm in positions, ± 4.5 mm/a in velocities) were included. Finally 50 local ties were selected, 37 between different techniques and 13 between different occupations of the same technique. Differently to the ITRF2000 procedure only a subset of reliable local ties were introduced and weighted by a-priori standard deviations according to the discrepancies obtained in the selection process. The standard deviations range for the positions from ± 1 mm (for differences less than 5 mm) to ± 5 mm (greater than 20 mm) and for the velocities from ± 1 mm/a (for differences less than 2.5 mm/a) to ± 5 mm/a (up to 4.5 mm/a). The complete set of selected co-location sites is shown in Color Fig. XLI on p. 311.

Another problem in the combination procedure is the equating of station velocities. If the velocities of the intra-technique combined solutions agree within a given limit (4.5 mm/a) they were considered as identical. These are all the selected co-location sites which also agreed with the local ties (see above). In addition there were about half of the remaining stations which have large discrepancies with respect to the local ties but in many cases velocities agree among each other. They were also equated. In total about 75 % of co-located stations were equated in their velocities.

The final inter-technique combination was done by accumulating the normal equations of the intra-technique combined solutions and by adding the weighted local ties and equated velocities constraints by pseudo observations. In the constrained normal equations from GPS and DORIS 14 parameters of a similarity transformation were included. The final datum was given for the origin (translation parameters and their velocities) from the SLR solution, for the scale and its rate from VLBI and SLR, and for the orientation by a no-net-rotation (NNR) condition with respect to ITRF2000. The final solution includes 412 occupations of 401 stations at 259 sites. It will be referred to as DGFI TRF03 in the sequel.

8 Validation of final results

The internal accuracy of the DGFI TRF03 is quite good. There are very few stations with large standard deviations (3 with position s $\geq \pm$ 5 dm, 1 with velocity s $\geq \pm$ 1 dm/a). Two third of the stations have got standard deviations $\leq \pm$ 1 cm in position and $\leq \pm$ 5 mm/a in velocity. This includes non-identical velocities at co-location sites, which are equated in ITRF2000. The average is below ± 5 mm and ± 1 mm/a for VLBI, SLR and GPS, and ± 7 mm and ± 3.5 mm/a for DORIS. A detailed analysis was done comparing the final solution with the individual intra-technique solutions. It shows features similar to the analysis of the formal errors.

The external validation bases on comparisons with other TRF realizations, in particular with the ITRF2000. The overall r.m.s. deviation of DGFI TRF03 with respect to ITRF2000 in 328 identical stations (without DORIS) is ±3 mm in X and Y, and ± 4 mm in Z. For the DORIS stations the r.m.s. deviation is ± 23 mm in X and Y, and ± 19 mm in Z. The r.m.s. deviation of velocities in 218 identical sites is ± 2 mm/a in all three components. A graph of the north, east and height comparison divided into different techniques is shown in Fig. 2. The graphic presentation of the comparison of velocities is given in Color Fig. XLII on p. 311. Color Fig. XLIII on p. 312 demonstrates some major discrepancies in detail.

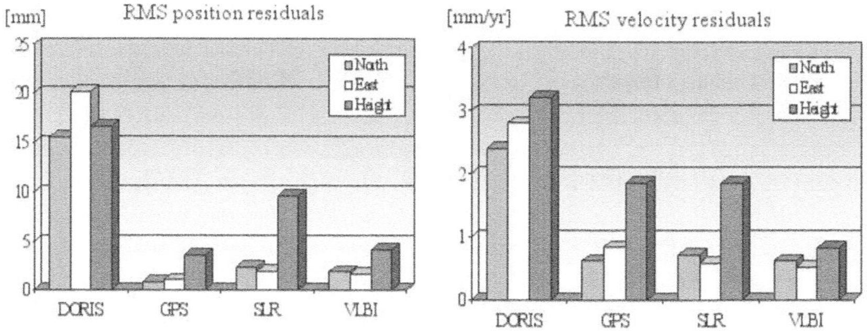

Fig. 2. r.m.s. deviations of DGFI TRF03 station positions and velocities w.r.t. ITRF2000 split into the individual techniques

The direct comparison with a recent cumulative IGS solution (IGS04P52) shows in 204 identical stations (after the transformation from the IGS epoch 1998.0 to the DGFI TRF03 epoch 1997.0) a r.m.s. deviation in the X and Y positions of ± 2.5 mm, and ± 2.7 mm in Z. The corresponding numbers for the velocities are ± 1.4 mm/a in X, ± 1.3 mm/a in Y, and ± 1.1 mm/a in Z.

9 Refined kinematic reference frame

The kinematic reference frame, i.e. the reference for the station velocities, is defined by a NNR condition with respect to the entire Earth crust (see Sect. 5). It is realized in ITRF2000 by transforming a set of selected stations supposed to represent a stable part of the crust to the geologic-geophysical model NNR NUVEL-1A. Thereby it is anticipated that NNR NUVEL-1A fulfils the NNR condition for the present-day motions of the Earth's crust.

There are two principal reasonings contradicting this anticipation:

- NNR NUVEL-1A is a rigid plate model computed from data of sea floor spreading rates, transform fault azimuths, and Earthquake slip vectors. The observations of the first two mentioned data are generated over geologic time scales (about 3 million years). It has been proven (Drewes, 1998) that they are not identical with present day plate motions. The third type of data represents motions in deformation zones. They may not be representative for the entire plate.
- The entire surface of the Earth is represented by rigid plates only with no deformation zones included. As there exist large deformation zones around the Earth (e.g., the Mediterranean orogenic belt, the Southeast Asian area, the Andes orogen) the NNR condition cannot be valid if they are excluded.

It is therefore necessary to compute a present-day model representing the motions of the entire Earth surface. This was done with the original input data of the DGFI TRF03 as the Actual Plate Kinematic and deformation

Model, APKIM2002 (Drewes and Meisel, 2003). Each station was assigned to a certain plate or deformation zone. As the deformation zones are not densely covered by the stations given in the input data, they were approximated as small rotating blocks. In total there were 12 rigid plates and 6 blocks. The geocentric rotation vectors of plates and blocks were estimates in a common adjustment of all velocities from the different techniques simultaneously with the techniques' kinematic datum parameters. The integral over all plates and blocks provides the NNR condition. The adjusted motions are shown exemplarily in Color Fig. XLIV on p. 312.

It can clearly be seen that there are significant discrepancies between the NNR NUVEL-1A and the APKIM2002 motion, not only in the deformation zones, where they are extremely large, but also inside the plates. The rotation of APKIM2002 with respect to NNR NUVEL-1A corresponds to a maximum of 2 mm/a in the equator of the spherical rotation. This rotation enters as a bias into the ITRF2000 realization affecting mainly the station velocities.

10 Conclusion and outlook

From the analysis of methodology and results of the TRF computation we may draw some conclusions for the procedure of future ITRF realizations. The principle shortcomings of present day methods may be summarized as follows (Angermann et al., 2005a):

- The input data in terms of multi-year solutions for epoch station positions and linear velocities are not optimal. The linear velocities do not always represent the motions of the Earth's crust. There are undetected episodic motions, e.g. caused by smaller seismic events, periodic motions, e.g. annual oscillations caused by atmospheric and hydrologic loading, and changes in velocities, e.g. after earthquakes (Angermann et al., 2005b).
- Most of the solutions are deformed by direct or indirect constraints (Drewes and Angermann, 2003). From the variance-covariance matrices or the SINEX files of the solutions one may not see these constraints, a removal is problematic, i.e., deformations enter into the reference frame.
- The combination of various individual technique solutions multiplies identical data and suffers from unclear internal conventions.
- The orientation of the ITRF is presently given by the orientation of the previous ITRF realization. It is not consistent with the Earth orientation parameters (EOP) which are computed independently at the IERS Earth Orientation Product Centre nor with the International Celestial Reference Frame (ICRF) computed independently at the ICRF Product Centre.

From these deficiencies we derive our recommendations for the realizations of the ITRF in the future, e.g., the ITRF2004:

- The input data should be provided in terms of time series of station coordinates (e.g. weekly). This allows immediately the detection of irregular position changes or instrumental variations, e.g., antenna changes, (Meisel et al., 2005).

- The observation data should be provided as unconstrained, i.e. singular, normal equations. The removal of constraints causes always troubles. All relevant parameters have to be included (e.g. loading parameters instead of loading corrections with unclear physical models).
- The intra-technique combination should be done by the technique services which should provide one unique technique data set.
- The computation of the ITRF should be done as a consistent adjustment of all parameters involved: ITRF, EOP, and ICRF.
- The ITRF network should include reliable stations only, e.g., permanent observations longer than three years with highest data quality.

Acknowledgement. This is publication no. GEOTECH-163 of the programme Geotechnologien of BMBF and DFG, Grant IERS(03F0336C).

References

Altamimi Z, Sillard P, Boucher C (2002) ITRF2000: A new release of the International Terrestrial Reference Frame for earth science applications. J Geophys Res 107 (B7), 2214, doi:10.1029/2001JB000561

Angermann D, Drewes H, Krügel M, Meisel B, Gerstl M, Kelm R, Müller H, Seemüller W, Tesmer V (2004) ITRS Combination Center at DGFI: A terrestrial reference frame realization 2003. Deutsche Geodätische Kommission, Reihe B, Heft Nr. 313

Angermann D, Drewes H, Gerstl M, Kelm R, Krügel M, Meisel B (2005a) ITRF combination - status and recommendations for the future, Sanso F (Ed): A Window on the Future of Geodesy, IAG Symposia, Vol. 128, Springer

Angermann D, Krügel M, Meisel B, Müller H, Tesmer V (2005b) Time evolution of the terrestrial reference frame, Sanso F (Ed): A Window on the Future of Geodesy, IAG Symposia, Vol. 128, Springer

Boucher C, Altamimi Z (1990) Evaluation of the realizations of the Terrestrial Reference System done by the BIH and IERS (1984-1988). IERS Technical Note No. 4

Boucher C, Altamimi Z, Sillard P, Feissel-Vernier M (2004) The ITRF2000. IERS Technical Note No. 31, Bundesamt für Kartographie und Geodäsie, Frankfurt am Main

DeMets C, Gordon R, Argus D F, Stein S (1994) Effect of recent revisions to the geomagnetic reversal time scale on estimates of current plate motions. Geophys Res Lett (21) 2191-2194

Drewes H (1998) Combination of VLBI, SLR and GPS determined station velocities for actual plate kinematic and crustal deformation models. IAG Symposia, Vol. 119, 377-382, Springer

Drewes H, Angermann D (2003) Remarks on some problems in the combination of station coordinate and velocity solutions. Proc. of the IERS Workshop on Combination Research and Global Geophysical Fluids, Richter B, Schwegmann W,

Dick WR (Eds), IERS Technical Note No. 30, 30–32, Bundesamt für Kartographie und Geodäsie, Frankfurt am Main

Drewes H, Meisel B (2003) An actual plate motion and deformation model as a kinematic terrestrial reference system. Geotechnologien Sciene Report No. 3, 40-43, Koordinierungsbüro Geotechnologien, Potsdam

Krügel M, Angermann D (2005) Analysis of local ties from multi-year solutions of different techniques. Proc. of the IERS Workshop on site co-location, Richter B, Dick W, Schwegmann W (Eds), IERS Technical Note No. 33, 32–37, Bundesamt für Kartographie und Geodäsie, Frankfurt am Main

McCarthy D, Petit G (2004) IERS Conventions (2003), IERS Technical Note No. 32, Bundesamt für Kartographie und Geodäsie, Frankfurt am Main

Meisel B, Angermann D, Krügel M, Drewes H, Gerstl M, Kelm R, Müller H, Seemüller W, Tesmer V (2005) Refined approaches for terrestrial reference frame computations, Adv Space Res, Elsevier (in press)

Zhu SY, Massmann FH, Reigber C (2003) Satellite antenna phase center offsets and scale errors in GPS solutions. J Geodesy (76) 668-672

Combination of VLBI Analysis Results

Axel Nothnagel, Dorothee Fischer, Christoph Steinforth, and Markus Vennebusch

Geodetic Institute of the University of Bonn, Nussallee 17, D-53115 Bonn, Germany, nothnagel@uni-bonn.de

Summary. The quality and reliability of products of geodetic Very Long Baseline Interferometry (VLBI) observations, as those of any other space geodetic technique, are greatly enhanced if results of different analysis software packages with their individual models are brought together using a suitable combination process. In the framework of the *Geotechnologien*-project "Integration of space geodetic techniques - IERS", tools and procedures have been developed for the combination of VLBI results taking into account the peculiarities of the VLBI observing technique. They have found their use in the routine operation of the International VLBI Service for Geodesy and Astrometry (IVS) which produces earth orientation parameters (EOP) and terrestrial reference frame (TRF) realisations from session-wise geodetic VLBI observations. The operational combination itself has several levels of latency requirements. Therefore, different types of combined EOP series based on EOP input series alone or on the full variance/covariance information of the EOP and the underlying TRF are generated.

Key words: Combination, Earth rotation, VLBI

1 Introduction

Geodetic Very Long Baseline Interferometry (VLBI) belongs to the group of modern space geodetic techniques which are necessary for the determination of celestial and terrestrial reference frames as well as for monitoring the earth's variable rotation. All techniques have certain strengths and weaknesses which can only be exploited in an optimal way if observations and results are intertwined (Rothacher et al., 2005).

The analyses of VLBI observations are carried out with several different analysis software packages using special models and analysis strategies on a best-effort basis. However, no solution can claim for itself that it is superior to the other. Under pure accuracy and reliability considerations the combination of individual solutions of similar quality has proven to be the ultimate solution to this dilemma (Kouba et al., 1997). The first step on the way to link all space geodetic techniques is the combination of all suitable VLBI solutions to a single set of technique-specific results. The combination of pure VLBI results should guarantee that the peculiarities of the VLBI observing technique are taken into account properly and that the output is suitable for an integration of all space geodetic techniques. This challenge has been taken up in the framework of the *Geotechnologien*-project "Integration of space geodetic techniques - IERS" under sub-project "Development of Methods for the Integration of VLBI Results into the Combination of Geodetic Space Techniques as a Contribution to the IERS".

2 VLBI and Earth Rotation Determinations

Geodetic VLBI is the only observing technique which realises a direct link between the quasi-inertial reference frame of radio source positions and the earth-fixed (terrestrial) reference frame. The analysis of geodetic VLBI observing sessions of 24 hours duration, thus, permits a direct, hypothesis-free determination of all components of earth rotation.

Although a transformation of the earth-fixed system into the quasi-inertial reference system would require only three angles, e.g. Eulerian or cardanic angles (Richter, 1995), the concept of a rotation axis and five angular components to describe the rotation of the earth is widely used for practical reasons and easier interpretation.

The motion of the earth's axis of rotation with respect to the crust is represented through the two components of polar motion (x_p and y_p). Polar motion is a free oscillation and has to be observed routinely. The motion of the earth's axis of rotation in the space-fixed frame is described using the concept of precession and nutation. Since this behaviour is forced predominantly through torques, recent models provide a fairly high level of accuracy. However, unmodeled contributions of the earth's elasticity and especially the free core nutation (FCN) still generate significant deviations of the observations from the models. Therefore, regular VLBI observations of the motion of the earth's axis of rotation in space are always necessary for further progress in the understanding of "System Earth".

The phase of the earth's diurnal rotation is measured as the angle between the dynamical equinox and Greenwich longitude, commonly called Greenwich Apparent Sidereal Time (GAST) from which the time argument UT1 is deduced. In order to describe the variable rotation of the earth as the difference with respect to atomic time, the quantity UT1–UTC is normally reported. The time derivative is called length-of-day (LOD).

Quite recently a major revision has taken place in the paradigm of describing the motion of the earth's rotation axis in inertial space (precession and nutation). The concept in use during the last century has been based on a dynamical equinox (Lieske et al., 1977). Furthermore, the two-dimensional motion has been split up in a radial and in a tangential component with respect to the precession cone around the celestial ephemeris pole. The results of VLBI observations are reported as offsets relative to the IAU1980 model in longitude ($d\psi \times \sin \varepsilon_0$) and obliquity ($d\varepsilon$). In the new paradigm according to the IAU2000 resolutions (IERS, 2002), non-rotating origins have been introduced for the celestial as well as for the terrestrial reference systems (Capitaine et al., 1986). In addition, the two-dimensional motion of precession and nutation has been separated into two Cartesian components (X, Y) in order to be consistent with polar motion.

The new paradigm has mainly been introduced for a clearer concept of the description of the earth's rotation around its spin axis, both in phase angle and in velocity. At the current level of accuracy this change does not have a significant effect yet but long term time series will have a slightly reduced noise level. However, this change had a serious impact on the operational reporting of nutation parameters by the analysis centers (see Sect. 3.4).

Current geodetic VLBI observations analysed with up-to-date analysis software packages yield EOP accuracies at the 100 μas level for x_p, y_p, $d\psi \times \sin \varepsilon_0$, $d\varepsilon$ and 6 μs for UT1–UTC. The accuracy of the TRF results (coordinates at a reference epoch and velocity components) heavily depends on the station's stability of the electronics and its observing history which should be several years long with participation in a sufficient number of observing sessions. Good stations can be located with ± 1 - 2 mm and ± 0.1 mm/y (std. dev.).

The combination itself has several levels of latency requirements. Generally, new EOP data points should be made available to the users as soon as possible. However, geodetic VLBI requires transportation of recording media (tapes or disk packs) and subsequent correlation which generate a natural delay in the delivery cycle. Therefore, 'as soon as possible' in geodetic VLBI always means a delay of at least between seven and ten days after the observations. Due to logistics and priorities some of the sessions may even have a much longer delay. In the latter case EOP results are published in batches of sessions at preset delivery times. These can be once every quarter of a year in the case of complete EOP time series (see Sect. 3.3) or eight weeks after the observations for the IERS Combination Pilot Project (see Sect. 3.4). In all cases the resulting combined EOP series may be based on EOP input series applying only their standard deviations alone or on the full variance/covariance information of the EOP and the underlying TRF.

3 Combinations

3.1 Combination Strategies

Since all analyses of VLBI observations have deficiencies in both accuracy and precision, a rigorous combination of several sets of results will reduce the impact of these imperfections on the final result (Cannon, 2001). In the combination process for a time series without time critical delivery requirements the individual solutions mostly just complement and control each other leading to a higher accuracy and a higher reliability. Another aspect of these combinations is that more objective error bounds can be established which is of particular importance where formal errors of contributing time series are not representative.

Generally, users require EOP products with a certain quality and long term consistency at a fixed recurring date. Under operational considerations where the latency has to be kept low, the combination of the results of different solutions may serve further purposes as well. There is always a danger that, for various reasons, the constellation of analysis centers may change or that errors in data and product submissions occur. Therefore, one would not like to rely on just one analysis center but rather establish some redundancy. In this case, the more analysis centers contribute the smaller the effect of taking out or adding a contributor. The combination of many EOP series, thus, yields a product of optimal long term stability. In addition, the residuals w.r.t. the combined series can be analysed to further improve the analysis process.

Although one of the fundamental rules of statistics is to never use an observation more than once, it is still appropriate to produce combined results from submissions of different VLBI analysis centers. One of the reasons is simply that analysis centers use different weighting schemes and may reject different data points as outliers. Even more important is the fact that by using different software packages with different models and corrections as well as different analysis strategies, the initial observations are to a very large extent not identical to the original ones any more. It may, thus, be safely assumed that the observations used are independent enough to permit a combination process.

In the course of the project's lifetime two different classes of VLBI analysis output have been generated by the VLBI analysis centers. The first one consists of the EOP results of each VLBI observing session containing x_p, y_p, UT1–UTC and their time derivatives as well as the two nutation offsets $d\psi \times \sin \varepsilon_0$ and $d\varepsilon$ together with their standard deviations and a few selected correlation coefficients. The combination of these results is described in section 3.2. Another approach uses VLBI output containing also information on the underlying TRF with the full covariance in order to carry this over to the combination phase. So-called SINEX files are used for this purpose (see Sect. 3.4).

3.2 Combination of Time Series

Since the beginning of space geodetic combination activities in the framework of the Bureau International de l'Heure (BIH) and later of the International Earth Rotation and Reference Systems Service (IERS) EOP results have been reported by analysis centers in the form of time series together with their formal errors (standard deviations). In the case of VLBI analyses these time series consist of the EOP results of each VLBI observing session containing x_p, y_p, UT1–UTC and their time derivatives as well as the two nutation offsets $d\psi \times \sin\varepsilon_0$ and $d\varepsilon$ together with their standard deviations. In addition, the correlation coefficients between x_p, y_p and UT1–UTC and between $d\psi \times \sin\varepsilon$ and $d\varepsilon$ are being reported. For historical reasons the number of correlation coefficients had been limited to only these four values since it was assumed that the other elements of the full correlation matrix were of minor importance. No information about the TRF used in the determination of the EOP series is carried over for further use.

In preparing the input data sets for combination one item of concern is the reference epoch of the EOP. From an operational point of view reporting of EOP should either be referred to 0.00 h UT (preference of the IERS EOP Product Center) or 12.00 h UT (for combination with other space geodetic techniques). However, VLBI analysts tend to use the middle epoch of the sessions since this is the best choice in terms of error propagation. So, due to the fact that geodetic VLBI observing sessions of 24 hours duration seldom run from 0 to 24 hours UT but for organisational reasons mostly start at about 18.00 h UT, VLBI analysts mainly use the middle epoch of the sessions at around 6.00 h UT. In addition, small deviations between the reference epochs of the individual series may occur. Therefore, a suitable interpolation method has to be applied to produce EOP at 0.00 h UT or 12.00 h UT reference epochs.

The next preparatory consideration has to deal with the fact that each EOP determination from VLBI observations is linked to a terrestrial and a celestial reference frame (CRF). The frames can either be fixed for estimating the EOP alone or be estimated together with the EOP. The latter, however, requires a certain degree of constraining. Owing to the fact that each EOP series is based on a slightly different TRF and/or CRF, and sometimes caused by differences in the models used, the EOP series may show offsets relative to each other. In addition, linear drifts may also occur depending on the drift rates of the telescopes assumed by the analysis centers. Offsets and drifts have to be subtracted prior to the combination process.

The combination of the input series has to take into account the standard deviations as they have been reported. However, it has to be considered that the formal errors, though representative and correct within a single series, may not display the correct relative level between individual series. Before carrying out the combinations it is, thus, necessary to establish a correct relative weighting of the input series. This can be done on the basis of the whole series or, as deemed necessary for an operational procedure, on a data point by data point basis.

In the first case the procedure of variance component estimation (Koch, 1988) can be used to establish proper relative weights. In an operational process, however, where only a single epoch has to be combined at a time, a different strategy has to be found which takes into account that the overall situation changes quite rapidly from session to session and may not produce convergent estimates of variance components (see Sect. 3.3).

Another issue is the use of the correlations between the reported EOP to construct the proper variance/covariance matrix. This can only be done to a certain extent considering the restricted input information of the correlations as described above. A further refinement of the covariance matrix originates from correlations between EOP from different analysis centers. Owing to the fact that the analysis centers use the same raw observations and sometimes also identical software packages, although with different analysis philosophies, it has to be expected that non-negligible correlations exist between the different EOP series. These have to be taken into account in a proper way based on empirical investigations or on apriori estimates to properly construct the covariance matrix for the EOP.

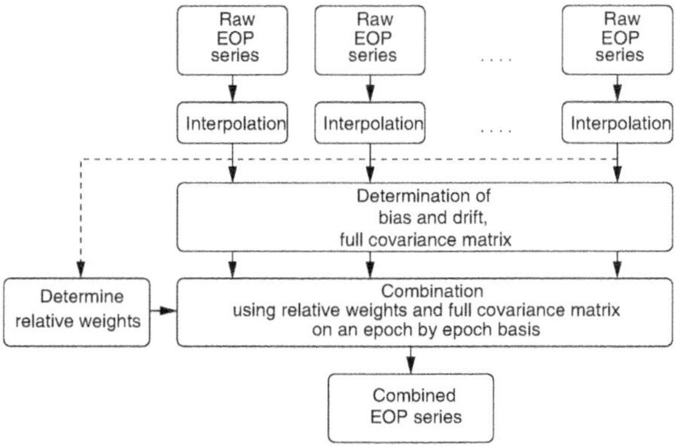

Fig. 1. Overview of combination flow

Summarising, the first step of the combination process is, thus, the interpolation to a common reference epoch and the computation of biases and drifts with respect to a chosen EOP reference series in order to maintain long term consistency with a well established EOP series. Relative weights for each of the input series have to be determined only from time to time since rapid changes in the quality of the input series have not been encountered so far (Fig. 1). In the second step, the EOP series are combined into one set of EOP results by applying the biases, the weighting factors and the correlations between the analysis centers using also the initial variance/covariance information to the extent available. This is done on an epoch by epoch basis computing a least squares average with full covariances (Koch, 1988).

3.3 Operational Time Series Combinations

The combination procedures developed within the *Geotechnologien*-project "Integration of space geodetic techniques - IERS" are being applied routinely in the operational EOP generation of the International VLBI Service for Geodesy and Astrometry (IVS). Here, five analysis centers (see Table 1) regularly submit EOP results as described above. Two types of observing sessions have to be distinguished since they are treated in a different manner in the combination and dissemination process. The first group covers all IVS observing sessions which are designed for rapid turn-around of recording media, correlation and data analysis. These run under the acronyms IVS-R1 and IVS-R4 every Monday and every Thursday with EOP results to be delivered as quickly as possible after the observations which, at present, is roughly seven to ten days later (IVS, 2005). All other geodetic and astrometric VLBI sessions without delivery time requirements and with processing delays of as long as two months form the second group of sessions.

Table 1. List of solutions and software used

Abbrev.	Institution/Software
AUS	Geoscience Australia, Belconnen, Australia OCCAM, Kalman Filter
BKG	Bundesamt für Kartographie und Geodäsie, Leipzig, Germany Calc/Solve, Gauß-Markov Least Squares Adjustment
GSFC	NASA Goddard Space Flight Center, Greenbelt MD, USA Calc/Solve, Gauß-Markov Least Squares Adjustment
IAA	Institute of Applied Astronomy, St. Petersburg, Russia OCCAM, Kalman filter
USNO	U.S. Naval Observatory, Washington D.C., USA Calc/Solve, Gauß-Markov Least Squares Adjustment

Considering these operational circumstances the EOP combination and dissemination procedures have been devised to optimally project this situation into the respective products. The first series, called *IVS Rapid Series*, is generated as soon as possible, i.e. when at least a certain number of analysis centers has submitted its results for combination. Within this process only results of new sessions are added to the series while existing EOP data points are fixed at the values of their first publication.

The second series is produced once every quarter of a year with all input data available at the time of combination, i.e. with all sessions processed and analysed to this date. Here, each EOP value may end up with a slightly different result compared to the previous realisation since the general analysis or combination process may have been improved or additional analysis centers may have submitted EOP for this session. These EOP series, called *IVS Quarterly Series*, are always state-of-the-art realisations of VLBI EOP series.

For both series almost the same procedures apply. Thus, as the first steps interpolation to a common epoch, computations of relative weights and determination of biases and rates have to be carried out. In the beginning of the IVS combination activities, it had been decided to maintain long term consistency with the IERS C04 series (IERS, 1996). For this purpose, polar motion and UT1–UTC biases relative to IERS C04 had been computed for each data set using only the data in the period from the beginning of 1999 to September 30, 2000 which is the date immediately prior to the first epoch of official IVS EOP combinations. By subtracting these biases from the individual series prior to the combination, the combined series had received just some small initial guidance but afterwards was free to evolve along with the VLBI observations.

Nowadays, consistency is sought with respect to a well established TRF. Currently, VTRF2003 (Nothnagel, 2003) is chosen for its consistency with the VLBI scale and its close relationship with the axis and origin definition of ITRF2000. The link with this TRF is achieved through forming a weighted mean of those EOP series which use VTRF2003 as their reference TRF, either as fixed TRF or constrained to this realisation (currently AUS, IAA and BKG). The apriori offsets and possible rates are then computed and applied with respect to this mean series (see Table 2).

Table 2. Current average statistics

	x_p		y_p		$UT1-UTC$		$d\psi \sin \epsilon_0$		$d\epsilon$	
	bias [μas]	wrms [μas]	bias [μas]	wrms [μas]	bias [μs]	wrms [μs]	bias [μas]	wrms [μas]	bias [μas]	wrms [μas]
AUS	56.2	130.5	114.7	136.8	-0.1	6.7	86.2	195.0	11.3	62.6
BKG	7.3	124.7	45.4	121.5	4.1	3.2	0.3	101.2	11.3	41.0
GSFC	-99.4	77.1	18.9	42.5	-0.1	1.6	-8.3	54.4	-13.5	18.7
IAA	139.5	79.6	-28.8	74.4	-1.0	3.0	-1.3	25.4	6.2	33.9
USNO	-22.3	76.7	-30.6	66.1	-1.1	1.6	36.5	110.2	17.3	32.6

The situation is slightly different for the nutation offsets and the underlying CRF. Here, long term consistency of the combined nutation series with other existing series is maintained through the fact that all analysis centers use the source positions of the International Celestial Reference Frame (ICRF) (Ma et al., 1998) as the defining frame in their VLBI analyses. Even with a few sources observed today which do not belong to the ICRF, this guarantees that the individual nutation series show only small relative biases (Table 2). The input series of the nutation offsets, thus, do not need a special procedure to ensure alignment with a well defined reference frame and can be used directly.

For the relative weight factors to be applied to all EOP components a special scheme has been devised and thoroughly tested for routine operations. A single weight factor for all EOP components of each input series is computed from the relationship of the scatter of the nutation angle series, $d\psi \times \sin \varepsilon_0$ and $d\varepsilon$. The background is again that all analysis centers use the ICRF as their celestial reference frame of choice. The ICRF as such is very stable at the sub-milliarcsecond level (IERS, 1999) and deficiencies of individual sources average out in the large number of sources used in geodetic VLBI observing sessions. Therefore, most of the scatter to be found in the individual nutation offset series can be attributed to the way the analysis is carried out, e.g. which software or which parametrisation is used etc. (analyst's noise) (Nothnagel and Steinforth, 2001). As a consequence, the scatter in the nutation results ideally represents the overall quality of each solution.

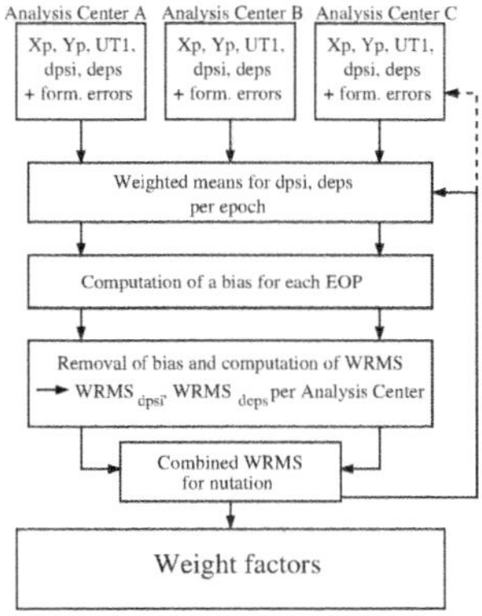

Fig. 2. Combination procedure (1st step)

The determination of the weight factors is an iterative process (Fig. 2). Initial mean values of the nutation offsets ($\overline{d\psi \times \sin\varepsilon_0}$ and $\overline{d\varepsilon}$) are computed for each epoch where the input data is only weighted according to the formal errors assigned by the analysis centers. In the next step, biases for each nutation offset component and analysis center are computed and subtracted. These are relatively small but significant nevertheless. From the bias-free series a weighted root mean squared error (WRMS) is then computed for each analysis center combining both nutation components in a root-sum-squared (RSS) sense. The combined WRMS of the nutation offsets are used to derive the weight factors f_i for each of the analysis centers dividing the center's WRMS $wrms_i$ by the mean WRMS of all analysis centers $\overline{wrms}$:

$$f_i = wrms_i / \overline{wrms}. \tag{1}$$

The apriori weights of each individual input series including those of polar motion and UT1–UTC are then multiplied with the respective weight factors. For fine tuning, the process is repeated by applying the new weights to the input data until a certain convergence threshold is met (Nothnagel and Steinforth, 2001).

Table 3 gives an impression of the current level of agreement of the weight factors which are used to increase (if > 1) or decrease (if < 1) the initial weights of the individual solutions. The weight factors do not vary dramatically but moderately balance the deficiencies of the formal errors of the input series. Variance component estimation has not proven to be robust enough for day-to-day operations since convergence is often not achieved with the small sample of only five analysis centers and eight EOP components.

Table 3. Weight factors

AC	weight factor
AUS	0.83
BKG	0.91
GSFC	1.03
IAA	1.31
USNO	1.05

The final combination of polar motion and UT1-UTC is calculated using the full variance and covariance information according to (Koch, 1988)

$$x_{j,combi} = \frac{\sum_{i=1}^{n} f_i p_{x_{ij}} (x_{ij} - (rate_{i,x} \times \Delta MJD_j - b_{i,x}))}{\sum_{i=1}^{n} f_i p_{x_{ij}}} \quad (2)$$

with x_{ij} = EOP components; $p_{x_{ij}}$ = input weight; f_i = weight factor; $rate_{i,x}$ = rate relative to reference series (cf. Table 2); ΔMJD_j = epoch difference w.r.t. bias reference epoch; $b_{i,x}$ = y-axis intercept of straight line fit; j = epoch; n = number of analysis centers. For the combination of the nutation offsets no rates and biases are applied. The covariances between the EOP components are applied in the setup of the normal equations which are used for the final estimation step. The results of the operational combinations are available both graphically and numerically on the IVS Analysis Coordinator's web page via the IVS Home page (*http://ivscc.gsfc.nasa.gov*) or directly from *http://giub.geod.uni-bonn.de/vlbi/IVS-AC*.

3.4 Combination with Full TRF and Full Covariance Information

The combination of EOP time series as described above completely neglects the direct interaction between EOP and the TRF used to estimate the EOP. In order to overcome this deficit the information about the TRF may be carried over to the combination process in two different ways. One option is to report the estimated parameters, i.e. TRF and EOP, together with their full covariance matrix and to use these for the combination. For this purpose the SINEX file format (Solution INdependent EXchange format) has been invented (e.g. IERS Web page). One of the drawbacks of this scheme is that the solutions of TRF and EOP together always need constraints to eliminate the datum defect in some hard or loose way. For the combination, however, the datum itself should be taken out again since it cannot be guaranteed that all analysis centers or, in the case of inter-technique combinations, even other techniques have used the same TRF realisation. This requires additional information and further processing steps. For this reason the IVS, at an early stage, decided to use the direct approach of extracting and reporting complete datum-free normal equation systems including the right hand side which is the second option to carry over the information of the TRF to the combination process. SINEX format Version 2.0 has been established for this very purpose.

By using pre-reduced normal equation matrices, i.e. reducing all non-TRF and non-EOP parameters, all relations between the TRF and EOP parameters are carried over to the combination step which is also known as "adjustment of groups of observation" (Mikhail, 1976). This approach may be used for the combination of several solutions of the same VLBI session generated by different IVS Analysis Centers under the assumption that the observations processed by different analysis centers are sufficiently independent of each other (cf. Sect. 3.1). Compared to the combination on the basis of covariances this method can be regarded as an enhancement since the normal equations are clean of any hidden residual datum effect which may otherwise remain in the data when the constraints are not removed properly.

The combination process itself mainly consists of stacking the input normal equation systems by adding the respective elements. Through this method all the relationships between the parameters (correlations) are carried over and are taken into account correctly. In the next step, the datum is applied through one of several options converting the normal equation system into a regular one. The final inversion steps then produce the combined parameters.

In order to carry out the combination in a proper way preparatory steps have to be carried out before the actual combination which are rather similar to those of the time series combination (cf. Sect. 3.2). The normal equation systems have to be transformed to be based on the same apriori values for the parameters, the reference epochs have to be changed to identical epochs, and the normal matrices have to be scaled for numerical consistency.

For the VLBI intra-technique combination the DOGS-CS combination software developed at the German Geodetic Research Institute (DGFI), Munich, forms an ideal tool. In addition, further software elements have been developed within the *Geotechnologien*-project for quality checks and for controlling DOGS-CS. The following tasks are carried out:

- The rank deficiency of datum-free normal equation matrices from VLBI observations has to be six. Quality checks have been developed to control this fact applying different techniques for rank-determination based on eigenvalue and QR-decompositions (Lay, 2000).

- Outliers have to be rejected before combination. After imposing a common datum definition to all the solutions, EOP or site positions can be estimated and a comparison can be carried out. Solutions exceeding a certain threshold are identified as outliers and the input data is rejected until further checks have been carried out.
- In order to account for different levels of magnitudes of the elements of the normal equation matrices, scaling of the input matrices has to be applied (Kelm, 2003).
- Before performing the final combination step it is sometimes necessary to reduce auxiliary parameters like clock or atmosphere parameters (Brockmann, 1997) if analysis centers also provide normal equation elements for these parameters.

3.5 Operational Combination with Datum-free Normal Equation Systems

Routines and procedures have been developed to carry out these combinations within the activities of the IVS to generate the input for the next realisation of the International Terrestrial Reference Frame, the ITRF2004, in the form of combined datum-free normal equation systems. Seven IVS Analysis Centers have contributed normal equations in SINEX format on a session-by-session basis for the intra-technique combination (see Table 4).

Table 4. List of solutions and software used

Abbrev.	Institution/Software
AUS	Geoscience Australia, Belconnen, Australia OCCAM, Kalman Filter
BKG	Bundesamt für Kartographie und Geodäsie, Leipzig, Germany Calc/Solve, Gauß-Markov Least Squares Adjustment
DGFI	Deutsches Geodätisches Forschungsinstitut (DGFI), Germany OCCAM, Gauß-Markov Least Squares Adjustment
GSFC	NASA Goddard Space Flight Center, Greenbelt MD, USA Calc/Solve, Gauß-Markov Least Squares Adjustment
MAO	Main Astronomical Observatory, Kiev, Ukraine SteelBreeze, Square-root information filter
SHA	Shanghai Astronomical Observatory, Shanghai, P.R. China Calc/Solve, Gauß-Markov Least Squares Adjustment
USNO	U.S. Naval Observatory, Washington D.C., USA Calc/Solve, Gauß-Markov Least Squares Adjustment

Each analysis center generates one file for each VLBI session containing normal equation coefficients at least for site coordinates and EOP. Coefficients for auxiliary parameters, e.g. atmosphere and clocks, may also be included.

The combination itself along the lines as described in Sect. 3.4 is not yet in an operational stage since a number of steps still have to be carried out semi-automatically.

- A conversion program also accounting for different versions of the format of the SINEX files (ASCII) into binary DOGS-CS input files had to be developed and is being applied.
- All parameters not belonging to the list of EOP and station coordinates have to be extracted by applying the respective reduction steps (Brockmann, 1997).
- In the period of transition between the old and the new IAU paradigm for nutation, the IVS Analysis Centers tend to use different nutation models. Since these cannot be combined reduction steps for the nutation coefficients have to be applied as well. Nutation parameters are, thus, not included in the combination process at the moment.
- Since most analysis centers use different apriori values in setting up the normal equations, the input solutions have to be transformed to an equal set of apriori values for the TRF and the EOP. At this step, a first check for outliers can be carried out when an identical datum definition is applied to all normal matrices and EOP are estimated. If the residuals do not exceed a certain threshold (e.g. 0.3 mas for polar motion and 0.02 ms for UT1) the solution can safely be included in the subsequent combination process.

The final step starts with the accumulation of datum-free normal equation matrices for each observing session. Then, the resulting normal equation system can be solved after imposing a datum definition as mentioned above. For the generation of the IVS input to ITRF2004 the analysis centers re-processed almost all available VLBI data from 1980 onwards (see statistics in Table 5). The IVS column indicates the number of sessions which have been combined successfully using input data of at least four analysis centers. An example of combined EOP results for a 24h-session is depicted in Fig. 3.

Table 5. Sessions processed by IVS Analysis Centers for ITRF2004

IVS Analysis center:	AUS	BKG	DGFI	GSFC	MAO	SHA	USNO	IVS
Processed sessions:	2481	3107	2642	3940	3441	3508	1220	2038

4 Conclusions

The combination of VLBI data from different analysis centers is a pre-requisite for an integration of the VLBI results into the combination of geodetic space techniques. Methods and procedures have been developed and successfully implemented for operational combinations of EOP time series.

SINEX – Combination (Fri Jul 1 2005 – 12:19:18) 1 / 2

Session: **Database: 99JUN29XE**

settings for aprioris: APRIORI SINEX BKG

settings for REF_EPOCH: MEAN

settings for datum definition: COND FIX 'S*' SIG=0.0001

comment:

AC	St.	# obs	# par.	VAR.–FAC.	weight for datum	scaling factor	weight factor	variance fac. (s0)
AUS	5	1069	46 / 45	1.000	1.000	1.064	1.000	0.996
BKG	5	1038	282 / 23	1.000	1.000	1.508	1.000	1.017
DGFI	5	1277	23 / 23	1.000	1.000	0.887	1.000	0.732
GSFC	5	1069	542 / 23	1.000	1.000	0.898	1.000	1.324
MAO	5	1060	168 / 20	1.000	1.000	1.568	1.000	1.176
SHA	5	1069	522 / 23	1.000	1.000	0.723	1.000	1.426
USNO	5	1060	545 / 23	1.000	1.000	0.881	1.000	1.354
COMBI					0.000			1.176

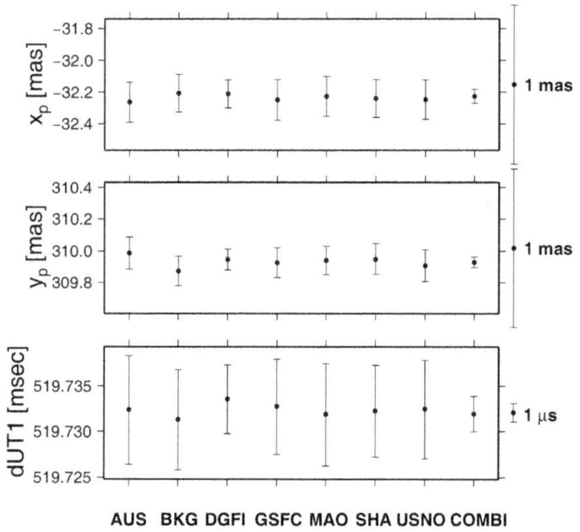

Fig. 3. Example for combined EOP results

The combination of EOP and TRF together in the form of datum-free normal equations has reached a certain level of maturity. The methods and procedures developed here can be applied to input data which has been generated for this very purpose regarding special conditions and configurations. Further developments have to be carried out in order to apply the procedures to all types of input configurations. The theory for the transformation of nutation parameters based on different nutation models into one common set has still to be developed. Furthermore, improved strategies for outlier detection, scaling and weighting algorithms for the input matrices and procedures for a fully automated combination procedure covering all possible input scenarios need to be devised. Only then can the combination and generation of EOP be routinely carried out on the basis of joint EOP and TRF information.

Acknowledgement. This is publication no. GEOTECH-165 of the programme GEOTECHNOLOGIEN of BMBF and DFG, Grant 03F0336B. The combination of genuine VLBI data would not have been possible without the many individuals working at IVS Analysis Centers and regularly producing solutions according to our specifications. We are thankful for all contributions.

References

Brockmann E (1997) Combination of Solutions for Geodetic and Geodynamic Applications of the Global Positioning System (GPS), Geodätisch-geophysikalische Arbeiten in der Schweiz, Schweizerische Geodätische Kommission, Band 55

Cannon W H (2001) Comments on VLBI as a Fundamental Geodetic Positioning Technique and IVS Combined VLBI Products; Memorandum available electronically from http://ivscc.gsfc.nasa.gov under Analysis Coordinator.

Capitaine N, B Guinot, J Souchay (1986) A Non-rotating Origin on the Instantaneous Equator: Definition, Properties and Use; Celstial Mechanics, **39**, 283 – 307

IERS (2002) IAU Resolutions Adopted at the 24th IAU General Assembly (B1.1 – B1.9, B2); IERS Technical Note No. 29, Frankfurt a.M., 117 – 128

IERS (1996) 1995 IERS Annual Report, Observatoire de Paris, Paris

IERS (1999) 1998 IERS Annual Report, Observatoire de Paris, Paris

IVS (2005) International VLBI Service for Geodesy and Astrometry 2004 Annual Report, Behrend and K. Baver (eds.), NASA/TP-2005-212772

Kelm, R (2003) Rank Defect Analysis and Variance Component Estimation for Inter-Technique Combination, Proceedings of the IERS Workshop on Combination Research and Global Geophysical Fluids. Bavarian Academy of Sciences, Munich, Germany, 18 - 21 November 2002. (IERS Technical Note; 30) Verlag des Bundesamts für Kartographie und Geodäsie, Frankfurt am Main

Koch, K-R (1988) Parameter Estimation and Hypothesis Testing in Linear Models, Springer, Berlin

Kouba J, Y Mireault (1997) Analysis Coordinator Report; IGS Annual Report 1996, JPL Pasadena CA, 55 – 100

Lay D (2000) Linear Algebra and its Applications, Second Edition, Addison-Wesley

Lieske J H , T Lederle, W Fricke, B Morando (1977) Expression for the Precession Quantities Based upon the IAU 1976 System of Astronomical Constants; Astronomy & Astrophysics, **58**, 1 – 16

Ma C, E F Arias, T M Eubanks, A L Fey, A-M Gontier, C S Jacobs, O J Sovers, B A Archinal, P Charlot (1998) The International Celestial Reference Frame realized by VLBI; Astron. J., **116**, 516 – 546

Mikhail E M (1976) Observations and Least Squares, IEP-A Dun-Donnelley Publisher

Nothnagel A (2003) VTRF2003: A Conventional VLBI Terrestrial Reference Frame; Proceedings of the 16th Working Meeting on European VLBI for Geodesy and Astrometry, held at Leipzig, May 09 10, 2003, edited by W. Schwegmann and V.

Thorandt, Bundesamt für Kartographie und Geodäsie, Frankfurt/Leipzig, 195 - 205

Nothnagel A, C Steinforth (2001) Analysis Coordinator Report 2000, In: International VLBI Service for Geodesy and Astrometry 2000 Annual Report, NASA/TP–2001-209979, N. R. Vandenberg and K. D. Baver (eds.), 55–64

Richter B (1995) Die Parametrisierung der Erdorientierung. Zeitschrift für Vermessungswesen, **120**, 109 – 119

Rothacher M, R Dill, D Thaller (2005) IERS Analysis Coordination, this issue

Towards a Rigorous Combination of Space Geodetic Observations for IERS Product Generation

Detlef Angermann[1], Rainer Kelm[1], Manuela Krügel[1], Barbara Meisel[1], Horst Müller[1], Volker Tesmer[1], Daniela Thaller[2], and Robert Dill[2]

[1] Deutsches Geodätisches Forschungsinstitut, München, angermann@dgfi.badw.de
[2] Forschungseinrichtung Satellitengeodäsie, Technische Universität München, daniela.thaller@bv.tum.de

Summary. This paper focusses on some relevant aspects when proceeding towards a rigourous combination of space geodetic observations that are addressed by the IERS Combination Research Centers at DGFI and FESG in the frame of the IERS-GEOTECHNOLOGIEN project. These are in particular investigations related to the datum realization for the space geodetic solutions, the analysis and combination of SLR and VLBI data, and research activities within the IERS SINEX Combination Campaign and the IERS Combination Pilot Project.

Key words: IERS products, geodetic datum, combination methods, SLR, VLBI, SINEX Combination Campaign, Combination Pilot Project

1 Introduction and motivation

The core products of the International Earth Rotation and Reference Systems Service (IERS) comprise the International Terrestrial Reference Frame (ITRF), the International Celestial Reference Frame (ICRF), the Earth Orientation Parameters (EOPs), the global geophysical fluids, and the standards and constants given in the IERS Conventions (see Terms of Reference, http://www.iers.org/iers/about/tor). The contributing space geodetic observation techniques are Very Long Baseline Interferometry (VLBI), Satellite and Lunar Laser Ranging (SLR/LLR), the Global Positioning System (GPS) and the Doppler Orbitography and Radiopositioning Integrated by Satellite (DORIS). Each of these space techniques has its strengths and weaknesses and contributes in a different and unique way to the determination of geodetic parameters (e.g., site positions and velocities, EOPs, quasar positions).

The goal of the combination is to make optimal use of the complimentary properties of the different techniques. Besides this, it has to be considered that the accuracy achieved today is mainly limited by technique-related and/or solution-related systematic effects, which are often poorly characterized or quantified. This can lead to highly optimistic precision estimates that provide too optimistic accuracy expectations. Both, the different characteristics of the various space techniques for determining geodetic parameters and the differences between space techniques observations (solutions) strongly require the development of rigorous integration and combination methods. This is of vital importance to exploit the full potential of the space techniques and to provide highly accurate and consistent results.

The analysis and combination of data (solutions) of the same observation technique is one of the major tasks of the geodetic services, the International GNSS Service (IGS, see http://igscb.jpl.nasa.gov), the International Laser Ranging Service (ILRS, see http: //ilrs.gsfc.nasa.gov), the International VLBI Service for Geodesy and Astrometry (IVS, see http://ivscc.gsfc.nasa.gov), and the International DORIS Service (IDS, see http://ids.cls.fr). During the last few years these services gained significant experiences and made a lot of progress regarding the intra-technique combination.

The IERS is in charge of the combination of different observation types, so-called inter-technique combinations. However, at present the IERS products (ITRF, ICRF and EOPs) are computed (combined) separately by different product centers. Consequently, the results are not consistent, e.g., different ITRF realizations produce offsets and drifts in the EOP series (Rothacher, 2000). The results of the IERS Analysis Campaign to align EOPs to ITRF2000/ICRF reveal that significant biases between EOP series exist (Dill and Rothacher, 2003; Gambis and Bizouard, 2003; Nothnagel et al., 2003). The discrepancies are observed not only between different techniques, but also between various analysis centers of the same technique. This means that there are clear deficiencies in the space geodetic solutions and in the present IERS product generation. The accuracy is limited by, e.g.: (a) systematic differences in the reference frames realized by different techniques resulting in offsets and drifts in individual EOP series, (b) offsets in geocenter positions between different techniques and solutions, (c) systematic effects in the time series of station positions, (d) unmodeled nonlinear effects (e.g. seasonal variations, seismic deformations) in site motions.

Up to now, combinations are primarily performed on the solution level. This strategy may lead to deformed solutions if individual solutions with badly or not clearly documented constraints are included in the combination, as it was the case for some of the ITRF2000 contributions. Furthermore, the input data provided by various analysis centers must be consistent concerning modeling and parameterization. This requires the adoption of common standards and models according to the most recent set of conventions (McCarthy and Petit, 2004), which is currently not always fulfilled by the different processing software packages. Other deficiencies regarding the combination methodology include, e.g., the weighting, handling of biases and the datum realization.

This paper focusses on some relevant aspects when proceeding towards a rigorous combination of space geodetic observations that were addressed by the IERS Combination Research Centers (CRC) of the Forschungseinrichtung Satellitengeodäsie of the Technical University Munich (FESG) and DGFI in the framework of the joint project "Integration der geodätischen Raumverfahren und Aufbau eines Nutzerzentrums im Rahmen des Internationalen Erdrotationsdienstes (IERS)" within the "Sonderprogramm GEOTECHNOLOGIEN" of BMBF and DFG. The topics addressed in this paper are closely related to the other CRC contribution of FESG and DGFI published in this volume (Thaller et al., 2005).

2 Datum realization

The TRF datum is defined by the Earth's center of mass (geocenter) as the origin, a mean Earth rotation vector for the orientation, and a scale given by the velocity of light, as well as the time evolution of these parameters.

The individual space geodetic observations do not contain the complete information to realize the TRF datum. Satellite methods, such as SLR, GPS and DORIS, are (more or less) sensitive to geocenter motions relative to the TRF, because they use the geocenter as dynamical origin for computing the satellite orbits. VLBI is quasi-independent of the gravity field and does not contribute at all to the realization of the geocenter. Thus, it is necessary to introduce no-net-translation (NNT) conditions for VLBI.

Since the orientation of the frame is attached to a mean rotation vector which can only be defined with respect to an external frame, and since the computation of the satellite orbits needs an external inertial frame, and since both external frames are supposed to coincide in the ICRF, we have to solve for the EOPs connecting the TRF with the adopted external quasi inertial frame. The separation of station positions and velocities from the EOP is achieved by appropriate condition equations, the no-net-rotation (NNR) conditions, minimizing the common rotation of the TRF solution w.r.t. its approximate values for the orientation at the reference epoch, and minimizing the horizontal velocity field over the whole Earth for the time evolution of orientation.

Each individual space technique has its own characteristic influence on the geodetic datum of the combined solution. Therefore it is an important issue to study the strengths and influence of the geodetic datum contribution of each technique. For that purpose we apply a method suggested by (Sillard and Boucher, 2001), where the covariance information of the datum parameters is calculated by

$$C_{datum} = (G^T C^{-1} G)^{-1}$$

C is the network covariance obtained by applying loose constraints on station positions and G is the matrix of a differential similarity transformation.

The investigations are performed on the basis of weekly and daily solutions for GPS, SLR, DORIS and VLBI. Examples for the formal errors and correlation matrix of the datum parameters of the different techniques are displayed in Color Fig. XLVI on p. 314.

The origin of the SLR solution shows a small standard deviation as SLR is sensitive to the geocenter. The z-component has a higher standard deviation than the x- and y-component. This result is consistent with theoretical considerations, as the origin in z is determined by the disturbed satellite motion due to the influence of the Earth's gravity field only, whereas the origin x and y is additionally determined by Earth rotation.

The rotations in the DORIS solution are fixed to ITRF2000, which is shown by standard deviations close to zero and small correlations with other datum parameters. The standard deviations for the translation parameters and the scale are larger compared to SLR, but the correlations are smaller probably due to a better DORIS network geometry.

The GPS solution is free w.r.t. rotations. The standard deviations of the translations and scale show a similar behaviour as the other satellite techniques. The small correlation of the z-translation and the scale might be caused by the weaker network density in the southern hemisphere.

As expected, the origin and the rotations are free in the VLBI solution which is shown by the standard deviations that are in the same order of magnitude as the loose constraints. According to theory, the scale is well determined and not correlated to the other datum parameters.

To gain further insight into the characteristics and the contributions of the different space techniques to the realization of the datum, we analysed time series of scale and translation variations obtained from weekly SLR, GPS and DORIS solutions as well as from daily VLBI session solutions (Angermann et al., 2005; Meisel et al., 2005). The time series of scale and translation (origin) variations w.r.t. ITRF2000 derived from various individual solutions are shown in Fig. 1 and 2.

In principle, the VLBI and SLR scales are in good agreement with the ITRF2000 scale. The VLBI scale variations of the daily session solutions have a higher noise level than the weekly solutions of the other techniques, mainly due to the fact that the solutions span only one day, and due to the relatively poor network geometry of a 24 h VLBI session. The DORIS scale has an offset of about 4 ppb w.r.t. ITRF2000. The three GPS series (CODE, JPL and SIO) agree well (within 1 ppb) during the last two years, whereas before 2000 larger discrepancies and some irregularities exist. The significant jump of about 2 ppb in the SIO scale in early 2000 was probably caused by a change of the elevation cut-off angle.

The most stable results for the translation parameters were obtained from SLR. The SLR solutions computed by two ILRS analysis centers, one at ASI (Agenzia Spatiale Italiano) with the software GEODYN, and one at DGFI with the DOGS software agree well. The time series for the translation parameters derived from the GPS and DORIS solutions show larger variations than SLR, in particular for the z-component.

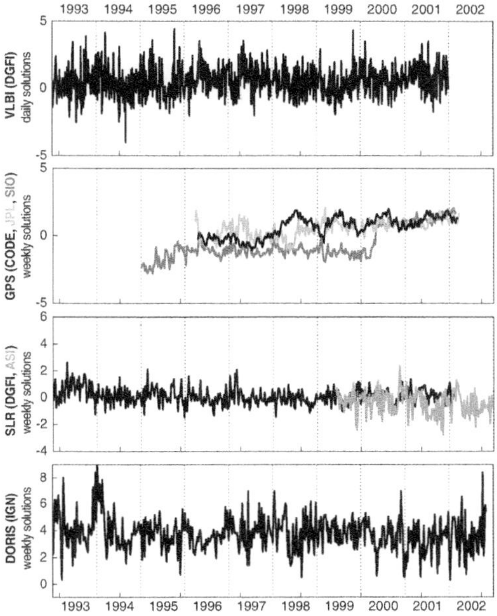

Fig. 1. Time series of scale variations in parts per billion [ppb] of different solutions

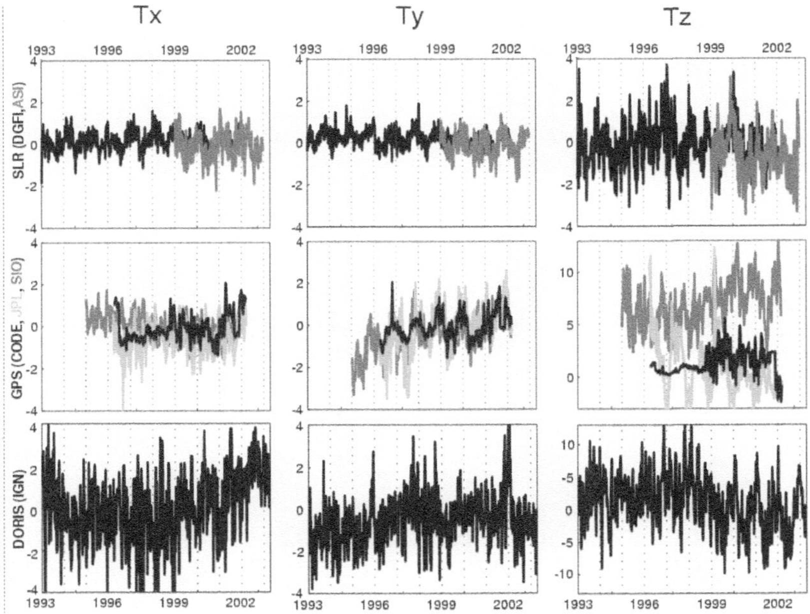

Fig. 2. Time series of weekly translation variations [cm] of different solutions

The amplitudes of the annual signals in the translation variations obtained from the different solutions are summarized in Table 1. Both SLR solutions show amplitudes of a few millimeters for the translation components and the observed differences are within their standard deviations. The amplitudes of the GPS solutions differ in particular for the z-translation, which is obviously by far too large in case of the JPL solution. The annual amplitude in the z-component of the DORIS solution is not significant.

Table 2 shows the results of the SLR solution of DGFI in comparison with geophysically derived model results (Chen et al., 1999; Dong et al., 1997) and with GPS results derived from a degree-one approach (Blewitt et al., 2001; Dong et al., 2003). The SLR results are in good agreement with the model results, whereas the z-component of the GPS-derived results is obviously too large by a factor two. Furthermore, the discrepancies between the two GPS solutions are significantly larger than the estimated standard deviations, in particular for the annual amplitudes of the z-component.

Table 1. Amplitudes of annual signals of translation variations derived from different space geodetic solutions

Technique	AC	Software	Annual Signal (Amplitude)		
			X [mm]	Y [mm]	Z [mm]
SLR	ASI	GEODYN	4.1 ± 1.2	4.5 ± 1.1	3.7 ± 2.3
SLR	DGFI	DOGS	3.1 ± 0.6	3.0 ± 0.6	4.5 ± 1.3
GPS	CODE	Bernese	2.3 ± 1.1	3.6 ± 0.9	5.0 ± 2.6
GPS	JPL	GIPSY	3.5 ± 1.5	7.8 ± 1.7	19.1 ± 4.1
GPS	SIO	GAMIT	0.5 ± 0.9	5.2 ± 1.4	9.8 ± 9.2
DORIS	IGN/JPL	GIPSY/OASIS	5.9 ± 1.8	4.7 ± 1.4	1.1 ± 0.7

Table 2. External comparison of translation (origin) variations

Source	Data	X		Y		Z	
		A [mm]	Φ [deg]	A [mm]	Φ [deg]	A [mm]	Φ [deg]
DGFI	SLR	3.1 ± 0.6	222 ± 12	3.0 ± 0.6	316 ± 11	4.5 ± 1.3	245 ± 12
Dong 1997	model	4.2	224	3.2	339	3.5	235
Chen 1999	model	2.4	244	2.0	270	4.1	228
Blewitt 2001	GPS	3.3 ± 0.3	184 ± 3	4.8 ± 0.3	285 ± 3	$11. \pm 0.2$	214 ± 1
Dong 2003	GPS	2.1 ± 0.3	224 ± 7	3.3 ± 0.3	297 ± 3	7.1 ± 0.3	232 ± 3

3 Analysis and combination of SLR data

SLR data to LAGEOS-1 and LAGEOS-2 are a fundamental basis for the realization of the origin (centre of mass) and scale of the terrestrial reference frame. Hence a continuous evolution and improvement of the SLR station coordinates is necessary. The most recent standards and models (McCarthy and Petit, 2004) allow a more precise modelling for the processing of SLR observations. Therefore we have reprocessed all LAGEOS tracking data back to 1981 using the latest version of the software package DOGS (DGFI Orbit and Geodetic parameter estimation Software) developed at DGFI. More information on the DOGS software is available from http://ilrsac.dgfi.badw.de/dogs/index.html.

The basis for the computations were weekly single satellite arcs. Figure 3 shows the r.m.s. fit of the weekly arcs over the entire data time span. During the first years (1981-1984) the tracking precision improved rapidly. Until 1993 the accuracy level was about 3 cm. Since the launch of LAGEOS-2 the 1 cm level was nearly reached, but for some weeks the accuracy is worse, probably due to tracking problems of some stations. Since about 2000 almost all stations have reached a high tracking performance, so the weekly r.m.s. is below 1 cm.

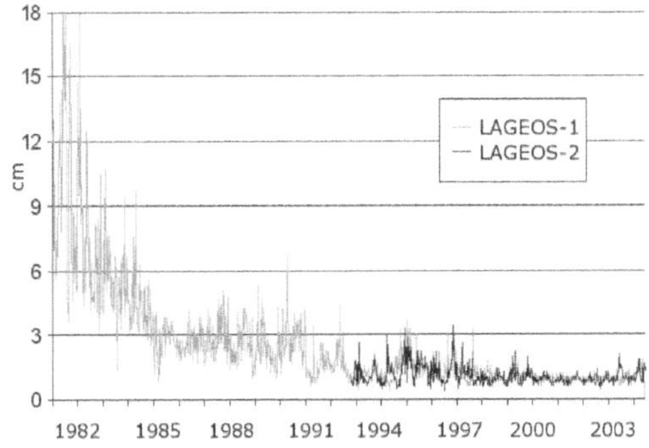

Fig. 3. RMS fit of weekly SLR solutions

The weekly arcs were accumulated to a consistent multi-year SLR solution (Müller et al., 2005). In addition to station positions, velocities and Earth Orientation Parameters (EOP) we solved also for low-degree spherical harmonic coefficients of the Earth's gravity field. As an example Fig. 4 shows the weekly J_2 values, estimated independently for LAGEOS-1 and LAGEOS-2. The higher scatter before 1985 is in accordance with the worse data quality. In general there is a good agreement between the estimates of both satellites. However, the higher noise after 2000 for LAGEOS-1 and the discontinuity in the year 2000 is not yet clear and subject of further investigations.

The multi-year SLR solution serves as reference for various issues, such as the bias estimation for the tracking stations, the operational weekly compu-

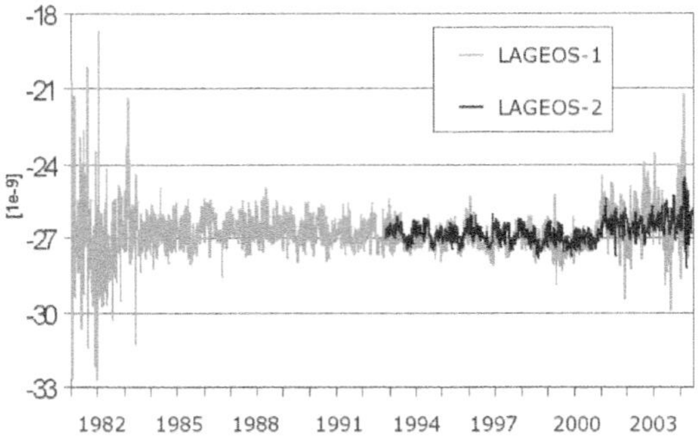

Fig. 4. J_2 values relative to 1.0826e-03

tations and combinations of SLR solutions within the ILRS, the weekly inter-technique combination in the frame of the IERS Combination Pilot Project, and for the computation of a refined terrestrial reference frame (Angermann et al., 2004; Drewes et al., 2005; Meisel et al., 2005).

Within the ILRS, the Analysis Working Group (AWG) has initiated a number of pilot projects on "positioning and Earth orientation", which aimed at the development of a unique and official ILRS product on station positions and EOPs. Besides this, a pilot project aims at the benchmarking of software and analysis procedures in use by the various analysis groups. During the ILRS AWG Meeting on April 22-23, 2004, in Nice, France, five groups (that have passed the benchmarking) have been nominated as official ILRS Analysis Centres: ASI (Agencia Spatiale Italiano), GFZ (GeoForschungsZentrum Potsdam), JCET (Joint Center for Earth Systems Technology, USA), NSGF (Natural Environment Research Council, NERC, Space Geodesy Facility, UK) and DGFI. The ILRS Analysis Centres are processing on an operational weekly basis SLR data to Lageos-1/2 and Etalon-1/2 and provide weekly loosely constrained solutions (SINEX files) with station positions and Earth orientation parameters (x-pole, y-pole and length of day).

Since June 2004, DGFI serves as official ILRS Backup Combination Centre. The function of the ILRS Primary Combination Centre has been taken over by ASI, Italy. Both centres are obliged to weekly compute a combined SLR solution as official product of the ILRS. The products are stored at the data centres of CDDIS and EDC. ASI combines the individual SLR solutions on the level of loosely constrained solutions (Bianco et al., 2003). DGFI performes the SLR intra-technique combination on the level of unconstrained normal equations. The processing flow and combination methodology applied at DGFI is described for example in (Kelm, 2003). The official weekly products are: (1) a combined solution for station coordinates and EOP, and (2) a combined solution for EOP aligned to ITRF2000.

As an example for the combined ILRS products Color Fig. XLV on p. 313 displays the weekly translation parameters of the five individual SLR solutions and the combined solutions of ASI and DGFI. In general there is a rather good agreement between the two combined solutions, but there is an offset of a few millimeter in the x-component and there are some weeks with too large discrepancies which need further investigation. For all three translation components the combined solutions are much smoother than the individual contributions, which proves the quality of the combined ILRS products.

4 Simultaneous estimation of a TRF, the EOP and a CRF with VLBI data

VLBI is the unique observation technique to provide the direct link between the celestial and terrestrial frame, including fully consistent time series of parameters to transform between the frames (pole coordinates and their first derivatives, dUT1 and LOD, as well as daily corrections to a precession-nutation model). Furthermore, the several million VLBI observations of the last 24 years can easily be reprocessed in one common solution, which ensures consistent results over the whole time period.

A VLBI solution with simultaneous estimation of station positions and velocities (TRF), celestial coordinates of the radio sources (CRF) and the full set of Earth orientation parameters (EOP) was computed at DGFI. First results of such a solution, still containing some deficiencies, are presented in Tesmer et al. (2004). The datum is realized by NNR and NNT conditions for the terrestrial (ITRF2000) and NNR for the celestial (ICRF-Ext1) frame (Ma et al., 1998). With this method, biases can be avoided which are due to fixed reference frames or other relevant parameters of the observation equations.

For 2614 seesions between 1984 and 2004, normal equations were set up with the VLBI software OCCAM 6.0 (Titov et al., 2004), including a total of 49 telescopes (of which 45 are well determined also in the ITRF2000) observing 1955 sources (of which 561 are part of the ICRF-Ext1). The auxiliary parameters (for troposphere and clocks) are reduced for each session. All pre-reduced session-wise normal equations are then accumulated to one normal equation system with the DGFI software DOGS-CS and solved with an appropriate datum, namely NNR and NNT for 25 stable stations w.r.t. ITRF2000 and NNR for 199 "stable" sources w.r.t. ICRF-Ext1.

A major focus in this paper is the comparison of this completely undistorted VLBI solution with the official IERS products, namely the IERS C04 series for the EOP parameters, the ITRF2000, and the ICRF-Ext1. For a validation of the consistency of the IERS reference frames, three solution types were computed: (1) the standard solution with a simultaneous adjustment of EOP, TRF and CRF; (2) a solution fixing station positions and velocities to ITRF2000 (for 45 stations) and estimating EOP and quasar positions; and (3) a solution fixing the CRF to the coordinates of ICRF-Ext1 (for 561 stable sources) and estimating EOP, station positions and velocities.

The numerical results of a comparison of the EOP from solution (1) and IERS C04 are displayed in Tables 3 and 4, a graphical representation of the data after 1998 is provided in Fig. 5. It has to be considered, that the VLBI data used at DGFI also contain sessions with few observing telescopes and rather short baselines (e.g. European sessions) which do not allow for precise EOP estimation. The RMS and WRMS values in Table 4 indicate that the different accuracy level of the VLBI EOP estimations is rather well reflected by the formal errors of the adjustment. Assuming that this VLBI solution is free of (systematic) errors, the results indicate inconsistencies between the IERS C04, ICRF-Ext1 and (the VLBI part) of ITRF2000. These inconsistencies, such as offsets (a rotation of station positions), drifts (a rotation of station velocities) and periodic effects (especially for the early years up to 1990) are obvious and vary over the entire period of 20 years. Inconsistencies between the IERS C04 EOP series and ITRF2000 were also observed in the framework of the IERS EOP Alignment Campaign (Dill and Rothacher, 2003; Gambis and Bizouard, 2003; Nothnagel et al., 2003). The results clearly show, that there is an urgent need for the development of rigorous combination methods for the generation of consistent IERS products, as for example envisaged within the IERS Combination Pilot Project (Rothacher et al., 2005).

Table 3. Offsets and rates of the EOP between the solution (type 1) and IERS C04 for different time spans. The offsets are related to the mean epochs of the intervals

Period	Offsets			Rates		
	Xpol [μas]	Ypol [μas]	dUT1 [μs]	Xpol [μas/y]	Ypol [μas/y]	dUT1 [μs/y]
84-04	-159 ± 6	28 ± 5	0.4 ± 0.5	71 ± 1	27 ± 1	0.5 ± 0.1
94-04	-123 ± 4	135 ± 4	3.4 ± 0.4	-13 ± 1	32 ± 1	0.1 ± 0.1
98-04	-139 ± 4	216 ± 4	5.1 ± 0.3	-30 ± 2	18 ± 2	-1.4 ± 0.2

Table 4. RMS and WRMS values for the EOP differences between the solution (1) and IERS C04 for different time periods (after applying the offsets and rates of Table 3). This table also contains values for the session-wise estimated EOP rates

Period		Xpol [μas]	Ypol [μas]	dUT1 [μs]	Xpol dot [μas/d]	Ypol dot [μas/d]	dUT1 dot [μs/d]
84-04	RMS	577	521	30.2	1752	1778	85.7
84-04	WRMS	185	151	13.6	454	436	26.5
94-04	RMS	344	253	20.9	801	826	49.5
94-04	WRMS	150	130	12.8	383	369	24.6
98-04	RMS	281	202	14.9	638	688	40.2
98-04	WRMS	121	107	9.2	344	345	23.9

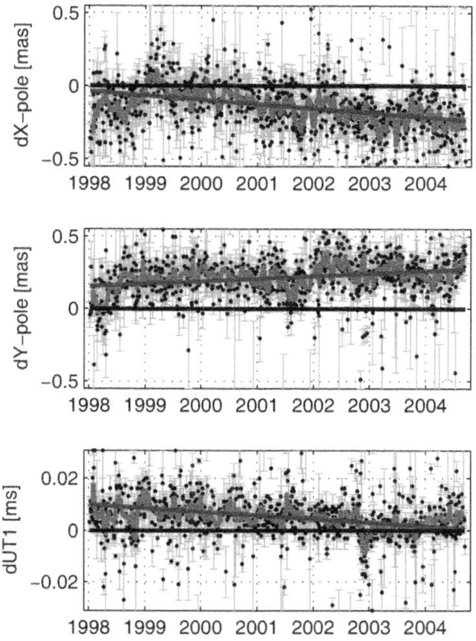

Fig. 5. Differences between the EOP of solution (1) and IERS C04 from 1998 until 2004 (solid curves show the median of 10 values and a best-fitting linear function)

The source positions obtained from the undistorted standard solution (type 1) are compared with those of solution type 2 (ITRF2000 fixed) to assess how the parameters of the terrestrial and celestial VLBI reference frames interact. Figure 6 shows the differences for the estimated source positions in declination (DE) and right ascension (RA × cos(DE)) between both solution types as a function of the declination. A systematic behaviour can be observed especially in the declination. These systematic effects show very clear resemblance with results of (MacMillan and Ma, 1997), except for the magnitude, which was up to almost 0.5 mas in that study, compared to 0.15 mas here. In the study of (MacMillan and Ma, 1997) two solutions were compared, where horizontal gradients for the tropospheric parameters were estimated in one case and in the other not. This suggests, that at least some of the ITRF2000 contributions handled the effect of the horizontal gradients differently than our VLBI solution (in which we assume this affect is handled properly).

Finally, the station positions of the standard solution type (1) are compared with those obtained from solution type 3 (ICRF-Ext1 fixed) to estimate the effect of the celestial on the terrestrial reference frame. The RMS values of the differences of the station positions and velocity estimates between the two solutions are 0.7/0.5 mm and 0.15/0.07 mm/yr for the north/east, and 1.3 mm and 0.16 mm/yr for the height component of the 45 stations compared. Although these values are quite small, some systematics become very clear:

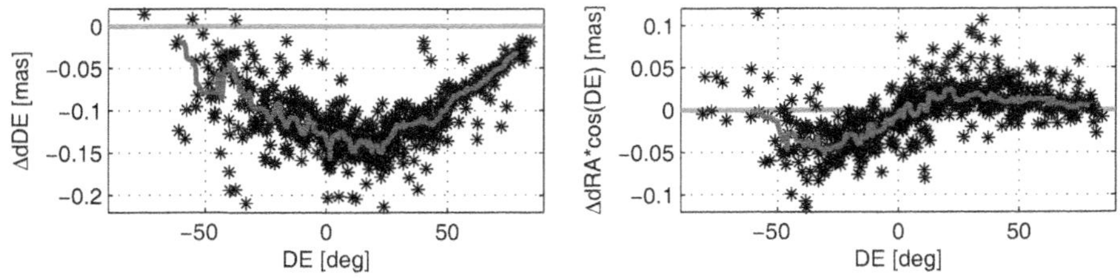

Fig. 6. Differences between source position estimates of the undistorted VLBI solution type (1) compared to type (2) (ITRF2000 fixed)

The largest differences become apparent for all stations in the south (which all have differences in the north component of the coordinates with the same sign larger than 1.3 mm), and for many of the stations, which did not observe regularly. This suggests, that parameters (coordinates and velocities) of VLBI stations that are subject to an observing constitution that provides a merely partial redundant information (e.g. because they observe mostly in the same networks, the same sources with the same topocentric observing geometry, etc.) are not completely uncoupled frame parameters of the VLBI-CRF. In this context, a combination with observations of other techniques will help fundamentally.

5 IERS SINEX Combination Campaign and Combination Pilot Project

Among other activities, the IERS Analysis Coordination has initiated the IERS SINEX Combination Campaign in 2002 to analyse and combine solutions of the different space techniques, to assess systematic biases and to develop suitable methods for a rigorous combination of the IERS products. At the IERS Retreat 2003 in Paris it was decided that the IERS SINEX Combination Campaign should be converted into a pilot project, namely the IERS Combination Pilot Project, to be started in spring 2004, and to prepare the IERS product generation on a weekly basis. The general scope and the objectives of these IERS projects are presented in (Rothacher et al., 2005). In this paper we focus on major activities concerning data analysis and combination performed at DGFI and FESG as CRCs.

In the frame of the SINEX Combination Campaign DGFI has provided SLR, VLBI and (regional) GPS solutions and/or unconstrained normal equations containing station positions and EOPs for the entire year of 1999. The available SINEX solutions are visible at the SINEX file archive, `http://tau.fesg.tu-muenchen.de/~iers/web/sinex/datapool.php`, see also (Angermann et al., 2003). Among other combination groups, DGFI and FESG were involved in the analysis and combination of the weekly SLR, GPS, DORIS and daily VLBI solution files. For this purpose the combination software packages DOGS-CS (at DGFI) and Bernese (at FESG) were updated and suitable combination methods were developed. Major tasks include the analysis of the individual solutions concerning various aspects (e.g., SINEX format, parameter definition, modelling, rank deficiencies, suitability for a rigorous combination), the selection of suitable local ties to connect the different techniques, the weighting of solutions and the datum definition of the individual and the weekly combined solutions. Results of the weekly combination (Krügel and Meisel, 2003; Thaller and Rothacher, 2005) prove the potential of such a weekly combination, though some details need to be further studied.

The IERS Combination Pilot Project (CPP) aims at more consistent, routinely generated IERS products. "Weekly" SINEX solutions, which are available from the different technique services contain station positions, EOPs, and, possibly, quasar coordinates. These solutions shall be rigorously and routinely combined into consistent IERS products. The procedure for the combination and validation of the "weekly" SINEX solutions is presented in (Rothacher et al., 2005). In the frame of the CPP, DGFI provides individual SLR and VLBI solutions and combined SLR solutions for the intra-technique combination (step 1) and has been accepted by the IERS as a combination center for the inter-technique combination (step 2). The presently available SINEX files were analysed regarding the suitability for a rigorous combination. The methodology for the weekly inter-technique combination of station positions and EOPs, including input data check and validation of the results, has been developed and implemented in the DGFI software DOGS-CS.

During the CPP and within the IERS Working Group on Combination it was recognized that the weekly SINEX solutions now routinely generated by the Technique Centers are not sufficient to generate combined inter-technique solutions over longer time periods. It was decided that the most recent IERS realization of the terrestrial reference frame, the ITRF2000, does not fulfil all the needs for the CPP, and thus a refined TRF realization is an essential prerequisite for the weekly rigorous combination of space geodetic observations. In December 2004 a Call for long time series of "weekly" SINEX files for ITRF2004 and a supplement of the IERS CPP was released. The ITRF2004 will be based on the combination of time series of station positions and EOPs. Weekly or (daily VLBI) contributions will allow better monitoring of nonlinear motions and other kinds of discontinuities in the time series. The ITRS Combination Centers, namely DGFI, IGN, and NRCan, led by the ITRS Product Center (IGN), are in charge for the generation of the ITRF2004 solution.

6 Conclusions and outlook

It is obvious, that there are clear deficiencies in the present IERS products, which are computed (combined) almost independently from each other. This paper addressed various issues that are essential for a rigorous combination of space geodetic observations. Now, for the first time, weekly (and daily VLBI) SINEX files with station positions and EOPs became available for the different space techniques. These input data provide the basis for a new ITRF2004 with station positions, velocities and EOP, as well as for the development and implementation of rigorous combination methods for the generation of consistent set of IERS products for ITRF, EOPs and ICRF, as envisaged within the IERS Combination Pilot Project (CPP). Highly accurate and consistent results with highest long-term stability are of vital importance, especially in respect to the challenges the IERS is facing with the new and upcoming satellite missions (gravity, altimetry, astrometry), and for the IAG's Global Geodetic Observing System (GGOS).

Acknowledgement. This is publication no. GEOTECH-161 of the programme Geotechnologien of BMBF and DFG, Grant IERS(03F0336C).

References

Altamimi Z, Sillard P, Boucher C (2002) ITRF2000: A new release of the International Terrestrial Reference Frame for earth science applications. J Geophys Res 107 (B7), 2214, doi:10.1029/2001JB000561

Angermann D, Thaller D, Rothacher M (2003) IERS SINEX Combination Campaign, Proceedings of the IERS Workshop on Combination Research and Global Geophysical Fluids, B. Richter, W. Schwegmann, W.R. Dick (eds), IERS Technical Note 30, 94–101, Bundesamt für Kartographie und Geodäsie

Angermann D, Drewes H, Krügel M, Meisel B, Gerstl M, Kelm R, Müller H, Seemüller W, Tesmer V (2004) ITRS Combination Center at DGFI: A terrestrial reference frame realization 2003. Deutsche Geodätische Kommission, Reihe B, Heft Nr. 313

Angermann D, Krügel M, Meisel B, Müller H, Tesmer V (2005) Time evolution of the terrestrial reference frame, Sanso F (Ed): A Window on the Future of Geodesy, IAG Symposia, Vol. 128, Springer

Bianco G, Devoti R, Luceri C (2003) Combination of loosely constrained solutions. Proceedings of the IERS Workshop on Combination Research and Global Geophysical Fluids, B. Richter, W. Schwegmann, W.R. Dick (eds), IERS Technical Note 30, 107–109, Bundesamt für Kartographie und Geodäsie

Blewitt G, Lavallee D, Clarke P, Nurutdinov K (2001) A new global mode of Earth deformation: Seasonal cycle detected. Science, 294 (5550), pp. 2342–2345

Chen J, Wilson C, Eanes R, Nerem S (1999) Geophysical interpretation of observed geocenter variations. J. Geophys. Res., 104, B2, pp. 2683–2690

Dill R, Rothacher M (2003) IERS analysis campaign to align EOP's to ITRF2000/ICRF, GEOTECHNOLOGIEN Science Report No. 3, 36–39, Koordinierungsbüro Geotechnologien, Potsdam

Dong D, Dickey J, Chao Y, Cheng K (1997) Geocenter variations caused by atmosphere, ocean and surface ground water. Geophysical Res. Lett., Vol. 24, No. 15, pp. 1867–1870

Dong D, Yunck T, Heflin M (2003) Origin of the International Terrestrial Reference Frame, J. Geophys. Res., Vol. 108, No. B4, 2200, doi:10/1029/2002JB0022035

Drewes H, Angermann D, Gerstl M, Krügel M, Meisel B, Seemüller W (2005) Analysis and Refined Computations of the International Terrestrial Reference, this issue

Gambis D, Bizouard C (2003) Consistency analysis between EOP series and reference frames, Proceedings of the IERS Workshop on Combination Research and Global Geophysical Fluids, IERS Technical Note 30, 57–62, Bundesamt für Kartographie und Geodäsie

Kelm R (2003) Automated combination of SLR solutions within ILRS. GEOTECHNOLOGIEN Science Report No. 3, 89–91, Koordinierungsbüro Geotechnologien

Krügel M, Meisel B (2003) DGFI results of IERS SINEX Combination Campaign. GEOTECHNOLOGIEN Science Report No. 3, 96–100, Koordinierungsbüro Geotechnologien, Potsdam

Ma C, Arias EF, Eubanks TM, Fey AL, Gontier AM, Jacobs CS, Sovers OJ, Archinal BA, Charlot P (1998) The International Celestial Reference Frame as realized by Very Long Baseline Interferomety, Astronomical Journal 116, pp. 516

Meisel B, Angermann D, Krüegel M, Drewes H, Gerstl M, Kelm R, Müller H, Seemüller W, Tesmer V (2005) Refined approaches for terrestrial reference frame computations, Adv. Space Res. (in print)

McCarthy D, Petit G (2004) IERS Conventions (2003): D. McCarthy, G. Petit (eds)., IERS Technical Note, 32, Verlag des Bundesamtes für Kartographie und Geodäsie, Frankfurt am Main

MacMillan D, Ma C (1997) Atmospheric Gradients and the VLBI Terrestrial and Celestial Reference Frames. Geoph. Res. Letters, Vol. 24, No. 4, 453–456

Müller H, Angermann D, Meisel B (2005) A multi-year SLR solution. Proceedings of the ILRS Workshop, San Fernando (in print)

Nothnagel A, Dill R, Feissel-Vernier M, Ferland R, Noomen R, Willis P (2003) EOP alignment campaign, IDS/IGS/ILRS/IVS combinations, systematic errors. Proceedings of the IERS Workshop on Combination Research and Global Geophysical Fluids, IERS Technical Note 30, Bundesamt für Kartographie und Geodäsie

Rothacher M (2000) Towards an Integrated Global Geodetic Observing System. In: R. Rummel, H. Drewes, W. Bosch, H. Hornik (eds.), Towards an Integrated Global Geodetic Observing System (IGGOS), International Association of Geodesy Symposia, Vol. 120, 41–52, Springer

Rothacher M, Dill R, Thaller D (2005) IERS Analysis Coordination, this issue

Sillard P, Boucher C (2001) Review of algebraic constraints in terrestrial reference frame datum definition. J. Geod. 75, 63–73

Tesmer V, Kutterer H, Drewes H (2004) Simultaneous estimation of a TRF, the EOP, and a CRF. IVS General Meeting Proc, 311–314, NASA/CP-2004-212255

Titov O, Tesmer V, Böhm J (2004) OCCAMv6.0 software for VLBI data analysis, IVS General Meeting Proceedings, 267–271, NASA/CP-2004-212255

Thaller D, Rothacher M (2003) Comparison and combination of GPS, VLBI and SLR solution series. GEOTECHNOLOGIEN Science Report No. 3, 176–180, Koordinierungsbüro Geotechnologien, Potsdam

Thaller D, Dill R, Krügel M, Steigenberger P, Rothacher M, Tesmer V (2005) CONT02 analysis and combination of long EOP series, this issue

CONT02 Analysis and Combination of Long EOP Series

Daniella Thaller[1], Robert Dill[1], Manuela Krügel[2], Peter Steigenberger[1], Markus Rothacher[1,3], and Volker Tesmer[2]

[1] Forschungseinrichtung Satellitengeodäsie, Technische Universiät München, daniela.thaller@bv.tum.de, dill@bv.tum.de
[2] Deutsches Geodätisches Forschungsinstitut München, kruegel@dgfi.badw.de
[3] GeoForschungsZentrum Potsdam

Summary. This report is divided into two parts: the first part gives an overview of the combination studies performed by the Forschungseinrichtung Satellitengeodäsie TU München (FESG) and the Deutsches Geodätisches Forschungsinstitut (DGFI) based on the data of the continuous IVS campaign CONT02. The close cooperation of the two institutions established the basis for a detailed adaption of the GPS and the VLBI software concerning models and parameterization to avoid systematic differences between the technique contributions. Special attention was payed to parameters with a high temporal resolution; in this study tropospheric parameters and Earth rotation parameters (ERP) are considered. Including the troposphere parameters offered a good possibility to study the correlation between troposphere parameters and station coordinates. It was found that this interaction can deliver a very important contribution to validate the available local tie information. For comparison of the troposphere results derived for the 14-days campaign CONT02, long time series for VLBI and GPS were used as well, and it turned out that the results are in good agreement. Regarding the sub-daily Earth rotation parameters it can be shown that a combination of the space techniques improves the results compared to single-technique solutions. Furthermore, it is illustrated that UT1-UTC can be combined from VLBI together with the satellite techniques. All in all, the presented results demonstrate the high potential of a combination of VLBI, GPS and SLR data. The second part is devoted to the combination of long sub-daily EOP time series from VLBI and GPS. Space geodetic techniques like the Global Positioning System (GPS) and Very Long Baseline Interferometry (VLBI) can provide Earth Orientation Parameter (EOP) time series with very high sampling rates. This offers the opportunity to study sub-daily tidal excitations and the influences of high-frequency or episodic geophysical effects on Earth rotation. Therefore we need sub-daily time series as consistent and homogeneous as possible. Based on the Combined Smoothing method of Vondrak and Cepek (2000), we developed a new combination scheme for sub-daily EOPs to obtain a new sub-daily time series which benefits from the longterm stability of VLBI and the continuity of GPS. Furthermore we can remove the weakness of UT1 estimations of the satellite techniques.

We analysed the combination using spectral imaging methods and the results from sub-daily tidal harmonic estimation.

Key words: rigorous combination, co-location, tropospheric parameters, Earth rotation, Combined smoothing, sub-daily variations, tides, earthquakes

Part I: Combination Studies Using the CONT02 Campaign

(D. Thaller, M. Krügel, P. Steigenberger, M. Rothacher, V. Tesmer)

1 Introduction

This report summarizes the results of a common combination study of the IERS Combination Research Centres at the Forschungseinrichtung Satellitengeodäsie TU München (FESG) and the Deutsches Geodätisches Forschungsinstitut (DGFI). This study is based on the data of a continuous VLBI campaign and data of a global GPS and a SLR network covering the same time span.

The International VLBI Service for Geodesy and Astrometry (IVS) set up a 15-day campaign of continuous VLBI observations, named CONT02, that took place in October 16-31, 2002. Altogether eight VLBI stations participated in this campaign, all of them are co-located with GPS. Due to a broad spectrum of common parameters for GPS and VLBI the studies are mainly concentrated on the combination of these two techniques. As both techniques make use of microwave signals, identical tropospheric parameters, i.e., zenith delays and horizontal gradients, can be estimated from the VLBI and GPS data for each observing station and can be combined together with the station coordinates and Earth orientation parameters (EOP). With respect to other combination studies, one big advantage is the continuous availability of observations of eight VLBI stations for the whole time span.

2 Preparations Concerning Data and Software Packages

For each technique daily normal equations were generated for the time span of CONT02. The VLBI data, usually managed in sessions starting and ending at 17 h UTC had to be concatenated to files from 0 h to 24 h UTC according to the GPS sessions, optimizing the compatibility. The analysis of the VLBI data was done at DGFI using the software OCCAM (Titov et al., 2004). The GPS network, consisting of 153 globally distributed stations, was evaluated at FESG using the Bernese GPS Software 5.0 (Hugentobler et al., 2004). The networks are displayed in Fig. 1. Additionally, a SLR solution was processed with the DOGS-OC and the Bernese GPS Software 5.0, including all Lageos-1 and Lageos-2 data.

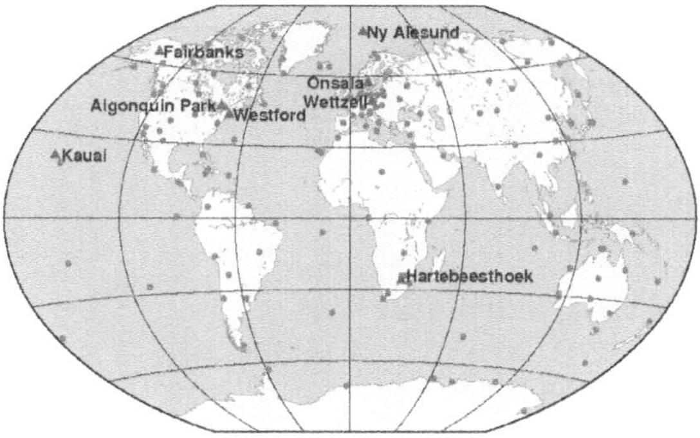

Fig. 1. Station networks: dots: GPS only; triangles: co-location VLBI and GPS

To perform a combination as rigorous as possible the functional models of the techniques, especially VLBI and GPS, were adapted carefully. Therefore, identical a priori models were used (so for: solid Earth tides, pole tide, ocean loading, tropospheric delays, a priori EOP values, as well as their interpolation to the observation epoch, sub-daily EOP model, nutation model). Furthermore, the unknown parameters were set up identically. Table 1 summarizes the temporal resolution and the type of parameterization that were chosen for the estimated parameters. This results in a large amount of parameters, especially due to the sub-daily estimation of Earth rotation parameters (ERP: xpole, ypole, UT1-UTC) and tropospheric zenith delays (ZD). A detailed statistical summary of parameters and observations is given in Thaller et al. (2005b). At the beginning of the CONT02 analysis, a lower temporal resolution, i.e., two hours, was chosen for the ERP and tropospheric zenith delays. The corresponding analyses and results are documented in Thaller et al. (2005a) and Krügel et al. (2004). Based on these results, the temporal resolution was doubled in order to enable the detection of faster changing features, especially in the tropospheric delays. Additionally, a GPS and a VLBI time series computed the same way as the CONT02 solutions were compared over a time span of 11 years (see Steigenberger et al. (2005)). These results are used for validating the tropospheric parameters derived from the CONT02 campaign (see Sect. 3.4, paragraph Long-Time-Series)

Table 1. Temporal resolution and parameterization

Station coordinates	daily	constant
ERP	1 hour	piece-wise linear
Nutation angles	daily	piece-wise linear
Tropospheric ZD	1 hour	piece-wise linear
Tropospheric gradients	daily	constant

3 Combination Results

3.1 Co-location Sites and Local Ties

Terrestrial measurements at co-location sites, so-called local ties, are needed for the combination of the station networks. For all eight co-location sites listed in Table 2 local ties are available.

Table 2. Co-locations between GPS and VLBI during the CONT02 campaign

Site	VLBI	GPS	Source
Ny Alesund	10317S003	10317M003	(2)
Onsala	10402S002	10402M004	(1)
Wettzell	14201S004	14201M010	(1)
Hartebeesthoek	30302S001	30302M004	(1)
Algonquin Park	40104S001	40104M002	(1)
Fairbanks	40408S002	40408M001	(1)
Kokee Park	40424S007	40424M004	(3)
Westford	40440S003	40440S020	(1)

(1) Altamimi et al. (2002)
(2) Steinforth et al. (2003)
(3) Altamimi et al. (2002) and IGSMAIL#4151

A first validation of the local ties was done by comparing them with the coordinate differences derived from the space techniques. At first a GPS solution computed with relative antenna phase center corrections (PCV) was used. Changing to absolute PCV corrections, the height component of the local ties fits much better to the space techniques. Some further modifications, especially the use of new axis offsets for the VLBI antennas (see Homepage IVS Analysis Coordinator[1]), lead to a still better agreement. The corresponding values are given in Table 3.

Looking at the horizontal components, three stations are remarkable: Fairbanks, Westford and Hartebeesthoek. They show differences of about 10 mm or more in one of the two components. The influence of the differences in Hartebeesthoek on the combination results is very small, because this station is very isolated and has large formal errors compared to the other VLBI stations. However, the bias in the Fairbanks and in the Westford tie becomes visible

[1] http://giub.geod.uni-bonn.de/vlbi/IVS-AC/data/axis-offsets.html

Table 3. Differences between space techniques and terrestrial measurements with relative and absolute PCV for GPS and old and new antenna axis offsets for VLBI

	relative PCV old axis offsets			absolute PCV new axis offsets			ITRF2000		
Site	north [mm]	east [mm]	up [mm]	north [mm]	east [mm]	up [mm]	north [mm]	east [mm]	up [mm]
Ny Alesund	-7.0	6.4	-32.8	-2.1	1.4	4.8	0.8	-0.8	-11.8
Onsala	0.6	6.4	-14.2	6.8	-1.2	-2.7	3.6	-1.8	-0.4
Wettzell	-3.4	6.2	3.0	4.9	-1.2	8.9	-0.2	-0.4	1.2
Hartebeesthoek	1.4	21.4	45.2	-1.5	14.4	22.9	-9.2	14.4	11.8
Algonquin Park	2.0	2.6	18.4	-1.2	-3.9	1.7	-3.0	-3.0	15.4
Fairbanks	-9.4	-0.8	38.8	13.1	2.1	-31.8	-16.8	5.4	20.0
Kokee Park	9.8	-6.0	13.0	5.3	2.6	0.5	-4.0	-3.4	14.6
Westford	2.6	14.8	35.2	1.9	9.8	8.9	-2.6	4.6	18.4

in the results as will be shown below. For comparisons the differences as they occure in the ITRF2000 are given additionally: the size of the differences are comparable for some stations (e.g., Fairbanks, Hartebeesthoek), whereas others differ significantly more. However, these values let us assume that a part of the differences identified in CONT02 may result from the chosen reference frame.

3.2 Station Coordinates

The station coordinates are combined using the local ties listed above. Daily repeatabilities (RMS) are chosen to analyze the effect of a combination on the time series of station coordinates. As the formal errors are unrealistically small, in particular in the case of a GPS solution, they are not considered in the analysis. The repeatabilities of the single solutions (GPS- and VLBI-only) and the combined daily solutions are compared in Fig. 2.

For 80% of the VLBI and the GPS coordinate components the repeatabilities improve as a consequence of the combination, although the benefit is visible especially for the VLBI stations. This can be explained by the poorer configuration of the VLBI network compared to GPS. Additionally, it can be recognized that the repeatabilities of the north component of the VLBI stations are improved more than those of the east component. This can be traced back to the sparse distribution of VLBI stations in north-south direction. Stations which do not benefit from the combination in all components are Fairbanks and Westford. For these stations the local tie does not fit well to the space technique solutions (Fairbanks: especially the north component, Westford: especially the east component) and was not introduced in the combination.

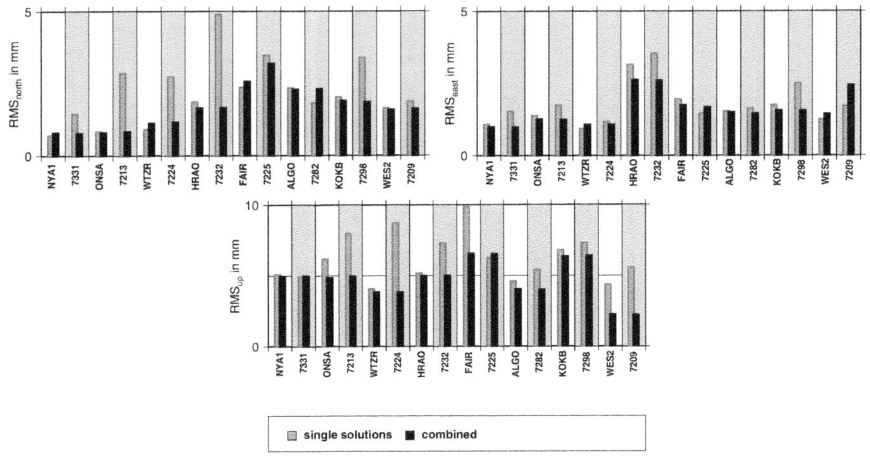

Fig. 2. Repeatabilities of station coordinates derived from single and combined solutions (VLBI stations with a grey background)

3.3 Earth Orientation Parameters (EOP)

The hourly resolution of the ERP allows to analyze sub-daily variations in the ERP time series. Fig. 3 shows the combined x-pole estimates in comparison to an a priori model (C04 and sub-daily pole model IERS2003 derived from altimetry (McCarthy, Petit, 2004)). The very good agreement demonstrates, that the estimation of high-resolution ERP gives very reasonable results.

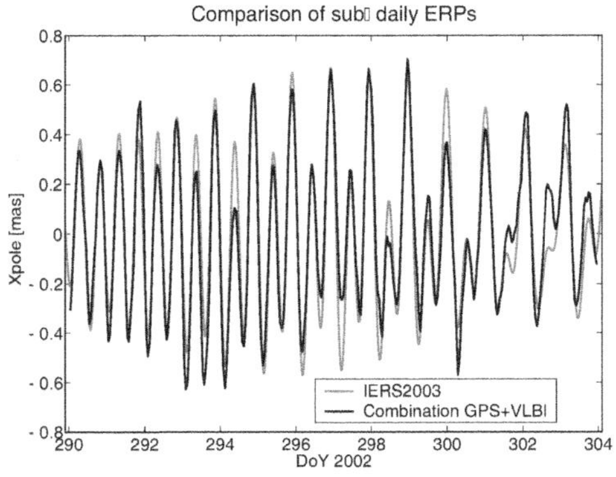

Fig. 3. Comparison of estimated x-pole and the a priori model (C04/IERS2003)

In the sub-daily estimates of combined ERP some periodical differences to the a priori model are visible. These variations exist also in the GPS and VLBI single solutions and agree very well for the two techniques. Hence, we conclude that the combination of the ERP should be beneficial. The results summarized in Table 4 point up that the ERP estimation of both single technique solutions can be improved by a combination.

Table 4. RMS values of the remaining differences to C04/IERS2003 (offset removed)

	GPS	VLBI	Combination
X-pole [mas]	0.143	0.259	0.120
Y-pole [mas]	0.144	0.253	0.130
UT1-UTC [ms]	-	0.015	0.011

The essential contribution to the determination of UT1-UTC and the nutation time series comes from VLBI, because VLBI is the only technique to determine the offsets. From observations of satellite techniques only the corresponding rates can be derived. As an example, the hourly UT1-UTC values estimated by VLBI and GPS w.r.t. the official C04-series and the IERS2003 sub-daily model are shown in Fig. 4. In spite of the big drift in the GPS estimates the combined UT1-UTC is perfectly aligned to the VLBI estimates. The RMS of the remaining differences to C04/IERS2003 can even be reduced from 0.015 ms for the VLBI-only solution to 0.011 ms for the combined solution, clearly demonstrating the advantage of an inter-technique combination.

The capability of VLBI to determine UT1-UTC and the nutation offsets is sustained in the combination and not disturbed by the satellite techniques as it was often alleged. The reason can be seen in the a posteriori formal errors of the estimated parameters displayed in Fig. 5a and 5b for UT1-UTC and the nutation in obliquity, respectively. Whereas the VLBI estimates are more or less of equal accuracy during the entire time span, the UT1-UTC and nutation estimates from the satellite techniques heavily degrade with time. This can be shown for GPS and SLR similarly. Daily values for UT1-UTC were generated for these two figures in order to show reasonable SLR results as well. Comparing only GPS and VLBI, the same behavior is visible for the hourly estimates.

3.4 Tropospheric Parameters

CONT02

The estimation of tropospheric zenith delays was performed in the same way for GPS and VLBI. The zenith delay was divided into two parts, a dry part, modelled a priori using the Saastamoinen model as it is implemented in the Bernese GPS Software (Hugentobler et al., 2004) and a wet part, estimated in the adjustment. Both contributions are mapped from the direction of observation to the zenith using the dry and wet Niell mapping function NMF

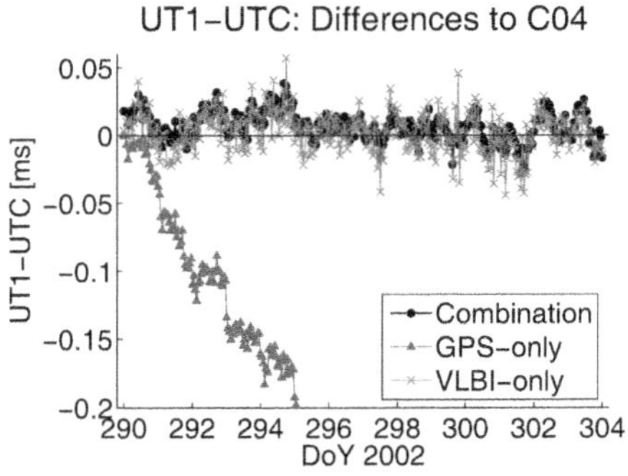

Fig. 4. Hourly UT1-UTC estimates w.r.t. C04/IERS2003 derived from VLBI, GPS and a combined solution

a)

b)

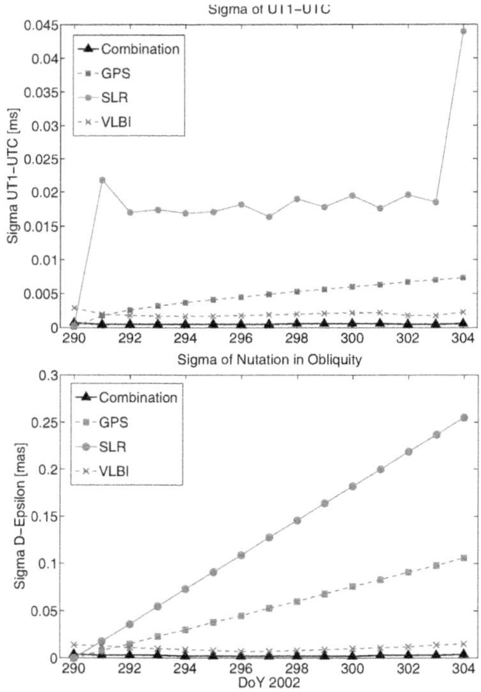

Fig. 5. Formal errors of daily EOP estimates: UT1-UTC, Nutation in obliquity

(Niell, 1996). For both techniques we used the same a priori delay, referred to the GPS reference point. Thus, the VLBI estimates contain the effect of the height difference between the GPS and the VLBI reference point. In general it must be stated that the zenith delays derived from the two techniques agree very well as it is shown exemplarily for the station Onsala in Fig. 6. The mean of the difference (lower figure) contains the effect of a height difference of 13.7 m.

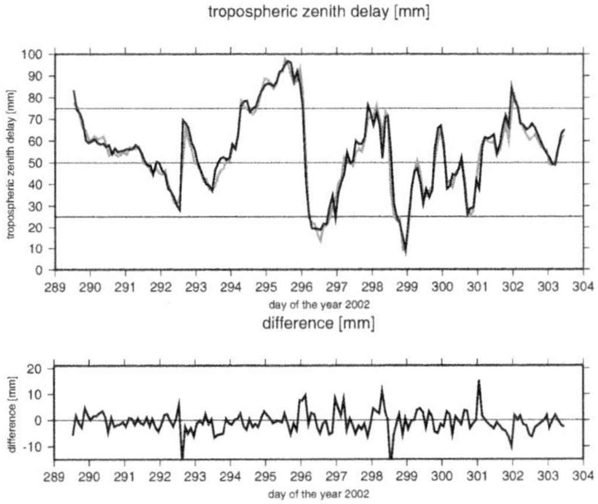

Fig. 6. Tropospheric zenith delays estimates from GPS (grey) and VLBI (black) of station Onsala (Sweden)

The height differences between the corresponding GPS and VLBI reference points have to be corrected for, if the ZD estimates of both techniques are compared or combined. For the studies presented here, the Saastamoinen model with mean surface meteorological data[2] was used for this purpose. The modeled ZD differences to be expected theoretically ("tropospheric ties") are compared to the differences between the space techniques averaged over 14 days (see Table 5). For Algonquin Park, Wettzell and Hartebeesthoek the agreement between the estimation and the model is quite good. The biases for Onsala, Fairbanks and Ny-Alesund might be caused by unmodeled phase center variations of the radomes installed on the GPS antennas, whereas the biases for Kokee Park and Westford cannot be explained at the moment. Although the errors in the zenith delay estimates due to a radome seem to be systematic, a conclusive statement about this topic cannot be given because the number of eight stations is too small and not enough studies were done up to now.

[2] In the case of long time series no meteorological data are available for several stations. Therefore, parameters of a standard atmosphere are used. The computed values show only small differences to those derived from meteorological data.

Additionally, it should be mentioned that the agreement of theoretical and space technique derived differences improved significantly when changing from relative to absolute antenna phase center corrections for GPS (see Schmid et al. (2005)).

Table 5. Comparison of tropospheric zenith delay differences [mm] from the Saastamoinen model and the space geodetic techniques for CONT02 and Long-Time-Series (see below)

Site	CONT02				Long-Time-Series	
	Model	ΔZD GPS-VLBI	ΔZD_{cor} GPS-VLBI	Radome	ΔZD_{cor} GPS-VLBI	Radome(s)
Ny-Alesund	0.96	−0.50	−1.46	SNOW	−1.41	NONE/SNOW
Onsala	4.53	1.17	−3.36	OSOD	−3.55	DUTD/OSOD
Wettzell	0.98	1.26	0.28	NONE	−1.09	NONE
Hartebeesthoek	0.46	−0.40	−0.86	NONE	−0.89	NONE
Algonquin Park	7.33	7.28	−0.05	NONE	−0.55	NONE
Fairbanks	3.90	0.74	−3.16	JPLA	−4.53	NONE/JPLA
Kokee Park	3.04	8.40	5.36	NONE	5.14	JPLA/NONE
Westford	0.57	4.38	3.81	NONE	4.21	NONE

It is well known that the tropospheric zenith delays are highly correlated with the station heights. Combining the station networks, the discrepancies between local ties and space technique measurements in the height component (see Table 3) will be absorbed by the ZD estimates. To avoid this problem the potential of the combination of ZD estimates was tested. The tropospheric ties are introduced using an a priori standard deviation of 1 mm. The results displayed in Fig. 7 demonstrate, that this alternative method works well. However, as expected, the combination of zenith delays cannot fully replace the combination of the height components themselves.

The correlation between tropospheric parameters and station coordinates can also be shown for the tropospheric gradient estimates. In general, the gradients estimated independently by VLBI and GPS agree quite well as shown in Fig. 8a for the north-south gradient of the station Fairbanks. Combining the station networks introducing all available local tie information the resulting gradients for VLBI and GPS (gradients not yet combined) show a clear bias (Fig. 8b). As can be seen in Fig. 8c, the bias can be removed if the north component of the Fairbanks local tie is not used in the combination. Together with comparably large discrepancies between the coordinate differences and the local tie values seen before, this investigation leads to the conclusion that the north component of the local tie for Fairbanks should be disregarded in a combination. Very similar results are obtained for the east component of the station Westford. This sensitivity of the tropospheric gradients to discrep-

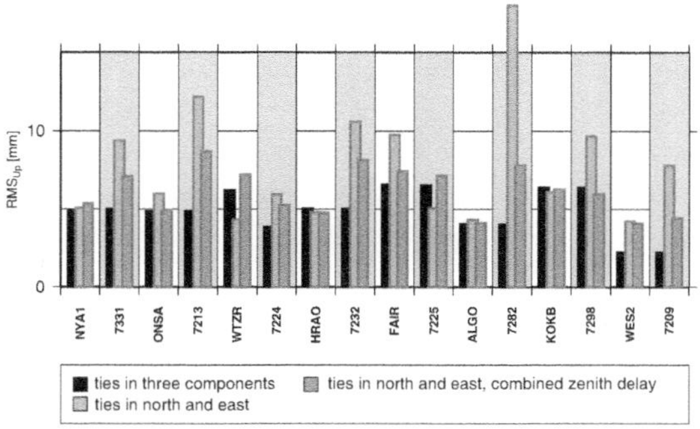

Fig. 7. Repeatabilities of station heights for different types of combination (VLBI stations with a grey background)

ancies between terrestrial measurements and space techniques makes them suitable to validate local tie information.

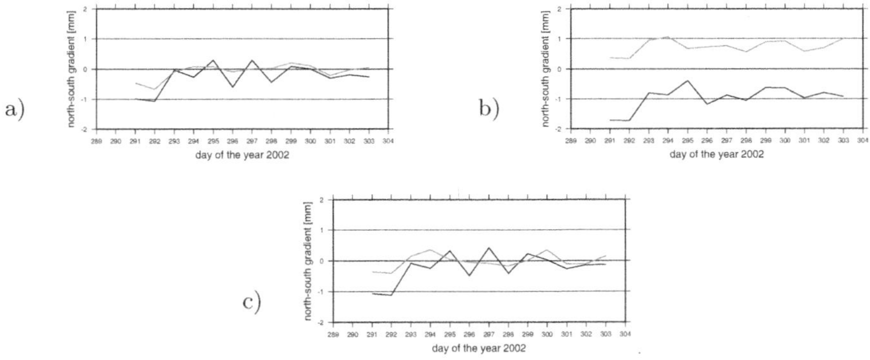

Fig. 8. North gradient of Fairbanks estimated from VLBI (black) and GPS (grey): (a) single technique solutions, (b) combination with all local ties, (c) combination with all local ties except for north component of Fairbanks

For the validation of the tropospheric zenith delays independent measurements from water vapor radiometers (WVR) are available for the stations Wettzell and Onsala for the period of the CONT02 campaign. As an example, the wet tropospheric zenith delay (ZWD) from the WVR and from a combined GPS/VLBI-solution (after removing the dry delay using pressure measurements) are displayed in Fig. 9 for the station Wettzell. In view of the large offset (about 20.6 mm) and the high variability of the difference between the two techniques (14.5 mm), the usefulness of WVR for validating the tropospheric zenith delays of the space techniques is limited. For comparison, the RMS of the differences between GPS and VLBI zenith delay estimates is only about 6.9 mm, which means that these results are more stable by a factor of two. Comparisons for the station Onsala are more problematic because both antennas (GPS and VLBI) are covered by a radome evoking changes in the tropospheric parameters that cannot be modelled yet. Nevertheless, comparisons with WVR were performed (see Elgered and Haas (2003); Schmid et al. (2005)) and in spite of radomes, the results are much more encouraging.

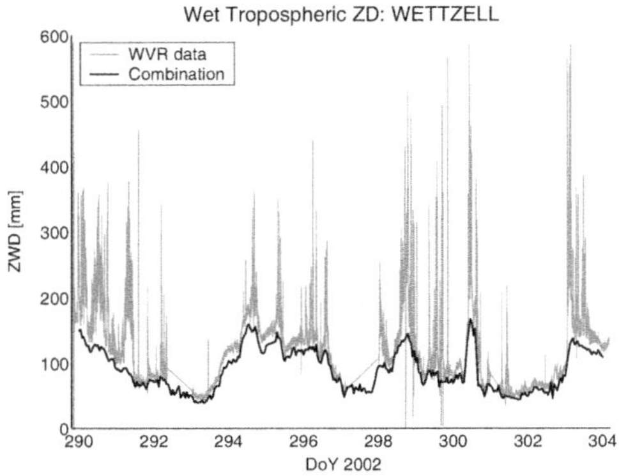

Fig. 9. Comparison of wet zenith delays (ZWD) from space techniques and WVR at Wettzell

Long-Time-Series

In order to study the suitability of GPS and VLBI time series for a rigorous combination covering several years, troposphere parameters of homogeneously reprocessed GPS and VLBI long-time series were compared. The VLBI solution is (except the longer time period) identical to the one mentioned in Sect. 2. The GPS solution has been computed by FESG in cooperation with the Technical University in Dresden (Steigenberger et al., 2004) and covers the time interval from 1 January 1994 till 31 December 2004. The tracking network is slightly different from the one used for CONT02, but the processing scheme is pretty much the same except the time resolution of the troposphere zenith delays: due to the huge computation load of the completely reprocessed GPS solution, the parameter interval has been reduced from one to two hours. For comparisons with the hourly VLBI parameters, the GPS values have been interpolated linearly. Altogether 36 co-located GPS antennas nearby 27 VLBI telescopes have been used for comparisons.

The RMS differences of the two ZWD time series are on the level of 5 to 10 mm, the correlation is larger than 0.95 for most stations. Only stations with very few observations (some sporadically observing VLBI stations) or technical problems (degraded GPS receiver or antenna performance) show larger discrepancies. Fig. 10 shows the estimated zenith wet delay for both

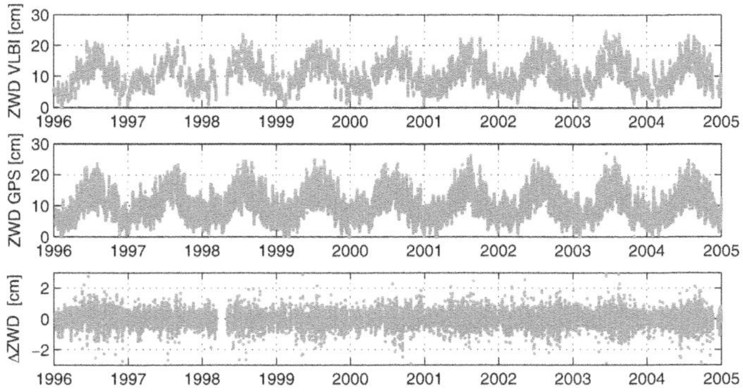

Fig. 10. Estimated zenith wet delay from VLBI and GPS for station Wettzell. The correlation of the two time series is 0.994, the RMS of the difference 4.65 mm

techniques and their difference for station Wettzell. The differences between GPS and VLBI are larger in the summer where also the variability of the troposphere is larger. The ZWD biases for the long-time series of the eight stations also used in CONT02 are included in Table 5. The consistency of the biases from the only 15 days of CONT02 and the long-time series of up to eleven years is remarkably high and on the level of several tenth of a millimeter. Only for Fairbanks (where the radome configuration has changed) and Wettzell the differences are on the level of one millimeter. Although more stations with radomes are involved in this comparison than in the CONT02 analysis, a conclusion about the effects of radomes on ZWD biases is still difficult, as for many stations the radome configuration changes with time (changes of radome types, time periods without radomes).

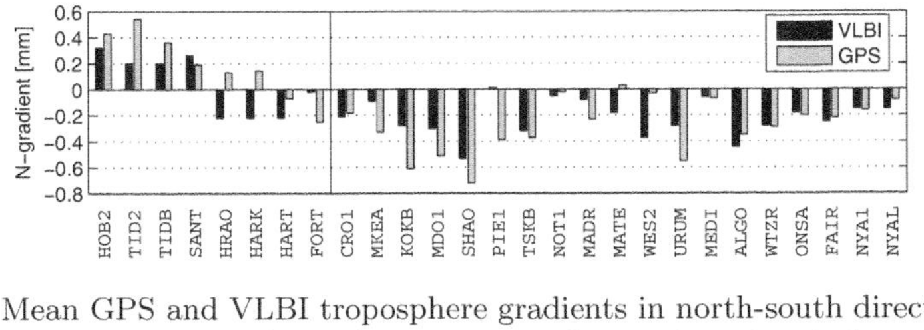

Fig. 11. Mean GPS and VLBI troposphere gradients in north-south direction. The stations are sorted by their latitude, starting in the southern hemisphere. The vertical line marks the equator

The mean GPS and VLBI troposphere gradients in north-south direction are shown in Fig. 11 (stations with few observations have been excluded). The stations are sorted by their latitude and the systematic differences between the northern and southern hemisphere can clearly be seen. The agreement of the two techniques is in general good, although the VLBI gradients are systematically smaller due to constraints of 2 mm on these parameters (GPS gradients have been estimated without any constraints). As in the case of the mean gradients, the gradient time series of both techniques show the same signals, see Fig. 12 as an example for station Wettzell. Especially the agreement in the years 2003 and 2004 is excellent. Concluding, one can say

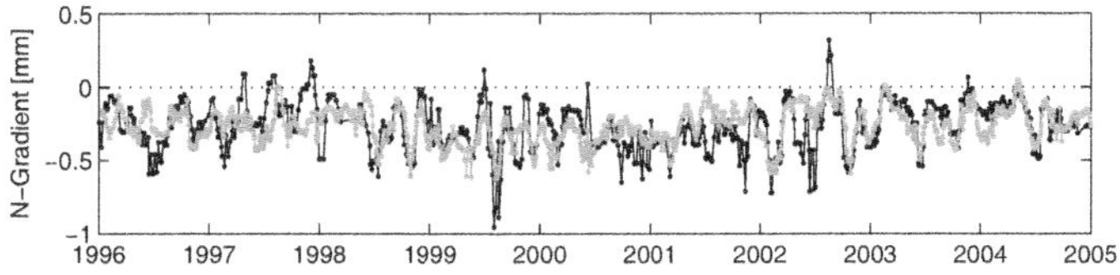

Fig. 12. GPS (gray) and VLBI (black) troposphere north gradient time series for station Wettzell. Both series have been smoothed with a 30-day median filter.

that the troposphere parameters of the reprocessed GPS and VLBI long-time series agree very well and are stable even for a time span of a decade: the combination of such series, which remains a goal for the future, seems to be promising.

4 Conclusions

The data set used in these studies is excellent and unique to perform a rigorous combination. In contrast to the usual VLBI sessions schedule continuous VLBI observations are available for a time span of 14 days. The generation of homogeneous normal equations was possible due to a careful adaption of the software packages based on the intensive cooperation of the two institutions involved. The combination of GPS and VLBI data delivers promising results for all parameters common to both techniques considered here. Looking at the station coordinates and the ERP, it could be shown that both techniques benefit from a combination. As one important result, it was demonstrated that the systematic effects present in LOD and nutation rates derived from satellite techniques do not necessarily result in a deterioration of UT1-UTC estimates and nutation offsets contributed to an inter-technique combination by VLBI.

The tropospheric parameters derived from the single technique solutions show a surprisingly good agreement that can be traced back mainly to the adaption of the functional models. Furthermore, it could be demonstrated that tropospheric parameters are useful to detect problems in the local ties due to their strong correlation with the station coordinates. The analyses also showed that the combined solution can be improved by combining the tropospheric parameters. However, there are remaining offsets in the tropospheric parameters for some of the stations that cannot be explained until now and have to be analyzed in more detail.

Part II: Combination of Long Sub-daily EOP Time Series from VLBI and GPS

(R. Dill and M. Rothacher)

1 Introduction

Today space geodetic observations of Earth rotation with high accuracy and high time resolution open up the opportunity to study the influence of sub-daily tidal excitation, short-periodic and episodic geophysical effects on polar motion and the variation of length of day (ΔLOD). Up to now there are more than 8 years of almost continuous sub-daily EOPs available from GPS and VLBI. Harmonic analyses of these time series provide sub-daily tidal excitation amplitudes matching very well the estimations from geophysical models Rothacher et al. (2001). 60 - 70% of the sub-daily variations can be explained by sub-daily tide models. The remaining variations in the GPS and the VLBI data are quite different and show outliers, offsets and short-periodic, non-tidal signals. Combining sub-daily EOPs from GPS and VLBI could yield a new sub-daily time series matching the requirement of a more consistent and homogeneous series to study high frequencies or episodic geophysical effects like earthquakes. Vondrak and Cepek (2000) developed the so-called "combined smoothing" using a Lagrange polynomial representation of the unknown combined curve and introducing a smoothness constraint in the least squares estimation. This method is also capable of using the first time derivative for one input time series, like ΔLOD from GPS, instead of UT1-UTC. This method works well as long as the smoothing is high enough to bridge the data gaps between VLBI sessions and to suppress high-frequency noise. Applying the combined smoothing method on sub-daily EOPs, we would have to use very high smoothing coefficients eliminating most of the signal with frequencies higher than two days. In order to keep as high frequencies as possible we extended the procedure to use it for the combination of high-frequency data. Designed as a multi-step scheme, we divided the sub-daily time series in a low-frequency and a high-frequency part and adapted two different sets of smoothing and combination coefficients to both frequency parts.

2 Multi-step Combination Approach based on the Combined Smoothing Method

Combined smoothing as described by Vondrak (2000) is based on the smoothing of observational data developed originally by Whittaker and Robinson (1946). The advantages of this method are the possibility to use unequally spaced input data with different uncertainties of the measurements and the suppression of high-frequency noise in the observations. Combined smoothing tries to find a compromise between closeness of the derived smoothed curve to the original values and smoothness of the result. Furthermore the method allows for not knowing the values of the function itself but only their first time derivatives, like the Earth rotation rates from satellite techniques. We start with two time series of observations - one with measured function values of a certain quantity whose analytic expression is unknown, like UT1-UTC of VLBI, and the other with measured time derivatives of the same quantity, like ΔLOD from GPS. Both series are given at unequally spaced epochs that do not necessarily need to be identical, and the individual observations are given with different precision, defined by their formal uncertainties. We assume that both series are independent of each other, and also that the individual observations of the same series are uncorrelated. The combination of both individual series should fit well to each of the original series and should be sufficiently smooth. Using the two series

$$y^{GPS}(i = 1...n_{GPS}) \quad \text{and} \quad y^{VLBI}(i = 1...n_{VLBI})$$

we define three quantities constraining the combination:

Smoothness of the combined curve:

$$S = \frac{1}{x_N - x_1} \int_{x_1}^{x_N} \dot{\varphi}^2(x) dx \tag{1}$$

In the equation, x_i denotes all epochs from 1...N with $N \leq n(GPS) + n(VLBI)$ and $\varphi(x)$ is the unknown function of the combined curve $y(i = 1...N)$ which is estimated numerically as third-order Lagrange polynomial at the four subsequent epochs $i, i+1, i+2, i+3$.

Fidelity of the combined curve y to the original values y^{VLBI}:

$$F_{VLBI} = \frac{1}{n} \sum_{i=1}^{N} w_i^{VLBI} (y_i^{VLBI} - y_i)^2 \tag{2}$$

Fidelity of the combined curve y to integrated first derivatives y^{GPS}:

$$F_{GPS} = \frac{1}{n} \sum_{i=1}^{N} w_i^{GPS} (y_i^{GPS} - y_i)^2 \tag{3}$$

The combination is done by a simultaneous adjustment of all three constraints using a linear equation system which minimizes the condition

$$Q = S + f_{VLBI} \cdot F_{VLBI} + f_{GPS} \cdot F_{GPS} = min \tag{4}$$

where the combination coefficients $f_{VLBI} \geq 0$ and $f_{GPS} \geq 0$ control the relative emphasis of smoothing and fidelity.

The numerical solution and some properties of the choice of the smoothing coefficients are given by Vondrak and Cepek 2000.

If we want to use the combined smoothing method for long time series of sub-daily EOPs, we have to deal with more than 10,000 data points from VLBI and over 50,000 data points from GPS. Even modern computers fail to solve the linear equation system in one step. In order to save memory storage space we split the input data into overlapping subsets of 1800 points, one subset every 600 points. After applying the combined smoothing method on one subset, we used only the middle 600 points as result because the margins are contaminated from the beginning and the end of the subsets. Shifting this procedure continuously over the whole input series and concatenating all 600 point-subsets we produced the entire combination result.

The nature of the combined smoothing method as adjustment between smoothness and fidelity implicates that we have to decide whether we want to emphasize the smoothness and accept the suppression of higher frequencies or whether we want to emphasize the fidelity to keep as high frequencies as possible and loose the capability to filter the noise. Unfortunately, when analyzing sub-daily EOPs one is interested in the highest frequencies in the series. This leads to combination coefficients $f > 10^7$. J. Vondrak mentioned optimal coefficients in the range of 1 to 100 for the combination of daily EOPs. Accordingly, the suppression of noise will not be very effective, but the problem of a worse signal-to-noise ration can be counterbalanced by using long time series and the fact that any combination amplifies the common signals of both input series and mute irregular features. The almost negligible smoothing results in a best fit between both input series. This is the desired behavior for the combination if both input series are smooth enough on the long-periodic time scale. If one input series is only available as first time derivative, it is imperative that the second time series provides enough sampling points to stabilize the integration of the Lagrange polynomials representing the first time derivative. Most of the time series of polar motion fulfill this requirement after removing an a priori pole model and the most important tidal constituents. The set of UT1-UTC data from VLBI and ΔLOD from GPS normally does not satisfy this condition because the data gaps in the VLBI series are often too large. If we use almost no smoothing to preserve the sub-daily variations we obtain an additional signal due to the utilization of the Lagrange polynomial representation. Even if we use coefficients lower than $f < 10^5$, suppressing periods lower than 1 day, or weaker constraints on the VLBI series, we see this undefined behavior of the Lagrange polynomial induced by the data gaps of the VLBI series.

To separate the low-frequency part, containing the unintential polynomial signal, we used a truncated version of the combined smoothing with high smoothing and adapted it to each single input series individually. This means to omit one input series by setting one coefficient equal to zero in Eqn. 4. In this way we were able to determine and remove for each input series a very smooth low-frequency curve before the combination step. This pseudo-smoothed curve represents most of the unrequested oscillations coming from the loose determination of the integrated Lagrange polynomial function. After the combination step, we have the possibility to resubstitute the mean or filtered low-frequency part.

3 Combination Result

The original time series contain EOPs with a 1h sampling interval. The GPS data contain 52584 samples from 2 January 1995 to 31 December 2000 provided by the IGS Analysis Center CODE. Due to the session-wise observation of VLBI, the VLBI series from the IVS Analysis Center GSFC only contain 12464 hourly samples from 3 January 1995 to 29 December 1999. From these two input series we obtained 63008 different epochs, with 2040 epochs common in VLBI and GPS. Further information on the original GPS and VLBI input series can be found in Rothacher et al. (2001). We limited the following combinations and analysis to the three parameters of polar motion and UT1-UTC. Previous to any combination, the input data were reduced by an a priori ERP series, like IERS Bulletin A, and a reference sub-daily ERP model. The mean levels of the formal errors of both input series were equalized and converted into normal weights using relative variances. After these preprocessing steps and before the combination we inserted the determination of a pseudo-smoothed curve, see also the scheme of the combination approach in Fig. 13.

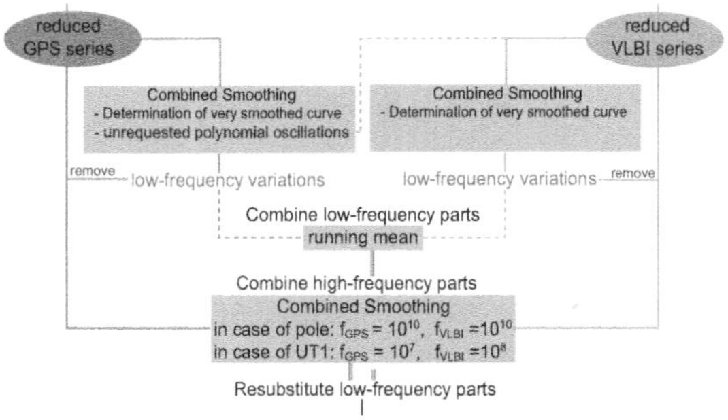

Fig. 13. Combination approach to combine UT1-UTC from VLBI with ΔLOD from GPS using the combined smoothing method in two steps

For the determination of the low-frequency part of UT1-UTC/ΔLOD we achieved the best results with combination coefficients between 1000 and 10,000 for both f_{GPS} and f_{VLBI}, using 5 times higher values for VLBI (e.g. $f_{GPS} = 1000$ and $f_{VLBI} = 5000$). This configuration yields an integrated smoothed curve of ΔLOD with almost the full Lagrange oscillation signal but also some "low"-frequency parts of the sub-daily UT1-UTC variations.

Implementing a low-pass filter as additional element, we can separate the Lagrange oscillations from the "low"-frequencies. The low-frequency smoothed curves of each input series were removed from the original data before the main combination step. For ΔLOD we removed the first time derivative.

The main combination was done similarly to the original combined smoothing method but with very high combination coefficients to keep as much high-frequency signals of both input series as possible. The a postiori RMS value of the ERP estimates from the original VLBI and GPS solutions, converted into relative weights, were used to combine the series in a weighted least squares sense. The standard deviation of the combined smoothed values was computed according to the general principle of propagation of variances and covariances of linear functions. If desired we can resubstitute any combination or mean of the smoothed curves after the combination. Additionally, we can also resubstitute the reference sub-daily model and the a priori ERPs.

4 Analysis

The quality of the new combined sub-daily EOP time series was analyzed by comparing the scatter, the standard deviation and the amplitude spectra of the two input series from VLBI and GPS with the combination and, in addition, by estimating tidal harmonic coefficients.

Table 6 summarizes the RMS and standard deviation statistics for the original and combined time series. In the polar motion components X-pole and Y-pole the combination shows a smaller RMS than the original time series. This is a result of the combination itself by reducing outliers and from the weak smoothing process within the combination. In order to obtain compatible weights for the combination the error levels of both input series were adjusted and transformed into relative weights. Therefore the mean sigma levels of the two input series are equal. As expected, the combination shows a smaller error level due to the increase observational data.

Table 6. Comparison of RMS and standard deviation of VLBI and GPS input data series with the new combined series

		VLBI	GPS	Combination
Xpol	RMS	0.430 mas	0.303 mas	0.267 mas
	mean (sigma)	0.084 mas		0.053 mas
	RMS (sigma)	0.042 mas	0.029 mas	0.021 mas
Ypol	RMS	0.512 mas	0.318 mas	0.273 mas
	mean (sigma)	0.083 mas		0.052 mas
	RMS (sigma)	0.053 mas	0.032 mas	0.021 mas
UT1 - UTC	RMS	0.028 ms		0.016 ms
	mean (sigma)	0.040 ms		
	RMS (sigma)	0.027 ms		
lod	RMS		0.237 ms	0.198 ms
	mean (sigma)		0.040 ms	
	RMS (sigma)		0.014 ms	

The same behavior is valid for UT1-UTC. The sigma values for the combined series are not given in Table 6 because, due to the integration of ΔLOD from GPS to UT1-UTC, the uncertainties of the combined series became very high and are no longer comparable to those of the input series.

Altogether the new combined series seems to be more homogeneous. But this property would be useless, if we achieve this homogeneity at the expense of the signal content by too much smoothing. Therefore we reviewed the amplitude spectrum of the combination comparing it with the amplitude spectra of the two input series.

In the spectra of the reduced sub-daily VLBI X-pole series we cannot detect any special periods. The GPS X-pole spectrum shows remaining signals at 8h, 6h, 4.79h, 3.43h, 2.67h and their harmonics, not contained in the a priori sub-daily tidal model. The combination process suppresses most of the GPS-specific periods, but keeps almost the same amplitude power in the whole series. Concerning the choice of the combination coefficients, the smoothing is weak enough to preserve even the highest frequencies.

Over 60% of the sub-daily variations in Earth rotation parameters are due to the influence of the tide generating gravitational potential of the sun, the moon and the planets. The dominant effects on the solid Earth's rotation come from the tidal deformations and mass redistributions in the atmosphere and ocean through the interchange of angular momentum. Modern tidal models consist of over 100 partial tides. The periods and the gravitational force can be predicted with very high accuracy. The sub-daily tidal excitation of Earth rotation variations can be directly modelled via atmosphere and ocean models or we can estimate the sub-daily tidal amplitudes from the observed sub-daily variations. The chances of success for this second method increase the longer and the more homogeneous the sub-daily Earth rotation time series are.

We started the estimation of tidal harmonic constituents with some GPS-only sub-daily time series to check and review the performance of our software, an extended version of T_TIDE from Pawlowicz et al. (2002) (The set of 148 frequencies and the definition of the rotation angle (GMST) was updated based on the IERS 2000 conventions, chapter 8.). The results provide the basis for comparisons of tidal estimations from the new combined time series with former results from other studies.

The results match generally the estimates from Rothacher et al. (2001), Chao (1996) and Egbert (2002). The estimates from Rothacher et al. were based on the same GPS series, whereas the other models were based on hydrodynamic ocean models assimilated with Topex/Poseidon seasurface height differences. Comparing the geodetic estimates with oceanic tide models is possible because the second major contribution coming from the atmospheric tides is very small. Brzezinski et al. (2002) determined the influence of atmospheric tides on Earth rotation parameters from the latest atmospheric angular momentum series. For the excitation of prograde polar motion, the influence of the atmospheric tides are one order of magnitude lower for the prograde diurnal tides and two orders of magnitude lower for the semidiurnal tides. For UT1-UTC the influence of the atmospheric tidal waves is more than 30 times smaller than the oceanic contributions.

Using the new combined time series leads to a noticeable improvement in the estimation of amplitudes and phases, reducing in particular the formal errors. The improvement is more pronounced for the UT1-UTC component (Table 7). Looking at the tidal constituents, estimated with more than 95% significance, the number increased from 21 (GPS-only) to 23 (combined series). The synthesized tidal prediction from the combined time series reproduced 77.4% of the sub-daily UT1-UTC variations and 64.3% of the sub-daily prograde polar motion. The tidal prediction from the VLBI-only analysis reflected 51.9% of UT1-UTC and 37.7% of polar motion. The tidal prediction from the GPS-only analysis reflected 66.9% of UT1-UTC and 58.4% of polar motion. The correlation can be even enhanced to 78.8% for tidal prediction of UT1-UTC, if we use a shortened combined series, where the last part with almost no VLBI observations is omitted.

Table 7. Tidal amplitudes, phase lags and their errors of UT1-UTC for GPS, VLBI and the combined sub-daily ERP series

Tide	Period [h]	Amplitude [µs]			Error [µs]			Greenwich Phase [°]			Error [°]		
		GPS	VLBI	Comb.	GPS	VLBI	Comb.	GPS	VLBI	Comb.	GPS	VLBI	Comb.
2Q1	28.01	1.72		1.12	1.20		0.83	358.53		22.79	47.77		41.11
σ1	27.85	1.27		0.88	1.16		0.75	28.33		31.11	55.72		49.56
Q1	**26.87**	**5.25**	**4.15**	**5.41**	**1.34**	**1.01**	**0.86**	**26.70**	**23.95**	**26.36**	**13.73**	**14.32**	**8.79**
ρ1	26.72			0.97			0.90			36.09			50.86
O1	**25.82**	**21.22**	**14.47**	**21.35**	**1.31**	**1.05**	**0.61**	**37.37**	**27.88**	**38.54**	**3.26**	**4.78**	**1.78**
β1	24.98		2.63			1.01			39.48			22.10	
NO1	24.83	1.26		1.05	0.86		0.60	9.90		14.04	40.48		34.92
π1	24.13	1.79		1.12	0.99		0.69	136.43		153.23	32.73		32.53
P1	**24.07**	**6.62**	**4.25**	**5.96**	**0.95**	**1.16**	**0.53**	**19.91**	**39.41**	**26.68**	**7.70**	**16.48**	**5.28**
S1	24.00	2.85		1.68	1.38		0.85	317.55		304.69	29.03		30.55
K1	**23.93**	**20.00**	**11.29**	**20.29**	**1.11**	**1.16**	**0.67**	**22.62**	**36.75**	**25.40**	**3.15**	**6.07**	**1.71**
ψ1	23.87	1.47		1.16	0.97		0.62	331.66		335.62	39.74		34.53
J1	23.10	1.16		1.37	1.12		0.83	49.65		42.11	63.51		32.21
S01	22.42		1.41			0.99			60.55			41.54	
2N2	12.91	0.93	2.13	0.87	0.62	0.73	0.42	243.59	258.11	255.56	38.55	22.21	27.37
μ2	12.87	0.66	1.15	0.80	0.56	0.80	0.43	22.46	167.93	236.04	47.76	30.90	31.17
N2	**12.66**	**3.95**	**3.50**	**4.02**	**0.52**	**0.72**	**0.45**	**239.91**	**247.62**	**243.37**	**8.52**	**13.03**	**6.76**
ν2	12.63	0.82		0.72	0.61		0.43	250.34		259.39	41.24		33.72
M2	**12.42**	**17.22**	**17.89**	**17.41**	**0.52**	**0.73**	**0.39**	**242.91**	**237.37**	**244.60**	**1.95**	**2.49**	**1.39**
H2	12.40		1.08			0.73			5.12			41.16	
L2	12.19			0.60			0.46			232.32			46.89
S2	**12.00**	**6.98**	**7.22**	**7.13**	**0.60**	**0.65**	**0.45**	**260.63**	**263.09**	**259.62**	**5.29**	**5.18**	**3.76**
R2	11.98	0.51	1.45		0.49	0.59		104.54	319.80		57.51	22.63	
K2	**11.97**	**1.70**	**1.99**	**1.98**	**0.79**	**0.65**	**0.61**	**217.39**	**293.91**	**223.42**	**31.23**	**17.93**	**16.79**
SO3	8.19		0.59			0.36			53.70			36.43	
SK3	7.99	0.45	0.55	0.31	0.42	0.41	0.29	70.31	24.62	89.15	44,55	50.52	54.20
S4	**6.00**	**5.42**		**0.96**	**0.32**		**0.24**	**236.35**		**215.00**	**3.17**		**14.40**
2SK5	4.80			2.00			0.17			312.37			52.50

The determination of the tidal components in sub-daily Earth rotation data allows, to a certain extent, the separation of tidal from non-tidal signals and leads to the non-tidal residuals of the sub-daily Earth rotation parameters containing some noticeable episodic signals coming mainly from the original GPS time series. An intensive comparison of earthquake events with the non-tidal sub-daily Earth rotation parameters showed no meaningful correlation between seismology and sub-daily Earth rotation. Further investigations of the impact of short-periodic and episodic geophysical effects on sub-daily Earth rotation excitation will follow. Additionally the new combination procedure should be applied to the latest sub-daily EOP time series.

Acknowledgement. This is publication no. GEOTECH-168 of the programme GEOTECHNOLOGIEN of BMBF and DFG, Grant 03F0336A. This research has made use of sub-daily EOP data provided by the International VLBI Service for Geodesy and Astrometry (IVS). The GPS data were produced from the International GPS Service (IGS) product center CODE.

References

Altamimi Z., Sillard P., Boucher C. (2002) ITRF2000: A new Release of the International Terrestrial Reference Frame for Earth Science Applications. J. Geophys. Res., 107 (B19): 2214, doi: 10.1029/2001JB000561

Brzezinski A., Bizouard C., Petrov S. (2002) Influence of the atmosphere on Earth rotation: What new can be learned from the recent atmospheric angular momentum estimates? Surveys in Geophysics 23:33-69

Chao B., Ray R., Gipson J., Egbert G., Ma C. (1996) Efficient inverse modeling of barotropic ocean tides. J. Geophys. Res. 101:20151-63

Egbert G. D., Erofeeva L. (2002) Diuarnal/semidiurnal polar motion excited by oceanic tidal angular momentum. J. of Atmospheric and Oceanic Technology 19

Elgered G., Haas R. (2003) The Geodetic VLBI Network Station at the Onsala Space Observatory - Activities During 2002. In: Schwegmann W. and Thorandt V.: Proceedings of the 16th Working Meeting on European VLBI for Geodesy and Astrometry, Bundesamt für Kartographie und Geodäsie, Leipzig, Germany

Hugentobler U., Dach R., Fridez P.(2004) (eds.) Bernese GPS Software Version 5.0. Astronomical Institute, University of Berne, Switzerland

Krügel M., Tesmer V., Angermann D., Thaller D., Rothacher M., Schmid R. (2004) CONT02 Campaign - Combination of VLBI and GPS. In: Vandenberg N.R., Baver K.D. (eds.): International VLBI Service for Geodesy and Astrometry 2004 General Meeting Proceedings, NASA/CP-2004-212255, Greenbelt MD

McCarthy D. and Petit G. (2004) IERS Conventions 2003, IERS Technical Note No.32 Verlag des Bundesamtes für Kartographie und Geodäsie, Frankfurt am Main

Niell A.E. (1996) Global mapping functions for the atmosphere delay at radio wavelengths, J. Geophys. Res., 101: 3227-3246

Pawlowicz R., Beardsley B., Lentz S. (2002) Classical tidal harmonic analysis including error estimates in MATLAB using T_TIDE. Computers & Geosciences 28:929-937

Rothacher M., Beutler G., Weber R., Hefty J. (2001) High frequency variations in Earth rotation from Global Positioning System data. J. Geophys. Res. 106, B7: 13,711-13,738

Schmid R., Rotacher M., Thaller D., Steigenberger P. (2005) Absolute phase center corrections of satellite and receiver antennas: Impact on global GPS solutions and estimation of azimuthal phase center variations of the satellite antenna, GPS Solutions, doi: 10.1007/s10291-005-0134-x (in press)

Steinforth C, Haas R., Lidberg M., Nothnagel A. (2003) Stability of Space Geodetic Reference Points at Ny-Alesund and their Excentricity Vectors. In: Schwegmann W and Thorandt V: Proceedings of the 16th Working Meeting on European VLBI for Geodesy and Astrometry, Bundesamt für Kartographie und Geodäsie, Leipzig, Germany

Steigenberger P., Rothacher M., Dietrich R., Fritsche M., Rülke A. (2004) Reprocessing of a Global GPS Network, European Geosciences Union, 1st General Assembly, 25-30 April 2004, Nice, Geophysical Research Abstracts, Vol. 6

Steigenberger P., Schmid R., Rothacher M., Tesmer V., Krügel M., Vey S. (2005) Homogeneous long-time series of GPS and VLBI troposphere parameters, EGU General Assembly, 25-29 April 2005, Vienna, Geophysical Research Abstracts, Vol. 7

Thaller D., Schmid R., Rothacher M., Tesmer V., Angermann D. (2005a) Towards a rigorous combination of VLBI and GPS using the CONT02 campaign. A Window on the Future of Geodesy. Proceedings of the International Association of Geodesy, IAG General Assembly, June 30 - July 11, 2003, Sapporo, Japan, Sanso‘ F. (Hrsg.), 2005, 128, 576-581. Springer-Verlag, Berlin Heidelberg, 2005. 3-540-24055-1.T

Thaller D., Krügel M., Angermann D., Rothacher M., Schmid R., Tesmer V. (2005b) Combination Studies Using the CONT02 Campaign, In: Int. VLBI Service for Geodesy and Astrometry 2004 Annual Report, edited by D. Behrend and K. Baver, NASA/TP-2005-212772

Titov O., Tesmer V., Böhm J. (2004) OCCAM v.6.0 software for VLBI data anlysis, In: Vandenberg N.R., Baver K.D. (eds.) International VLBI Service for Geodesy and Astrometry 2004 General Meeting Proceedings, NASA/CP-2004-212255, 311-314, Greenbelt MD

Vondrak J., Cepek A. (2000) Combined smoothing method and its use in combining Earth orientation parameters measured by space techniques. Astron. Astrophys. Suppl. Ser. 147:347-359

Whittaker E. and Robinson G. (1946) The Calculus of Observations (4th ed.). Blackie and Son Ltd., London, 303

Conventional and New Approaches for Combining Multi-Satellite Techniques

Rolf König[1], Chuang Shi[2], Karl Hans Neumayer[1], and Shengyuan Zhu[1]

[1] GeoForschungsZentrum Potsdam, Telegrafenberg, 14473 Potsdam, Germany, koenigr@gfz-potsdam.de
[2] Wuhan University, GPS Research Center, 129 Luoyu Road, Wuhan 430079, P.R. China

Summary. The combination of different space geodetic or satellite techniques offers the possibility to determine more complete, reliable, and accurate reference frame parameters. Two new approaches are proposed: the onboard collocation and the integrated method. The onboard collocation utilizes the availability of different techniques onboard a Low Earth Orbiting (LEO) satellite. For the CHAMP and GRACE satellites, and for the GPS-35 and -36 satellites, the radial distances between the Global Positioning System (GPS) phase centers and the Satellite Laser Ranging (SLR) retroreflectors are examined. The integrated method treats a multi-satellite configuration with measurements by various tracking techniques on the observation level. From three months of observations from the GRACE/GPS constellation, a series of low degree harmonics of the Earth's gravity field, representing dynamical geocenter location and axes orientation, is determined with high accuracy and daily resolution.

Key words: Reference Frames, Space Geodetic Techniques, Onboard Collocation, Integrated Method

1 Introduction

In the frame of the "Geotechnologien" program "Observation of the Earth System from Space", the Forschungsgruppe Satellitengeodäsie (FGS) and the GeoForschungsZentrum Potsdam (GFZ) cooperated in 2001 to set up a user center for the International Earth Rotation and Reference Systems Service (IERS, see IERS (2003)). One of the objectives was the development of strategies and algorithms for the rigorous integration and combination of space geodetic or satellite techniques in order to improve consistency and accuracy of IERS products.

The satellite techniques deliver observations of type Doppler Orbitography by Radiopositioning Integrated on Satellite (DORIS), Global Positioning

System (GPS), Satellite Laser Ranging (SLR), and Very Long Baseline Interferometry (VLBI). These techniques are represented in the International Association of Geodesy (IAG, see IAG (2005)) technique services, i.e. the International DORIS Service (IDS, see IDS (2005)), the International GPS Service (IGS, see IGS (2005)), the International Laser Ranging Service (ILRS, see ILRS (2005)), and the International VLBI Service for Geodesy & Astrometry (IVS, see IVS (2005)).

IERS reference frame products consist of site coordinates, quasar coordinates, Earth Orientation Parameters (EOPs), and nutation corrections being made public at regular intervals. Site coordinates have been combined from various space techniques into the International Terrestrial Reference Frame (ITRF) already since the advent of the first ITRF, the ITRF88 (Altamimi et al., 2002). The same was true for the generation of EOPs (e.g. Gambis and Bizouard (2003)), where however the contributing techniques were based on either the terrestrial (ITRF) or the celestial (the International Celestial Reference Frame ICRF) reference frame. All in all the ITRF, the ICRF, and the EOPs were combined independently from each other. Within the FGS/GFZ cooperation for "Geotechnologien", GFZ took the tasks of analyzing the quality of available IERS products, of developing models and methods to assess and mitigate the problems of the individual space techniques, and of conducting some pilot projects for the optimum combination of space techniques on the observation and on the normal equation level.

Meanwhile in 2004 the IERS called for participation in a combination pilot project (CPP, see IERS (2004)) for the routine generation of weekly IERS products that consistently comprise site coordinates, EOPs, and possibly quasar coordinates.

By combining various solutions from the same or from different space techniques, and based on different approaches by various institutions, the combined solution is expected to become more complete, more reliable, and more accurate. This fact was noted by Kouba et al. (1994) already in the early stage when the IGS went into its successful operations.

2 Single and Multi-Satellite Techniques and Problems

Reference frame products as generated within the IERS CPP reflect the strengths and limitations of a single space technique. SLR e.g. does not generate rates of EOPs from 7 day solutions due to the sparseness of observations in space and time. This is in contrast to DORIS and GPS where data coverage is highly dense. On the other hand, SLR provides scale in ITRF solutions what GPS can't due to nuisance effects on the unknown ambiguities of the phase observations.

A lot of factors could affect the definition of the reference frame or produce systematic errors respectively. These are model errors in dynamic approaches (gravity field model, thermospheric model, solar radiation pressure model etc.)

or systematic errors in the tracking observations (GPS phase center corrections for sender satellites or receivers derived or given as absolute or relative values).

The low degree harmonics, particularly the degree 1 and 2 coefficients of a spherical harmonic expansion of the gravity field, are actually part of the reference frame definition. They describe the dynamical location of the geocenter and the orientation of the main axes of inertia of the Earth's body. In the IERS CPP these parameters are not considered at all.

Before the IERS CPP, in many cases site positions and EOPs were generated separately. So the correlations between station positions and EOPs have been ignored. At this time, the IERS CPP holds reference frame products generated for each technique individually (with one exception: the Groupe de Recherche de Geodesie Spatial (GRGS) provides a product based on all techniques). In order to link the site positions of the individual techniques together, collocations are to be used. Then the normal equations of different techniques can be added and eventually be solved to obtain the final multi-technique reference frame solutions.

When doing a combination, quite some problems have to be faced. The number of collocation sites is generally small. And the quality of the collocation measurements is poor in many cases. The different space techniques suffer from different systematic errors. E.g. the dynamic model uncertainty used by SLR and DORIS may cause systematic errors in geocenter and Earth's orientation due to the effect of dynamic polar motion. The uncertainty in the phase center of the GPS receivers and transmitters may produce scale errors in the GPS reference frame (cf. e.g. Rothacher (2001), Zhu et al. (2003), Ge et al. (2005)). The detection of systematic differences and the ways to reduce them are important for the application of the reference frames. E.g. if the scale of a reference frame is changing with time (as in the case of GPS), determinations of sea level change based on such a reference will be contaminated.

3 New Approaches

Our aim is to strengthen the reference frame, to detect and reduce possible systematic errors, and to increase the homogeneity and consistency of the solutions. Two new methods are proposed: the so-called "onboard collocation" and the "integrated method".

3.1 Onboard Collocation

Most of modern Low Earth Orbiters (LEOs) are carrying instruments allowing for more than a single tracking or space technique. E.g. GPS, DORIS and SLR are found on the TOPEX/POSEIDON and JASON satellites, GPS and SLR on the CHAMP and GRACE satellites, etc. This onboard collocation is adopted here. The conventional ground station collocation is reversed, the local ties are replaced by the center of mass offsets of the instrument locations in the satellite body. In this approach firstly the onboard collocation information needs to be verified. So the phase center of the GPS receiver onboard the LEO is determined (Shi et al., 2003), as well as the position of the SLR reflector, both determinations checking the correctness of the information of the satellite manufacturer, if available. Secondly the GPS and LEO orbits are determined by using GPS tracking data. Finally all orbits are fixed, the SLR tracking data to the GPS-35 and GPS-36 and to the LEOs are used to solve for the SLR station coordinates (and velocities, if the number of data allows). The positions of the SLR stations refer then to the frame spanned by the GPS observations and ground station coordinates. By comparing them with ITRF coordinates (in the ITRF the SLR positions are given in a frame spanned solely by the SLR observations), the 7- or 14-parameter Helmert transformation parameters can be examined for systematic differences between the GPS and the SLR frame. Eventually these SLR station coordinate solutions can be used to strengthen the ITRF SLR reference frame and to improve the homogeneity and consistency between GPS and SLR frames. A detailed presentation of the onboard collocation is outlined in Neumayer et al. (2005).

3.2 Integrated Method

The EPOS-OC software of GFZ has been extended such that multi-satellites with various kinds of tracking observations or space techniques respectively can be combined at the observation level, instead of doing it at the normal equation level. This approach increases homogeneity and consistency among different satellite techniques.

Along with the coordinates and the EOPs, the gravity field should be part of the definition of a reference frame derived from satellite techniques. A lot of geophysical processes (mass movement etc.) cause deformations (hence a variation in the site position) as well as changes of the rotation axes and they produce gravity field variations. Conventionally station coordinates and EOPs are solved by fixing the gravity field model, vice versa for gravity field determinations station positions are fixed (e.g. to the ITRF values). Theoretically, the time variation of all the parameters should be solved simultaneously. This solution is not contaminated by fixing part of the parameters to certain constant values, therefore the results are more reasonable.

The integrated method is decribed in Zhu et al. (2004). The method links the ground station layer (low), the LEO layer (middle), and the GPS layer (high) together and strengthens the overall solution significantly. The layer concept has been introduced by Rothacher (2002). Applications of the integrated method for the improved solution of low degree harmonics of the Earth's gravity field are given in König et al. (2005) and Hu et al. (2005).

4 Results and Discussion

In the following some results when applying the newly proposed methods, the onboard collocation and the integrated method, are given and discussed.

4.1 Onboard Collocation

In conventional scenarios of SLR parameter adjustment, the ground station coordinates and the orbits are determined from SLR data only. For those space vehicles that are endowed with both onboard GPS receivers and SLR reflectors we have a reasonably good trajectory from GPS alone without making use of the laser observations at all. It is therefore possible to fix the GPS derived orbits and solve for the SLR ground station coordinates from SLR data only. The preparatory step to verify the instrument locations in the satellite body delivers the center of mass offsets.

Fig. 1 shows the adjusted georadial components for the eleven years of GPS-35 and GPS-36, together with the shorter time intervals for the LEOs CHAMP, and the two GRACE satellites. Yearly estimates are shown side-by-side along with a global one. Due to the larger number of observations (some 30000 SLR shots for either GPS satellite vs. some 140000 for CHAMP and 67000 for GRACE-A and 63000 for GRACE-B), and due to the denser coverage of the observations over time (the period for the GPS satellites is more than 3 times as large), the time series for the LEOs are more stable.

The suspicion that, in the case of GPS, the radial SLR reflector offset correction is due to a scale missmatch between ITRF2000 and the reference frame of the GPS satellites was tested via the following experiment: The above-mentioned global adjustment procedure, i.e. estimate SLR reflector coordinates in the spacecraft body, and fix everything else, was repeated with a series of copies of ITRF2000 coordinate sets that were scaled with common factors like 1 + 1 ppb, 1 + 2 ppb, ... etc. Fig 2 shows a plot of the solved-for radial offset vs. the corresponding scaling factors. For GPS-35 and GPS-36, the offsets disappear at scale differences of 13 ppb and 11 ppb respectively. Such an order of magnitude is unrealistic. The determined radial SLR bias, however, is real, for we have analogous statements in Springer (1999), where the same order of magnitude is established, and besides GPS for GLONASS as well. Keeping in mind that GPS and GLONASS are two independent systems, the latter findings indicate that the observed effect seems to originate from phenomena concerning the GPS reference frame and ITRF2000 rather than individual satellites.

Fig. 3 is the result of an experiment showing that an erroneous phase center of GPS sender satellites cannot account for any SLR reflector offset in the radial direction. For a fixed, arbitrary day (April 4, 2005), the GPS satellite / ground station network has been adjusted twice, the only difference between the two adjustement runs being a georadial 7-cm-shift of the sender

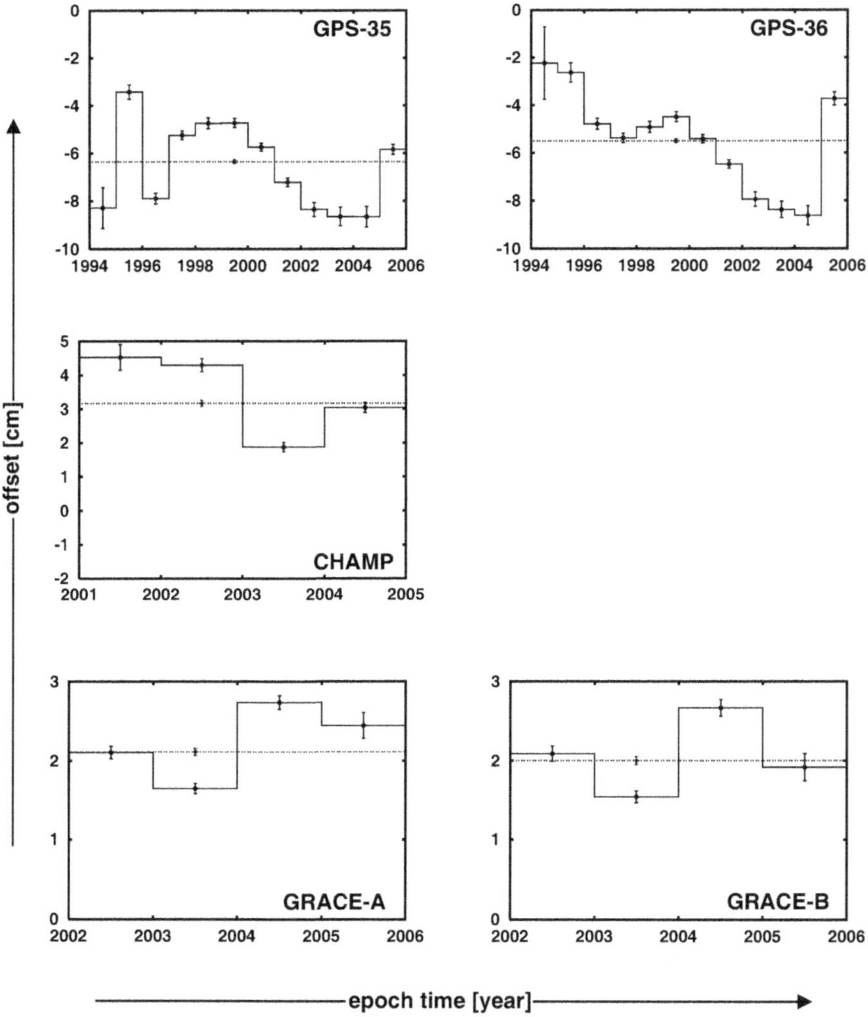

Fig. 1. Radial offsets of SLR reflectors for various satellites. In order to give some idea of the stability, yearly estimates of the offsets are plotted against the global value for the complete time horizon

phase centers on-board GPS-35 and GPS-36. The dense, spaghetti-like structures at the bottom of Fig. 3, close to zero, are the resulting coordinate-wise orbit differences in centimetres. The stable, almost horizontal lines at 7 cm offset are the sender clock differences respectively. It becomes obvious that the artificially introduced GPS phase center offset slips almost entirely into the sender clock estimates, not into the satellite position.

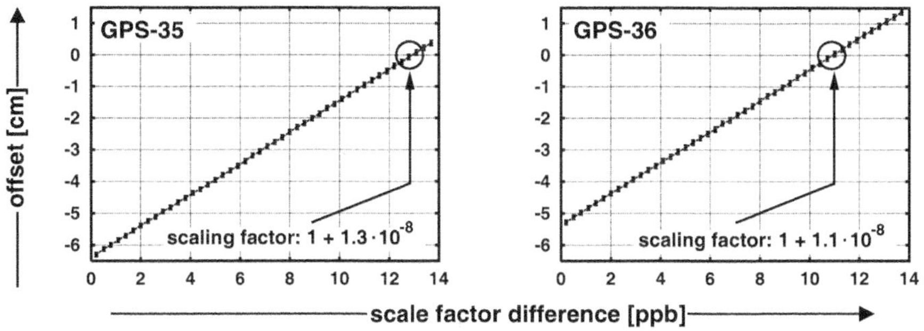

Fig. 2. A change of the fixed ground station coordinate scale of 13 and 11 parts per billion for GPS-35 and GPS-36, respectively, makes the georadial component of the solved-for SLR reflector offset disappear. Such a scale difference between traditional ITRF coordinates and the newly established SLR coordinates in the reference system of the GPS satellites, however, is not realistic

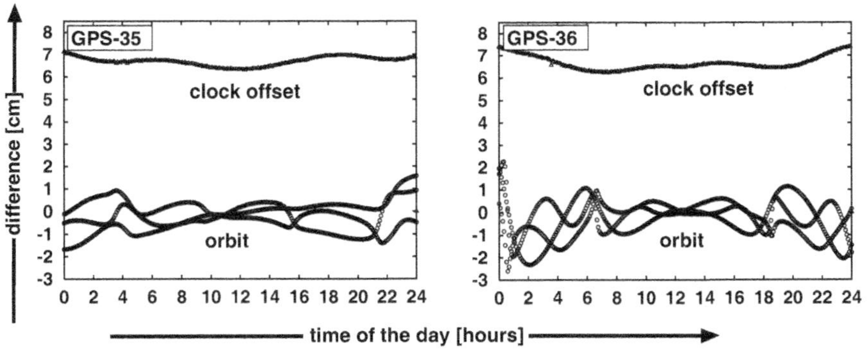

Fig. 3. This plot shows that phase center errors on-board the GPS satellites mainly correlate with clock offsets, not with the spacecraft position

After applying and fixing the above-mentioned offsets, station coordinates are established using GPS-35, GPS-36, CHAMP and GRACE. Due to the high number of observations and the observation density, coordinates from GRACE data alone are already of an impressive quality. For the 8 stations with the most SLR observations (Yarragadee, Graz, Herstmonceux, Zimmerwald, Washington, Riga, Changchun and Monument Peak), Helmert transformation parameters between the SLR coordinates in the GPS frame and ITRF2000 are computed as given in Tab. 1. No significant differences between the two reference frames can be detected. The root mean square (RMS) value of the differences of the positions of the stations is 1.7 cm before and 1.0 cm after applying the Helmert transformation. The drop in the RMS values is mainly caused by the relatively large, yet not significant Z-rotation. The RMS value after the Helmert transformation reveals more or less the accuracy that can be expected from SLR.

Table 1. Helmert transformation parameters between the SLR coordinates in the GPS frame and ITRF2000

	value	sigma	significant (3 sigma)
translations (mm)	−0.4	4.0	no
	−1.5	3.8	no
	−3.4	3.6	no
scale diff. (10^{-8})	0.006	0.060	no
rotations (mas)	−0.03	0.15	no
	−0.25	0.16	no
	−0.40	0.14	no

4.2 Integrated Method

The integrated method can advantageously be used to derive the low degree harmonics of the Earth's gravity field with high accuracy and resolution. König et al. (2005) adjusted degree 0 to 2 coefficients from the GPS/CHAMP constellation with 1.5-day resolution for nearly the full year 2002. Hu et al. (2005) computed daily coefficients up to degree 10 from GRACE data covering 3 months in 2003 and showed analyzes for the results referring to degree 2 and upwards. In the following results are given from the same series concentrating on the geocenter (i.e. C_{10}, C_{11}, S_{11}) and orientation coefficients (i.e. C_{21}, S_{21}).

Results for the geocenter fall in two categories. The Z-component (C_{10}) is less accurate than the X- and Y-components (C_{11}, S_{11}). Daily C_{10} values converted to meters are given in Fig. 4. For comparison Fig. 5 provides weekly, geometrical IGS solutions provided via the IERS Special Bureau for Gravity and Geocenter (SBGG, 2004). There seems to exist a bias of some 2 cm in both series. Formal errors of daily GRACE results are at a level of 6 mm where formal errors of the weekly IGS solution have a size of 9 mm. Maximum day to day variations reach 25 mm, which is as stable as the week to week variations of the IGS solution where at least 7 times more observations have been evaluated.

Fig. 6 gives the S_{11} series, C_{11} behaves similar. Maximum day to day variations are at the size of the C_{10} series, however formal errors are as good as 0.8 mm. Formal errors of the IGS weekly solution are given with 7 mm. All in all, the daily solutions seem to be of higher accuracy if compared to the IGS solution.

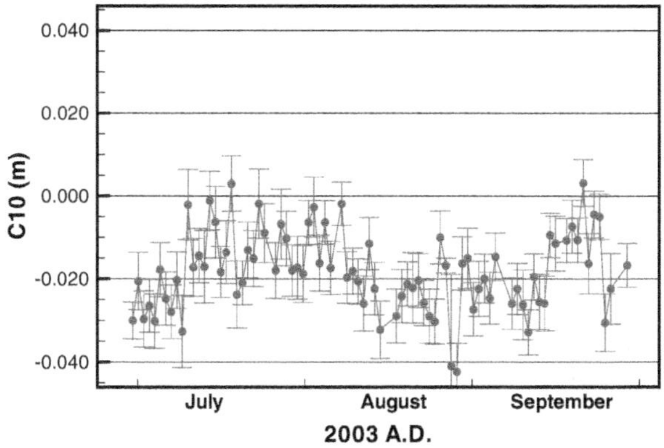

Fig. 4. C_{10} resolved daily

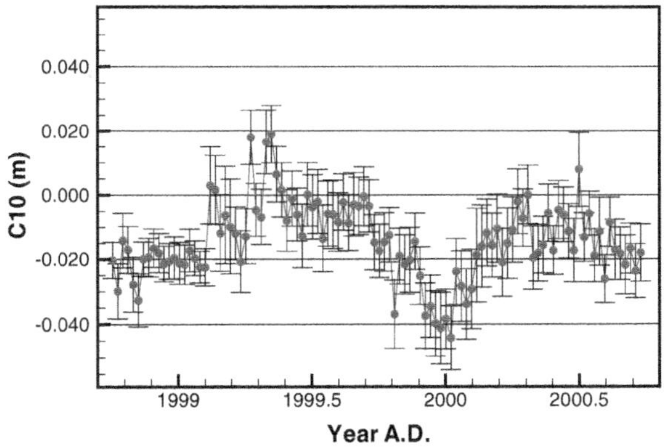

Fig. 5. C_{10} resolved weekly by IGS

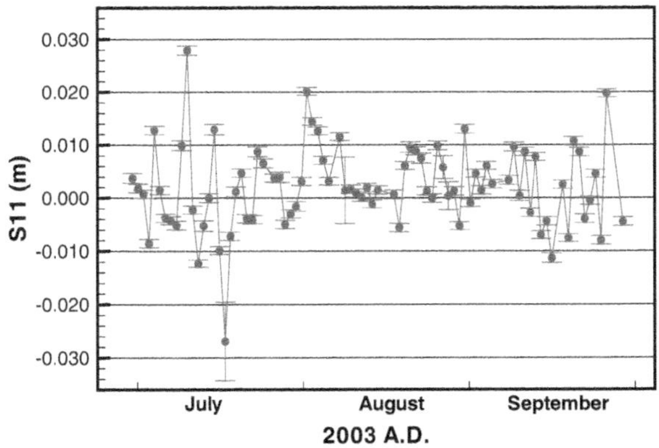

Fig. 6. S_{11} resolved daily

Results for the dynamic Earth orientation (i.e. the C_{21} and S_{21} coefficients) are given in Fig. 7 and 8. The daily series is compared to the GFZ monthly solution. The daily series follows nicely the general monthly behaviour. As the monthly solutions are more or less established by comparison to external solutions, the daily solution can be considered to be reliable in general. Formal errors are at the size of 0.7×10^{-11}, day to day variations maximally reach 15×10^{-11}. So formally the variations are significant.

On the whole we therefore really might see geophysical effects here. The quality achieved with this approach can hardly further be assessed as direct comparisons to external results are not possible at this time: time series end too early and resolution is too wide. E.g geocenter time series at the SBGG end in the year 2000 already. For a time series of orientation estimates that covers the period presented in the above one can refer to a recent publication by Chen et al. (2004), yet monthly solutions only are analyzed there.

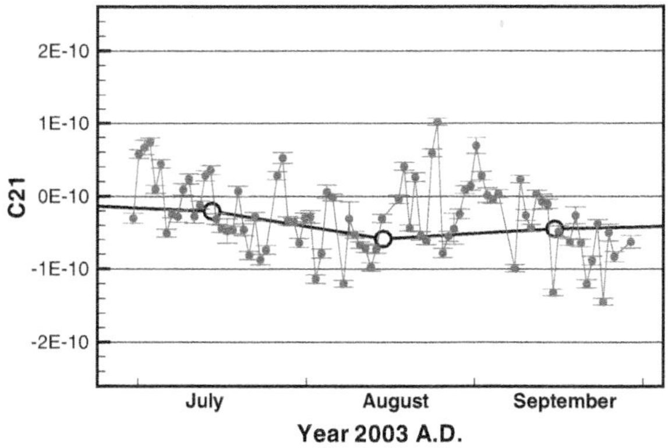

Fig. 7. C_{21} resolved daily versus monthly solution

5 Summary

Within the "Geotechnologien" program, GFZ developed two new approaches for the rigorous combination of space geodetic techniques: the onboard collocation and the integrated method. The onboard collocation needs in a first step the verification of the onboard locations of the various tracking instruments. Therefore the radial distances of the GPS phase centers and the SLR retroreflectors on GPS-35, -36, and on the CHAMP and GRACE satellites are determined. The resulting corrections for GPS-35 and -36 are likely not a problem of scale but rather some inherent problem of the GPS and ITRF2000 frames. Erroneous phase center settings for the GPS satellites do not affect SLR offset estimates, they move almost fully into the clock parameters. A

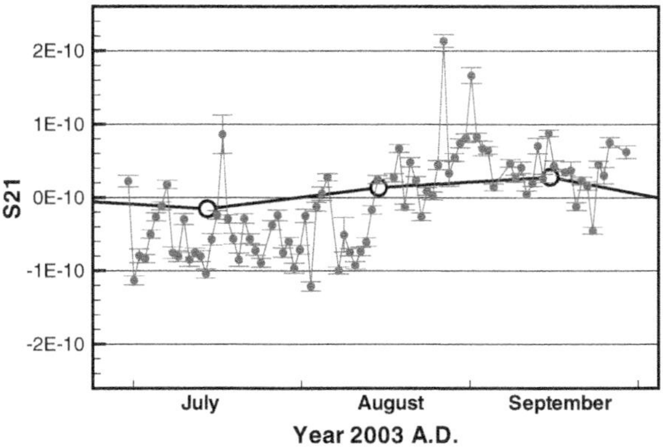

Fig. 8. S_{21} resolved daily versus monthly solution

Helmert transformation between eventually established SLR station coordinates in the GPS frame and ITRF2000 does not reveal systematic differences.

The integrated method combines various satellite techniques at the observation level. It leads to unprecedented high accuracies of the gravity field solutions, therefore the low degree harmonics of the Earth's gravity field, which should be part of the reference frame realizations, can be resolved with high accuracy and resolution. The series of daily C_{10}, C_{11}, and S_{11} coefficients, the Z-, X-, and Y-components of the dynamical geocenter location, from three months of GRACE/GPS observations, comes out with higher accuracy in comparison to IGS weekly solutions. The C_{21} and S_{21} series, the orientation of Earth's main axis of inertia, can reliably be resolved at daily intervals and seem to reveal geophysical effects.

Acknowledgement. This is publication no. GEOTECH-164 of the program GEOTECHNOLOGIEN of BMBF and DFG, Grant 03F0336E. Proxies for EOPs and ground station coordinates were provided by IERS and IGS, GPS ground data by IGS, SLR data by ILRS.

References

Altamimi Z, Sillard P, Boucher C (2002) ITRF2000: A new release of the International Terrestrial Reference Frame for earth science applications. J Geophys Res 107:B10, 2214, doi:10.1029/2001JB000561

Chen JL, Wilson CR, Tapley BD, Ries JC (2004) Low degree gravitational changes from GRACE: Validation and interpretation. Geophys Res Letters 31: L22607

Gambis D, Bizouard C (2003) Consistency analysis between EOP series and reference frames, Proceedings of the IERS Workshop on Combination Research and Global Geophysical Fluids, IERS Technical Note 30, 57–62, Bundesamt für Kartographie und Geodäsie, Frankfurt am Main

Ge M, Gendt G, Dick G, Zhang FP, Reigber Ch (2005) Impact of GPS satellite antenna offsets on scale changes in global network solutions. Geophys Res Letters 32: L06310

Hu XG, Shi C, Flechtner F, Koenig R, Schwintzer P, Schmidt R, Meyer U, Massmann F-H, Reigber C, Zhu SY (2005) High frequency temporal Earth gravity variations detected by GRACE satellites. In: this issue

IAG (2005) The International Association of Geodesy Site. Internet: http://www.iag-aig.org, modified June 1, 2005.

IDS (2005) The International DORIS Service Site. Internet: http://ids.cls.fr, modified March 2, 2005.

IERS (2003) The International Earth Rotation and Reference Systems Service Site. Internet: http://http://www.iers.org, modified July 23, 2003

IERS (2004) IERS Combination Pilot Project: Call for Participation and received proposals. Internet: http://www.iers.org/iers/about/wg/wg3/cfp_cpp_cont.html, modified August 16, 2004

IGS (2005) International GPS Service Site. Internet: http://igscb.jpl.nasa.gov, modified April 02, 2005

ILRS (2005) International Laser Ranging Service Site. Internet: http://ilrs.gsfc.nasa.gov, cited as of May, 2005.

IVS (2005) International VLBI Service for Geodesy & Astrometry Site. Internet: http://ivscc.gsfc.nasa.gov, modified January 1, 2005.

König R, Reigber Ch, Zhu SY (2005) Dynamic model orbits and Earth system parameters from combined GPS and LEO data. Advances in Space Research, In Press, Corrected Proof, Available online 27 April 2005

Kouba J, Mireault Y, Lahaye F (1994) 1994 IGS Orbit/Clock Combination and Evaluation. 1994 IGS Annual Report, pp 70–94

Neumayer KH, Massmann F-H, Meixner H, Raimondo J-C, Reigber Ch, Shi C, Yu Y, SY Zhu SY (2005) SLR Station Solutions in the GPS Reference Frame. In preparation to be submitted to the Journal of Geodesy

Rothacher M (2001) Comparison of absolute and relative antenna phase centers. GPS Solutions 4: 55–60

Rothacher M (2002) Towards a rigorous combination of space geodetic techniques. IERS Workshop on Combination Research and Global Geophysical Fluids, Bavarian Academy of Sciences, Munich, Germany, November 18-21, 2002

SBGG (2004) Special Bureau for Gravity/Geocenter Home Page. http://sbgg.jpl.nasa.gov/, modified October 30, 2004

Shi C, Grunwaldt L, Raimondo J-C, Massmann F-H, Zhu SY (2003) Determination of the CHAMP GPS Antenna with Respect to Satellite's Mass Center. In: Reigber C, Luehr H, Schwintzer P (eds) First CHAMP Mission Results for Gravity, Magnetic and Atmospheric Studies. Springer, Berlin Heidelberg

Springer TA (1999) Modeling and Validating Orbits and Clocks Using the Global Positioning System. PhD dissertation, University of Bern

Zhu SY, Massmann F-H, Yu Y, Reigber Ch (2003) Satellite antenna phase center offsets and scale errors in GPS solutions. J of Geodesy 76: 668–672

Zhu SY, Reigber Ch, König R (2004) Integrated Adjustment of CHAMP, GRACE, and GPS Data. J of Geodesy 78: 103–108

Part V

GEOsensor
Ring Laser for Seismology

The GEOsensor Project: Rotations – a New Observable for Seismology

Ulrich Schreiber[1], Heiner Igel[2], Alain Cochard[2], Alexander Velikoseltsev[1], Asher Flaws[3], Bernhard Schuberth[2], Wolfgang Drewitz[4], and Frieder Müller[4]

[1] Forschungseinrichtung Satellitengeodäsie der TU-München
`schreiber@wettzell.ifag.de`, `alex@wettzell.ifag.de`
[2] Department für Geo- und Umweltwissenschaften, Sektion Geophysik, Ludwig-Maximilians-Universität München
`igel@geophysik.uni-muenchen.de`, `alain@geophysik.uni-muenchen.de`
[3] Department of Physics and Astronomy, University of Canterbury, New Zealand
`acf28@student.canterbury.ac.nz`
[4] FMB Feinwerk- und Messtechnik GmbH, Berlin
`w.drewitz@fmb-berlin.de` `f.mueller@fmb-berlin.de`

Summary. Over the last 40 years ring laser gyroscopes became one of the most important instruments in the field of inertial navigation and precise rotation measurements. They have a high resolution for angular velocities, a very good scale factor stability and a wide dynamic range. These properties made them suitable for aircraft and autonomous submarine navigation. Over the last decade we have developed several very large perimeter ring laser gyroscopes for the application in geodesy and geophysics (Schreiber et al., 2001). Because of a substantial upscaling of these ring lasers, their sensitivity to rotations has been increased by at least 5 orders of magnitudes. At the same time the instrumental drift was reduced by about the same amount. This progress in rotational sensor technology led to the successful detection of rotational signals caused by earthquakes (Pancha et al., 2000) several thousands kilometers away. These observations stimulated the development of a highly sensitive ring laser gyro for specific seismological applications. The GEOsensor provides rotational motions along with the usual translational motions at a high data acquisition rate of at least 20 Hz. Observations of seismic induced rotations show that they are consistent in phase and amplitude with the collocated recordings of transverse accelerations obtained from a standard seismometer over a wide range of distances and frequencies.

Key words: ring laser, seismology, rotation measurements

1 Introduction

Currently there are primarily two types of measurements that are routinely used to monitor global and regional seismic wave fields. Standard inertial seismometers measure three components of translational ground displacement and provide the basis for monitoring seismic activity and ground motion. The second type aims at measuring the deformation of the Earth (strains). It is well known (Aki and Richards, 2002) that there is a third type of measurement that should be observed in seismology and geodesy in order to fully describe the motion at a given point, the measurement of ground rotation. In the past years, ring laser gyroscopes were developed primarily to observe variations in Earth's absolute rotation rate with high precision (Stedman et al., 1995; Stedman, 1997). The recording of the (complete) earthquake-induced rotational motion is expected to be useful particularly for (1) further constraining earthquake source processes when observed close to the active faults (Takeo and Ito, 1997); (2) estimating permanent displacement from seismic recordings (Trifunac and Todorovska, 2001); (3) estimating local (horizontal) phase velocities from collocated observations of translations and rotations (Igel et al., 2005a).

In standard analysis approaches in seismology earthquake induced rotations have been neglected in the past, because the corresponding magnitudes were thought to be small and no suitable instruments with the required resolution existed. The high sensitivity for rotations of large ring lasers along with their insusceptibility to linear translations makes the application of these instruments very attractive for seismological studies. The required range of angular velocities to be measured is expected to be 10^{-14} rad/s $\leq \Omega_s \leq 1$ rad/s and the required frequency bandwidth for the seismic waves is in the range of 3 mHz $\leq f_s \leq$ 10 Hz (Schreiber et al., 2004). Three such devices mounted in orthogonal orientations will eventually provide the quantitative detection of rotations from shear, Love and Rayleigh waves, thus providing the missing quantities for a complete 6 degrees of freedom measurement system.

2 Instrumental Section

Ring lasers are active Sagnac interferometers, where two laser beams are circulating around a triangular or square closed cavity in opposite directions (Aronowitz, 1971). If the whole apparatus is rotating with respect to inertial space one obtains a frequency splitting of the two counter propagating waves, which is proportional to the rate of rotations. The Sagnac frequency δf is

$$\delta f = \frac{4A}{\lambda P}\mathbf{n}\cdot\mathbf{\Omega}\ , \tag{1}$$

where A is the area, P the perimeter enclosed by the beam path and λ the optical wave length of the laser oscillation. $\mathbf{\Omega}$ is the angular velocity at which the instrument is turning and $\mathbf{n}$ is the normal vector to the laser beam plane. The resolution of a ring laser gyroscope is proportional to the ratio of the quotient of area and perimeter enclosed by the beam path. Therefore, an increased size of the laser cavity leads to an improved sensor sensitivity. For example, the 4 by 4 meters square ring laser G installed in Fundamentalstation Wettzell (Germany) has a sensor resolution of $\delta\varphi = 9 \cdot 10^{-11} \mathrm{rad}/\sqrt{\mathrm{s}}$. This outstanding sensitivity is good enough for the detection of both teleseismic waves and near source seismic signals. Typical seismic signals require a high sensor stability for up to one hour of continuous data acquisition. This requirement is much reduced from the long-term stability necessity of an instrument for the measurement of Earth rotation variations. Of more importance is the short-term precision and mechanical rigidness of the laser beam path for stable operation. The instrument must be capable of accurately recording seismic rotations while keeping the scaling factor (the quotient in (1)) constant. At the same time the whole system must be relocatable, cost effective and allow for a relatively simple installation. Given the fact that a ring laser is a highly sensitive optical interferometer, all these requirements are essentially contradicting design goals. For the data logging, a precise time stamping of the record is absolutely crucial for the desired application. A typical requirement for timing accuracy in seismology is to within 1 ms of UTC. As a consequence all data must be sampled using a stable reference and an accurate time such as the timescale represented by GPS.

2.1 GEOsensor Concept

In Sect. 1 we have listed the requirements for a ring laser in seismology. The frequency band of interest covers about 5 orders of magnitude, while the corresponding sensitivity for the measurement of rotations should cover 14 orders of magnitude, the range between strong motions during a local earthquake on one side and the signals of an earthquake more than 10000 km away on the other side. However, if we exclude strong motion domain from the immediate measurement interest, we obtain a viable measurement range for a prototype sensor of approximately 10^{-12} rad/s $\leq \Omega_s \leq 10^{-4}$ rad/s which still extends over 8 orders of magnitude.

The basic design of the GEOsensor is outlined in the block diagram of Fig. 1. A synchronously generated dataset contains the precise timestamping from a GPS system, the instantaneous Sagnac frequency from the ring laser component, tiltmeter recordings in two directions about the normal vector of the ring laser plane and the measured velocities of a seismometer.

In order to operate the GEOsensor, more signals such as the ring laser beam intensity are required. However, such signals are used in a feedback system to set the operation range of the ring laser and therefore are not logged for further use.

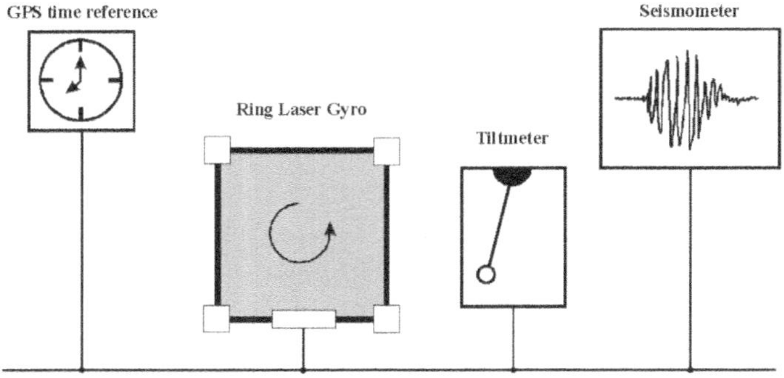

Fig. 1. Block diagram of the GEOsensor design

2.2 The Ring Laser Component

In order to obtain a stable interferogram of the two laser beams the cavity length has to be kept constant to within a fraction of a wavelength. Therefore, usually ring laser bodies are made from Zerodur, a glass ceramic which exhibits a very small relative thermal expansion of $\alpha = 5 \cdot 10^{-8}$ 1/K. Since a ring laser for seismic applications requires an enclosed area of more than 1 m^2, a monolithic ring construction would be both too expensive and not transportable.

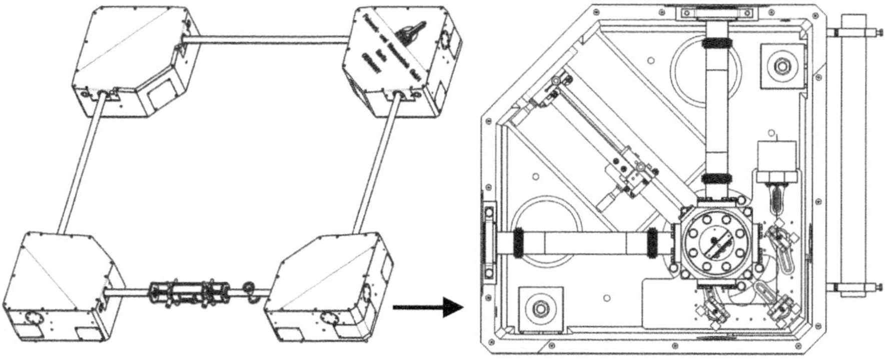

Fig. 2. Construction diagram of the GEOsensor ring laser

Figure 2 gives an impression of the actually realized ring laser hardware. The laser cavity has the shape of a square. The 4 turning mirrors are each located in a solid corner box. As shown in the right side of the plot, a folded lever system allows the alignment of each mirror to be within ± 10 seconds of arc. This high level of alignment is required to ensure lasing from an optically stable cavity. The mirrors are located inside steel containers which in turn are connected together with stainless steel tubes, forming an evacuated enclosure for the laser beams. In the middle of one side the steel tubes are reduced to a small glass capillary of 4 mm in diameter and a length of 10 cm, which is required for gain medium excitation. When operated, the ring laser cavity is first evacuated and then filled with a mixture of Helium and Neon reaching a total gas pressure of approximately 6 hPa. The left part of Fig. 2 gives an impression of the instrumental layout. The following two important considerations are unique for the GEOsensor design.

- Since the ring laser is constructed from several components, it requires a stable concrete platform base at the location of deployment. Such a pad is simple to specifiy and can be prepared totally independently of the actual GEOsensor deployment.
- The actual area of the ring laser component is not predetermined by the design. The instrument can be built according to the available space at the host observatory. Different GEOsensor realizations may therefore have different size and consequently different instrumental resolution. The length of the current instrument is 1.6 m on a side, which provides an area of 2.56 m^2.

In order to operate the GEOsensor, the cavity must be evacuated, baked and filled with a Helium/Neon gas mixture. This procedure requires a turbo molecular pump system and a manifold with a supply of ^{4}He, ^{20}Ne and ^{22}Ne. The pump system is not required during the operation of the GEOsensor but is necessary for the preparation of the instrument and once or twice during a year in order to change the laser gas. Laser excitation itself is achieved via a high frequency generator (Stedman, 1997), matched to a symmetrical high impedance antenna at the gain tube. A feedback loop maintains the level of intensity inside the ring laser and ensures monomode operation. When the ring laser is operated it detects the beat note caused by Earth rotation. The magnitude of this beat frequency is depending on $\sin(\Phi)$ with Φ the latitude of the ring laser location. Table 1 shows the value of the Earth's rate bias for a few locations of interest.

Table 1. Earth rotation bias for some GEOsensor locations

location	frequency [Hz]
Wettzell, Germany (49.145 N)	138
Pinon Flat, CA (33.6 N)	102
Tokyo (35.4 N), Japan	106
Cashmere (43.57 S), NZ	127

To date we have operated the GEOsensor at the first two locations. Since the Earth rotation acts like a rate bias on our ring laser measurements, any rotations caused by earthquakes will show up as an alternating frequency around the measured Earth rate.

2.3 Auxiliary Sensors

Ring lasers are rotation sensors that give very localized information on rotations. This is in contrast to an array of seismometers for example. To fully exploit the sensor potential and to investigate the properties of the GEOsensor instrumentation, other auxiliary data is measured along with the Sagnac frequency. The temperature at the monument of the ring laser installation provides informations on the scale factor variations due to the thermal expansion of the concrete foundation to which the ring laser hardware has been matched by design. Atmospheric pressure variations may also cause scale factor changes and are therefore measured too. For the purpose of studies on the general sensor behavior there is the option of logging the power of the high frequency generator and the beam power ratio. However this is not done routinely. The intensity of one laser beam is used to adjust the point of operation of the ring laser. This process is part of an automatic feedback loop arrangement but the data is currently not logged.

Apart from these direct ring laser related sensors we are recording variations in the orientation of the ring laser with respect to local **g** as well as all three components of translational seismograms. The tiltmeter measurements are used to reduce the measured rotation rate from orientation effects that enter the Sagnac frequency via the inner product in (1). The importance of this effect is discussed in Sect. 2.7 in more detail.

2.4 Logging System

The logging system has to satisfy demanding realtime conditions. This requires that the epoch of an observation can be timestamped to better than 1 ms with respect to UTC. The concept chosen for that purpose is outlined in Fig. 3. The GPS time-frequency receiver provides both the time stamping and the data sampling frequency. The incoming PPS (pulse per second) signal with an accuracy of 30 ns rms relative to UTC, triggers the acquisition. The exact epoch of the trigger is also taken from the GPS receiver. The data sampling is based on a receiver reference frequency output, which has a relative accuracy of better than 10^{-12} (while locked to GPS). To satisfy the 1 ms accuracy requirement for the application, the reference frequency (e.g. 10 MHz) output is divided down to 1 kHz, which is the actual data acquisition rate. This frequency is used as an external clock instead of the build-in computer clock. This ensures that the data logging is always started at the full second and the data samples are equidistant and phase locked to the GPS receiver clock.

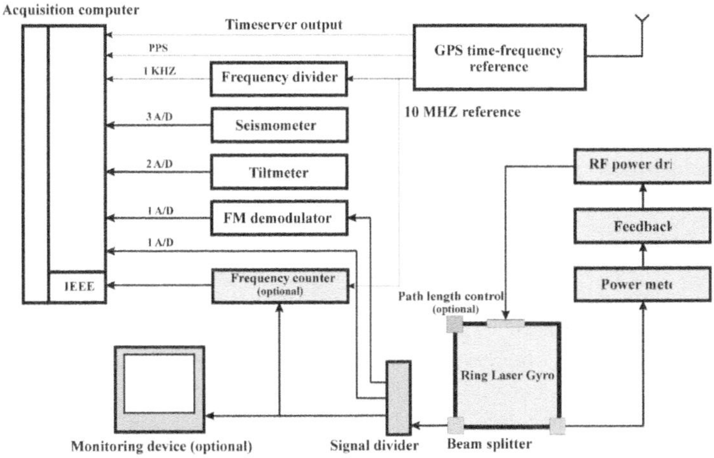

Fig. 3. Dataflow chart for the realtime logging in the GEOsensor

2.5 Deployment

After the development of the GEOsensor and including a test installation at the Fundamentalstation Wettzell, the instrument was shipped to the Scripps Institution of Oceanography in San Diego, California. In January 2005 the installation of the complete sensor took place at the seismological observatory Pinon Flat as shown in Fig. 4. The observatory is located between the San Jacinto and the San Andreas faults.

Fig. 4. The ring laser vault at the Pinon Flat (CA) observatory and the GEOsensor installation in one of the chambers

The goal of this installation is the measurement of a number of earthquakes at short distances from their epicenters. This will allow an extensive and systematic study of rotational motions with a particular emphasis on local and regional scales with source distances of up to 1000 km. From basic theoretical studies (Takeo and Ito, 1997) it became apparent that rotational motion information may contribute the most for local and regional earthquakes.

2.6 Sensor Performance and Verification

The validation of ring laser measurements as true rotational signals was an important item during the integration phase. For the assessment of the quality of the GEOsensor data it is necessary to distinguish true measured rotations from unknown sensor artefacts. Since there are two collocated ring lasers with identical orientation in the Cashmere Cavern in Christchurch, we used the simultaneous earthquake recording of both C-II and UG1 for the Fiji event on August 19, 2002. Figure 5 shows the Sagnac frequency as a function of time converted to rotation rate in nanoradians per second using (1) for the first 15 seconds of this earthquake.

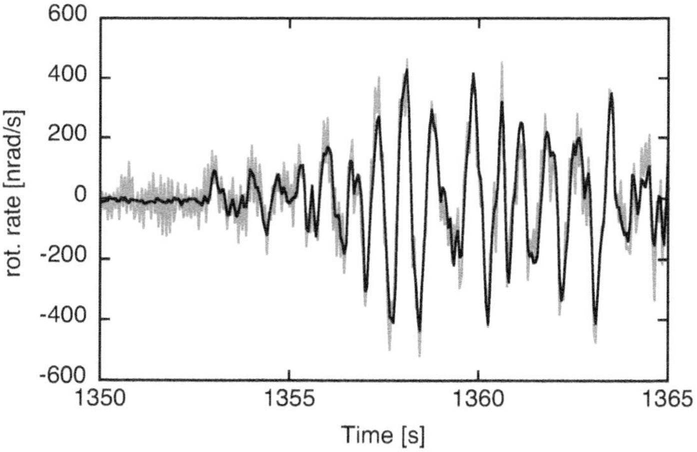

Fig. 5. Raw rotations obtained from an earthquake at Fiji (Aug. 19. 2002) with 2 collocated ring lasers C-II and UG1 located in the Cashmere Cavern near Christchurch (New Zealand)

Because ring lasers are purely optical sensors, they do not depend on mass inertia like ordinary seismometers. Therefore, there is no need for a restitution process of the measurement quantity. According to the ring laser equation the relationship between the obtained Sagnac frequency and the input rotation rate is linear over a wide dynamic range. This important property is reflected in Fig. 5. The data from the small C-II ring laser is much more noisy than the data from the very large UG1, because there is almost a factor of 20 difference in the respective scale factors. Nevertheless one can see that both ring lasers measure the same effect, in phase as well as in amplitude. It has to be noted that apart from the unit conversion we are comparing raw data for this measurement.

2.7 Sensor Orientation

Equation 1 shows another very unique property of ring lasers. They measure absolute rotations and this includes also changes of the instrumental orientation as one can see from the inner product between the rotation axis and the instrumental normal vector. Ring lasers may be subject to rocking motions in the presence of S- and Rayleigh-waves one could argue. In order to identify such motions and to correct for this, the GEOsensor was equipped with a sensitive tiltmeter. Figure 6 shows an earthquake in Northern Algeria on May 21, 2003, which was measured at the G ring laser in Southern Germany approximately 1600 km away. The top part of the diagram show the measured rotation rates from the ring laser, while the lower part of the diagram shows the corresponding contribution to the signal originating from variations of the ring laser orientation.

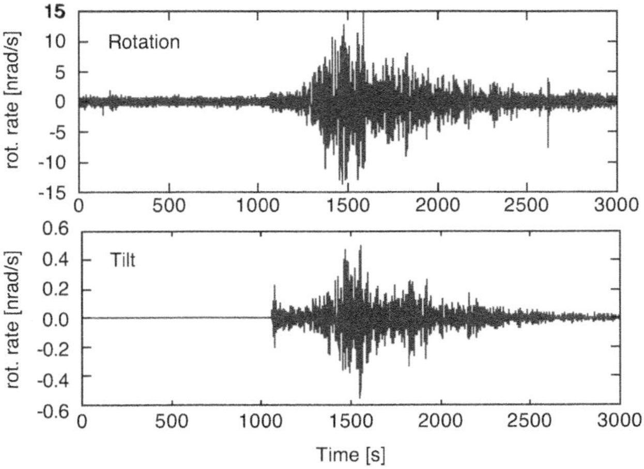

Fig. 6. Algerian earthquake real rotations and orientation induced variations of the Sagnac frequency expressed in the same units

This tilt equivalent rotation rate was computed by

$$\Omega_{tilt} = \Omega \cdot \sin(\Phi - T_{NS}), \tag{2}$$

where Ω is the actual rotation rate, Φ the latitude of the sensor and T_{NS} the North - South component of the locally measured tilt. The contribution of the East - West component is more than 2 orders of magnitude smaller, since it appears under a cosine function which is always very close to 1.

2.8 Detection Properties

Ring lasers provide optical interferograms where the external rate of rotation is proportional to the rate of change of the fringe pattern. This signal becomes available as an audio-frequency at the output of a photomultiplier tube. In seismology it is important to detect the rate of change of this frequency at 50 ms intervals (20 Hz) very accurately. Since frequency counting techniques do not provide a sufficient resolution at such short averaging intervals, a frequency demodulation concept has been developed. A voltage controlled oscillator is phase locked to the Sagnac frequency of the ring laser, exploiting the fact that Earth rotation provides a constant rate bias in the absence of any seismically induced rotation signals. In the event of an earthquake one obtains the rate of change of the Sagnac frequency at the feedback line of the voltage controlled oscillator. This voltage can be digitized and averaged at the required 20 Hz rate or higher.

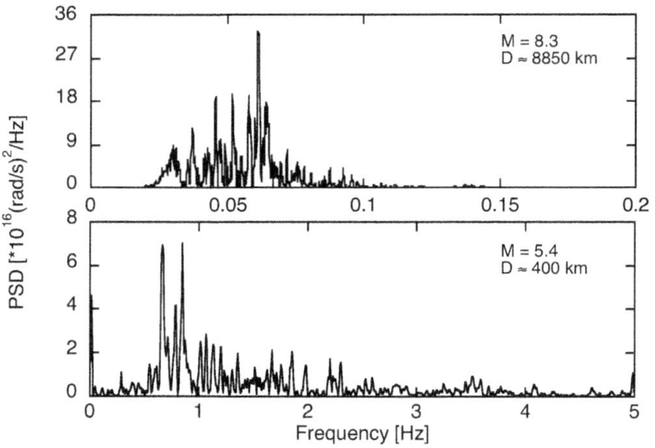

Fig. 7. Comparison of recorded rotation spectra from an teleseismic event (Hokkaido: Sept. 9, 2003) and a regional earthquake (France: Feb. 22, 2003). The much higher bandwidth of the rotational wave spectra requires data acquisition techniques such as the demodulator

To outline the importance of the frequency demodulation technique we compare two earthquakes with distinctly different properties. Figure 7 shows an example for a teleseismic event and an example from a much closer regional earthquake. While for the remote earthquake the spectral power density essentially drops off to zero above frequencies of 0.1 Hz, one can still see some signal signature up to about 4 Hz for the regional event. Frequencies with a rate of change above 2 Hz; however, are already outside the regime of reliable representation in phase and amplitude by conventional frequency counting and second order autoregression frequency analysis (McLeod et al., 2001).

2.9 Sensor Artefacts

Large ring lasers are currently the most sensitive operational devices for the measurement of geophysically induced rotations. The GEOsensor combines the high sensitivity of these gyroscopes with a technical design that allows a relative simple deployment at an observatory with some basic infrastructure, such as a concrete pad in a basement of a building, electricity and internet connectivity. However, in the presence of even moderate temperature fluctuations, one has to accept sudden mode jumps in the laser cavity, since thermal expansion or compression changes the length of the ring laser cavity enough that a neighbouring longitudinal mode starts to have more laser gain than the currently supported one. Another side effect of these mode drifts is a slow change of mode pulling (Aronowitz, 1971) as the laser mode passes through the gain curve. For frequencies below 0.01 Hz this may result in some ambiguities for the interpretation and some post processing may be required to distinguish between seismically induced rotations and internal biases from the ring laser.

Figure 8 shows a sample dataset from the GEOsensor measured at the temporary installation on the Fundamentalstation Wettzell. The upper diagram shows a section of the measured rotations from the magnitude 7.4 Hokkaido earthquake from Sept. 5, 2004.

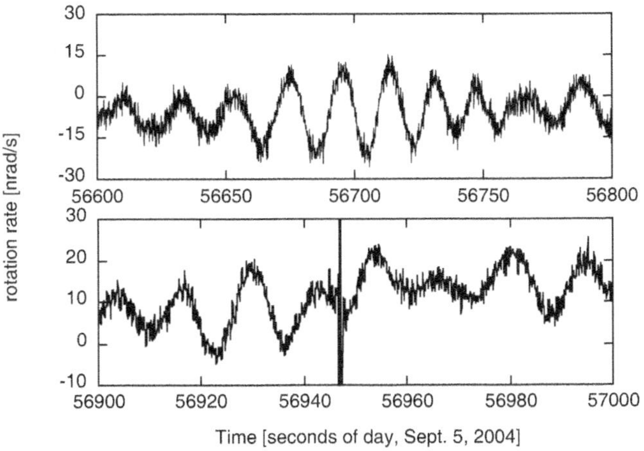

Fig. 8. Rotation measurements taken with the GEOsensor on Sept. 5, 2004. The top diagram shows obtained rotation rate, while the lower plot illustrates the effect of a mode jump in the data

In the lower diagram one can see the effect of a mode jump in the dataset. For about one second there is a gap in the data. This is the time it takes for the new mode to settle in and for the fringe pattern to become stable again.

For teleseismic events it is not difficult to interpolate the dataset and such a sensor behavior will not result in data loss. However for local earthquakes this may be different.

The effect of sensor drift is also apparent in Fig. 8. One would expect the measured rotations to appear symmetrically about the rate bias of Earth rotation. This is not the case for the displayed dataset because of variations of the laser frequency pulling in the ring laser cavity. For earthquakes like the one shown here this artefact can be fully removed.

3 Data Analysis

3.1 Observations of Rotations

In order to compare translations (measured by a standard seismometer) with the vertical component of the vector of rotation - which is what the G-ring is measuring - the horizontal components of seismic recordings were rotated into radial and transverse directions. Note that Rayleigh waves should not generate such a vertical rotation component, while Love waves are horizontally polarized hence generate rotations around a vertical axis only. To obtain transverse acceleration, the transverse velocity seismograms were differentiated with respect to time. Let us now assume a transversely polarized plane wave with displacement $\mathbf{u} = (0, u_y(t - \frac{x}{c}), 0)$, c being the horizontal phase velocity. The vector of rotation is thus given as $\frac{1}{2}\nabla \times \mathbf{u} = (0, 0, -\frac{1}{2c}\dot{\mathbf{u}}_y(t - \frac{x}{c}))$ with the corresponding z-component of rotation rate $\boldsymbol{\Omega}_z(x,t) = -\frac{1}{2c}\ddot{\mathbf{u}}_y(t-\frac{x}{c}))$. This implies that - under the given assumptions - at any time rotation rate and transverse acceleration are in phase and the amplitudes are related by $\ddot{\mathbf{u}}_y(x,t)/\boldsymbol{\Omega}_z(x,t) = -2c$. In practice, the phase velocities can be estimated by dividing best-fitting waveforms in sliding a time-window of appropriate length along the seismic signal and rotation rate. Thus, under the plane-wave assumption both signals should be equal in phase and amplitude (McLeod et al., 1998; Pancha et al., 2000). This assumption is expected to hold for a considerable part of the observed ground motion due to the large epicentral distance compared to the considered wavelengths and source dimensions. This property is exploited here to verify the consistency of the observations. Close to the seismic source this assumption no longer holds and may form the basis for further constraining rupture processes (Takeo, 1998; Takeo and Ito, 1997).

A data example (rotation rate and transverse acceleration) of the M8.1 Tokachi-oki event, September 25, 2003, and a time-dependent normalized cross-correlation coefficient (maximum in a 30s sliding window) is given in Fig. 9. The time window also contains an event (increase in cross-correlation at 3500 s) that was bearly visible in the seismograms without correlating the two signals. When the waveform fit between rotation rate and transverse acceleration is sufficiently good (e.g., a normalized correlation coefficient > 0.95), we estimate phase velocities by dividing the peak amplitudes of both

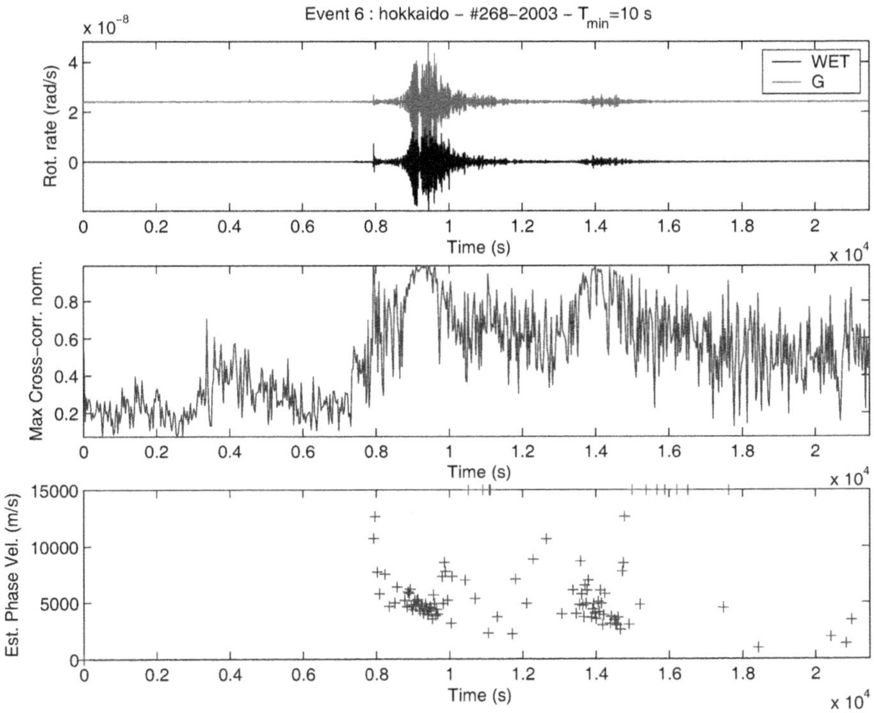

Fig. 9. Top: Observation of rotation rate (red) and transverse acceleration (black) after the M8.1 Tokachi-oki event, 29-9-03. Middle: The cross-correlation-coefficient in a 30 s sliding window. Note the increase in correlation during the main shock (>7500 s) and aftershock (>13800 s) to almost 1 (perfect match). Bottom: Estimates of horizontal phase velocities in time windows with good phase match (> 9500 s). Note the decreasing phase velocities in the Love wave train (e.g., 8000-10000 s, indicative of Love wave dispersion)

traces as explained above. These time dependent estimates of phase velocities are shown in Fig. 9 (bottom). Despite the scattering the phase velocities in the time window containing the Love waves (9000-10000 s and around 14000 s for the aftershock) the estimates are in the right range of expected phase velocities and the negative slope of the velocities with time suggest that the expected dispersive behavior (earlier longer periods have higher phase velocities) can be observed in the data using this processing approach. The lack of correlation in the time windows excluding the Love waves may indicate that either the (body-) wave fronts are not planar or that the energy comes from out-of-plane directions through scattering. The consequence of non-planar waves on structural interpretation was studied by (Wielandt, 1993). To demonstrate that we obtain such a fit consistently we further show in Color Fig. XLVII on p. 314 the standard seismograms (blue) and a comparison between translations and observations for another event out of a data base with now over 40 events with epicentral distances from 300-12000km.

3.2 Simulation of Rotational Motions in Realistic Earth Models

As there are basically next to no studies of rotational ground motions in the seismological literature several classic schemes to calculate synthetic seismograms have been extended to also output rotations. These are (1) the full space analytical solution of wave propagation due to a double couple point source e.g. (Aki and Richards, 2002) (with a surprising result at first sight: as expected, the rotation generated by the P wave-front is always zero - hence in particular in the far-field - yet the rotation derived from the far-field displacement alone is nonzero - as is obvious from the classical radiation pattern plot; one has to take all field terms into account. In other words, the usual separation between far-, intermediate-, near- displacement fields is not valid for rotations (Cochard et al., 2005)); (2) the Cartesian finite-difference algorithms with which we study near-source effects (Cochard and Igel, 2004); (3) the spherical finite-difference methods (Igel et al., 2002); and (4) the spectral-element code (Specfem) (Komatitsch and Tromp, 2002a; Komatitsch and Tromp, 2002b). The latter algorithm was used to specifically model individual earthquakes using finite fault models (Igel et al., 2004, 2005b). The M8.1, Sep. 25, 2003, Tokachi-oki event was modelled with the Specfem algorithm (Igel et al., 2005a; Schuberth et al., 2004) using a 3D tomographic model, crustal model and a finite source model provided by Ji Chen (CalTech). The results of this modelling exercise is shown in Color Fig. XLVIII on p. 315. The fit between observations and modelling for the shear-wave arrivals and the fundamental mode Love waves is excellent for both translations and rotations. The amplitude mismatch in the Love waves may be explained with the inaccurate finite source model leading to directivity effects that are not correctly modelled. However, it is important to note that the observed phase velocities match well with the modeled horizontal phase velocities (Igel et al., 2005a).

3.3 Long-period Rotations

How much long-period information is contained in the ring laser recordings? In Color Fig. XLIX on p. 315 the transverse accelerations and rotations are again superimposed and filtered in narrow frequency bands to highlight the fit towards lower frequencies. This also allows us to directly estimate best-matching phase velocities in the Love-wave train and to see whether the expected dispersion relation is visible. Color Fig. XLIX demonstrates that the ring laser rotations are matching the transverse accelerations down to periods of 150 seconds (and possibly beyond). As expected the maximum amplitudes occur within the Love waves. By dividing the peak amplitudes of acceleration and rotation rate we estimate the phase velocities for narrow-bandwidth surface wave signals. As surface waves are dispersive as a consequence of the seismic velocity structure with depth we expect increasing horizontal phase velocities with increasing dominant period. This is indeed what we observe at least in the band between periods of 20 seconds to 100 seconds. It is important to note that with the collocated recordings of rotations and translations one can estimate a quantity (phase velocity) that otherwise can only be determined with an array of seismometers or with collocated strainmeters (Mikumo and Aki, 1973). The results in Color Fig. XLIX on p. 315 are indeed promising. Even though this is work in progress the frequency-dependent amplitude ratio of rotations and translations suggests that a single-station measurement may allow estimates of Love-wave dispersion. These in turn would allow estimates of local 1D velocity models and - by combining several of them - eventually 3D tomography. However, whether the accuracy required for structural inversion is sufficient remains to be seen.

In summary, to the standard seismologist the current data base of rotations, array data, and collocated translations provides a realm of new opportunities. Even though the potential benefits to seismology, earthquake physics, and earthquake engineering still need to be further explored, the preliminary results indicate the many interesting new routes that can be taken.

Acknowledgement. This is publication no. GEOTECH-169 of the programme GEOTECHNOLOGIEN of BMBF and DFG, Grant 03F0325 A-D. The authors would like to thank Wolfgang Schlüter and Michael Wensauer from the Bundesamt für Kartographie und Geodäsie for their support of the project. We would also like to acknowledge the IQN Program by the German Academic Exchange Service and the Leibniz Computing Centre Munich for providing access to their supercomputers. We would like to thank Dr. Walter Zürn and Prof. R. Dunn for their constructive comments that improved the manuscript.

References

Aki, K., Richards, P. G. (2002) Quantitative Seismology, 2nd Edition, University Science Books

Aronowitz, F. (1971) The laser gyro. Laser applications, Vol. 1, edited by M. Ross, 133–200, Academic Press, New York

Bouchon, M., and Aki, K. (1982) Strain, tilt and rotation associated with strong ground motion in the vicinity of earthquake faults. Bull. Seismol. Soc. Amer., 72:1717–1738

Cochard, A., Igel, H. (2003) What can rotational measurements teach us about earthquake rupture histories? Eos Trans. AGU, 84(46), Fall Meet. Suppl., Abstract S42D–0200

Cochard, A., Igel, H. (2004) What can rotational measurements teach us about earthquake rupture histories? Geophysical Research Abstracts (EGU Meeting), 6, 06359

Cochard, A., Igel, H., Flaws, A., Schuberth, B., Wassermann, J., Suryanto, W. (2005) Rotational motions in seismology, in preparation, to be published in "Earthquake source asymmetry, structural media and rotation effects" eds. Teisseyre et al., Springer Verlag

Igel, H., Nissen-Meyer, T., Jahnke, G. (2002) Wave propagation in 3-D spherical section: effects of subduction zones, Phys. Earth. Planet. Int., 132:219–234

Igel, H., Flaws, A., Velikoseltsev, A., Cochard, A., Schreiber, K. U. (2004) Comparison of rotational and translational motions induced by distant large earthquakes, Geophysical Research Abstracts (EGU Meeting), 6, 06487

Igel, H., Schreiber, K. U., Flaws, A., Schuberth, B., Velikoseltsev, A. Cochard, A. (2005a) Rotational motions induced by the M8.1 Tokachi-oki earthquake, September 25, 2003, Geophys. Res. Lett., VOL. 32, L08309, doi:10.1029/2004GL022336

Igel, H., Cochard, A., Schuberth, B., Flaws, A. Velikoseltsev, A., Schreiber K. U. (2005b) Rotational ground motions: a new observable for seismology? Geophysical Research Abstracts (EGU Meeting), 7

Komatitsch, D., Tromp, J. (2002a) Spectral-element simulations of global seismic wave propagation, Part I: Validation, Geophys. J. Int., 149:390–412

Komatitsch, D., Tromp, J. (2002b) Spectral-element simulations of global seismic wave propagation, Part II: 3-D models, oceans, rotation, and gravity, Geophys. J. Int., 150:303–318

McLeod, D. P., Stedman, G. E., Webb, T. H., Schreiber K. U. (1998) Comparison of standard and ring laser rotational seismograms, Bull. Seism. Soc. Amer., 88:1495–1503

McLeod, D. P., King, B. T., Stedman, G. E., Schreiber, K. U., and Webb T. H. (2001) Autoregressive analysis for the detection of earthquakes with a ring laser gyroscope; Fluctuations and Noise Letters, Vol. 1, No. 1:R41–R50

Mikumo, T., Aki, K. (1973) Determination of local phase velocity by intercomparison of seismograms from strain and pendulum instruments, J. Geophys. Res., 69: 721–731

Pancha, A., Webb, T.H., Stedman, G. E., McLeod, D.P., and Schreiber, U. (2000) Ring laser detection of rotations from teleseismic waves. Geophys. Res. Lett, 27:3553–3556

Schreiber, U., Schneider, M., Rowe, C.H., Stedman, G. E., and Schlüter, W. (2001) Aspects of Ring Lasers as Local Earth Rotation Sensors. Surveys in Geophysics, Vol. 22: (5-6) 603–611

Schreiber, U., Velikoseltsev, A., Stedman, G. E., Hurst, R. B., Klügel, T. (2004) Large Ring Laser Gyros as High Resolution Sensors for Applications in Geoscience. Proceedings of the 11th International Conference on Integrated Navigation Systems, St. Petersburg, 326–331

Schuberth, B., Igel, H., Wassermann, J., Cochard, A., Schreiber, K. U.(2004) Rotational Motions from Teleseismic Events - Modelling and Observations, Eos Trans. AGU, Fall Meet. Suppl. Abstract S42D–0200

Stedman, G. E., Li, Z., Bilger, H. R. (1995) Sideband analysis and seismic detection in large ring lasers. Appl. Opt., 34:7390–7396

Stedman, G. E. (1997) Ring laser tests of fundamental physics and geophysics. Rep. Progr. Phys. 60:615–688

Takeo, M., Ito, H. M. (1997) What can be learned from rotational motions excited by earthquakes? Geophys. J. Int., 129:319–329

Takeo, M. (1998) Ground rotational motions recorded in near-source region of earthquakes, Geophys. Res. Lett., 25:789–792

Trifunac, M. D., Todorovska, M. I. (2001) A note on the usable dynamic range of accelerographs recording translation, Soil Dyn. Earth. Eng., 21(4):275-286

Wielandt, E. (1993) Propagation and structural interpretation of non-plane waves, Geophys. J. Int., 113: 45–53

Part VI

Airborne Gravimetry

Evaluation of Airborne Vector Gravimetry Using GNSS and SDINS Observations

Christian Kreye, Günter W. Hein, and Bernd Zimmermann

Institude of Geodesy and Navigation, University FAF Munich,
Werner–Heisenberg–Weg 39, D–85579 Neubiberg, Germany,
Christian.Kreye@unibw-muenchen.de, Guenter.Hein@unibw-muenchen.de and Bernd.Zimmermann@unibw-muenchen.de

Summary. Airborne gravimetry systems provide the most economical way to improve the spatial resolution of gravity data measured by satellite missions. So the paper deals with the presentation of a modern airborne gravitymeter designed, developed and tested at the University FAF Munich. The specific forces are measured by a high precision strapdown INS and the kinematical accelerations are derived using numerous differential GNSS observations.

The first part of the paper describes the system architecture and the aircraft installation. Then the data processing methods are mentioned including the filtering and derivation procedures, the computation of aircraft accelerations and different algorithms for providing the gravity profiles. Static tests in a laboratory environment approve the error budget of two sensor types on acceleration level.

The main part of the article contains the description of practical test flights carried out in the middle of Germany and the corresponding results. An evaluation of the current system performance is possible. Final remarks refer to the planned improvements until the end of the project.

Key words: Airborne gravimetry, acceleration determination, strapdown inertial navigation system

1 Introduction

Information about the earth gravity field is used for many applications in geophysics and geodesy dealing with figure and structure of our planet. In the context of current or planned satellite missions methods for determination of the gravity field are in discussion today. Caused by thresholds in possible spatial resolution (50-100 km), however, applications using data of satellite based systems are restricted to global or regional investigations. The observation of gravity with wavelengths up to 1 km in an efficient way, especially important for economical applications, is only possible with airborne methods.

Today airborne gravitymeters based on the platform design provide the amount of gravity in local to regional areas. Also strapdown systems are available today, but nevertheless the combination of modern INS technology basing on the strapdown principle and sophisticated processing methods should be able to improve these systems. Important advantages are the observation of the full gravity vector and the simpler system design that enables more efficient airborne gravimetry campaigns. Therefore it must be approved if it is possible to reach the often postulated goal: an accuracy of 1 mGal with a spatial resolution of 1 km.

2 System Design

The goal of airborne vector gravimetry is to provide the gravity disturbance vector $\delta\mathbf{g}$ as the difference between the measured gravity $\mathbf{g}$ and the normal gravity $\boldsymbol{\gamma}$ in the same observation point. Following Newton's second law of motion the gravity $\mathbf{g}^i$ in an inertial coordinate system can be calculated by the difference between the specific force $\mathbf{f}^i$ and the kinematical acceleration $\mathbf{a}^i$ (= second derivative of the position).
Therefore an implementation of an airborne gravitymeter must be able to provide both components as accurate as possible to fulfil the requirements of a large user community. Beside of resolution and accuracy, however, also the efficiency of a sensor system decides on its future applications. In case of an airborne gravimetry system this means that it must be as cheap, small and light as possible.
Following these arguments in the context of a new airborne gravimetry project we try to investigate the performance of an integrated sensor system based on GNSS receivers and a commercial high precision strapdown INS (SDINS). As it is demonstrated in Color Fig. L on p. 316 the specific forces are measured by a SAGEM Sigma 30 INS fitted with triads of ring-laser gyros and pendulous accelerometers. Using special interfaces both the raw data (specific forces $\mathbf{f}^b$ in body-frame and gyro rates $\boldsymbol{\omega}_{ib}^b$) and the navigation data (position, velocity and attitude) are available for the gravity calculation. The specific forces can be computed by

$$\mathbf{f}^i = \mathbf{C}_b^i \cdot \mathbf{f}^b \tag{1}$$

with

$$\mathbf{C}_b^i = \mathbf{C}_{b0}^i + \int \mathbf{C}_b^i \cdot \boldsymbol{\Omega}_{ib}^b dt \tag{2}$$

where $\boldsymbol{\Omega}_{ib}^b$ skew-symmetric matrix of gyro rates
$\mathbf{C}_b^i$ rotation matrix between body- and inertial frame

In order to calculate the kinematical acceleration of the aircraft $\mathbf{a}^i$ GNSS observations in a DGPS configuration are used. These measurements are mainly provided by an ASHTECH Z-Xtreme L1/L2 receiver with a data rate of 10 Hz.

Additionally L1-observations are generated by two NovAtel BeeLine multi-antenna systems. The corresponding four antennas are mounted on the fuselage and the wings of the test aircraft with fixed baseline lengths between 3 and 14 meters. The additional sensors are used to investigate the possible gain in performance in three fields: accuracy and reliability of derived kinematical accelerations, carrier ambiguity determination using fixed baseline informations and support of sensor orientation by GNSS attitude information. The results of practical flight tests (see Sect. 6.1) consider also these aspects.
A central PC provides the sensor controlling during the flight periods, the time synchronisation of GNSS and INS data sets and the storage capabilities for all observations. The used system design is adapted to the 19"-rack equipment and the 28 V power supply of the test aircraft. The implemented configuration has a total weight of 70 kg and a power consumption of 150 Watt. These values can be reduced to 50 kg and 120 Watt if the latest PC- and battery equipment is used. Therefore also small and relativly cheap aircrafts are able to carry out airborne gravimetry campaigns using the presented system design.
Beside of the aircraft installation GNSS reference stations on the ground along the flight trajectory guarantee differential GNSS observations.

3 Data Processing

Apart from sensor selection and system design the used data processing structure dominates the performance values of gravity information provided by airborne missions. Although the general principle of airborne gravimetry is very simple, the challenges consist of the bad ratio of occuring specific force measurements and the expected gravity signal (in general 10^5) and by adjoining or overlapping spectral properties of gravity field values and sensor errors.
Therefore the central aspects of a software tool for airborne gravimetry using GNSS and SDINS data should be highlighted in this chapter. Furthermore the used algorithms will be described in a more detailed way in order to understand the results presented in this article.

3.1 Filtering and Derivation

Neither the GPS measurements nor the inertial observations can ensure the required accuracy of airborne gravimetry without a reduction of high frequency noise effects by low-pass filtering methods. Thereby the corresponding cut-off frequency defines the spatial resolution of the derived gravity profile. In order to separate the gravity signal from error effects on the one hand a small transition band and small discrepancies in pass- and stopband in comparison to the nominal values are required, on the other hand the order of the filter should be possibly low, because edge effects caused by data gaps should be reduced. Furthermore a linear phase response is necessary to maintain the time synchronisation between GPS and INS data. Therefore only finite impulse response (FIR) filters are applicable. The higher filter order in comparison to IIR-Filters must be accepted. Based on the mentioned requirements of airborne gravimetry in (Hehl, 1992) it is pointed out that the design algorithms in frequency domain are superior to the space or time domain methods. Especially the so called 'REMEZ Exchange Algorithm' following the Tschebyscheff-criterion guarantees a small transition band using a relatively low filter order. According to some tests with 10 Hz DGNSS phase observations in a typical airborne gravimetry environment for a spatial resolution of 1 km (transition band between 0.01 and 0.03 Hz) a filter order of 3000 is required to guarantee a damping lower than -100 dB in the stopband. Figure 1 shows the amplitude response for the described filter implementation.

In contrast to the inertial observations the GPS measurements or position

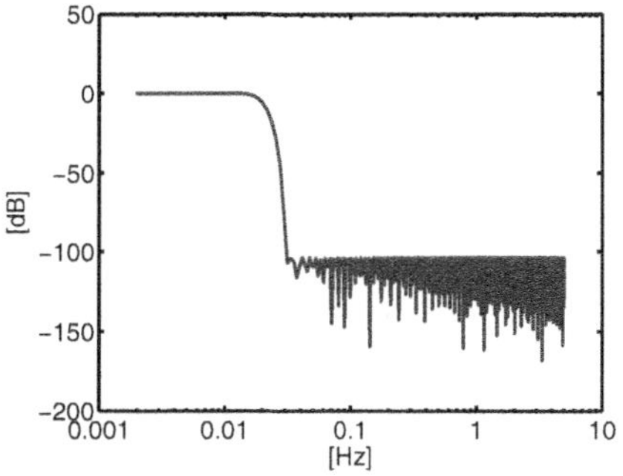

Fig. 1. Amplitude response of low-pass-filter

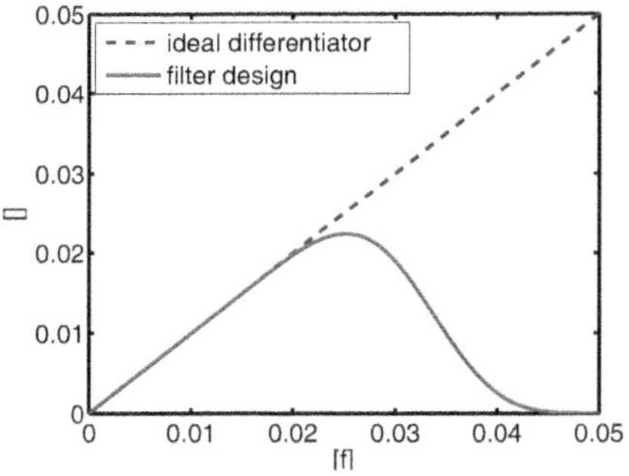

Fig. 2. Transfer function of differentiatior

solutions must be differentiated twice for the further processing on acceleration level. Following the literature of digital filters an ideal differentiator can be described using the following transfer function H:

$$H\left(e^{j\omega T}\right) = j\omega \qquad \text{for } 0 \leq |\omega| < \frac{\omega_s}{2} \tag{3}$$

ω : Angular frequency
ω_s : Data rate
T : Sampling period

The better the approximation of the real differentiator to this function the higher is its differentiation performance in regard to the investigated frequency area. Additionally the differentiators in airborne gravimetry must meet the same conditions for time synchronization (linear phase delay) and filter order as the used low pass filters. In (Bruton, 1999) several differentiation techniques are compared and assigned to static, low dynamic or high dynamic GNSS observations. It is stated, that already a differentiation algorithm designed by a first order Taylor model approximate the ideal differentiator at least in the low frequency area. The advantage of this type of algorithm is thereby the short filter length. Own investigations with real airborne gravimetry data sets show, however, that more sophisticated approaches, like the REMEZ-design provide a higher accuracy level. This is caused by the extended spectral range of the investigated measurements. Figure 2 demonstrates the transfer function of the implemented REMEZ-differentiator with an order of 2000 for 10 Hz data sets in comparison to an ideal design. The discrepancies are smaller than 10^{-8} in the spectral range of airborne gravimetry.

3.2 Calculation of Kinematical Accelerations

For applications of airborne gravimetry the kinematic accelerations of the aircraft must be derived using GNSS double differenced phase observations. This kind of processing can be done by various algorithms. The traditional approach is presented in the upper part of Fig. 3.

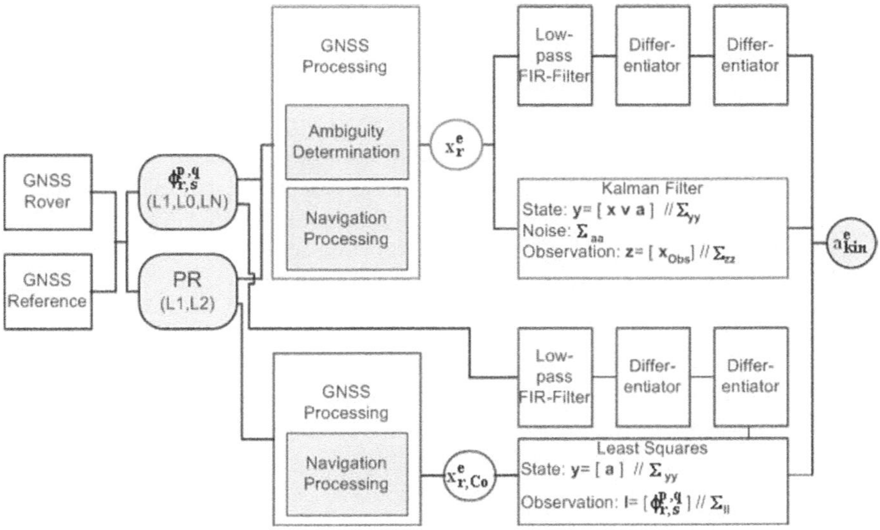

Fig. 3. GNSS processing methods

1. First of all standard software packages for kinematic GNSS processing are used to calculate the DGPS carrier phase solution including the ambiguity fixing and the navigation processing itself. Then the position is filtered

according to the required spatial resolution of the data set. The process of double differentiating finally leads to the kinematic accelerations.

2. Another technique is the Kalman filtering of the position data. Using a simple second order dynamic model containing position, velocity and acceleration the GNSS positions can be used as observations z to derive the current acceleration state. But also in this case the ambiguity terms have to be fixed. Additionally using this method a system design with a fixed frequency response and time delay is not possible. Considering the small frequency window of airborne gravimetry this fact cannot be accepted.
3. A more direct approach for derivation of acceleration using GNSS measurements is presented in (Jekeli, 1994). Assuming that the line of sight acceleration is a sum of the acceleration vector between satellite and receiver and a kind of centrifugal term caused by the dynamics of the baseline itself phase observations to three or more satellites can be used to calculate the kinematic acceleration without knowing the precise integer fixed phase solution of the aircraft.
   In this case the input data are satellite information derived from broadcast or precise ephemerides beside of the low-pass filtered and differentiated phase measurements of reference and rover, furthermore the position of the reference receiver and a standard DPGS code solution in order to derive the mentioned centrifugal term with a sufficient accuracy. Redundant information from additional satellite measurements are processed in a least-squares algorithm estimating the full acceleration vector $\mathbf{a}_{kin}$. The satellite geometry is taken into account in the stochastical model. This method is very interesting for the discussed application because the integer phase ambiguities are not required. Therefore it should be possible to substitute the generally used ionospheric free linear combination by the L1-phase measurement with a lower noise level. Furthermore the estimation process generating the position solution, which is not exactly defined from the frequency point of view, can be avoided.

For investigation purposes both the positioning approach (1.) and the raw data algorithm (3.) were implemented for the processing of the project data. Figure 4 shows the differences between both methods on acceleration level. If the low-pass filtering in (3.) is done after the least-squares estimation (modified) the differences are lower than 3 mGal during a full observation period.

3.3 Derivation of Gravity Information

The traditional way of determining the gravity disturbance components using GNSS/INS data is the integration on position level. Using precise GNSS positions both the INS errors and the gravity disturbance components can be estimated. The differentiation of GNSS data is replaced by the integration of

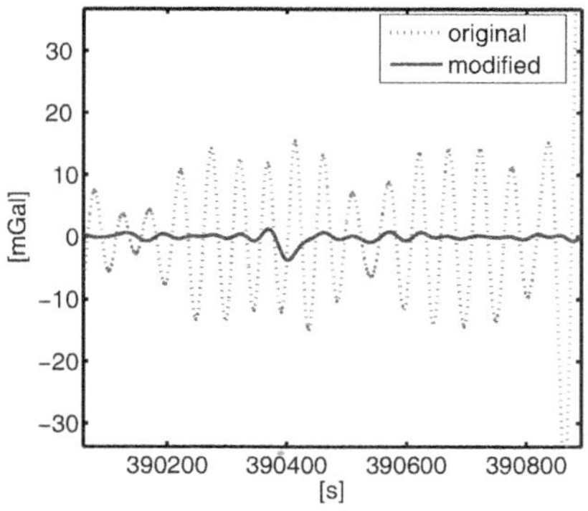

Fig. 4. Comparison of positioning and raw data approach

inertial observations. Caused by the instability of free inertial navigation in the down direction in this case only the horizontal components can be estimated. Furthermore additional stochastical information is required to seperate gravity information and accelerometer errors. Both restrictions can be avoided, if the accelerometer errors are neglegted. This simpler positioning approach is used for test purposes in the implemented data processing.
The central integration processing is carried out using the more direct algorithm on acceleration level (Jekeli, 1992). In the first approach the corresponding state vector only contains orientation errors, INS errors and the parameters of gravity disturbances. The Kalman filter updates are done by the low-pass filtered difference between kinematical accelerations and specific forces in an inertial coordinate system. Thereby the data preprocessing of both data streams must be completely harmonized (data rate, observation time, filter design) in order to reduce numerical errors. Also this approach needs a stochastical model of the gravity field in order to separate INS errors and gravity disturbances using their spectral properties. It must be noted that the choice of stochastic model influences the final gravity results. If this models is not approximated to the real distribution of gravity in the investigated area an additional error can be generated. In (Kwon, 2001) another data processing method is presented without any information of the expected gravity field. The gravity disturbances are not estimated in the state vector. Instead of this they are derived from the residuals corresponding to the adjusted observations. The functional model contains only orientation and INS errors. Therefore in a first approximation every difference between specific forces and kinematic accelerations are caused by these states. The disadvantage is that parts of the gravity signal are estimated as INS errors. But the same effect is caused by errors in the stochastical modelling of the gravity field. So this last processing method is preferred for the airborne gravimetry processing tool.
The corresponding software containing also different alignment procedures is

implemented in C++. Using a graphical user interface, the mentioned processing steps can be applied to the data sets. At the same time the results can be controlled by graphical and numerical outputs. An example for the output screen of this software is given in Color Fig. LI on p. 316.

4 Laboratory Tests

The theroretical error analysis of both sensor components is already done in (Kreye, 2003). Laboratory tests should additionally approve the different components for the application of airborne gravimetry. Inertial sensor errors caused by random effects remaining after the filtering process and bias and scale factor instabilities over time, especially if they are in the same spectral range as the gravity disturbances, are critical for airborne gravimetry. Figure 5 shows the frequency analysis of specific force measurements during a static period of 30 minutes. The amplitude of accelerometer errors in the relevant spectral area is at 1 mGal level. If (2) is taken into account also the gyro errors have an important influence on the measured specific forces based on the required transformation between body and inertial frame. In order to evaluate the total error budget of the inertial sensor, the variation of the specific forces **f** over a time of one hour in the down-component derived by (1) is presented in Fig. 6. Linear drift effects can be estimated by the KALMAN-Filter approaches described in Sect. 3.3. Using the residuals a RMS error of 1.5 mGal can be calculated. This is in the accuracy range of airborne gravimetry. Thus from this point of view the investigated INS should be able to provide gravity disturbances in the required accuracy range. In order to derive the kinematical

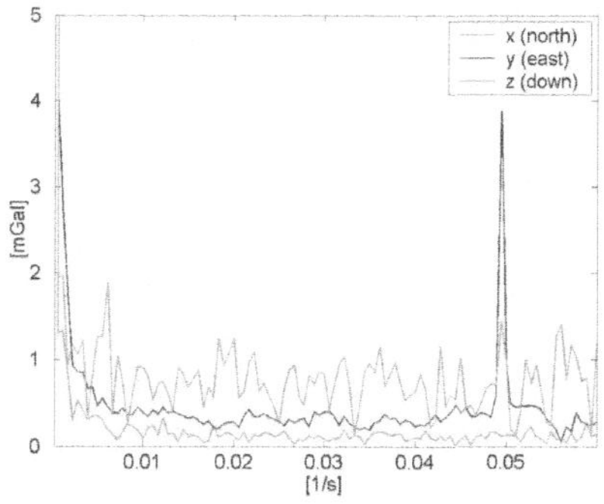

Fig. 5. Spectral analysis of specific force measurements

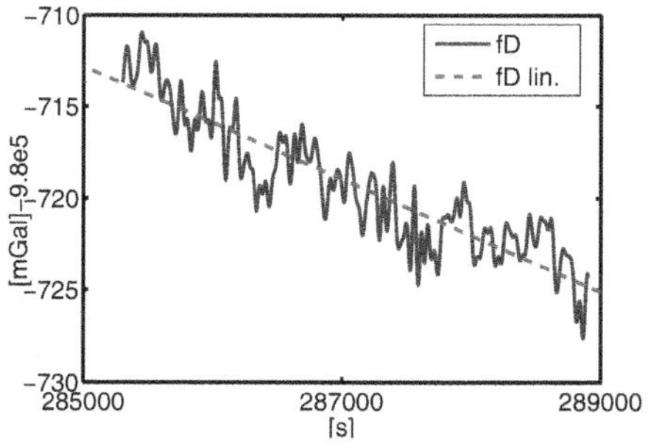

Fig. 6. Total error of specific forces in vertical direction

accelerations the carrier phase observations must be differentiated twice. This process amplifies the error components depending on increasing frequency.

Therefore in theory high dynamic influences like receiver phase noise, cycle slips, sudden changes in satellite configuration and multipath dominate the error budget in contrast to e.g. orbit errors and atmospherical conditions. The evaluation of specific error influences can be carried out using two geodetic GNSS receivers first in a zero baseline configuration, than with a static baseline of 50 m in a medium multipath configuration. The calculated acceleration error after the low-pass filtering for the zerobaseline configuration and the solution based on L1 measurements as well as the ionospheric free linear combination is shown in Fig. 7. Thus the effects of multipath can be seen significantly, but especially the noise of L1/L2 data increases the acceleration errors. So for airborne gravimetry the L1 solution must be preferred as often as possible to reduce the noise level.

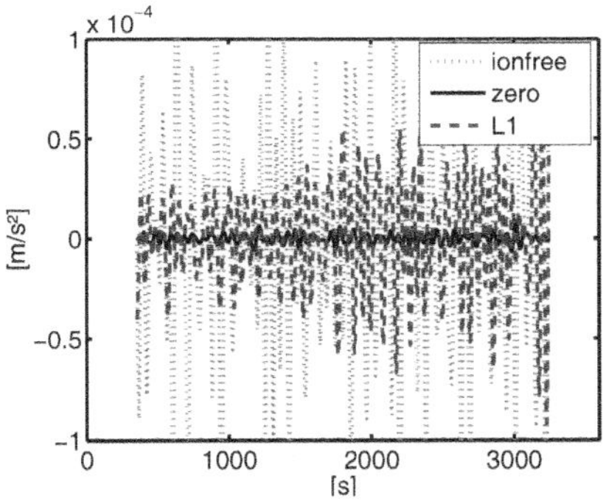

Fig. 7. Acceleration error

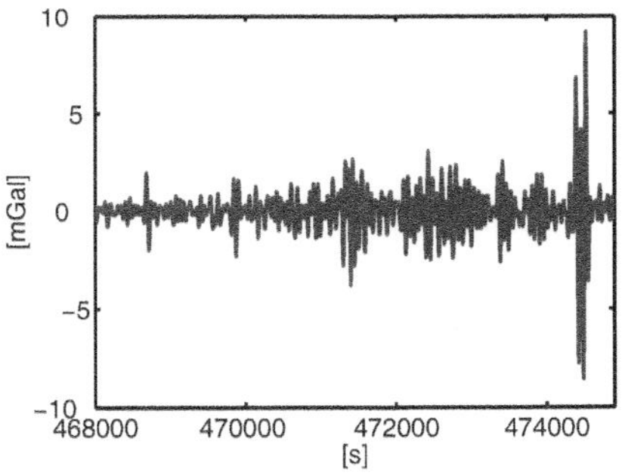

Fig. 8. Effect of changed satellite configuration

Figure 8 presents the differences of an acceleration profile in the vertical component if the data of the satellite with the highest elevation is eliminated from the processing. In general values of 2 mGal can be recognized. If the total number of availiable satellites is low then also differences of 10 mGal must be expected. Therefore the satellite configuration plays a significant role in accelerometry.

The influence of oscillator errors can be evaluated if the filtered carrier phase accelerations over time are computed in a static observation situation. Figure 9 demonstrates the acceleration profiles for satellites 3 and 15. The high frequent receiver oscillator errors superimpose the long term effects caused by the satellite dynamics. Single differencing eliminates this error influence as well as the acceleration based on double differences is free of satellite oscillator errors. In contrast to the effects described above the variation of ionosphere and troposphere assuming general atmospherical conditions is not significant for airborne gravimetry. In this context a GNSS signal simulator is used to reproduce a real observation flight of airborne gravimetry with defined error sources. The derived acceleration error including only atmospherical effects and receiver noise is smaller than 0.8 mGal (RMS)(Fig. 10). Therefore only in conditions with high atmospherical variations the ionospheric free linear combination should be used for airborne gravimetry purposes.

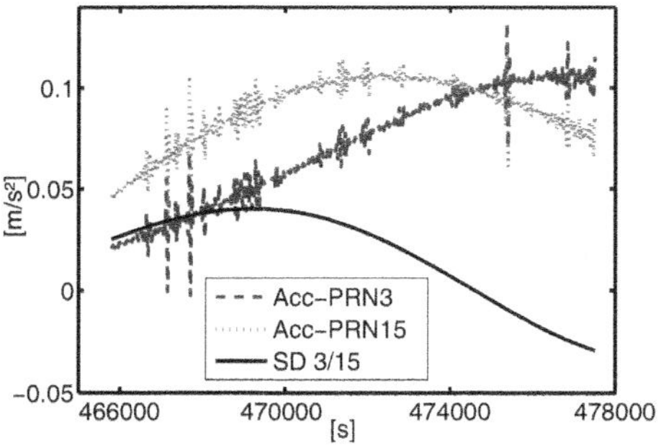

Fig. 9. Effect of receiver oscillator

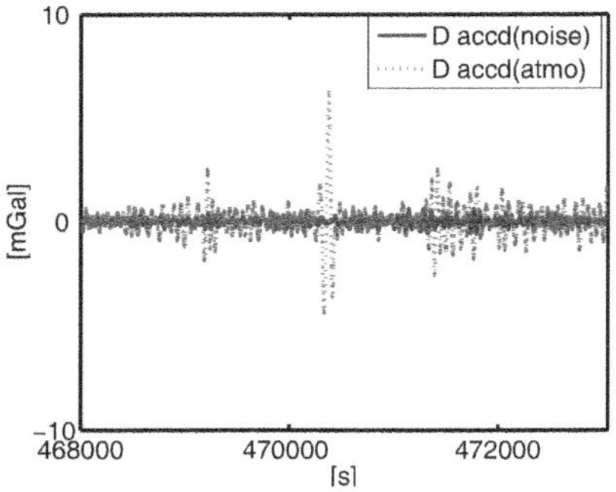

Fig. 10. Influence of atmospherical error

5 Flight Test Enviroment

In order to approve the performance of the described airborne gravimetry system pratical flight tests were carried out with a Do-128-6 aircraft in June 2003, March 2004 and November 2004 in the middle of Germany. The experiments were organized together with the other partners in the GEOTECHNOLOGIEN airborne gravimetry project. The flight trajectory in an area of 120 to 80 km is presented in Color Fig. LII on p. 317. Twelve GNSS reference stations were placed in the test area with a medium distance of 30 km. Some of them were provided by the German SAPOS DGPS service, others were temporary stations installed by the project partners. The coordinates of the temporary stations were estimated in the scope of a common network adjustment based on fixed positions of the SAPOS stations. Therefore the positions and also the accelerations derived by different reference stations are comparable to each other. The average speed of the aircraft was 70 m/s with a height over ground of 300 m. The measured kinematical accelerations reach maximal values of 5 m/s^2. Every flight was started and ended with characteristic calibration manoeuvres providing better observation conditions for the estimation of sensor errors. During the observation periods itself the aircraft dynamics were reduced as much as possible. So in the spectral range of airborne gravimetry the kinematical accelerations do not exceed 0.2 m/s^2. In the test area the gravity anomalies are known with an accuracy of 0.1 mGal and a spatial resolution of 500 m based on relative gravimetry on the ground. The structure of the gravity field is given in Color Fig. LII on p. 317. A large gravity anomaly of 70 mGal is superimposed by smaller structures. Therefore the test area is well suited to evaluate the performance of airborne gravitymeters.

6 Results

The results of the laboratory tests described in Sect. 4 give an impression of the total error budget regarding both sensor types. It could be approved that the sensors are able to fulfil the accuracy requirements of airborne gravimetry in static conditions. Statements concerning the real performance of the described airborne gravitymeter, however, can only be derived if real flight tests are carried out in an area with known gravity anomalies. Regarding the derived results in concrete observation situations and their repeatability the accuracy potential of spatial resolution of 1 km can be calculated. These results are presented in the next chapter, first concerning the determination of kinematical acceleration and then with regard to the derived gravity profiles.

6.1 Kinematical Acceleration

In the test area a GNSS reference station network was installed with a typical baseline length between the stations of 30 km. This density is not required for airborne gravimetry but allows investigations of baseline dependent error influences on the derived kinematical accelerations. If observations of different reference stations are used, the aircraft dynamics can be calculated on several ways. These solutions are not completely independent of each other, but every single result is influenced in a different manner by baseline variation and most of the receiver noise error effects. So far a comparison between these solutions can be used for the derivation of accuracy values for kinematical accelerations of the aircraft. In Fig. 11 the difference in kinematical acceleration is given for the reference stations Helmstedt and Haldensleben with a baseline length of 21 km. The standard deviation of the difference leads to a value of 1.9 mGal. Thereby the accuracy goal of airborne gravimetry can nearly be reached. A comparable investigation of two other reference stations (Kloetze and Stassfurt with a distance of 90 km) can be carried out. As it can be approved by Fig. 12 no significant loss in accuracy can be detected. This result emphasizes the statement of Sect. 4, that baseline dependent errors like atmospherical influences are not a dominant error source for acceleration determination because of their generally longperiodic properties. The occuring discrepancies are rather correlated with the sudden changes in the satellite configuration. Therefore it is possible to extend the baseline length between the reference stations to 100 km without a significant loss of accuracy. This is an additional advantage of airborne gravimetry in respect to the economical point of view.

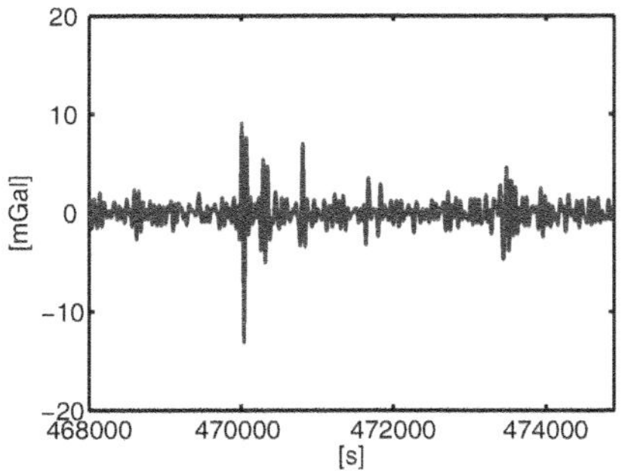

Fig. 11. Accuracy level of kinematical accelerations (21 km baseline)

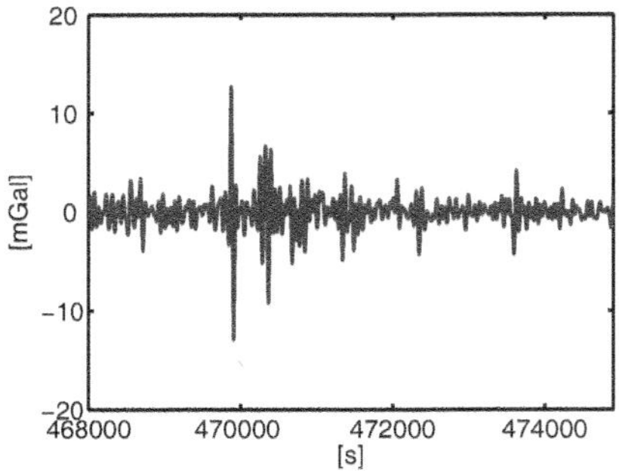

Fig. 12. Accuracy level of kinematical accelerations (90 km baseline)

The multi antennae configuration on the aircraft allows simultaneous 1 Hz observations of GNSS L1 measurements with fixed baseline length on the aircraft up to 14 meters. Using the INS gyro rates the lever arm effects can be corrected, so that four additional, comparable acceleration profiles are available. The difference of two kinematical acceleration results with respect to two antennae on the fuselage of the aircraft (baseline length 3.2 m) is given in Fig. 13. Medium values of 2 mGal indicate, that an averaging process between all theses single antenna solutions reduces the accuracy of the common acceleration profile. Unfortunately this cannot be confirmed by the collected test data, but the reliability of the final result is significantly increased caused by the better possibilites of error detection. An improvement in the integer ambiguity solution of the carrier phase can be detected only in some epochs. Mostly the additional baseline information do not lead to a higher reliability of the detected integer values, because the satellite configuration is identical for all antennae. The reduction of processing time caused by the decreased search space is equalised by the increased number of observations. Using the four antennae on the fuselage and the wings of the aircraft also an attitude solution can be derived following the interferometric principle. Based on comparisons with the inertial attitude angles an RMS error of these values at the level of 0.2°can be expected. This is not within the required accuracy level to support the orientation of the inertial sensor. Only initial values for the alignment process and observations to support the estimation of long periodic drift effects of the gyros can be provided by the GNSS attitude solution. In conclusion it must be pointed out, that a multi antennae configuration on the aircraft makes an important contribution to the reliability of the derived kinematical accelerations. Especially additional antennae on the fuselage of the plane increase the performance of airborne gravimetry.

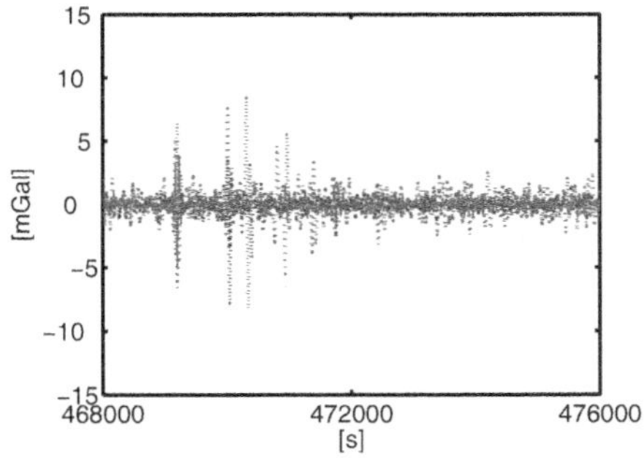

Fig. 13. Difference of accelerations between GNSS aircraft antennae

6.2 Gravity Profiles

Gravity profiles are the final product of airborne gravimetry missions. So this chapter should clarify the performance of the investigated system using measured gravity profiles and comparisons to the given reference field. This is done based on observation data of one test flight. First of all the sensibility of the sensor configuration in regard to the gravity field must be approved. For this purpose an alignement of the inertial sensor is carried out until the start epoch of the aircraft. Then the sensor is used as a free-running INS without any integration filter. The measured specific forces are low-pass filtered and subtracted from the derived kinematical accelerations to calculate the gravity information. Reductions concerning the normal gravity field, the free-air correction and the topography leads to the demanded Bouguer anomaly. A comparison between this calculated profile and the given reference field in down-component is presented in Fig. 14. A high correlation between both curves can be detected approving the sensibility of the sensor configuration. Small drift effects over time are caused by the neglected inertial errors. In the next step both data streams are integrated using a simplified positioning approach, were the accelerometer errors are neglected in order to have the capability to determine the whole gravity vector. A similar picture as in Fig. 14 can be derived. Investigations of the estimated error parameters and their accuracy values show the stability of the filter implementation. Graphs like Fig. 15 emphasize, however, the problems of vector gravimetry. Whereas the standard deviation of the vertical component is reduced very quickly to an accuracy level of 1.3 mGal, the horizontal elements are only decreasing, if turns in the flight path occur. Only during these epochs the observability is sufficient to seperate horizontal gravity values and yaw-angle errors. In a final approach the integration on acceleration level is used to realize a full INS error estimation together with a determination of the gravity anomalies based on provided residuals. The corresponding result in the vertical component is presented in Fig. 16. In comparison to Fig. 14 the long term drift can

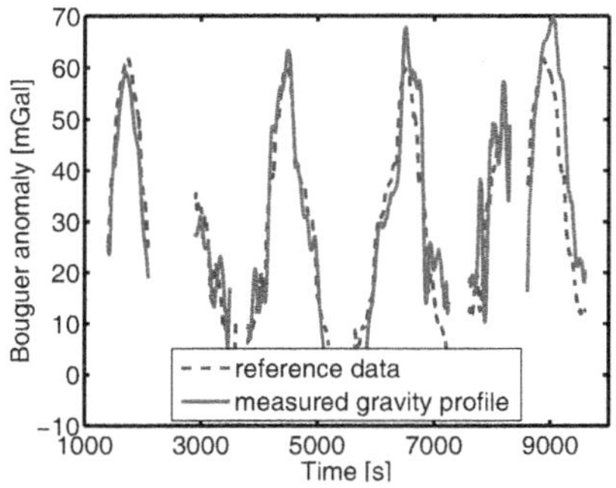

Fig. 14. Anomaly profile (raw data)

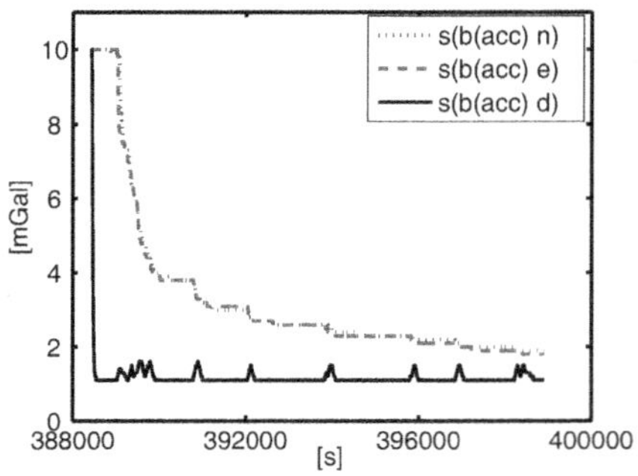

Fig. 15. Accuracy of derived gravity anomaly (positioning approach)

be eliminated. Figure 17 presents the estimated horizontal components of the gravity field. A reference field is missing in this case, but the small values are according to the general expectations. Finally also the repeatability of the derived gravity profiles can be shown using a detailed investigation of one observation track flown in forward and backward direction. This is demonstrated in Color Fig. LIII on p. 317 together with the reference profile. The particular differences between these lines leads to an accuracy of 6.8 mGal to the given reference field and 4.9 mGal between two measured data sets in regard to a spatial resolution of 1 km. These values also characterize the current performance of the investigated gravitymeter. For the horizontal components a decrease in accuracy between 10 and 15 mGal can be demonstrated.

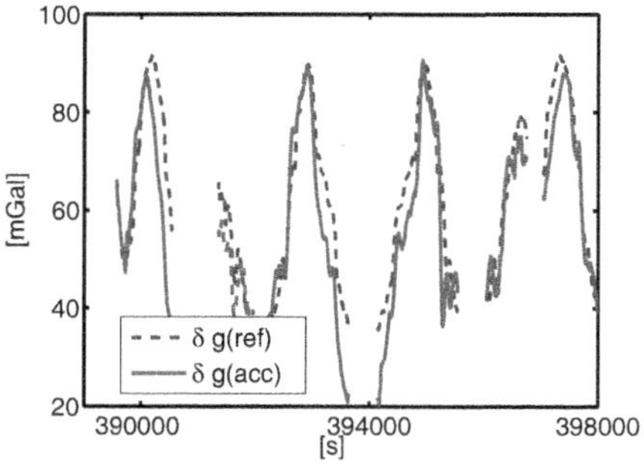

Fig. 16. Anomaly profile (acceleration approach)

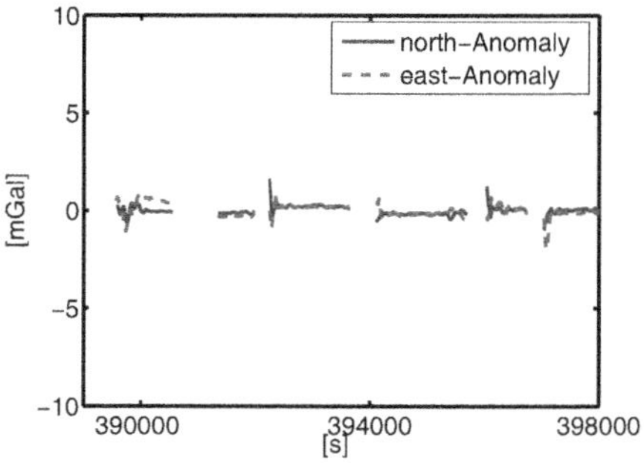

Fig. 17. Horizontal gravity anomalies

7 Summary and Outlook

Based on the current system performance given in the last chapter, during the last months of the project some further investigations are planned. In this context the benefits of shaping filters will be investigated, the potential of additional information (e.g. crossing points) will be evaluated and the methods of gravity downward continuation will be improved. Therefore a performance of 3 to 5 mGal in vertical and 10 mGal in horizontal gravity components with a spatial resolution of 1 km seems to be possible with this economical airborne gravitymeter based on a commercial strapdown INS.

Acknowledgement. This is publication no. GEOTECH-171 of the programme GEOTECHNOLOGIEN of BMBF and DGF, Grant 03F0341A

References

Bruton A M, Glennie C L, Schwarz K-P (1999) Differentiation for high precision GPS velocity and acceleration determination. GPS Solutions 1999 Vol.2, No.4

Bruton A M, Schwarz K P, Ferguson S, Kern M, Wei M (2002) Deriving acceleration from DGPS: Towards higher resolution applications for airborne gravimetry. GPS Solutions, Vol.5, No.3

Eissfeller B, Spietz P(1989) Basic filter concepts for the integration of GPS and an inertial ring laser gyro strapdown system. Manuscripta Geodetica 14:166-182

Hehl K (1992) Bestimmung von Beschleunigungen auf einem bewegten Traeger durch GPS und digitale Filterung. Schriftenreihe des Studiengangs Vermessungswesen, University FAF Munich, Heft 43

Jekeli C (1992) Vector gravimetry using GPS in free-fall in an earth fixed frame. Bulletin Geodesique 66:54-61

Jekeli C (1994) On the computation of vehicle accelerations using GPS phase measurements. International Symposium on Kinematic Systems in Geodesy, Geomatics and Navigation, Banff, Alberta, Canada, 1994

Kreye Ch, Hein G W (2003) GNSS Based Kinematic Acceleration Determination for Airborne Vector Gravimetry - Methods and Results. Proceedings of ION GPS 2003, Portland, Oregon, 2003

Kwon J H, Jekeli C (2001) A new approach for airborne vector gravimetry using GPS/INS. Journal of Geodesy 74, 690-700

SAGS4 – StrapDown Airborne Gravimetry System Analysis

Gerd Boedecker and Andrea Stürze

Bavarian Academy of Sciences and Humanities, Munich
boe@bek.badw-muenchen.de
stuerze@bek.badw-muenchen.de

Summary. A strapdown airborne gravimeter of a peculiar configuration has been developed and is nearly operational for observing total acceleration. Precision high sampling rate GPS receivers provide the kinematic acceleration. The overall system and its hardware is analysed with focus on signal flow.

Key words: Airborne gravimetry, kinematic GPS positioning

1 Introduction

Airborne gravimetry as an observation technique for the gravity field of the Earth has emerged in the last decades and still undergoes vivid development because high resolution gravity information is extremely useful to mineral exploration, precision navigation, geodesy, geophysics. Because the spatial resolution of satellite methods is limited to >70 km, airborne gravimetry with a resolution of currently some 4 km and target 1 km at 1 mGal[1] is indispensable. Conventional gravity ground observations are by far not as efficient.

The basic idea is to recover gravity – vector or scalar – from the difference of total observed acceleration and kinematic acceleration:

$$\mathbf{g} = \mathbf{a} - \mathbf{b}, \text{ where} \tag{1}$$
$$\mathbf{g} = \text{gravity vector}$$
$$\mathbf{a} = \text{vector of total acceleration}$$
$$\mathbf{b} = \text{vector of kinematic acceleration}$$

Among airborne gravimetric observation devices, platform scalar gravimeters for the vertical component were the first to become operational as modifications of ship gravimeters. The most ambitious is a strapdown vector gravimeter with its inherent advantages such as full vector information **a**, robustness, small size, ease of operation. In the past, approaches to utilise off the

[1] $1 \text{ mGal} = 1 \cdot 10^{-5} \frac{\text{m}}{\text{s}^2}$

shelf strap down inertial navigation instruments which include the proper sensors, showed some promising results but were not successful on the long run. For this reason, our approach aims at the optimisation of configuration and takes into account some good experiences from classical gravimetry by using temperature control, duplicate sensors, high vibration isolation. Likewise, the kinematic acceleration **b** has to be determined primarily by GNSS, currently: GPS to an adequate quality; this is at least as difficult (Schwarz, 2001).

We may consider the data fusion expressed in the above equation $\mathbf{g} = \mathbf{a} - \mathbf{b}$ as a system with subsystems for **a** and **b**. The performance for the determination for **g** may be studied by the study of the subsystems for **a** and **b** and the sub-subsystems and/or by input/output analysis. The user of gravity information is interested in spatial resolution; this quantity can be transformed to time resolution at a given aircraft speed. For this reason, it is important to study the performance of the system not only in terms of e.g. accuracy or geometric transformation among intermediate reference frames, but also in terms of time resolution or (time) spectral properties.

2 Methodology of Description

A system like SAGS4 (StrapDown Airborne Gravimetry System prototype no. 4) may be described from various views. We selected to view it from

- the hardware side to some extent and
- from the system signal structure.

As to the latter, we shall use partly time domain representation, partly signal spectral domain representation. The description will remain incomplete because of space limited, e.g., we shall not study the full acceleration vector. Rather, we shall try to present the overall structure and to focus on a few components.

A system usually is understood as an entity that responds to some input signal $x(t)$ with some output $y(t)$, t = *time*; in a discrete system, we are dealing with series of input $x[n]$ and output $y[n]$ (Fig. 1). For more details

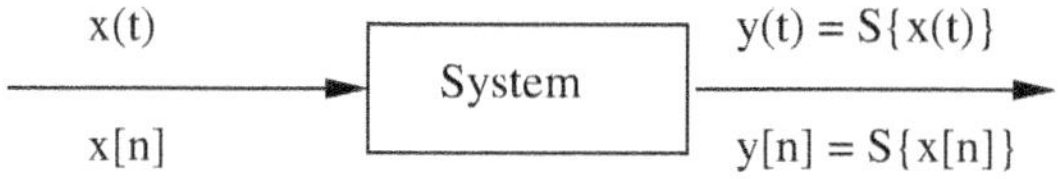

Fig. 1. System: input-output-relation

of system properties, see Kiencke and Jäkel (2002), Kronmüller (1991), Lathi (2002), Lathi (2005), Oppenheim et. al (2004), Unbehauen (1996), Eykhoff (1974). The following treatise will not be self-sufficient but requires the above background literature.

In the sequel, we shall limit ourselves to linear time invariant systems and structures of systems, mainly of single input single output type; we shall consider these primarily as continuous systems knowing that a good deal are discrete. The conception of a system in this context is that of a physical device with some useful properties such as mechanical systems – e.g. dampers –, electronic systems – e.g. analog filters –, filters to be described mathematically, and discrete digital systems, e.g. an ARMA process or a FIR filter. Often, system behaviour is described by the transfer of the system input spectrum to the system output spectrum. In the case of a continuous causal system, a system function or transfer function may be formulated by:

$$\begin{aligned} Y(s) &= G(s) \cdot X(s), \text{ where} \\ G(s) &= L\{g(t)\} = \int_0^\infty g(t)e^{-st}dt, \; s = \sigma + j\omega \end{aligned} \tag{2}$$

where $L\{g(t)\}$ is the one-sided Laplace transform of weight function $g(t)$, $G(s)$ is the transfer function and $X(s)$ and $Y(s)$ are the Laplace transforms from input and output respectively. Typically, systems form structures such as a sequence or parallel or hierarchy (Fig. 2).

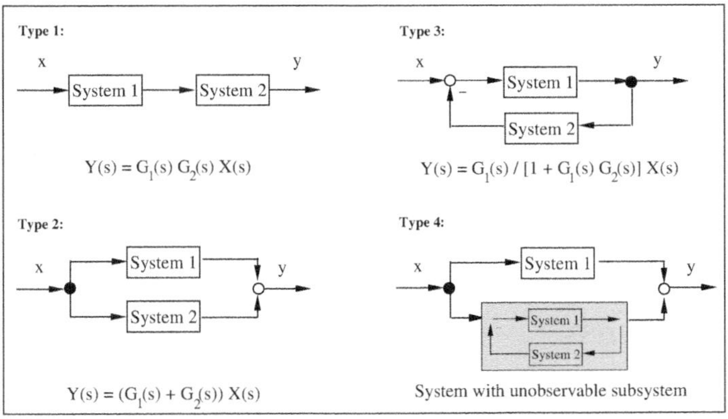

Fig. 2. Types of systems

The last example shows that systems may include subsystems that may or may not be accessible to analysis. If not, they are sometimes called uncontrollable or unobservable (Lathi, 2005). In any case, we are usually dealing with different levels of system hierarchy. In our case, we shall select one useful level of system analysis being aware that there are subsystems not analysed. As to spectral representation, we shall make use of:

$$
\begin{aligned}
&\text{Fourier Transform: } X(f) = F\{x(t)\} = \int_{-\infty}^{\infty} x(t)e^{-j\omega t}dt,\ \omega = 2\pi f \\
&\text{Laplace Transform: } X(s) = L\{x(t)\} = \int_{0}^{\infty} x(t)e^{-st}dt,\ s = \sigma + j\omega
\end{aligned}
\tag{3}
$$

One significant difference between the two transformations is the different integration range, making the one-sided Laplace transform adequate for causal signals where $x(t) = 0$ for t < 0; the Fourier transform enables more illustrative signal spectra. Transfer function estimate is possible from the input-output relation (2) above:

$$G(s) = \frac{Y(s)}{X(s)} \tag{4}$$

Another approach makes use of a statistical transfer function estimate (Hütte, 1996):

$$\hat{G}(\omega) = \frac{\hat{S}_{xy}(\omega)}{\hat{S}_{xx}(\omega)} \tag{5}$$

where $\hat{S}_{xy}(\omega)$ and $\hat{S}_{xx}(\omega)$ are the estimated auto- and cross power spectral functions.

The overall SAGS4 system is a system of type 2 above (Fig. 2) with the two parallel main subsystems for total acceleration and for kinematic acceleration. These subsystems and their sub-subsystems will be studied in the sequel. The total acceleration subsystem will directly make use of the above system structures. The kinematic acceleration subsystem will make use of the ARMA type system description.

3 The System

The general idea is to solve the above fundamental equation of kinematic gravimetry (1) by observation in the simplest possible way, i.e. observe **a** and **b** and take the difference in a common reference frame to yield **g**. The basics of the practical realisation of the strapdown airborne gravimetry system were published e.g. in Boedecker et. al (1999-2001).

A system hardware overview is given in Fig. 3. The total acceleration vector **a** is sensed by a triad of four accelerometers with one duplicate in the vertical channel. The attitude is sensed by a triad of gyroscopes aided by a multi antennae GPS receiver (MA-GPS).

For the determination of **b**, two GPS receivers of high sampling rates are available in the GPS assembly. Some details of these major components will be explained in the corresponding subsections.

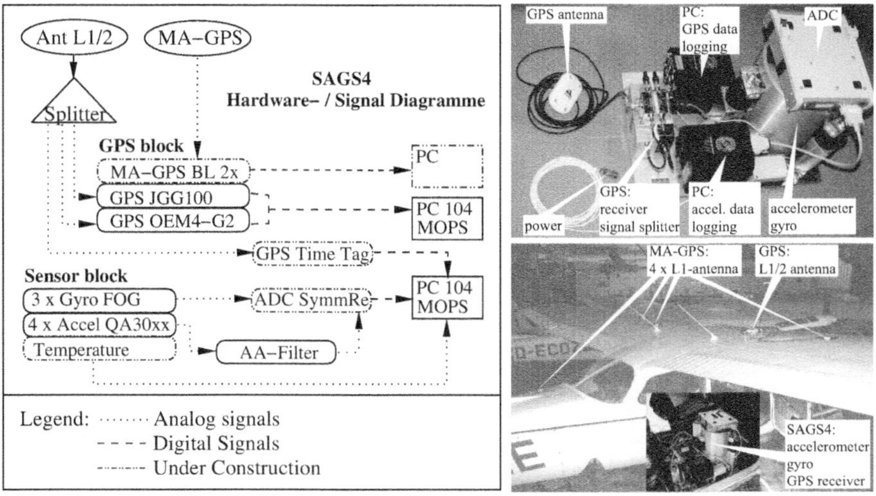

Fig. 3. SAGS4 system.

The subsystems are all strictly time tagged, the data of all sensors are recorded in the PCs for post processing. The SAGS4 total weight is some 30 kg. It does not require an operator. The total equipment may be mounted to a light aircraft within a few hours. The subsystems proved their expected performance, the whole system was approved safe by the German air traffic authority for use in aircraft. It has been flight tested frequently in various aircraft types. Currently (spring 2005), fine tuning of thermal control is underway, afterwards it will be calibrated and flown operationally.

The various components of SAGS4 are mounted to different parts of the aircraft (Fig. 3, right bottom). It has been tested that the mechanical structures are sufficiently rigid, i.e. the geometric lateral deformations are below 3 mm, probably much less. As to the torsion between MA-GPS, these are small and are properly handled by means of the gyro-MA-GPS data fusion. As to engine driven vibrations with small geometric amplitude, these are currently being studied.

3.1 Total Acceleration

The total acceleration in our approach to strapdown airborne gravimetry is sensed at the seat rails of the co-pilot's seat of a light aircraft, where the SAGS4-system is mounted. As mentioned above, this is part of a consistent frame of the various components.

One important question is now, to what extent is the true total acceleration vector of the aircraft, represented by the seat rails, sensed by the SAGS4 total acceleration sensor. What is the transfer function in an important frequency band? We shall not be dealing with calibration problems, nonorthogonalities etc. Rather, our interest focusses on the acceleration signal spectral properties.

For a number of reasons, e.g. for anti-aliasing at reasonable sampling rates and for mechanical stress reduction to improve accelerometer performance, we are using mechanical dampers and electronic analog filters. We shall study the characteristics of these elements both from a synthetic approach and from a signal input-output-analysis. The elements significant for this analysis are shown in Fig. 4. The standard signal track is shown left, an experimental track

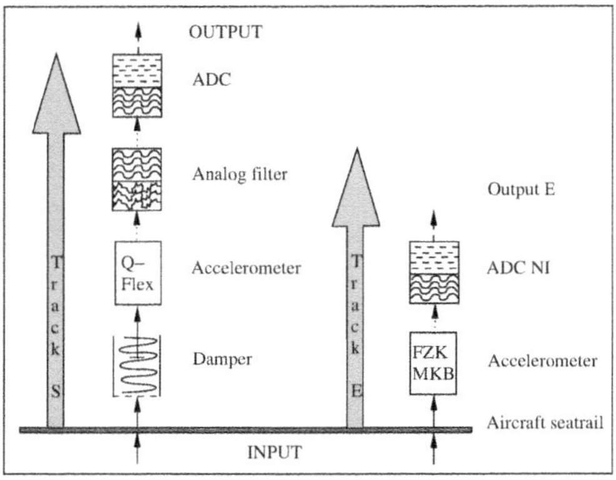

Fig. 4. Total acceleration system

right. We realise, that the total acceleration observation subsystem structure (track S) is of sequential type, see type 1 in Fig. 2.

Components

Track S – Standard: components

Damper for vibration isolation by Barrymount type SLM 1-A: The shock-mounts (Fig. 5) are a combination of elastomer and air damping with a natural frequency around 10 Hz depending on air pressure and load. The load for each is about 5 kg. The height change at flight dynamic conditions is less than 1 mm.
Accelerometers Q-Flex QA-3000-030: The Q-Flex accelerometer's resonance frequency is beyond 1000 Hz. The response remains within 5 dB up to 2000 Hz, within 0.5 dB up to 300 Hz and within 0.01 dB up to 10 Hz. (Fig. 6).
Analog filter Kemo 1608: Low pass 8 pole -48dB/Octave Bessel type, cut off (-3 dB) at 10 Hz (Fig. 7).
Analog-Digital-Converter ADC Burr Brown ADS 1210. The ADC is used at 100 S/s (samples per second), i.e. at a resolution of 22 to 23 bits (see company information) or, at the conversion factors implemented by the hardware,

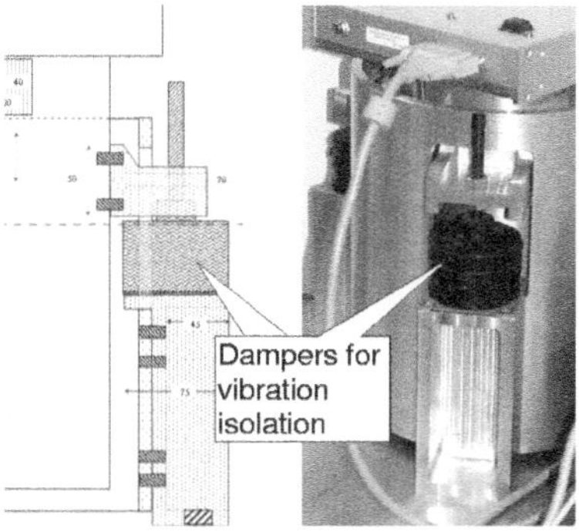

Fig. 5. Damper

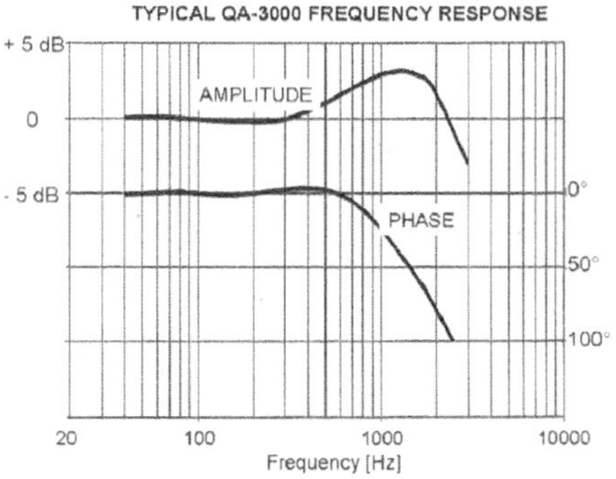

Fig. 6. Q-Flex transfer function (Sundstrand Data Control, 1986).

about $1 \cdot 10^{-5} \frac{m}{s^2}$ or 1 mGal. For experiments, we also used sampling rates up to 500 S/s (Fig. 8, upper curve).

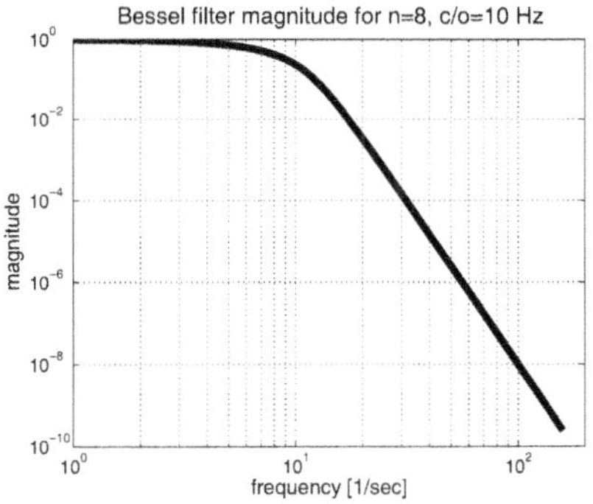

Fig. 7. Analog Bessel filter

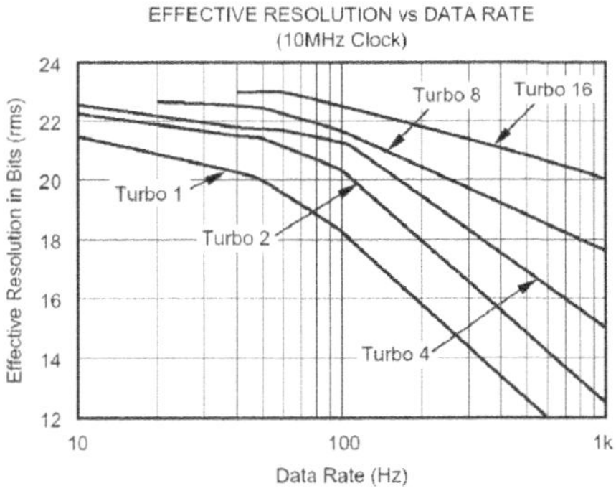

Fig. 8. ADC resolution (upper curve)

Track E – Experimental: components

Accelerometer FZK-MKB: The 3D accelerometer is characterised by a damped resonance frequency of > 500 Hz. Up to this frequency, the linear coefficient is fairly constant (Wüstling, 1997).
Analog-Digital-Converter ADC: DAQPad-MIO-16XE-50 of National Instruments. The analog values are sampled at sampling rates of 500 S/s and 16 bit.

Analysis of Total Acceleration Observations Vertical Channel

The transfer function of the total acceleration system, see track S above, may be studied by different approaches, for which we make use of the above general and specific tools and information. The analysis is based on vertical acceleration data.

Approach 1: Impulse response function

The transfer function of a system is equal to the frequency response of a Dirac impulse. We have exerted an impulse in the form of a stroke with a hammer to the SAGS4 ground plate, which was put on loose sand. The signal took track S and likewise track E, that was also installed for this experiment. Sampling rate for both tracks was 500 S/s. Figure 9, up, shows the power spectral density function for track E, i.e. the spectrum without damping, which is considered as representing the true input. As it should be, the spectrum of an impulse is a constant. The lower curve shows the effect of damping particularly in the higher frequencies.

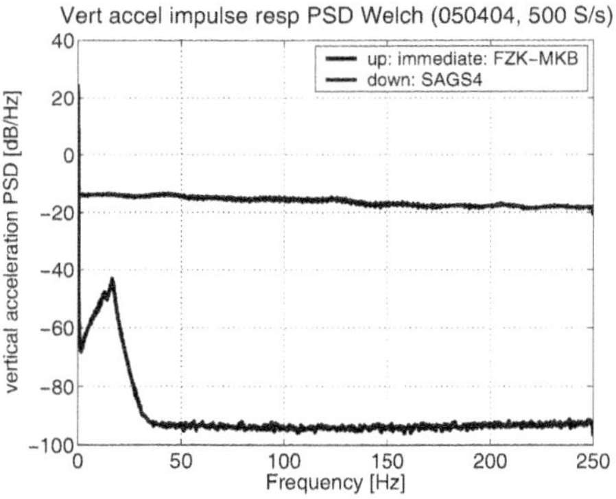

Fig. 9. Vertical acceleration: impulse response (up: track E, down: track S)

The left part of the lower curve deserves our special interest: Let us remember that the system of track S primarily constitutes a sequence of subsystems – see Sect. 2; hence, the transfer functions of the subsystems add up. The roll off between some 17 and 40 Hz is due to the analog Bessel filter, see above, which acts as designed. In the frequency range between 2 ... 17 Hz, the transfer function of the mechanical damper adds up. Its transfer function is somewhat unexpected, because it has a bit of high pass characteristics. It turned out that under the conditions of the experiment, the natural frequency of the damper was close to 20 Hz. Above 40 Hz, the filter roll off has achieved the general noise floor.

Approach 2: Statistical frequency domain system identification

Figure 10 (left) shows the PSD spectra taken during a flight with the light aircraft. The upper curve is the PSD of the vertical channel of track E, sampled at 500 S/s, taken as true input. The lower curve depicts the PSD of the vertical channel of track S of SAGS4. Again, we see very nicely the filter effect in the SAGS4-data. The transfer function estimate using (5) above, we obtain the graph in Fig. 10 (right).

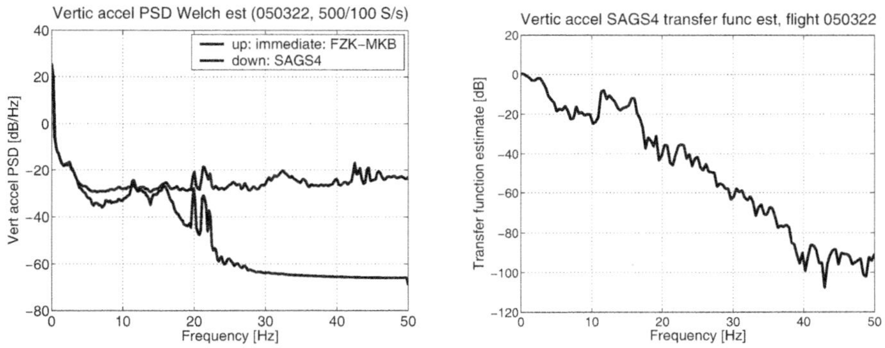

Fig. 10. Vertical acceleration: PSD's (left) and transfer function (right)

We recognise the similarity with the impulse response function above; please note the different frequency ranges of the two figures. In both figures we see a relative maximum around 15 Hz and a rolloff to noise floor above 40 Hz.

Approach 3: Subsystem synthesis

If we integrated our information on the subsystems transfer functions of track S, we should end up with the same transfer function as shown above. As a matter of fact, our knowledge on the transfer functions of the subsystems of track S is poor. For a part of these, we could try to establish the transfer function by a system analysis of the sub-subsystem as indicated in Fig. 4 above. However, this is in vain because also in this case we do not have sufficient knowledge. For this reason, the sparse information on the subsystems of track S such as the analog Bessel filter transfer function support our interpretation of the analyses under Approach 1 and Approach 2, but they are not sufficient for a subsystem synthesis.

Total Acceleration Subsystem Analysis Summary

We learned that the two approaches 1 and 2 above show good agreement for the total acceleration observations system analysis. The discrepancies can be explained by different conditions of the practical experiments. The standard sampling rate of 100 S/s is above the Nyquist rate and appropriate in view of the pre-filtering. The behaviour of the mechanical damping was somewhat unexpected. The dampers will be tested further with different pressures and optimised for a target transfer function that permits a full spectral transfer up to about 10 Hz because the flight kinematics spectra of a light aircraft may go that far. Because of lack of information on some components, we could not carry out a full subsystem synthesis. For similar reasons, we did not study discrete systems transfer functions, as would have been adequate. Nevertheless, the information on the subsystems helped us to understand and interpret the total acceleration spectral transfer function. With minor modifications for the mechanical dampers, we should be able to transfer a bandwidth of 10 Hz signal and limit the noise above 45 Hz to -95 dB.

3.2 Kinematic Acceleration

Measuring gravity in airborne mode requires the separation of gravitational from non-gravitational acceleration (1). To obtain gravity with an accuracy of 1 mGal, the aircraft acceleration must be measured within the same accuracy. Because of the amplification of measurement noise by time differentiation, the acceleration determination still shows errors of several hundred mGal using the individual measurements. Since gravity changes slowly in horizontal direction, filtering over pre-determined time periods deliver a contribution to reduce the noise (Schwarz, 2001). Another contribution will be described below.

Hardware

The GNSS observation system consists of two receiver boards and a signal splitter that divides the incoming GPS signal to both:

- *Javad JNS100* single frequency board (L1), claims to provide raw data 50 times per seconds (S/s) and a carrier phase precision of 0.1 mm,
- *Novatel OEM4-G2* dual frequency board (L1/L2), enables observations at 20 S/s raw data output rate with a specified carrier phase precision of 0.75/2.0 mm for L1/L2. A special firmware enables the use of 50 S/s.

Signals

Theory and experiment

The precision of code and carrier phase measurements largely depends on how much noise accompanies the signals in the receiver's tracking loops. This noise either comes from the receiver electronics itself or is picked up by the antenna along with GPS signals in the form of naturally produced electromagnetic radiation. The most basic kind of noise is that produced by the random movement of electrons in any conductor with a temperature above absolute zero. This thermal noise occupies a broad frequency spectrum and its power in a given passband is independent of the passband's centre frequency. Such noise is referred to as white noise whereas all errors, which vary with time, definitely are not white noise (Spilker, 1996). For characterisation of the code and phase lock loops the bandwidth of the respective loop filter is a common parameter. It describes the passband of a linear system for white noise (Eissfeller, 1997) and is chosen to balance noise performance against authentic signal dynamics. Additionally, the order of the loop filter affects the dynamic capabilities of the receiver. Hence, it could play a key role for the separation of gravitational from non-gravitational acceleration in airborne gravimetry.

For the *Javad* single frequency board *JNS100* it is possible to select bandwidth and order of the delay lock loop (DLL) and phase lock loop (PLL) filter so that these theoretical relations can be investigated by experiment. One way to study the receiver's performance is the comparison of the computed GPS trajectory with measurements of an independent positioning device of superior quality. Therefore, a lift with photoelectric scanning of a glass scale enables a resolution of 50 S/s in time and 0.02 mm in length, has been employed. Using this lift it was tried to simulate the dynamics of light aircraft. So several tests using different receiver settings for the phase lock loop filter have been carried out. Because the *JNS100* code tracking loop is aided by the use of an estimate of the dynamics from the carrier tracking loop, the DLL noise bandwidth needs only be wide enough to track the ionospheric divergence between carrier phase and pseudorange and can be set to 1 Hz only. Thus the PLL noise bandwidth must be wide enough for the tracking loop to follow the authentic changes in the satellite user geometry.

All investigations in this section are based on the vertical component h of the receiver trajectories, because ground truth H exists for this component only. The effect of different receiver settings became visible by graphical comparison of the GPS trajectory and ground truth, so that the theory is first confirmed by experiment (Fig. 11).

Model building

Only the common studies of measurement noise and dynamic capabilities result in a realistic assessment of the receiver's performance. Therefore a representation of the essential aspects of the existing system in a usable form is searched. Such a model does not need to be a description of the actual mechanism of the system. It is also possible to mimic the system behaviour (Eykhoff, 1974).

The GPS time series, represented by the vertical component h, are analysed using a statistical method that leads at first to a separation of the deterministic and the stochastic part. After trend elimination, the stochastic structure of the resulting time series can be approximated by a model. The idea is, that a time series consisting of highly correlated observations may

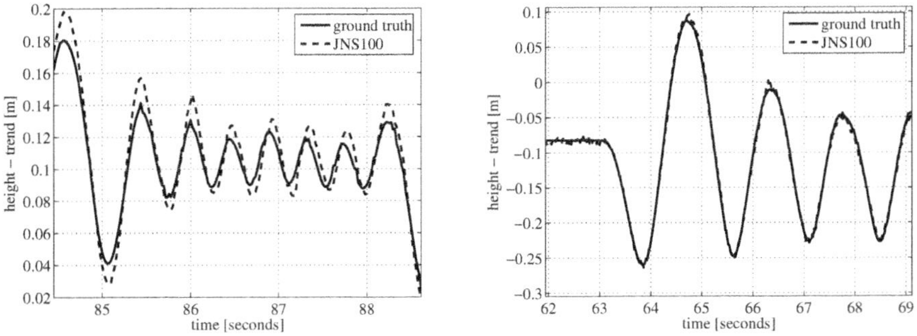

Fig. 11. Graphical comparison of the GPS height component and ground truth using different settings for the *JNS 100* receiver

be generated by applying a series of independent 'shocks' to a linear system. If white noise ϵ_t is an input to a linear filter in order to produce a correlated stochastic process, such a model is named autoregressive-moving average model (ARMA). Based on the fact that measurements are not perfect, the deterministic trend elimination results in the sum y_t of the stochastic part x_t and measurement noise v_t. Therefore, the stochastic model is complemented by white measurement noise to an enhanced ARMA model (EARMA):

$$y_t = x_t + v_t = \epsilon_t + b_1\epsilon_{t-1} + \cdots + b_q\epsilon_{t-q} - a_1x_{t-1} - \cdots - a_px_{t-p} + v_t \quad (6)$$

where $a_1, \ldots, a_p$ and $b_1, \ldots, b_p$ are called ARMA parameters. The order of the EARMA process is (p, q). Only for this special case it is possible to estimate the variances of measurement and system noise σ_v and σ_ϵ simultaneous (Chen, 1996). The noise analysis using such a model in consideration of the selected loop bandwidth reveals the receiver's noise performance (Fig. 12, left).

Because the order of the loop filter affects the dynamic capabilities of a receiver, the comparison of ground truth with the GPS time series in consideration of the selected loop order will be used for the assessment of the receiver's dynamic capabilities (Fig. 12, right).

At last, the PLL noise bandwidth must be chosen to balance noise performance against authentic signal dynamics. In total, it is an analysis tool for a realistic evaluation of the receiver's overall performance. For more details of this type of model building see Stürze and Boedecker (2004).

Because of the double difference GPS processing strategy, all results are affected by the configuration of the reference receiver. Hence, no absolute values are available. The raw data update rate of the *JNS100* receiver and the reference receiver configuration stayed unchanged during all experiments. Also the dynamics of experiments no. 1 and 3 to 6 was nearly the same. So the changes of the rms values for measurement noise σ_v mostly depend on the PLL noise bandwidth of the *JNS100* receiver and are immediately visible (Table 1). Consequently, the smallest PLL noise bandwidth of 10 Hz leads apparently

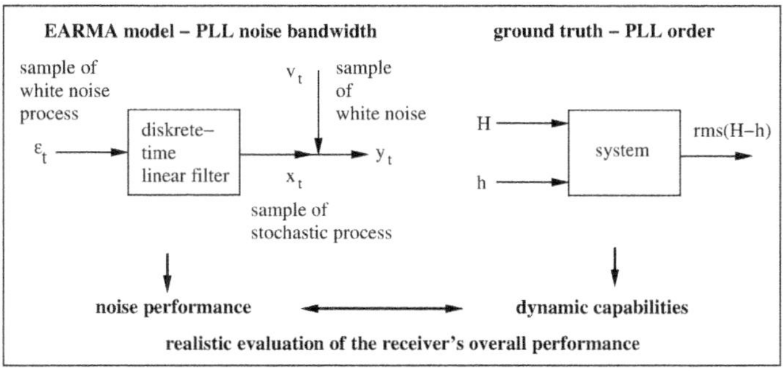

Fig. 12. Strategy of model building for evaluation of the receiver's performance

to the best noise performance. However, the additional contemplation of the rms value of the difference between ground truth and GPS height component $rms(H - h)$ shows that the best quality of phase measurements is achieved with a PLL order of 3 and a PLL noise bandwidth of 30 Hz. A comparison of experiments no. 1 and 2 reveals that movements of the aviation antenna are barely visible in the values of σ_v whereas the $rms(H - h)$ reflect the movements. This means that the variance of measurement noise, estimated during the EARMA modeling procedure, contains white noise only as defined and a PLL order of 2 is not sufficient to follow all movements of the lift.

Table 1. Configuration of the *Javad* GPS board *JNS100* during different experiments and results of model building in form of the rms values of measurement noise and of the difference between ground truth H and GPS height component h

Experiment no.	1	2	3	4	5	6
static/kinematic mode	kin	stat	kin	kin	kin	kin
raw data update rate [S/s]	50	50	50	50	50	50
PLL order	2	2	2	3	3	3
PLL bandwidth [Hz]	25	25	50	10	30	50
σ_v [mm]	1.4	1.8	2.6	0.6	1.6	2.6
$rms(H - h)$ [mm]	6.5	4.7	6.2	8.3	5.2	5.4

As shown, the strategy of model building mentioned above leads to a realistic evaluation of the receiver's overall performance. With the same procedure the influence of the signal splitter was investigated; no noise amplification was detected. After all, the investigation of the *Novatel OEM4-G2* dual frequency board indicates the PLL noise bandwidth in case of the standard firmware is a wider one than in case of the special firmware, which enables observations at 50 S/s.

Completing, it should be mentioned that the system analysed in this case includes both, reference and rover receiver with their parallel channel subsystems (type 2 in Fig. 2). Therefore, the individual channel subsystem is an unobservable system (type 4 in Fig. 2).

Signal Processing

The kinematic acceleration shall be determined now from discrete-time signals of a single receiver, which include only the noise contributions of this one. Starting point of signal processing, which is pictured exemplary for the velocities in Fig. 13, are the best quality observations regarding the dynamics of lift experiment no. 5.

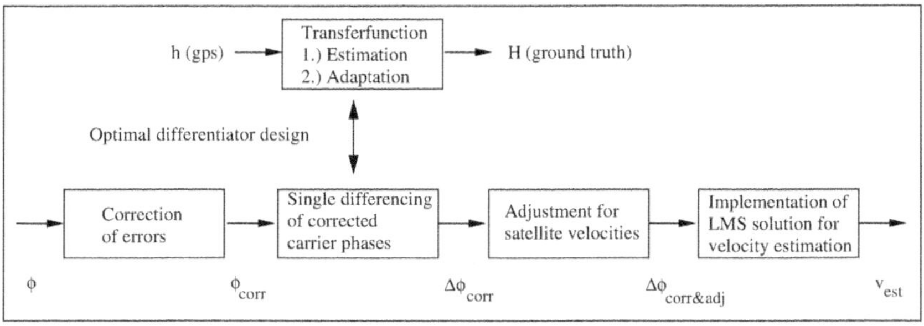

Fig. 13. Experimental signal processing strategy, exemplarily for the determination of velocities using GPS carrier phase observations ϕ

Since pseudorange measurements for the position solutions and carrier phases for deriving range rates and accelerations are used, some errors in the raw observations ϕ can be eliminated. They are the errors in satellite clock and propagation effects in the ionosphere and troposphere.

The optimal differentiator design is important for an accurate velocity and acceleration determination. An intuitive approach to estimate a derivative of a noisy discrete-time signal is by fitting a curve to it and differentiating this curve fit. The amount of smoothing will be a function of the order of polynomial and the length of the sliding window. The attempt to find a suitable filter design consists of two steps. At first, a transfer function estimation (5) shall represent the relation between observation and ground truth in the frequency domain. Because the GPS height determination is mostly affected by atmospheric effects and geometric deficiencies and ground truth exists only for the vertical component, both are acting as input and output respectively. Second, in order to find a compromise between the noise level reduction and the spectral resolution of the differentiated signal, this real input-output relation will be approximated using a Savitzky-Golay smoothing filter. The resulting differentiator design shall preserve the pertinent high frequency components of the carrier phase observations during differentiation (Fig. 13 top).

Both, the satellite velocities and accelerations can be obtained from the derivatives of Chebyshev-approximated precise ephemerides.

The formulation of the stochastic model is based on the assumption that the relationship between satellite zenith distance z and system noise can be modelled by the weighting function $w(z) = \cos^2 z$.

This procedure is performed using GPS observations at both, 10 and 50 S/s. The resultant vertical accelerations compared with the derivatives of ground truth heights H are displayed in the time and the frequency domain (Fig. 14).

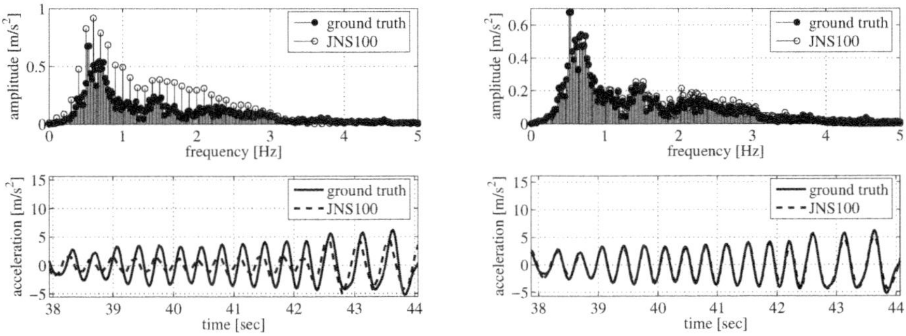

Fig. 14. Graphical comparison of the vertical acceleration component, derived from ground truth and from GPS carrier phase observations at different sampling rates (left: 10 S/s; right: 50 S/s); time and frequency domain

The potential of this procedure is not utilised in this experimental way, because the individual error behaviour isn't considered. Nevertheless it is apparent that increasing of the raw data update rate from 10 to 50 S/s alone leads to an improvement of the kinematic acceleration resolution.

3.3 Signal Fusion

After passing the total acceleration and the kinematic acceleration through the referring sensor subsystems (Sect. 3.1, 3.2) they will undergo data fusion according to $\mathbf{g} = \mathbf{a} - \mathbf{b}$. The further procedure is beyond the scope of this paper.

Acknowledgement. This is publication no. GEOTECH-170 of the programme GEOTECHNOLOGIEN of BMBF, Grant 03F0340C.

References

Boedecker G, Leismüller F, Neumayer K H (1999-2001) Device and Method for measuring Gravitation. Dgl. Patent USA #5,924,056 (1999), Dgl. Patent Australien #700890 (1999), Dgl. Europ. Patent EP 0 778 955 (1999), Dgl. Patent Russland #2144686 (2000), Dgl. Patent Canada #2,198,925 (2001)

Chen G (1996) Robuste Verfahren zur Analyse linearer stochastischer Prozesse im Zeitbereich. Doctoral Thesis, Institute of Geodesy and Navigation, University FAF, Munich

Eissfeller B (1997) Ein dynamisches Fehlermodell für GPS Autokorrelationsempfänger. Doctoral Thesis, Institute of Geodesy and Navigation, University FAF, Munich

Eykhoff P (1974) System Identification. Parameter and State Estimation. John Wiley & Son, London New York Sydney Toronto

Hütte, Czichos H (Hrsg.) (1996) Die Grundlagen der Ingenieurwissenschaften. Springer, Berlin Heidelberg New York

Kiencke, U, Jäkel H (2002) System Identification. Signale und Systeme. 2. Aufl., R. Oldenbourg Verlag, München Wien

Kronmüller H (1991) Digitale Signalverarbeitung. Springer, Berlin Heidelberg

Langley R B (1997) GPS Receiver system noise. GPS World June: 40-45

Lathi B P (2002) Linear Systems and Signals. Oxford University Press, Oxford New York

Lathi B P (2005) Linear Systems and Signals. 2nd edition, Oxford University Press, Oxford New York

Oppenheim A V, Schafer R W, Buck J R (2004) Zeitdiskrete Signalverarbeitung. 2. Aufl., Pearson Studium, München

Schwarz K P (2001) The impossible dream, thoughts on the development of airborne gravimetry. Wissenschaftliche Arbeiten der Universität Hannover, Nr. 241, Hannover

Spilker JJ, JR (1996) Fundamentals of Signal Tracking Theory. In: Parkinson, B W und Spilker, J J JR: Global Positioning System: Theory and Applications. Volume I, American Institute of Aeronautics and Astronautics.

Stürze A, Boedecker G (2004) High Precision Kinematic GNSS Observations Up to 50 S/s for Airborne Gravimetry. Proceedings of the ION GNSS 2004, Long Beach, September 21-24

Sundstrand Data Control (1986) Inc.: Sundstrand Data Control's Q-Flex Accelerometers. Instruction manual

Unbehauen R (1996) Regelungs- und Steuerungstechnik. In: Czichos (Hrsg). Hütte: Die Grundlagen der Ingenieurwissenschaften. 30. Aufl., Springer, Berlin Heidelberg

Unbehauen R (2002) Systemtheorie I. 8. Aufl., R. Oldenbourg Verlag, München Wien

Wüstling S (1997) Hochintegriertes triaxiales Beschleunigungssensorsystem. Forschungszentrum Karlsruhe. Wissenschaftliche Berichte, FZKA 6003, Karlsruhe

Further Development of a High Precision Two-Frame Inertial Navigation System for Application in Airborne Gravimetry

Tim H. Stelkens-Kobsch

Institute of Flight Guidance, Technical University of Braunschweig,
Hermann-Blenk-Strasse 27, D-38108 Braunschweig, Germany
t.stelkens@tu-bs.de

Summary. The Institute of Flight Guidance (IFF) of the Technical University of Braunschweig (TU BS) is involved in the development of airborne gravimetry since 1985. Fundamental examinations of airborne gravimeters were carried out between 1991 and 1993. In 1998 a high-precision two-frame inertial platform and a gravimeter sensor were purchased and modified for airborne application in cooperation with the Russian manufacturer ELEKTROPRIBOR.

Successful flight tests have been executed in the recent years. So far the resolution achieved is 2 km with a standard deviation of 3 mGal[1] (resp. 5 km, 1 mGal).

The two-frame inertial platform was extended to a three-frame INS by mounting a ring laser gyro on top of the platform. The gyro provides an additional degree of freedom (yaw) around the vertical axis.

Since 2001 the IFF was involved in the programme GEOTECHNOLOGIEN funded by the German Federal Ministry of Education and Research. The goal of the project, which ended in 2005 was to develop an airborne gravimetry system with a resolution of 1 mGal for wavelengths of 1 km.

Key words: Airborne gravimetry, gravimeter, inertial platform

1 Introduction

The variation of local gravitational acceleration is an indication for changes in density of the Earth's crust. These changes in density are both scientifically and economically very interesting for geophysics, geodesy and for exploration purposes (Fig. 1). Methods of gravimetry are usually applied in order to specify reference heights for land surveying, to investigate the geological structure of the subsurface or to explore occurrences of drinking water, oil, natural gas and other resources. In the high resolution range from wavelengths between 1 to 70 km the knowledge of gravitational anomalies is generally very small.

[1] 1 mGal $=10^{-5}$ ms^{-2}

Today, large areas of the Earth are still hardly explored regarding this wavelength band, as for example parts of Africa, Asia, South America (especially the Amazon area) or the Antarctic region. Especially for this accuracy range airborne gravimetry is an effective and economical tool.

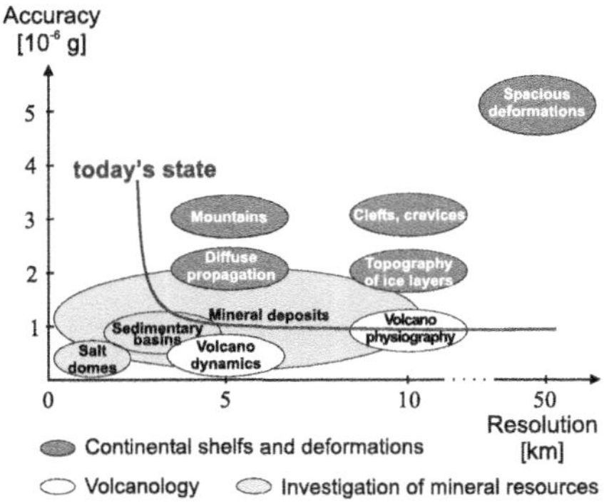

Fig. 1a. Required accuracy of geophysics (Hein, 1995)

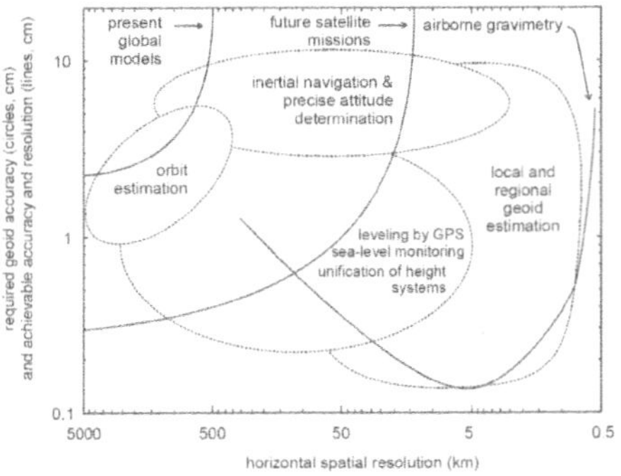

Fig. 1b. Required accuracy of geodesy (Bruton, 2000)

Airborne measurements are especially of interest for regions which cannot at all (or only with substantial difficulties) be attained by land vehicles or ships. A further crucial advantage of airborne gravimetry is the possibility of being able to measure large surfaces fast and at low cost.

Potential customers for airborne gravimetry can be essentially divided into two groups. These are national and international

- institutions/research establishments as well as
- exploration companies.

The task of land surveying authorities is, among other things, to determine the surface of a country by a sufficient large number of coordinates of fixed points. In order to achieve the demanded accuracy and resolution the spacious curvature conditions of the Earth and their exterior gravity field must be considered. The research institutions cover problems of basic research about the structure of the Earth and important questions for the precaution of existence (e.g. investigation of geological structures in cultivated areas, volcanic dynamics, geoid for measurement and cartography, ...). Thus extensive international programmes for the gravimetric determination of the geoid are (and will be) carried out in order to permit meaningful transfer of altitude information with GNSS systems. For these purposes different satellite-based Earth observation programmes (CHAMP, GRACE, GOCE) were started, which, however, can resolve only anomaly wavelengths over 70 km for physical reasons. In order to evaluate the data from satellite missions, reference surfaces are necessary, which can be provided by airborne gravimetry.

Terrestrial gravimetry is very expensive and time consuming, thus not practicable for this application. In contrast airborne gravimetry supplies an extremely economic and competitive alternative.

The situation in regard to the exploration of raw materials is similar. Considering the increasing scarceness of resources, a large, world-wide need exists for economical investigation of new deposits (e.g. oil, copper, other resources). Here, a profitable operational area opens up for airborne gravimetry.

The IFF is involved in the development of a gravity measurement system using an inertial platform, a gravity sensor, a barometric sensor and kinematic differential carrier phase based GPS positioning.

The two basic approaches for specific force measurement use strapdown inertial systems (Wei and Schwarz, 1998) or platform inertial systems (Olesen, Forsberg and Gidskehaug, 1997). Damped inertial platforms are typical for commercial airborne gravimetry systems. The vertical stabilisation of the gravity sensor is one of the fundamental problems, which has to be solved in airborne gravimetry based on inertial platforms. The accuracy of vertical specific force measurement depends on the performance of the vertical accelerometer and the levelling accuracy of the inertial platform. Aircraft dynamics and flight manoeuvres affect the accuracy of vertical specific force measurement via platform levelling errors.

To carry out airborne gravimetry research at the IFF, a Russian gravimeter (CHEKAN-A) was acquired, modified and successfully put into operation onboard the experimental twin-engine turbo-prop aircraft Dornier Do 128-6. The platform was originally designed to be installed in submarines and was developed at the Central Scientific & Research Institute ELEKTROPRIBOR.

In this study the evolution of the gravimetry system will be described.

2 Platform Description

Beside the sensors for altitude measurement the inertial platform is the decisive item of the sensor package necessary for airborne gravimetry. The platform supplies both the acceleration signal necessary for determination of gravity anomalies and data necessary for different corrections. The measuring system supplied from the Russian manufacturer ELEKTROPRIBOR consists of two main parts: a control unit and a gyro-stabilised platform. In Fig. 2a the two components are shown mounted in the experimental aircraft of the IFF.

The inertial platform ensures stabilisation of the gravity sensor's sensitivity axis in vertical direction. It is designed as a two-axes gyro stabiliser with a two-degree-of-freedom gyro and a gearless servo drive. Outer and inner rotating gimbals are supported by precise bearings and their axes are aligned parallel to longitudinal and lateral vehicle axis correspondingly. A set of sensors is installed on the platform. A highly sensitive accelerometer (gravity sensor, Fig. 2b) is located in the centre and measures vertical acceleration. Two further accelerometers are mounted in the horizontal plane and provide data for platform levelling. Each gimbal axis is provided with an angular position transducer and a brushless torque motor.

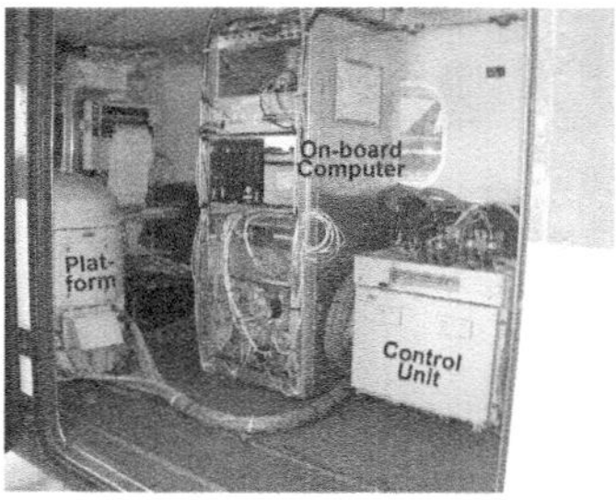

Fig. 2a. Gravimeter in test aircraft

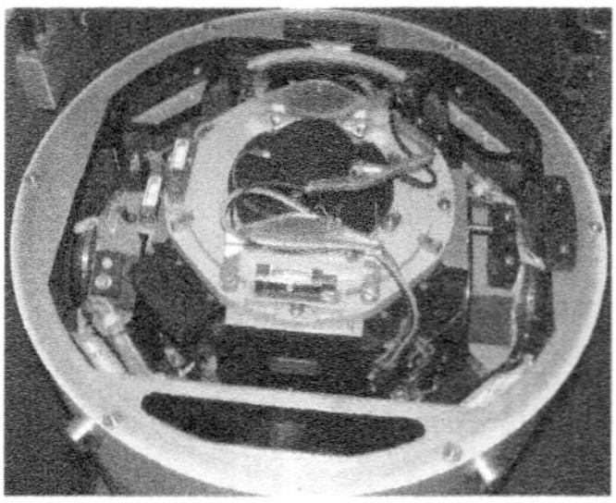

Fig. 2b. Inertial gravimeter platform

To compensate the influence of Earth rotation Ω_E and the relatively large drift ($D = 3^\circ \, \mathrm{h}^{-1}$) of the stabilising gyro, the platform was originally controlled with an analogue internally damped stabilisation loop. However, this kind of levelling loop does not meet the requirements for airborne gravity measurement systems. After high dynamic course manoeuvres like turning flights, the analogue controller needs unacceptable long to realign the platform. Therefore this control was upgraded with a digital controller, which creates an additional degree of freedom (yaw) for the platform around the vertical axis. The azimuth information is acquired from a ring laser gyro, which is installed on top of the platform. In other words, by use of GPS-measurements and a heading gyro a three-axis-platform is established with a strapdown calculation.

3 Modelling of the Gravimeter

The central element within the system is the gravity sensor (Fig. 3a). The sensing element consists of two identical torsion frames with pre-stressed quartz filaments. Each filament holds a pendulum with a mass and a reflector. The pendulums are inversely arranged to minimize cross coupling effects.

The torsion frames are mounted on a carrier which also holds a reference prism. A change of gravity causes a change of the pendulum angle according to $\xi = \mathbf{C} \times \mathrm{d}g$. $\mathbf{C}$ describes the sensitivity of the quartz filament.

To enhance attenuation of the vibratory pendulum system the entire construction is installed in a case which is filled with a viscous fluid.

Due to the viscosity the occurrence of friction between pendulums and fluid has to be considered. The basic approach is the Newtonian law of friction (Kaufmann, 1963). Besides the friction also buoyancy of pendulums in the fluid occurs. Steady state equilibrium results from the torsional moment of quartz filaments, the weight of masses and buoyancy inside the fluid. The angle ξ_0 is the steady state deflection of the pendulum.

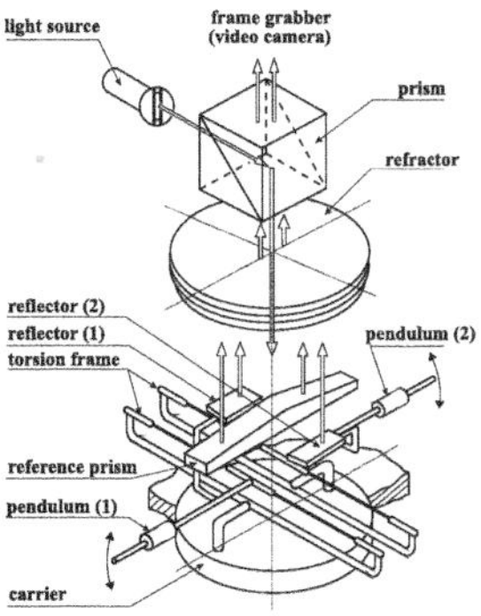

Fig. 3a. Construction of gravity sensor

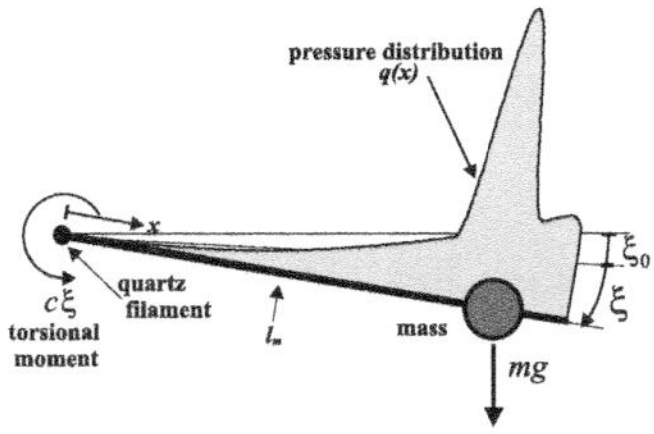

Fig. 3b. Modelling of gravity sensor

The mathematical description for a pendulum in viscous fluid (Fig. 3b) is:

$$
\begin{aligned}
ml_m\ddot{\xi} &= -c\xi - \int_0^l q(x)\,x dx \\
&+ m(\delta g - a_z)\,l_m \cos(\xi + \xi_0) + ma_x l_m \sin(\xi + \xi_0).
\end{aligned} \tag{1}
$$

Here δg describes a change in gravity and a_z is the vertical kinematic acceleration. According to the specification of viscous fluids follows:

$$\int_0^l q(x)\,x dx = \left(d_1\mu + d_2\left|\dot{\xi}\right|\right)\dot{\xi}. \tag{2}$$

In (2) parameter d_1 describes friction and parameter d_2 describes drag. With (2), (1) can be written as:

$$\ddot{\xi} + \frac{1}{ml_m^2}\left(d_1\mu + d_2\left|\dot{\xi}\right|\right)\dot{\xi} + \frac{c}{ml_m^2}\xi = \frac{\cos(\xi+\xi_0)}{l_m}(\delta g - a_z) + \frac{\sin(\xi+\xi_0)}{l_m}a_x. \tag{3}$$

Experiments demonstrated that pressure drag can be neglected in comparison to friction drag. This means $d_2\left|\dot{\xi}\right| \ll d_1\mu$ and therefore $d_2\left|\dot{\xi}\right|$ will be neglected in the following.

Equation (3) can now be written as:

$$\ddot{\xi} + \frac{d_1\mu}{ml_m^2}\dot{\xi} + \frac{c}{ml_m^2}\xi = \frac{\cos(\xi+\xi_0)}{l_m}(\delta g - a_z) + \frac{\sin(\xi+\xi_0)}{l_m}a_x. \tag{4}$$

The system has two cut-off frequencies. As the system cannot sample very high frequencies the influence of $\ddot{\xi}$ can be neglected. This leads to a simplified equation which describes the sensor as a low pass:

$$\frac{d_1\mu}{c}\dot{\xi} + \xi = \frac{ml_m\cos(\xi+\xi_0)}{c}(\delta g - a_z) + \frac{\sin(\xi+\xi_0)}{c}a_x. \tag{5}$$

In Fig. 4 the observed output signal of the gravity sensor is shown in comparison to the developed model.

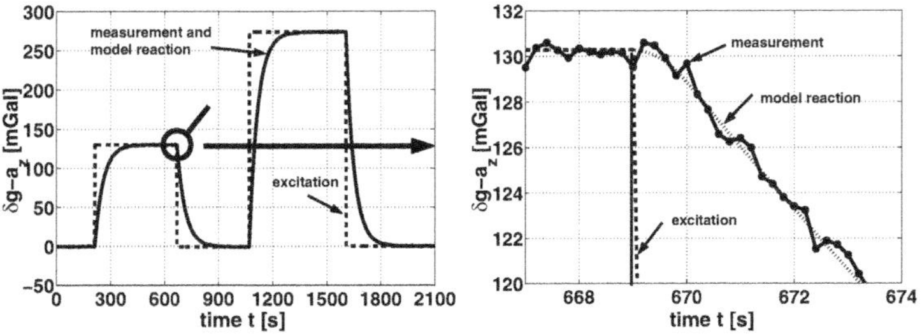

Fig. 4. Gravity sensor output signal

4 Platform Stabilisation

To achieve a gravity measurement accuracy of 1 mGal onboard a light fixed-wing aircraft under the influence of horizontal accelerations (turbulence of air) a sophisticated platform levelling is required. This requirement can hardly be met by commercially available inertial platforms. Even the magnitude of horizontal acceleration is about $1\,\mathrm{m\,s^{-2}}$, dependent on air turbulence conditions. For a Schuler-tuned inertial platform accelerometer biases must be less than $5{\times}10^5\,\mathrm{m\,s^{-2}}$ in order to achieve the demanded levelling accuracy. Gyro drifts must be less than $0.001°\,\mathrm{h^{-1}}$ (Zhang, 1995). This grade of inertial sensors is either not commercially available or too expensive to be used in civilian applications. To solve this problem, new techniques of using GPS measurements for damping the platform oscillation have been developed at the IFF in the recent years.

4.1 Velocity-Aided Platform

Inertial platform levelling errors can be physically damped by feeding reference measurements into the platform levelling loop. The advantage of this approach over internally damped platforms is to provide the vertical accelerometer with a more stable environment and to physically reduce influences of horizontal accelerations on the measurement of vertical specific force.

In Fig. 5 the diagram of a levelling loop is shown, which is physically stabilised by GPS-velocity since there is no better reference system to date. Dashed lines in the diagram represent a physical feedback path of the gravity signal due to platform mislevelling and a forward path of local level frame rotation due to system movement and Earth rotation. The other part of the diagram is feeding velocity back into the platform levelling loop. The levelling error ε_x is defined as the platform mislevelling with respect to reference ellipsoid. The Laplace transformation of the levelling error is (Fig. 5):

$$\hat{\varepsilon}_x = \frac{\hat{D}s - \frac{\hat{B}}{R}\left(1+F\left(s\right)\right) + \frac{s}{R}F\left(s\right)\delta\hat{V}_{ref}}{s^2 + \frac{g}{R}\left(1+F\left(s\right)\right)}, \quad F\left(s\right) = k\frac{T_D s+1}{T_I s+1}, \tag{6}$$

where s is the complex frequency of the Laplace transformation, g is the magnitude of gravity and R is the average Earth radius.

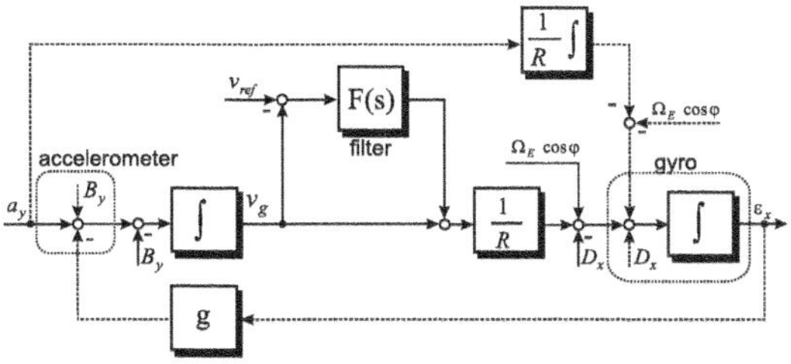

Fig. 5. Levelling loop of the platform stabilised by GPS-velocity (East-channel)

Equation (6) shows that the levelling error is not affected by system acceleration and the levelling loop can be stabilised by choosing values of T_D, T_I and k. The characteristic polynomial $n(s)$ can be written in dependency of controller gains and Schuler frequency ω_S:

$$n(s) = T_I s^3 + s^2 + \omega_S^2 (T_I + kT_D) s + (k+1) \omega_S^2. \tag{7}$$

This equation has three Eigenvalues. The optimal behaviour of the platform is achieved, if the three poles of the characteristic polynomial are located on the left part of a circle with the radius ω_0.

The levelling loop is a low pass filter for accelerometer biases and a band-pass filter for gyro drift and GPS velocity. Errors within the frequency range close to natural frequency have major effects on platform levelling errors.

4.2 Position-Aided Platform

In normal operation GPS velocity shows a time delay in relation to INS velocity due to complex signal processing in the GPS receiver. An inaccurate determination of this delay causes an excitation of the platform, which in turn causes an enlargement of platform misalignment.

The time delay problem is basically treated, as the position information of GPS is used as reference for the platform. The position information received from GPS has a time delay of an order of magnitude smaller than the velocity information. Furthermore it is exactly assignable.

In order to carry out a control based on position information, the control loop must be changed as indicated in Fig. 6 (north channel). The problem of platform control on position basis is the stabilisation of the control loop. Now three integrations have to be performed and therefore it tends to be instable. The influence of the additional integrator must be compensated by a suitable selection of the filter $G(s)$. For the represented block diagram the following transfer function results for levelling errors of the platform:

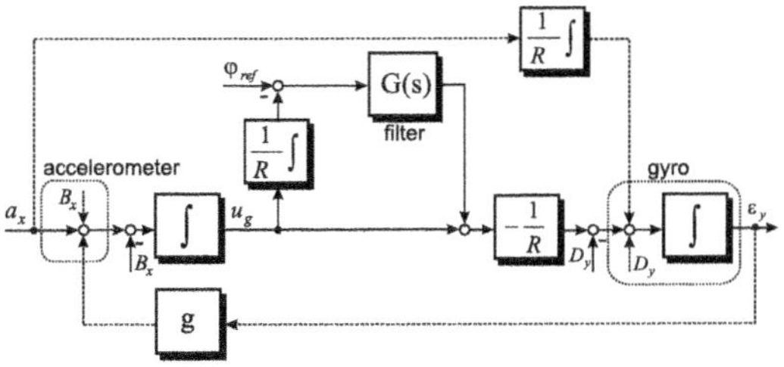

Fig. 6. Levelling loop of the platform stabilised by GPS-position (North-channel)

$$\hat{\varepsilon}_y = \frac{\hat{D}s^2 - \frac{\hat{B}}{R}\left(s + \frac{G(s)}{R}\right) + s^2\frac{G(s)}{R}\delta\hat{\varphi}_{ref}}{s^3 + \frac{g}{R}\left(s + \frac{G(s)}{R}\right)}. \tag{8}$$

The characteristic polynomial $n(s)$ becomes:

$$n(s) = (s + \omega_0)\left(s^2 + 2\zeta_1\omega_0 s + \omega_0^2\right)\left(s^2 + 2\zeta_2\omega_0 s + \omega_0^2\right). \tag{9}$$

The conditions regarding the critical band close to the natural frequency are almost the same as for velocity based control, but the delay of the system is now much smaller. Therefore the control on basis of position information outperforms the velocity based control.

4.3 Initial Alignment

The task of initial alignment is to level the platform and to find the azimuth angle. In addition, gyro drifts can be calibrated during alignment. After initial alignment, the platform is switched to navigation mode and is ready for gravity survey.

To obtain accurate alignment results, fine alignment and gyro calibration have to be performed. Velocity can be determined using (10):

$$\begin{aligned} u &= \frac{g\left(\Omega_E \cos\varphi\left(\sin\psi - \sin\psi_s\right) + D_y\right)}{s^2 + \frac{g}{R}\left(1 + F(s)\right)} \\ v &= \frac{g\left(\Omega_E \cos\varphi\left(\cos\psi - \cos\psi_s\right) - D_x\right)}{s^2 + \frac{g}{R}\left(1 + F(s)\right)}, \end{aligned} \tag{10}$$

where φ is the geodetic latitude, Ω_E is the Earth rotation, u and v are the horizontal velocity components, D_x and D_y are the components of gyro drift.

In steady state and under assumption of constant gyro drift during fine alignment velocity signals are recorded as:

$$\begin{aligned} u_\infty &= \tfrac{R}{1+k}\left(\Omega_E \cos\varphi\left(\sin\psi - \sin\psi_s\right) + D_y\right) \\ v_\infty &= \tfrac{R}{1+k}\left(\Omega_E \cos\varphi\left(\cos\psi - \cos\psi_s\right) - D_x\right), \end{aligned} \tag{11}$$

where ψ and ψ_s are true azimuth and estimated azimuth in coarse alignment, respectively.

After a measurement the platform is turned by a certain angle $\Delta\psi$ in azimuth and the levelling step is repeated. Using velocity data obtained from the two orientations ($u_{1\infty}$, $v_{1\infty}$ and $u_{2\infty}$, $v_{2\infty}$), both the azimuth at the two orientations and the constant components of gyro drift can be estimated. The azimuth of the second orientation can be computed using (12):

$$\begin{bmatrix} \sin\psi \\ \cos\psi \end{bmatrix} = \begin{bmatrix} 1 & \frac{-\sin\Delta\psi}{1-\cos\Delta\psi} \\ \frac{\sin\Delta\psi}{1-\cos\Delta\psi} & 1 \end{bmatrix} \begin{bmatrix} b_1 \\ b_2 \end{bmatrix}, \tag{12}$$

where

$$\begin{bmatrix} b_1 \\ b_2 \end{bmatrix} = \frac{1+k}{2R\Omega_E\cos\varphi} \begin{bmatrix} u_{2\infty} - u_{1\infty} \\ v_{2\infty} - v_{1\infty} \end{bmatrix}. \tag{13}$$

If constant components of gyro drift were determined beforehand and then compensated, the azimuth can be obtained after setting the first orientation:

$$\begin{aligned} \sin\psi &= \frac{(1+k)u_\infty}{R\Omega_E\cos\varphi} + \sin\psi_s \\ \cos\psi &= \frac{(1+k)v_\infty}{R\Omega_E\cos\varphi} + \cos\psi_s \end{aligned} . \tag{14}$$

Using the azimuth obtained from (14) in (11) the constant gyro components can be calculated with:

$$\begin{aligned} D_x &= \Omega_E\cos\varphi\,(\cos\psi - \cos\psi_s) - \frac{(1+k)v_{2\infty}}{R} \\ D_y &= \Omega_E\cos\varphi\,(\sin\psi - \sin\psi_s) - \frac{(1+k)u_{2\infty}}{R} \end{aligned} \tag{15}$$

The levelling accuracy is mainly limited by accelerometer biases. If gyros and accelerometers are well calibrated, the platform can be levelled within 2" in steady state. The constant components of gyro drift can be calibrated with an accuracy of $0.03°\,\mathrm{h}^{-1}$ and the azimuth angle can be calculated with an accuracy of 30".

5 Further Development of the System

During the expired project in the programme GEOTECHNOLOGIEN the demand for an even better platform levelling came up in order to achieve the high goal of 1 mGal accuracy. On this account a Kalman filter for platform levelling was developed. Furthermore a possibility for increasing the accuracy in determination of flight altitude was investigated.

5.1 Platform Levelling with a Kalman Filter

As the platform has been extended by an additional degree of freedom around the vertical axis a levelling loop similar to those used in integrated navigation systems can be set up. For this reason a simulation of the entire system was developed at the IFF. This system includes a Kalman filter with 13 state variables. The levelling loop principle is shown in Fig. 7.

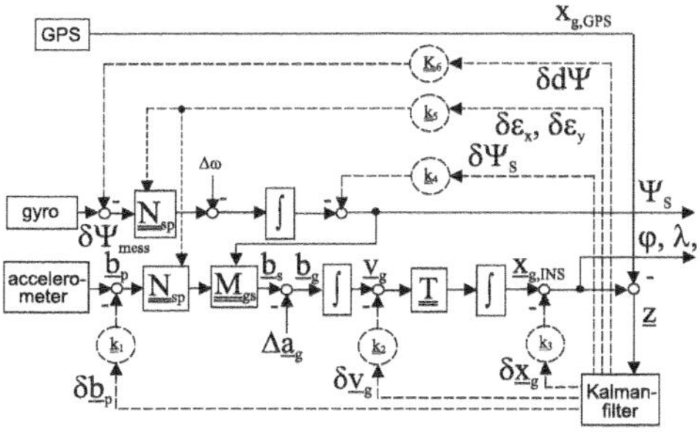

Fig. 7. Scheme of the Kalman filter

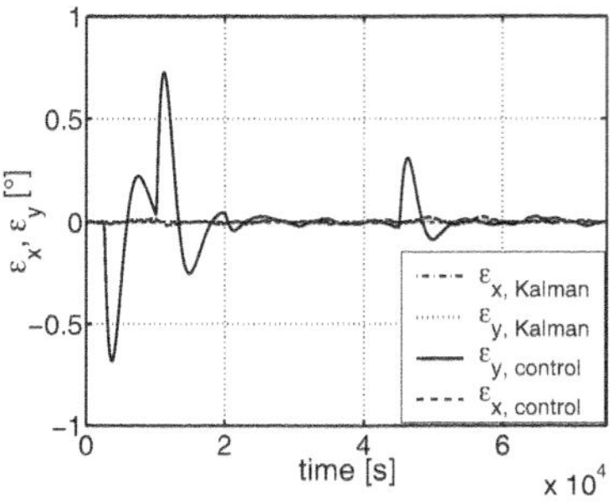

Fig. 8. Comparison of levelling methods

Simulation results exhibit much better performance for a levelling algorithm on base of a Kalman filter. A comparison between the control described in Sect. 4.1 and the Kalman filter control is shown in Fig. 8. Levelling errors of the Kalman filter are magnitudes smaller.

5.2 Aiding the Determination of Altitude with a Statoskop

All gravimetry systems have in common that influence of air turbulence makes a determination of gravitational anomalies almost impossible, even in calm air. The wavelengths of anomalies to be determined lie in the same wavelength scope as the disturbing turbulences. A simple example shall clarify this.

Atmospheric turbulence with an intensity of $\sigma_w = 0.5\,\mathrm{m\,s^{-1}}$ is called 'calm' by the pilot. The characteristic wavelength L_w particularly depends on the stability of atmosphere (Richardson number) (Hinze, 1975) and flight altitude.

Vertical acceleration of an airplane as a reaction on vertical turbulence corresponds to a band-pass filter, where the typical wavelength of the aircraft $L = TV$ (T, L: airplane-typical time constant and wavelength, respectively) and the characteristic wavelength L_w of the turbulence constitute the cutoff frequencies. The variance of airplane vertical acceleration may be calculated by Parseval's theorem (Schänzer, 1985):

$$\sigma_{\ddot{h}}^2 = \frac{\left(\frac{V}{L}\right)^2}{1+\frac{L_w}{L}}\,\sigma_w^2 . \qquad (16)$$

For small turbulence $\sigma_w = 0.5\,\mathrm{m\,s^{-1}}$ and the example airplane Do 128 with an airspeed $V = 50\,\mathrm{m\,s^{-1}}$, vertical accelerations like shown in Table 1 result in dependence of turbulence wavelength r.

Table 1. Vertical accelerations in dependence of turbulence wavelength r

r [m]	$\ddot{h}$ [m s^{-2}]	$\ddot{h}$ [mGal]
2	5×10^{-3}	500
20	5×10^{-2}	5000
200	5×10^{-1}	50000
2000	5×10^{-2}	5000
20000	5×10^{-3}	500
200000	5×10^{-4}	50

The largest vertical accelerations arise at turbulence wavelengths between L and L_w (Fig. 9). Depending on type and configuration of the aircraft, flight altitude and stability of the atmosphere this results in the range between 150 m and 400 m (Schänzer, 2000).

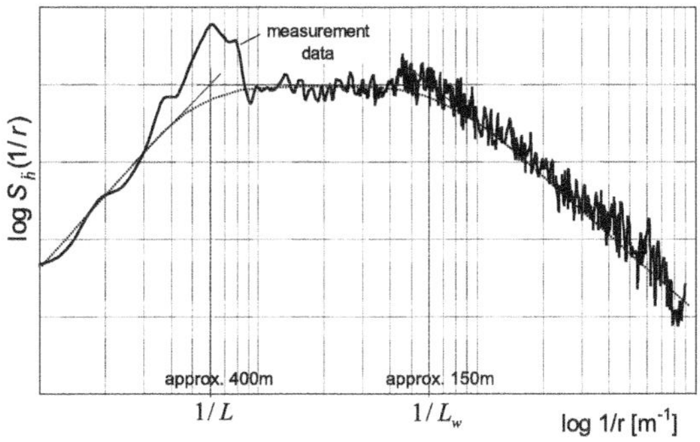

Fig. 9. Power density spectrum of aircraft vertical acceleration in turbulent air

Even at a turbulence wavelength of 20 km, the disturbing accelerations arise to 500 mGal. For a measuring accuracy of the gravitational anomalies of 1 mGal this amounts to a signal to noise ratio of 500:1. With conventional low-pass filters, even if digitally realized, gravitational anomalies cannot be determined reliably. In order to improve this, further actions have to be taken.

The undesirable influence of turbulence can be reduced (but not eliminated) by flying at calm air (daily edge times, zero wind), in low altitude ($H < 200$ m), with low airspeed and with a high surface load of the airplane.

The necessary altitude resolution in order to dissolve gravity anomalies of 1 km wavelength can be estimated, if sinusoidal turbulence is varied with different wavelength. The vertical acceleration of the aircraft along the flight path x is then likewise sinusoidal:

$$\ddot{h} = \ddot{h}_0 \sin\left(2\pi \frac{V\,t}{\lambda}\right). \tag{17}$$

This results in a change in altitude:

$$\Delta h = \left(\frac{\lambda}{2\pi V}\right)^2 \ddot{h}. \tag{18}$$

If the permissible error of the accelerometer is in the range of $\Delta\ddot{h} = 1\,\text{mGal}$ the accuracy of altitude determination has to be as shown in Fig. 10. Following this, extreme demands are made on measurements of altitude (1 mGal for 1 km $\Rightarrow \Delta h = 0.1\,\text{mm}$). This cannot be achieved at present with DGPS phase measurements or barometric determination of altitude. Thus, possibilities must be found to substantially increase the resolution of altitude determination.

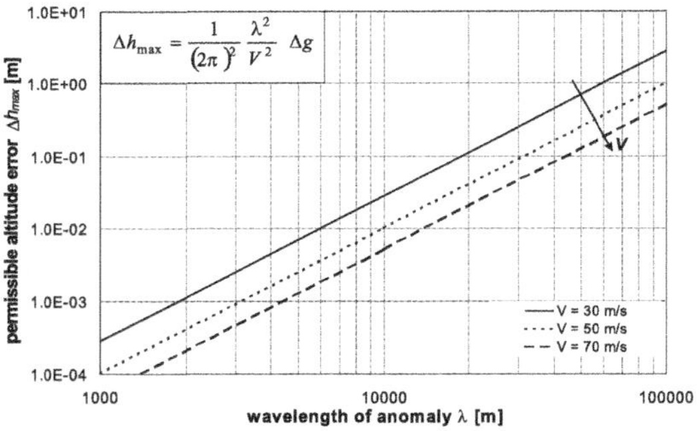

Fig. 10. Necessary resolution of an altitude sensor for 1 mGal accuracy

All systems used in airborne gravimetry consist of two sensors. One sensor determines the position the other kinematic acceleration. For the combination of signals correct phasing is extremely important, since a phase shift of sensor signals makes a determination of anomalies impossible. Furthermore sensor characteristics change during measurements which typically last 2 to 3 hours. This must be considered and appropriate countermeasures applied.

The main research topic for the high precision barometric altitude determination was the improvement of the pressure measurement itself which included the modelling of the sensors. For the differential pressure transducer both a static error model as well as a model for the dynamic behaviour was set up. Steady temperature changes of the reference volume cause the largest error in the sensor system. To compensate the drift following these changes an active volume control was installed in the Statoskop. The modelling of the thermodynamic behaviour of the reference volume was a major topic in the work. It included the heat transfer as well as the dynamic model of the temperature sensors. A real time use of the temperature signal was impossible due to the high time constant. Therefore a model for the temperature inside the reference volume was build which is based on the measurement of ambient temperature and other parameters.

Using these models a significant improvement of the pressure measurement was reached. A reduction of the signal noise and an improvement of the resolution by two orders of magnitudes was reached. Due to the measurement principle the absolute accuracy of the Statoskop corresponds to the accuracy of absolute pressure transducers. The influence of variation of meteorological parameters such as humidity and temperature was not investigated yet, but has significant impact on the accuracy of altitude measurements.

In Fig. 11 altitude measurements of the Statoskop, a common absolute pressure transducer and GPS are shown.

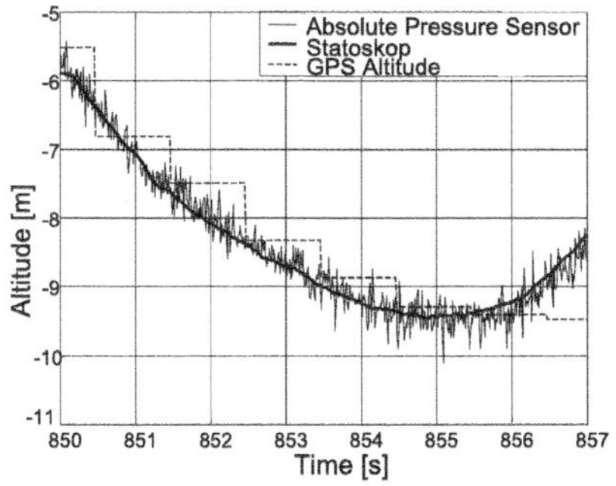

Fig. 11. Comparison of Statoskop, absolute pressure sensor and GPS

5.3 Aiding the Platform Control with Phase Based Kinematic DGPS

For applications of airborne gravimetry systems based on gimballed inertial systems an exact horizontal alignment of gimbals is of essential importance. Even attitude control errors of a few arc seconds invoke erroneous gravity measurements. At the IFF, a control with a Kalman Filter is investigated, augmented by GNSS data based on phase measurements in real-time. Therefore correction data has to be received in-flight and processed by the navigation algorithm. In the FRG, the satellite positioning service of the German Survey Department (SAPOS) is suitable in this context. The correction data is transmitted via the GSM network and can be received with modified commercial off-the-shelf (COTS) mobile phones. For applications outside of GSM networks a transportable reference station may be used.

6 Results

At the IFF, investigations of airborne gravimetry have been carried out for approx. 25 years. In the course of these investigations the concept of a refed complementary airborne gravimeter was developed (patent no. 40 13 570, announced to 27th April 1990) and examined systematically in extensive simulations (Schänzer, 2000), (Abdelmoula, 1998a), (Abdelmoula, 1998b). In the DFG project Scha 334/6-1 this new concept of a gravimeter was simulated in combination with a real airplane model (5-point model), air-turbulence and simple analytic gravitation models. During the SFB 420, subproject B4 'Fluggravimeter', a highly exact, platform-based gravimeter was procured, installed and tested in the experimental aircraft Do 128-6 of the IFF under use of the patented concept. The attainable resolution of anomalies was increased to 5-6 km with a standard deviation of expectation errors of 1 mGal.

With funds of the SFB 420 an azimuth gyro was acquired, which was incorporated into the system in the context of the GEOTECHNOLOGIEN-project 'Entwicklung der Fluggravimetrie unter Nutzung von GNSS-Satellitenbeobachtungen' (Development of Airborne Gravimetry Using GNSS Satellite Observations). Thus a three-frame-platform could be created from the two-frame-platform with a strapdown calculation. The additional rate information permits a substantial shortening of the time needed for levelling of the platform after dynamic flight manoeuvres. Furthermore, a self-sufficient operation of the gravimeter is possible, which improves its operational serviceability.

In comparison with static references, the obtained results of airborne gravity measurements correspond well (Fig. 12).

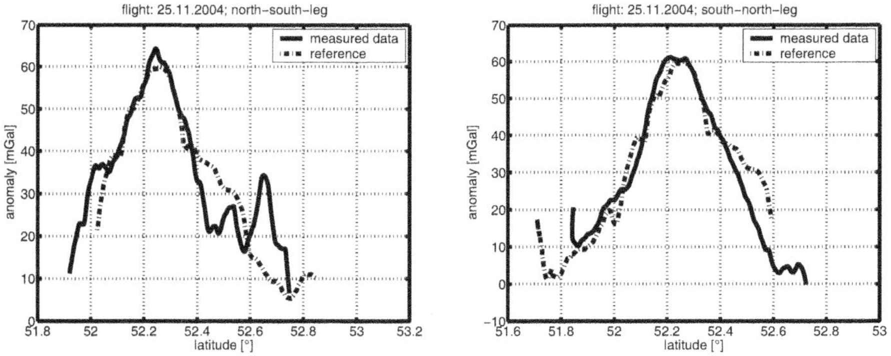

Fig. 12. Comparison of measurement data with reference data

The pressure measurement with the Statoskop could be improved during the preceding GEOTECHNOLOGIEN-programme. These improvements were reached by detailed modelling of the thermodynamic procedures in the reference volume and refer mainly to the pressure measurement itself.

The investigations showed that variation of meteorological parameters such as air temperature and humidity have substantial influence on attainable accuracy of barometric altitude determination. For this reason further investigations of meteorological parameters are essential.

Only after correction of these meteorological influences on measurements the barometric altitude may be filtered appropriately with GPS altitude.

The principal purpose of the preceding joint project (1 mGal for 1 km) was not fulfilled and is still an extremely fastidious task. The demand of the geodetic community was, however, not fulfilled internationally by any research group or institution so far. During the project it became clear that the target may be achieved by an increased quality of platform levelling, by optimisation of evaluation algorithms and, particularly, by an extraordinary increase in the accuracy of altitude determination (Fig. 10).

Acknowledgement. This is publication no. GEOTECH-172 of the programme GEOTECHNOLOGIEN of BMBF and DFG, Grant MGS/F0340B

References

Abdelmoula F (1998a) Neue Konzepte und Lösungsansätze bei der Fluggravimetrie. Progress in geodetic science at GW98, 9-16, Shaker Verlag, Aachen

Abdelmoula F (1998b) Probleme und Lösungsansätze bei der Fluggravimetrie. 1. Braunschweiger Symposium für Flugmesstechnik, Braunschweig

Abdelmoula F (2000) Ein Beitrag zur Bestimmung der Erdbeschleunigungsanomalien an Bord eines Flugzeuges. Dissertation. Shaker Verlag, Aachen

Bruton A M (2000) Improving the Accuracy and Resolution of SINS/DGPS Airborne Gravimetry. UCGE Reports Number 20145. University of Calgary, Calgary, Alberta

Hein G W (1995) Progress in Airborne Gravimetry: Solved, Open and Critical Problems. Proc. of the IAG Symposium on Airborne Gravity Field Determination. Boulder, Colorado, USA

Hinze J O (1975) Turbulence. 2nd Edition. McGraw-Hill Book Company, New York

Kaufmann W (1963) Technische Hydro- und Aerodynamik. Springer-Verlag, Berlin Göttingen Heidelberg

Olesen A V, Forsberg R, Gidskehaug A (1997) Airborne Gravimetry Using the LaCoste & Romberg Gravimeter - an Error Analysis. In: M.E. Cannon and G. Lechapelle (eds.). Proc. Int. Symp. on Kinematic Systems in Geodesy, Geomatics and Navigation, Banff, Canada, June 3-6, 1997, Publ.Univ. of Calgary, pp. 613-618

Schänzer G (1985) Abschätzung von stochastischen Böenlasten unter Berücksichtigung instationärer Luftkräfte. Zeitung Flug-wiss.Weltraumforschung 9, Heft 3

Schänzer G, Abdelmoula F (1999) Fluggravimeter; Zwischenbericht Sonderforschungsbereich Flugmesstechnik SFB 420

Schänzer G (2000) An Approach to Utilising In-Flight Gravimetry in Geodesy and Geophysical Exploration. Aerospace Science and Technology, subm. February 2000

Schänzer G (2000) Fluggravimeter hoher Auflösung. Geodätische Woche, Potsdam

Wei M, Schwarz K P (1998) Flight Test Results from a Strapdown Airborne Gravity System. Journal of Geodesy 72:323–332

Zhang Q J (1995) Development of a GPS-aided Inertial Platform for an Airborne Scalar Gravity System. PhD Thesis. University of Calgary, Calgary, Alberta